ERGONOMIC DESIGN FOR PEOPLE AT WORK

VOLUME 2

Suzanne H. Rodgers

Principal Author and
Technical Editor

Deborah A. Kenworthy, Editor
Elizabeth M. Eggleton, Editor

Commercial Publications
Eastman Kodak Company

Contributing Authors

David M. Kiser
*Thomas J. Murphy
*Waldo J. Nielsen
*Suzanne H. Rodgers

Ergonomics Group
and Human
Factors Section
Eastman Kodak Company

Artwork by William Sabia

*No longer associated with Eastman Kodak Company

ERGONOMIC DESIGN FOR PEOPLE AT WORK

Volume 2

The Design of Jobs, including Work Patterns, Hours of Work, Manual Materials Handling Tasks, Methods to Evaluate Job Demands, and the Physiological Basis of Work

A Source Book for Human Factors Practitioners in Industry including safety, design, and industrial engineers; medical, industrial hygiene, and industrial relations personnel, and management.

by: The Ergonomics Group
Health and Environment Laboratories
Eastman Kodak Company

VNR VAN NOSTRAND REINHOLD
_____ New York

I(T)P™ Van Nostrand Reinhold is a division of International Thomson Publishing, Inc.
The ITP logo is a trademark under license

Printed in the United States of America

For more information, contact:

Van Nostrand Reinhold
115 Fifth Avenue
New York, NY 10003

Chapman & Hall GmbH
Pappelallee 3
69469 Weinheim
Germany

Chapman & Hall
2-6 Boundary Row
London
SE1 8HN
United Kingdom

International Thomson Publishing Asia
221 Henderson Road #05-10
Henderson Building
Singapore 0315

Thomas Nelson Australia
102 Dodds Street
South Melbourne, 3205
Victoria, Australia

International Thomson Publishing Japan
Hirakawacho Kyowa Building, 3F
2-2-1 Hirakawacho
Chiyoda-ku, 102 Tokyo
Japan

Nelson Canada
1120 Birchmount Road
Scarborough, Ontario
Canada M1K 5G4

International Thomson Editores
Seneca 53
Col. Polanco
11560 Mexico D.F. Mexico

96 97 98 99 RRD HB 15 14 13 12 11 10

Library of Congress Cataloging-in-Publication Data
Main entry under title:

Ergonomic design for people at work.

Includes bibliographies and indexes.
Contents: v.1. Work place, equipment, and environmental design and informa-
tion transfer—v.2. The design of jobs including work patterns, hours of work, manu-
al materials, handling tasks, methods to evaluate job demands, and the physiological
basis of work.
1. Human engineering—Handbooks, manuals, etc.
I. Eastman Kodak Company. Human Factors Section.
T59.7.E714   1983   620.8'2   83-719
ISBN 0-534-97962-9 (v.1)
ISBN 0-442-22103-7 (v.2)
ISBN 0-442-23939-4 (v.2ppr.)

The Ergonomics Group of Eastman Kodak Company acknowledges with gratitude permission received from the following companies and organizations to use material reprinted from their publications in the development of this book. Specific references are noted in the text, and complete citations follow in the bibliography at the end of each chapter.

Academic Press, Inc., New York, New York
American Academy of Orthopaedic Surgeons, Chicago, Illinois
American Congress of Rehabilitation Medicine, Chicago, Illinois
American Industrial Hygiene Association, Akron, Ohio
American Physical Therapy Association, Fairfax, Virginia
British Association of Rheumatology and Rehabilitation, London, England
Butterworth Scientific Ltd., Guildford, England
Flournoy Publishers, Inc., Chicago, Illinois
Gower Press, Industrial Society, London, England
Human Factors Society, Santa Monica, California
Karolinska Institutet, Stockholm, Sweden
Lea and Febiger, Philadelphia, Pennsylvania
Alan R. Liss, Inc., New York, New York
McGraw-Hill, Inc., New York, New York
Methuen, Inc. New York, New York
NASA Scientific and Technical Information Office, Yellow Springs, Ohio
National Institute for Occupational Safety and Health, Washington, D.C.
Pergamon Press, Inc., Elmsford, New York
S. P. Medical & Scientific Books, Jamaica, New York
W. B. Saunders Co., Philadelphia, Pennsylvania
Scientific American, Inc., New York, New York
Taylor & Francis, Ltd., Basingstoke, England
George Thieme Verlag, Stuttgart, Germany and New York, New York
C. C. Thomas, Springfield, Illinois
Van Nostrand Reinhold Co., Inc. New York, New York
John Wiley and Sons, Inc., New York, New York
Wright Air Development Center, Wright-Patterson AFB, Ohio
Year Book Medical Publishers, Inc., Chicago, Illinois

NOTICE

We believe the information provided in this work is reliable
and useful, but it is furnished without warranty of any
kind from either the author, Eastman Kodak Company, or
the publisher, or the editors and contributing authors.
Readers should make their own determinations of the
suitability or completeness of any material or procedure for
a specific purpose and adopt such safety, health, and other
precautions as may be necessary. Of course, no license
under any patent or other proprietary right is granted or to
be inferred from our provision of this information.

Contents

Chapter 17. Shift Schedules   **304**

**Chapter 18. Alternative Work Schedules and Guidelines
      for Shift Workers**   **326**

List of Illustrations

Chapter 27.

Chapter 28.

List of Tables

Chapter 27.

Chapter 28.

Related Titles from Van Nostrand Reinhold:

Ergonomic Design for People at Work, Vol. 1
by Eastman Kodak Company

Fundamentals of Biomechanics:
Equilibrium, Motion and Deformation
by Nihat Ozkaya and Margareta Nordin

Engineering Physiology, Second Edition:
Bases of Human Factors/Ergonomics
by K.H.E. Kroemer, H.J. Kroemer, and K.E. Kroemer-Elbert

Ergonomics in Back Pain:
A Guide to Prevention and Rehabilitation
by Tarek M. Khalil, Elsayed M. Abdel-Moty, Renee Steele Rosomoff,
and Hubert Rosomoff, M.D.

The Ergonomics Edge
by Dan MacLeod

Hand-Arm Vibration:
A Comprehensive Guide for Occupational Health Professionals
by Peter L. Pelmear, M.D., William Taylor, M.D., and Donald E. Wasserman

On the Practice of Safety
by Fred A. Manuele

Physical and Biological Hazards of the Workplace
by Peter Wald and Gregg M. Stave

MANPRINT:
An Approach to Systems Intergration
by Harold R. Booher

Preface

This book is the second volume of a comprehensive survey of industrial ergonomics and human factors written by members of the ergonomics and human factors staffs at Eastman Kodak Company. The first volume considered the design of workplaces, equipment, information, and the environment to improve worker performance. Its emphasis was on the design of the interfaces between the worker and objects or conditions in the work environment. Proper design of these interfaces will improve worker performance on the job, and increase productivity, but all of the benefits can be lost if the job itself is not well designed. This book provides guidelines on the design of jobs based on information about people's capacities for work.

Although there are published studies about acceptable work loads and how to design work, we have often found them to have limited practical application in a production area. This is the case if the published guidelines were based on studies of healthy young men, but the industrial population is made up of men and women, young and old, and in varying degrees of health. In other instances, the published guidelines were based on short-duration tasks, whereas in industry the work is often sustained for eight or more hours a day, five or more days a week for several years. Other published approaches for matching people to jobs, such as extensive training or selection to prevent overexertion injuries, can simply be impractical to use in many departments. They can require significant amounts of time to obtain benefits that are often ill-defined; the requirements for validation and delays in getting the person into the job will act to discourage the use of these techniques.

In the past twenty-six years, Kodak's ergonomics and human factors personnel have devoted a substantial amount of time to the analysis of existing industrial tasks and jobs. Our focus has been on tasks where performance problems or injuries indicated potential design problems. In addition, we have assisted with the design of new jobs as production facilities were expanded. The guidelines and data found in this book are drawn from that experience. Each guideline has been tested in a manufacturing operation. Published guidelines that did not pass the practical test have been discarded or adjusted to fit our experience. In all instances, we have tried to accommodate as many women and older workers in our design guidelines as is possible. Further information is provided in the section entitled "Whom to Design Jobs For" in the Introduction.

Eastman Kodak Company

Acknowledgments

Members of the Ergonomics Group and Human Factors Section who contributed to the research and writing of each section of this book are listed opposite the title page and at the beginning of each chapter.

Lending support to the project were Harry L. Davis, Supervisor, Human Factors Section; Kenneth T. Lassiter, Publications Director, Consumer/Professional and Finishing Markets; and Alexander Kugushev, Lifetime Learning Publications. Elizabeth M. Eggleton and Deborah Kenworthy, Technical Editors, Consumer/Professional and Finishing Markets, provided editorial services. The following people reviewed part or all of the manuscript and offered valuable suggestions for its improvement:

Thomas S. Ely, M.D., Director, Occupational Health Laboratory, Health and Environment Laboratories—the whole manuscript.

David M. Caple, Ergonomics Consultant, Victoria, Australia—Parts IV and VI.

Robert H. Jones, M.D., Rehabilitation Consultant, Eastman Kodak Company—Parts IV and VI.

Thomas J. Armstrong, Ph.D., University of Michigan School of Public Health—Part IV.

Thomas Bernard, Ph.D., Westinghouse Corporation—Part IV.

W. Monroe Keyserling, Ph.D., University of Michigan Industrial and Operations Engineering Department—Part VI.

Stover H. Snook, Ph.D., Liberty Mutual Insurance Company—Part IV.

Harry Snyder, Ph.D., Virginia Polytechnic Institute & State University—Part IV.

We would like to thank especially Carol McCreary of the Ergonomics Group, who checked and compiled the reference materials, and Paul Champney, Human Factors, and Don Buck, Photographic Illustrations, who staged and took the photographs. The preliminary artwork was drawn by Leigh Ann Smith of Albany, New York, and by Anne Wilkinson and Gerry Bommelje of Photo Services; Bill Sabia of Rochester, New York, did the final art. Leslie B. Norton, Julie L. Witt, and Shari Revell are thanked for typing the draft manuscript.

The Ergonomics Group
Health and Environment Laboratories

PART **I**

Introduction

CONTRIBUTING AUTHORS

Suzanne H. Rodgers Ph.D., Physiology

CHAPTER 1

Introduction

CHAPTER 1. INTRODUCTION

A. The Contents of This Book

B. The Industrial Need for Ergonomic Job Design
   1. Improve Worker Productivity
   2. Increase Available Work Force by Reducing Problematic Tasks
   3. Provide More Jobs for Older Workers and Women

C. Possible Impacts of Poorly Designed Jobs

D. Options Other Than the Ergonomic Design of Jobs for Addressing Difficult Jobs

E. Whom to Design Jobs For

References for Chapter 1

Job design includes an understanding of how much a person can do, how long a given level of effort can be sustained, how work can be organized or patterned to reduce the possibility of accumulating fatigue, and how external pressures, such as machine pacing, can influence the worker's perception of job difficulty. This book covers these subjects in depth in discussions of the demands of jobs and how to design them ergonomically. The physiological responses of a person doing work have been used here in the development of guidelines for the design of short- and long-duration work of light, moderate, or heavy intensity. Psychological measurements of ''acceptable'' workloads have been used especially in the development of guidelines about lifting tasks. Although the areas of job satisfaction and job enrichment are certainly a part of job design, they will not be discussed in this book. The interested reader is encouraged to read texts devoted to these topics (Gellerman, 1960; Herzberg, 1968; Maslow, 1943).

This volume extends the information in Volume 1 of *Ergonomic Design for People at Work* by including techniques for analyzing and designing jobs, especially physically demanding ones, in order to:

- Improve worker productivity by eliminating unnecessary effort and designing tasks within the capabilities of most workers.
- Reduce opportunities for overexertion injuries.
- Curtail the development of fatigue over the shift.
- Utilize workers' skills optimally, thereby increasing job satisfaction and individual fulfillment.

From our studies we are persuaded that well-designed jobs are more satisfactory to both worker and manager. When the worker's effort is efficiently directed to the production process and little is wasted on unnecessary activity, productivity will be high. Likewise, when the job demands are brought within the capabilities of more people and the worker has control over the way in which tasks are safely accomplished, job satisfaction will be high.

This chapter reviews the contents of the book and discusses the benefits to industry of using ergonomic principles in task and job design. The impacts of poor job design are included and some options for matching people to jobs are covered briefly. A final section indicates who should be considered in job design; the guidelines in this book are based on the capabilities of that population.

A. THE CONTENTS OF THIS BOOK

Among the topics covered in this book are: scheduling of work and recovery periods; the amount of force that can be exerted and weight lifted as a function of task duration and frequency of occurrence; the design of shift work and overtime schedules; and the design of highly repetitive and machine-paced jobs. To support the guidelines and to assist the industrial human factors or ergonomics practitioner in solving problems, this book also contains methods for evaluating existing job demands, data on human strength and endurance capabilities, and information on the energy requirements of industrial and personal activities. Background information on work physiology and biomechanics has been included to aid the reader who is not familiar with biology. A brief summary of Parts II through VII follows.

Part II, "The Physiological Basis of Work," provides information on the way work affects the body. Specifically, it discusses the use of energy by muscles, how muscles produce movement, how the body's circulation and respiration support muscular work and define its limits, and the role of the nervous system in coordinating and controlling the body's response to increased work requirements. A section on biological, or circadian, rhythms of the body that can be affected by shift work or extended hours of work is also included. Part II, then, provides the reader with the basic physiological information on which to build an understanding of the job design guidelines in the rest of the book.

Part III, "The Evaluation of Job Demands," describes techniques for evaluating the workload of existing industrial jobs. A survey approach is outlined, as well as specific methods that are noninvasive and do not require significant capital expenditures. Use of these techniques allows the ergonomics practitioner to quantify the problem so that its seriousness can be evaluated.

Part IV, "Patterns of Work," gives guidelines for the design of work and recovery periods for physically demanding jobs and for lighter jobs that are highly repetitive or machine-paced. The pattern of work can determine how much worker fatigue will develop over a shift, influencing the potential for overexertion injuries. These guidelines are intended for use in the design of new jobs, but information is also provided to assist in the redesign of existing jobs with suboptimal work patterns.

Part V, "Hours of Work," contains information about the impact on workers of alterations in the length of time worked per shift and in the time of day worked. Examples of shift schedules are given, as well as information about other factors that may affect worker productivity on overtime schedules. The guidelines for selecting shift work or overtime schedules should assist compensation or production personnel when increased production time requires additional labor or extended hours.

Part VI, "Manual Materials Handling," presents guidelines for the design of common industrial handling tasks, such as lifting, carrying, pushing, and grasping tasks. The length of time the activity has to be sustained and its frequency of occurrence are included in the recommended task designs. This information is drawn from published studies, internal research, and from wide experience in problem solving in the plant.The ergonomics practitioner should be able to use the guidelines to design new jobs and to help in the evaluation of "problem" handling tasks.

The Appendices in Part VII include information on human strength, motion, and endurance capabilities, job and task energy demands, and methods for evaluating human capabilities such as strength and aerobic work capacities. In addition, problems are presented with examples of how material in the book can be used to design or redesign jobs. These data should help the more experienced ergonomics practitioner solve problems that are not specifically addressed in this book. The sample problems are ones common to industry, and the solutions are ones that have been found successful in practice; because each situation is different, there can be more than one solution to a general problem.

Finally, a Glossary and an Annotated Bibliography of books, monographs, and journals in the fields of ergonomics and job design are included. The interested reader should consult the Glossary for clarification of unfamiliar terms and the Annotated Bibliography for additional information about the topics covered in this book.

B. INDUSTRIAL NEED FOR ERGONOMIC
   JOB DESIGN

The design of jobs using physiological and psychological principles and information about human capabilities and capacities for work is still not very widespread in American industry. There are a number of possible reasons for this:

- Some lack of recognition of how poorly designed jobs can reduce productivity or affect worker health and safety.
- A lack of practical guidelines than can be applied to job design.
- A lack of awareness of the information available or how to apply it.
- The unavailability of documented studies demonstrating the impact of the ergonomic redesign of jobs on productivity, accident rates, quality performance, and so on.
- An acceptance of the existence of ''hard'' jobs, and a willingness to recognize the difficulty of finding people to do these jobs by increasing the compensation for them.
- The existence of an informal ''natural selection'' process on these jobs, whereby people who find the job too difficult weed themselves out over time, leaving only people with the highest work capacities on the job.

This book addresses these points by presenting job design guidelines that have been applied successfully in manufacturing operations over the past 23 years. Table 1-1 summarizes some of the steps in the creation of a job in industry and gives examples of ergonomic information that can aid the process.

Three primary reasons for industry to incorporate ergonomic principles into job design are discussed below. They are:

- To improve productivity.
- To reduce the number of problematic tasks in jobs that make it difficult to staff them.
- To increase the number of jobs within the capabilities of women and older workers.

Table 1-1: Ergonomic Input to the Job Design Process   Column 3 gives examples of ergonomic information or studies that can assist members of industrial departments (column 2) responsible for designing jobs. Column 1 identifies specific activities in the job design process into which ergonomic information can be integrated very early on in order to ensure that the jobs fit people's capabilities.

Activity	Departments Usually Involved	Ergonomic Input to Assist the Departments
Develop new product or increase the capacity of an existing production system.	Research Production Engineering	Evaluate impact of increased capacity on existing line personnel.
Identify equipment and labor needs.	Production Planning Safety Industrial Hygiene Medical	Review equipment design or selection for ergonomic problems.
Identify costs of manufacturing the product.	Methods Engineering	Review labor requirements with respect to job demands.
Define tasks for people.	Industrial Engineering Industrial Hygiene Safety	Determine best utilization of people vs use of machines.
Determine hours of work and shift schedules.	Compensation Personnel Medical	Advise on shift work schedules and on work practices for shift work and overtime.
Identify skills needed to do tasks and structure tasks into jobs.	Production Compensation Personnel Industrial Hygiene	Define job demands from a physiological and psychological perspective.
Define compensation levels.	Production Compensation	Determine reasonable workload and environmental exposure levels.
Define productivity and quality standards for each job.	Industrial Engineering Medical Industrial Hygiene	Estimate workload, define recovery allowances, and identify performance capabilities where known.
Initiate new or expanded production process.	Production Safety	Evaluate job demands in relation to population capabilities.

1. IMPROVE WORKER PRODUCTIVITY

Productivity losses associated with jobs that are not designed according to ergonomic principles can be very significant. Several work tasks or situations contribute to worker fatigue or stress and, therefore, may result in decreased performance on the job over the shift. Job conditions such as machine-paced work or external pacing based on job standards that fail to recognize the variability in work styles and in individual capabilities can also affect productivity. Lack of control by the worker over the way the work is done may reduce job satisfaction and affect the quality of the work (Sen, Pruzansky, and Carroll, 1981). Table 1-2 summarizes some job conditions and describes the mechanism by which productivity is lost. Since many job standards are set after ergonomic problems

Table 1-2: Job Design Factors that Contribute to Losses in Productivity   Three job design factors (column 1) are listed with examples of industrial tasks that may include them (column 2). Column 3 explains the mechanisms through which these factors may be associated with losses in productivity. Examples of improvements in the workplace or task design that can overcome the productivity-reducing factor are given in column 4. Each of the factors listed has been observed to reduce productivity in studies carried out in manufacturing areas.

Job Design Factor	Example(s)	Mechanism of Lost Productivity	Design Improvement
Highly repetitive low lifts	Loading or unloading pallets on the floor.	Unnecessary effort is required with the "calisthenics" of raising and lowering the body. This increased workload results in less productive effort because more recovery time is needed.	Raise pallets to above 50 cm (20 in.).
Static muscle loading	Continuous holding of an object or tool. Elevation of arms in overhead work. Stooping or crouching repeatedly.	Local muscle fatigue can limit continuous work time on a task to less than a minute with several minutes needed for recovery.	Provide holding fixtures. Avoid work tasks that require frequent or constant overhead work and low clearances that require awkward postures.
High-paced work	Inspection of fast-moving product on conveyors whose speed cannot be controlled by the operator.	Increased waste and quality control problems can be associated with nonoptimal conditions. Lack of control of conditions can stress inspectors when product quality problems require adjustments in conveyor speed.	Provide some control for the inspector by unlinking the work station from the direct conveyor line. Conveyor "spurs" to take product off and put it back on the line have been used successfully.

like these are designed into the job, demonstration of the productivity losses is often difficult until the redesign process is completed.

2. INCREASE AVAILABLE WORK FORCE BY REDUCING PROBLEMATIC TASKS

An industry seldom has large numbers of jobs that exceed the capabilities of most of the work force for the duration of a shift. In one study, only 4 percent of the jobs, involving less than 1 percent of all production workers, were identified as very difficult (Rodgers, Eastman Kodak Company, 1979). Far more jobs may include a task that exceeds the strength of many workers, requires extended reaches or overhead work that cannot be sustained for long periods by many workers, or has time pressure that is difficult for many people to adjust to or sustain. By using ergonomic principles to design these tasks, more people should be able to perform the job without risking injury or serious fatigue.

3. PROVIDE MORE JOBS FOR OLDER WORKERS AND WOMEN

Two recent phenomena in industry should stimulate the application of ergonomic principles to job design. These are the increasing numbers of women entering the work force, many in nontraditional jobs where physical effort tasks may be moderate to heavy, and an increasing number of older workers, occasioned by later retirement ages and increased longevity and also related to more transient economic phenomena such as less expansion of the work force or the layoff of younger workers in a recession. Because women and older workers have, on the average, lower work capacities than young men, jobs designed with the latter in mind can be problematic. Use of job design guidelines that accommodate a large majority of the work force makes it easier to find people to do the job. It also reduces potential physical barriers for women and older workers who must do that job if they wish to progress in a work area.

C. POSSIBLE IMPACTS OF POORLY DESIGNED JOBS

The goal of the ergonomics practitioner in job design is the same as the goal of the industrial engineer or wage and salary analyst who usually determines job content: to provide for maximal productivity with minimal cost. For the ergonomist, "cost" is expressed as the physiological or psychological cost to, or stress on, the worker. These costs can be quantified when the following measurements are taken:

- Heart, or pulse, rate.
- Blood pressure.
- Energy expenditure.
- Local muscle fatigue.

- Individual perceptions of stress associated with attention requirements, time pressure, or perceptual demands, such as viewing small defects. These can be measured using psychometric methods.

For longer-duration stress or more intense effort, blood or urinary levels of stress hormones, such as catecholamines, may also indicate the cost of a given job design or shift work schedule. By using ergonomic principles to structure tasks, jobs, and hours of work, job designers can keep the stress levels within acceptable ranges for most people.

In every manufacturing operation there are jobs that are difficult to design for most people. In some instances the jobs are carried out in old facilities that were designed to accommodate healthy young male workers at a time when few women in the work force were in those types of jobs. In other cases, technology has not yet been developed to replace equipment that is difficult to operate. People who work in a production department where one of these jobs exists are usually aware of it because it is identified as a "problem job." The following list indicates some of the measurable impacts of a poorly designed job on the work force, productivity, and management goals (Rodgers, Eastman Kodak Company, 1979):

- Increased potential for accidents and overexertion incidents until "natural selection" has established the suitable work force.

- Increased turnover of people in the job as "natural selection" weeds out those who do not have the capacity to do the job.

- Difficulty in finding people to fill in on the job when the regular worker is on vacation or absent due to illness.

- Increased training or acclimatization time to the job, which is related to its high demands.

- Increased time spent by the worker in secondary or auxiliary tasks that serve as recovery periods from the more demanding primary tasks. This can result in decreased production per hour, for example, when hours per shift are increased.

- Pressure on the compensation department to increase the wage level of the job to reflect the difficulty of finding people who can do it.

- Reduced personal job satisfaction if the worker sees inefficiencies in the way he or she has to work but has no control over them. This is particularly true if a job is high-paced and the worker cannot vary task speed to accommodate differences in the quality level of parts, for example.

For physically demanding work, the following results are also possible:

- A possible reduction in the quality of the work over the course of the shift if the worker tires and makes more errors. For example, if a worker experiences local muscle fatigue, he or she may drop a product during a handling task, thereby increasing the possibility of damage and waste.

- Difficulty in returning people to work after an injury or extended illness, resulting in more people having medical restrictions from the job.
- Increased frequency of short-duration medical absences for mild ailments such as colds. These ailments can reduce the worker's capacity for heavy work, with the result that he or she stays home on those days instead of coming to work.
- Difficulty in getting women and older workers into these jobs.

The impacts of poor job design given above can also be used to help identify problem jobs in the workplace. Through analysis of safety, medical, industrial relations, and production department data bases, and through discussions with supervision and line personnel, ergonomics practitioners can easily identify jobs that merit redesign. Since many production departments have older workers and since some of these older workers may have developed chronic health problems such as heart disease or back pain, "problem jobs" can often be identified by asking which jobs would be difficult for people with those problems to perform.

D. OPTIONS OTHER THAN THE ERGONOMIC DESIGN OF JOBS FOR ADDRESSING DIFFICULT JOBS

It should be no surprise to the reader that the authors of this volume believe that designing jobs using ergonomic principles is superior to other ways of dealing with problem jobs in industry. However, two alternative approaches can be used successfully to deal with specific jobs that do not lend themselves easily to redesign. These are the selection of suitable people through testing based on the job requirements, and the improvement of individual physical fitness levels to meet the demands of heavy jobs. The latter can be done by providing conditioning programs for employees (D. Day, Eastman Kodak Company, 1979).

Some of the reasons why the above approaches are usually less satisfactory than job redesign using ergonomic principles are listed below:

- It is very time-consuming, expensive, and difficult to validate selection tasks for specific jobs. Each time the job changes, the validation should be repeated (Equal Employment Opportunity Commission, et al., 1978). One must show not only that people need a certain measurable capacity to do a job but also that people without this capacity invariably fail on the job. Because there is usually more than one way to accomplish a work task, it is often difficult to meet the latter validation requirement.
- Once a validated selection test has been developed, every person con-

sidered for that job has to be tested. If the job is one that people can be hired into directly, not one that requires prior experience in the department, a large number of people are likely to pass through the job, all of whom would need to pass the selection test. This means that substantial amounts of time can be tied up in the testing program. In the long run, redesign is often less expensive.

- Even if it is possible to find people to staff the difficult jobs, there are times when they will be on vacation, absent with illness, or needed in other work areas on temporary assignments. The people who fill in for them will also need to pass the selection test. The more difficult the job or task is, the harder it will be to find people who can "back up" the workers who are regularly assigned to these jobs.

- Making workers more physically fit to make them capable of sustaining heavy work may be an admirable goal; but, except in special circumstances like firefighting, it is difficult to accomplish from a practical standpoint. Unless the fitness program is built into the work schedule and provision is made for medical or physiological counseling of individual workers, physical fitness may not be sustained. Also, training for specific work tasks requires exercises that use the same muscles and joints in the same postures as they are used in the job. General fitness exercises may have limited value in developing the specific strengths and endurances needed on the difficult job (Mathews and Fox, 1976).

- The development of fitness for difficult jobs takes weeks and months. If the workers lose motivation to keep going through the fitness exercises at least two times a week after they are trained for the job, their fitness could fall below the recommended levels (Mathews and Fox, 1976). Other events that reduce job fitness on a temporary basis are: upper respiratory tract infections, sleep deficit, taking certain medications, recent recreational overexertion, emotional stress, physical environmental stressors such as heat and humidity, and temporary assignments to other jobs that have different requirements than those for which the fitness program has been developed.

Although the above examples of the difficulties of using selection tests and fitness programs have been directed to physically demanding work, similar problems exist if the tests and training programs are for lighter jobs. For example, identifying workers who will be less likely to aggravate muscle or joint problems in highly repetitive tasks, or who will be able to tolerate continuous work on machine-paced jobs over which they have little control is also time-consuming, expensive, and difficult to validate. For these reasons, it is recommended that selection and training be reserved for dealing with job problems that are either technologically or economically unfeasible to redesign.

E. WHOM TO DESIGN JOBS FOR

In the first volume of this series, the goals for the design of workplaces and equipment were to accommodate people with the largest dimensions for clearances and to find the best compromise between people with long arms or legs and those with short limbs when designing workplace reaches and working heights. For information transfer and environmental stressors, one should design for as many people as possible, which generally means trying to include people from the lower capacity or tolerance end of the population's distribution of capacities. In job design, one has to be concerned about people with low capacities who risk overexertion injuries in heavy jobs, but one cannot underutilize people with higher capacities for the sake of accommodating the weakest potential workers. For light jobs with complex decision-making requirements, one should not remove all of the challenge of the job by breaking it up into simple steps, as this will reduce the motivation of the more skilled or capable workers to sustain quality performance.

It is not bypassing the question of who should be considered in design to say that each job situation has its own unique needs. In general, one should try to accommodate a majority of the potential work force in the area; this includes men and women, and older and younger workers. Figure 1-1 illustrates frequency distributions of handgrip, arm, and back maximum isometric muscle strengths for industrial men and women between the ages of 18 and 60 years. It is clear from the curves that there is much more variability in the strengths of men than of women, as seen by the broad ranges of the men's distributions and the narrow ranges of strength for the women. If one designs a task requiring handgrip strength to meet the maximum isometric strength of the average man, for instance, almost all of the women will have difficulty with it. If you try to accommodate 95 percent of the women, the requirements will be far below the capabilities of most of the work force. The compromise approach (depending upon other factors that the analyst would observe when studying the job) could be to design so that at least half of the female population could do the task.

Designs for shift work and for highly repetitive light work on which some people may experience joint and muscle problems usually try to accommodate most people, including all but the ones most susceptible to occupational illness. People who have a high tendency to develop medical problems on any of the types of jobs discussed in this book should be dealt with on an individual basis; accommodations should be made for them in the workplace, instead of trying to design the whole job to their level.

The identification of ways to accommodate people with lower work capacities or skills is recommended as one way of designing work so that people with higher capacities can remain challenged on the job. These accommodations may be: providing handling aids, providing visual aids, changing the work/rest cycle organization, making additional help available for specific tasks, or making other modifications of job demands. It is important to develop management policies about such accommodations, including how they will be dealt with when

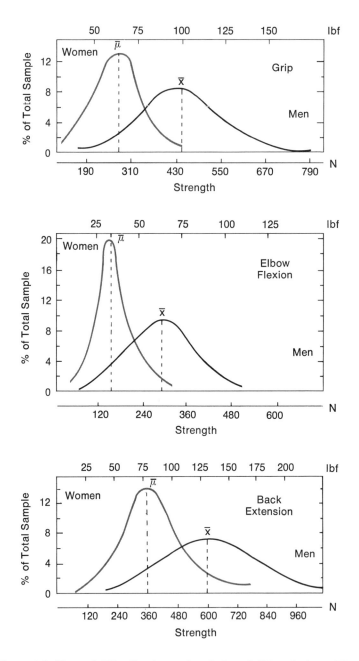

Figure 1-1: Strength Distributions of an Industrial Population   Three sets of frequency distributions of the maximum static strengths of hand-grip (*a*), arm muscles (*b*), and back muscles (*c*) are shown for industrial men and women. In each set of curves the horizontal axis shows the actual maximum strength in newtons (N) and in pounds of force (lbf), and the vertical axis shows the percentage of the population studied that had that much strength. Static strengths were measured by having each person pull or grasp maximally for five seconds against a stiff re-

job performance is evaluated, how other workers will react to them, and how their use and availability can be assured.

In summary, the goal of job design should be to accommodate as many people as possible while minimizing the number of people who will be either overstressed or understimulated by the work. The weaker or lower-capacity workers can often be accommodated as individuals on a difficult job, permitting the overall job design to match more closely the skills and capabilities of the somewhat stronger members of the population, such as those in the 25th percentile of a capability range.

REFERENCES

Equal Employment Opportunity Commission, Civil Service Commission, Department of Justice, and Department of Labor. 1978. "Uniform Guidelines on Employee Selection Procedures." No. 6570-06, Part 1607. *Federal Register, 43 (166)*: pp. 38290-38315.

Gellerman, S. W. 1960. *People, Problems and Profits.* New York: McGraw-Hill, 254 pages.

Herzberg, F. 1968. "One More Time: How Do You Motivate Employees?" *Harvard Business Review, 46:* pp. 53-62.

Kamon, E., and A. Goldfuss. 1978. "In-Plant Evaluation of the Muscle Strength of Workers." *American Industrial Hygiene Association Journal, 39 (10):* pp. 801-807.

Maslow, A. H. 1943. "A Theory of Human Motivation." *Psychological Review, 50:* pp. 370-396.

Mathews, D. K., and E. L. Fox. 1976. *The Physiological Basis of Physical Education and Athletes.* 2nd edition. Philadelphia: W. B. Saunders, 577 pages.

Sen, T. K., S. Pruzansky, and J. D. Carroll. 1981. "Relationship of Perceived Stress to Job Satisfaction." In *Machine Pacing and Occupation Stress,* edited by G. Salvendy and M. J. Smith. London: Taylor & Francis, Ltd., pp. 65-71.

sistance, represented by a Stoelting hand dynamometer. The frequency distributions of the women's strengths fall to the left of the men's values. In addition, they form a more pronounced peak while the men's distributions are broader. This indicates that the industrial women had less strength and that there was less difference among them than among the men. If a task is designed for an average man's strength ($\bar{x}$), most women will have difficulty performing it. If the task is designed to accommodate the strengths of at least 50 percent of the women ($\bar{\mu}$), about 75 percent of the total potential work force, men and women, should be able to do it. (*Adapted from Kamon and Goldfuss, 1978. Reprinted by permission of the American Industrial Hygiene Association Journal.*)

PART **II**

The Physiological Basis of Work

CONTRIBUTING AUTHORS

David M. Kiser Ph.D., Physiology

Suzanne H. Rodgers Ph.D., Physiology

The guidelines for job design presented in this book under work patterns, hours of work, and manual materials handling are based largely on information about people's capacities for work and their physiological and psychological responses to it. This part discusses the physiological basis of muscular effort and of a person's capacity for physical work in terms of the intensity and duration of effort needed to do a job. The circulatory and respiratory support of and limits to muscle work and the effects of environmental stressors are also reviewed. The underlying biological rhythms that affect people's tolerance of unusual work hours or affect their ability to sustain performance on a continuous task are presented in the final chapter.

Although individuals vary in their physiological work capacities, it is still possible to generalize about group responses to specific job stresses. Psychological responses to work, on the other hand, are highly individual and are not addressed in this part. It is difficult to generalize about how psychological stresses affect the work capacities of employees except when these stresses are quantified by interview or questionnaire data in specific industrial studies. This has been done in several studies included in both volumes of this book. For example, psychological data have been used in Volume 1 to establish the guidelines for population stereotypes, dials and displays, and controls (Chapter

III, Volume 1), for instruction design, coding, and inspection (Chapter IV, Volume 1), and for tolerance of noise, vibration, heat and humidity, as well as for proper illumination levels (Chapter V, Volume 1). Methods for collecting psychological or opinion data are included under questionnaire design in "Information Transfer" (Chapter IV, Volume 1) and under "Psychophysical Scaling Methods" (Appendix B, Volume 1) In this volume, psychological data are used extensively in the discussions of work pacing, repetitive work, and perceptual work stress (Part IV) and in the determination of appropriate shift and overtime schedules (Part V). The psychological data behind these guidelines are explained in each section, where the study results are presented.

Information in this part will help the reader understand why the analysis of work is not simple or easily modeled by computer analysis. Techniques to evaluate the demands of jobs are given in Part III; the reader's interpretation of data gathered using those techniques will be more cautious if he or she first understands the underlying physiological principles of work presented in this part. The material covered here goes from the muscle cell, where energy is made available to the muscle, to the coordination of muscle activity, to the oxygen and nutrition support systems of circulation and respiration, and, finally, to overall control systems represented by biological rhythms.

CHAPTER 2

Muscular Contraction and Movement

CHAPTER 2. MUSCULAR CONTRACTION AND MOVEMENT

A. Metabolism, Energy Transformation, and Energy Utilization

    1. Aerobic and Anaerobic Combustion of Nutrients
    2. The Formation and Utilization of High-Energy Compounds

B. Anatomy of Muscle and Muscular Contraction

    1. Anatomy of Muscle
    2. Muscular Contraction

C. Motor Units

D. Muscle Mechanics and Movement

    1. Muscle Mechanics
    2. Movement

E. Motor Control

    1. Proprioceptors
    2. Supraspinal Control
        a. Motor Control Areas of the Brain
        b. Ascending and Descending Pathways for Motor Control

F. Motor Learning and Skill Acquisition

G. Fatigue and Performance Impairment

References for Chapter 2

T he use of muscles to effect movement is the basis of all industrial tasks. Whether the tasks are physically demanding and require large muscle groups or are less demanding and mainly use the muscles of the fingers and eyes, muscles must contract to do the job. Frequently asked questions in industry are: "How much is too much to lift?" and "How often can it be lifted?" The intensity, duration, and frequency of muscle effort, plus other factors, determine whether a task is designed for most people, or whether it is difficult for all but a few very strong or fit people. Table 2-1 summarizes some of the types of muscle effort in common industrial tasks.

This table is a simplification of the muscle activity involved, but it demonstrates the complexity and degree of interrelation of the variables of intensity, duration, and frequency. The need to identify which factor limits the amount of work a person can sustain productively for a full shift is further discussed in the survey methods section of Part III and also in Part IV.

Table 2-1: Types of Muscle Effort in Several Work Tasks Nine examples of industrial tasks are given in column 1, ranging from heavy to very light physical effort. The intensity, duration, and frequency of muscle effort for each task are indicated in columns 2-4. *Intensity* is categorized as heavy, moderate, or light according to the oxygen demands or strength requirements of the task. *Duration* is the continuous period of muscle contraction or the total amount of time per minute the muscles are active. *Frequency* defines the number of work cycles per minute, hour, or shift and defines the pattern of work. The muscle groups most actively involved are shown in column 5. ATP usage by the working muscle will be greater with more intense contractions, longer strenuous effort, and more frequent repetitions. Each of these aspects of work should be evaluated when assessing the appropriateness of a given work task.

Task	Muscle Effort			Muscles Involved
	Intensity	Duration	Frequency	
Lifting a 23 kg (50 lbm) box from a 76-cm (30-in.) high conveyor to a 102-cm (40-in.) high table for inspection or labeling.	Heavy	<3 sec per lift	Up to 12 per minute	Arms and shoulders
Lifting a pallet off the floor and putting it on a stack of 5 other pallets.	Heavy	<6 sec	6 or 7 times per shift	Back, arms, shoulders, legs

Table 2-1 (*Continued*)

Task	Muscle Effort			Muscles Involved
	Intensity	Duration	Frequency	
Pushing a loaded cart down a long corridor.	Moderate to heavy	About 1 to 2 min, continuous	A few times per hour	Back, arms, shoulders, hands, legs
Packing boxes with product weighing 5–10 kg (11–22 lbm); standing work at 90-cm (35-in.) high table.	Moderate	<3–18 sec per lift	Up to 6 per minute	Arms, hands, shoulders
Painting overhead, arms elevated.	Moderate	Continuous, until person takes a recovery break; probably only a few hours a shift.	>20 brush strokes per minute	Shoulders, arms, hands
Activating foot pedal on a production machine, such as a hand press.	Moderate	<1 sec	Several times per minute	Foot, leg, lower trunk
Assembling a small appliance weighing less than 1 kg (<2 lbm) at a seated workplace.	Light	>1 min	Usually 0.5 to 1 per minute	Arms, hands
Inspection of small parts moving by workplace on conveyor belt.	Light	Continuous except for rest breaks	Visually inspect thousands of parts per shift. Move hands a few times a minute to remove bad parts or record data.	Eyes, hands arms
Entering text into a word processor.	Light	Continuous except for rest breaks	>250 strokes per minute	Fingers, arms shoulders

Information about the structure and function of muscle, including its energy usage, and about how it is linked to nerves and to skeletal structures helps explain how movement is produced and why it is important to design strength-requiring jobs using ergonomic principles. This chapter covers the mechanisms by which maximum muscle strength is developed and the factors that influence muscle strength. Coordination, skill acquisition, and training are included to indicate how a person learns to perform an industrial task successfully. A short discussion of neuromuscular fatigue and how it relates to the design of work in industry concludes the chapter.

A. METABOLISM, ENERGY TRANSFORMATION, AND ENERGY UTILIZATION

In order for a person to do work, he or she has to use muscles. Muscles work only when there is energy to permit muscular contraction to occur. This energy, in the form of high-energy phosphate compounds such as creatinine phosphate (CP) and adenosine triphosphate (ATP), is transformed from nutrients by oxidative and nonoxidative processes known as metabolism. The creation of high energy phosphate compounds is called phosphorylation; all body cells are capable of forming and using them to sustain life. Muscle cell metabolism and energy transformation are discussed below since these processes are needed for muscle contraction and, thus, for work.

1. AEROBIC AND ANAEROBIC COMBUSTION OF NUTRIENTS

The two nutrients most involved in providing energy for muscular work are cabohydrates, or sugars and starches, and fats, or free fatty acids (Issekutz and Miller, 1962). These nutrients come initially from the food we eat. They are broken into less complex molecules by the digestive system, absorbed into the blood stream from the intestines, and transported to the cells of the body. The liver plays a special function in storing and breaking down the nutrients to provide a source of energy for work that lasts more than a few minutes. The liver and muscle cells store the carbohydrates as glycogen, a starch built up from glucose delivered by the blood supply. This becomes one important source for high-energy compound formation for the working muscles.

Nutrients can be burned to provide energy for muscle work both with oxygen (aerobic) and without (anaerobic). The aerobic combustion of nutrients refers to the oxidation of glucose or glycogen and fatty acids. Small amounts of oxygen are stored bound to the muscle protein myoglobin and can be drawn on in short intense work periods. Most of the oxygen is delivered to the muscle cells from the circulatory system, with blood flow to the muscles matching the muscular demands for oxygen after a few minutes of work.

Because it takes up to one minute for blood flow to respond to a sudden increase in workload similar to that seen in many industrial tasks, it is not always possible to have adequate amounts of oxygen available at the beginning of work. The muscle can contract without oxygen through two mechanisms: it can use the high-energy phosphates stored in the muscle, and it can generate a limited amount of energy through anaerobic glycolysis (the breakdown of glucose to lactic acid). Work cannot be sustained for very long if glycolysis is the only energy source, because lactic acid will accumulate in the active muscles and, along with declining levels of the high-energy phosphate compounds ATP and CP, bring about rapid fatigue (Margaria, 1972). When enough oxygen is available, oxidation can occur and provide additional high-energy phosphate compounds for muscle work. A byproduct of oxidation is carbon dioxide, which must be removed by the blood in order to keep the muscles working efficiently.

Through the processes of glycolysis and the oxidation of glucose and fatty acids in the muscle, energy is released. The high-energy phosphate system traps this energy during the metabolic combustion of nutrients. Adenosine triphosphate (ATP) then provides energy to the cell as needed to fuel chemical reactions in the body, which then lead to growth, activity, secretions, and the generation of body heat.

By measuring the amount of oxygen a person uses during work, we can determine the amount of aerobic metabolism taking place. Energy expenditure, in kilocalories (kcal) per minute, is calculated by multiplying the oxygen use, in liters per minute, by 5. (A watt equals 70 kcal per minute.) The amount of anaerobic metabolism used during work can be estimated by measuring the excess oxygen used, compared to resting values, during recovery from a physically demanding task.

A goal in designing industrial tasks requiring moderate or heavy effort is to keep the demands low enough to permit aerobic metabolism to supply the necessary energy. If the demands are very high, anaerobic metabolism can result in a buildup of lactic acid. If the work comes in short, very intense bursts, such as those seen in firefighting or other emergency responses, the amount of ATP and CP available for sustained work will be reduced. The worker can fatigue more rapidly in these conditions and will take longer to recover between the tasks. This means that industrial job designers should reduce the intensity and duration of peak demands and provide adequate recovery allowances for the full shift's workload.

2. THE FORMATION AND UTILIZATION OF HIGH-ENERGY COMPOUNDS

The chemical reactions of anaerobic and aerobic metabolism both provide energy for the formation of high-energy phosphates and need energy to proceed. The adenosine triphosphate (ATP) and creatinine phosphate (CP) compounds are energy carriers and are present in each cell. If a reaction needs energy to proceed, such as the relaxation of muscle after its contraction, ATP will be broken down by the enzyme adenosine triphosphotase (ATPase) to yield the necessary energy. When energy is released, as it is in several steps of the oxidation of nutrients, adenosine diphosphate (ADP) will pick up another phosphate molecule and become ATP again, thereby trapping the energy. This is illustrated by the following equation:

$$\text{ATP} \overset{\text{ATPase}}{\rightleftharpoons} \text{ADP} + \text{P}_i + \text{Free Energy}$$

where ATP and ADP are as defined above, P_i is the inorganic phosphate split off of ATP or combined with ADP, and the free energy is available for use in muscle contraction and relaxation.

The total supply of ATP at any one point in a 75 kg (165 lbm) person with 20 kg (44 lbm) of muscle is about 4 kilojoules, or 1 kilocalorie of energy. Consequently, the ATP supply must be replenished constantly to permit muscular work to continue.

The role of CP in energy transformation in the cell is primarily as a storage system for ATP. It replenishes the ATP supply as the latter is used in muscle work or in other bodily activities (Gollnick and Hermansen, 1972). A person has about 15 kilojoules, or 3.6 kilocalories, of CP-stored energy available at any one time. This gives a pool of stored high-energy compounds of less than 20 kilojoules (<5 kilocalories), enough to support moderately heavy work for about one minute. If this supply drops to about 50 percent of its resting level, and more lactic acid is formed, the worker's muscles feel fatigued (Margaria, 1972).

Aerobic metabolism generates almost 20 times as much ATP as does anaerobic metabolism (Harper, 1959). Feelings of fatigue can be minimized by assuring that the workload intensity is low enough to be done with adequate oxygenation or that it only lasts a few seconds. Figure 2-1 illustrates how energy is supplied in the first minutes of moderately heavy work; for example, a production worker moving in quickly to keep a machine running by rapidly unloading product from it. The high-energy phosphate stores are exhausted within the first minute of work. Lactic acid will be formed until enough oxygen can be delivered by the muscle's circulation. If the work is extremely heavy and exceeds the ability of the circulation to deliver oxygen for ATP formation, the pool will be exhausted and muscle contraction cannot take place. Several hours may be needed to restore the energy supply after exhausting work.

The limits for work lasting more than a few seconds are called *maximum aerobic work capacities;* they vary according to the worker's age, sex, fitness level, and the number of muscles involved in the work. They are discussed in detail in Part IV under work/rest cycles and in the discussions of overtime in Part V and of repetitive lifting in Part VI. In addition, Appendix A includes normal values for maximum aerobic work capacities of industrial men and women. Appendix B includes methods of measuring the work capacities of the whole body and upper body.

High-energy phosphate compounds are required for the chemical reactions of many body processes; their role in muscle contraction and relaxation is of most interest to ergonomists. A more detailed discussion of the muscular contraction and movement processes is presented in the next section. ATP is involved in splitting apart the actin and myosin proteins and moving the calcium ions back out to the cell membrane where they await the next nerve impulse.

Since most muscle activity involves repetitive stimulation of muscle fibers and rapid use of ATP, oxidative combustion of nutrients is needed to replenish the high energy phosphate stores. ATP utilization is measured indirectly by measuring the amount of oxygen used (oxygen consumption) during both work

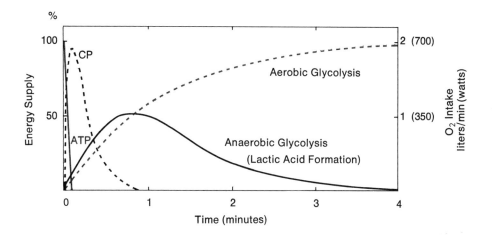

Figure 2-1: Energy Supplies in the First Minutes of Moderately Heavy Work   Four curves are shown representing the percent usage (vertical axis, left side) of high-energy phosphate stores, ATP and CP; of anaerobic glycolysis (lactic acid); and of aerobic metabolism to provide energy for moderately heavy muscle work. Only the first few minutes of exercise are shown (horizontal axis). The ATP stores are exhausted in a few seconds, while the CP stores last about 45 seconds at this workload (O_2 intake, vertical scale, right side). Anaerobic glycolysis supplies less and less of the energy required as the duration of exercise increases and aerobic metabolism takes over. The time until aerobic glycolysis provides 100 percent of the energy supply represents an oxygen debt that must be repaid during the subsequent recovery period. *(Adapted from Jones, et al., 1975.)*

and recovery periods. For several minutes after a heavy task is done, the worker's oxygen use remains elevated above resting levels. This excess oxygen use is refilling the stores of ATP and CP and increasing the myoglobin and blood oxygen stores so that the active muscles will be ready for the next work period. The heavier the work, the more time will be needed to fill the stores, so the longer will be the recovery period needed. The work and rest cycle guidelines in Part IV are based on this information.

B. ANATOMY OF MUSCLE AND MUSCULAR CONTRACTION

The performance of work requires energy. This is true when a secretary types a letter and when a laborer shovels slag in a cleanup operation. The effective utilization of energy in the human body is made possible through a coordinated effort among muscle, nerve, and bone. In order to understand clearly how the human body responds to and adapts to work, we need to take a closer look at this effector system.

1. ANATOMY OF MUSCLE

Skeletal muscle is a complex arrangement of thousands of contractile fibers bound by connective tissues into a functional unit (Figure 2-2a). These layers combine at each end of the muscle to form tendons. It is through these tendons that force developed by skeletal muscle contraction is transmitted to the levers of the body, the bones, to generate the forces required to perform work. The cross-sectional area of a muscle is measured for the whole unit within the sheathlike outer membrane, the epimysium. A table of cross-sectional areas for arm and leg muscle groups is given in Appendix A. These give some indication of the amount of tension a muscle can develop, greater areas being associated with higher strengths for similar muscle types. Individual muscle fiber cross-sectional areas are also associated with increased tension development.

The contractile unit of skeletal muscle is the myofibril. The myofibril is comprised of many protein filaments that, through complex associations, are responsible for muscle tension development. Two basic types of proteins are present in the myofibril—a thick filament called myosin and a thinner one called actin. These proteins interdigitate, giving the myofibrils alternating light and dark bands (Figure 2-2b). The light, or I, bands contain only the actin protein. The dark, or A, bands contain both the actin and myosin proteins. At rest, or during submaximal muscle contractions, an intermediate, or H, zone appears where myosin alone exists. This zone may disappear as the actin filaments totally overlap the myosin filaments in strong contractions. Each actin filament is also attached at one end to a transverse membrane, the Z-line or Z-disc. Actin filaments extend from both sides of the Z-line to interdigitate with the myosin filaments.

The portion of the muscle fiber that falls between two successive Z-lines is a sarcomere. In its resting state, the actin and myosin filaments are optimally associated for tension development. As the length of the sarcomere changes, such as when the muscle is stretched or contracted, its ability to develop tension changes. These changes are discussed under muscle mechanics later in this chapter. The capacity of the whole muscle to generate force will be determined by the changes in individual sarcomere lengths. This is illustrated in the muscle strength data in Appendix A where joint angle is shown to influence the maximum isometric forces developed during elbow flexion. The degree of flexion determines sarcomere length in the myofibrils and the amount of overlap of the actin and myosin proteins.

2. MUSCULAR CONTRACTION

Stimulation of muscle to develop tension is accomplished by complex electro-chemical processes (Layzer and Rowland, 1971; Winter, 1979). Although muscle will contract in response to a directly applied electrical stimulus, it is usually activated by an electrical signal transmitted from the central nervous system by the axon of its own motor neuron. This pathway is described later under motor control. There is no physical connection between the axon and the muscle fiber, so a chemical transmitter is used to activate the muscle fiber at the motor end-

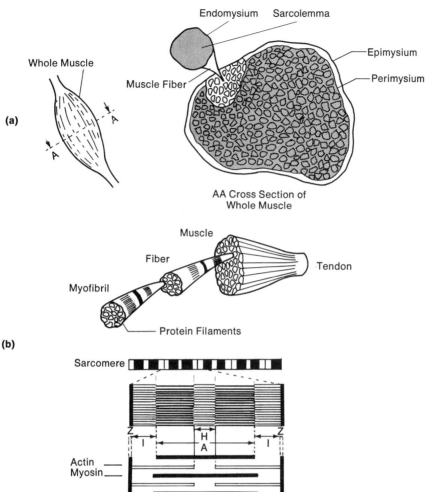

Figure 2-2: Gross and Microscopic Anatomy of Human Skeletal Muscle   The gross structure of a skeletal muscle, such as the gastro-cnemius muscle of the lower leg (calf), is shown in *a*. A cross section of the whole muscle is designated as AA and shows multiple bundles, or fasciculi, of muscle fibers. The membranes surrounding the individual fibers, fasciculi, and whole muscle are the endomysium, perimysium, and epimysium, respectively. They join to form the muscle tendon, which attaches the muscle to the skeleton. In *b*, the division of the muscle fiber into myofibrils is shown; these are made up of sarcomeres, the functional unit of muscle. The sarcomere is composed of filaments of actin and myosin, which combine to form actomysin in muscle contraction. The Z discs and the I, H, and A zones of the sarcomere are indicated. (*Part "a" adapted from The Physiological Basis of Physical Education and Athletics, Second Edition by Donald K. Mathews and Edward L. Fox. Copyright © 1975 by W. B. Saunders Company. Reprinted by permission of CBS College Publishing; Part "b" adapted from Textbook of Work Physiology 3rd Edition by P.-O. Astrand and K. Rodahl, copright 1986 by McGraw-Hill, also after Huxley, 1958.*)

plate, or myoneural junction. This area of the muscle is more sensitive to the chemical transmitter, acetylcholine, than are other parts of the muscle, and it is changed electrically, or depolarized. The result is muscle contraction and tension development, allowing work to be done. The enzyme cholinesterase breaks down the acetylcholine quickly, so the muscle can relax again. Thus, another impulse from the motor neuron is required in order to set up a new muscle contraction. Tension development in the muscle is controlled partially through this mechanism, resulting in a smooth increase in tension instead of a sudden jerk when muscle contraction occurs.

The pulling of the actin along the myosin during the active state occurs in thousands of locations. These many small forces summate and, through the functional arrangement of connective tissue and muscle, they are transmitted to the tendons and bones. The amount of tension developed is determined by the load on the muscle. For instance, for a person to lift a box with the elbow flexors, the amount of tension developed in the muscles will depend on the length of the lever (forearm) and the weight of the box. This is discussed in more detail in Part III under the biomechanical analysis of work. The duration of tension development depends on the activity of the motor neuron.

C. MOTOR UNITS

The sequence of muscle activation and movement initiation requires complex interactions between the nervous system and muscles. Skeletal muscle is a collection of muscle fibers arranged in parallel to produce force, through the muscle tendon, at a common point on a bone. For those fibers to contract together to produce motion, they must be activated by electrical signals from motor neurons; these fiber and neuron combinations form motor units. Motor units are the "final common pathway" for motor control of movement (Cooper, Denny-Brown, and Sherrington, 1926; Sherrington, 1947). Each unit includes an anterior motor neuron (AMN) and all of the muscle fibers innervated by its motor axon. A motor unit follows the "all-or-none" rule (Brown and Sichel, 1936). If the AMN is activated, all fibers supplied by it will contract.

The structure and function of individual motor units varies. The number of muscle fibers innervated by a single motor neuron also varies greatly depending on the function of the muscle (Hunt and Kuffler, 1954). For instance, the extrinsic eye muscles have about five fibers in a motor unit. This permits the very fine, graded control of eye movement that is important in visual tasks. Muscles acting on large masses, such as the leg muscles, can have several hundred muscle fibers in their motor units. These muscles are suited for gross movements requiring large forces.

The number of muscle fibers in a motor unit also varies within a specific muscle and is directly related to the size of the anterior motor neuron. The fibers of a specific motor unit are not segregated but intermingle with other motor unit fibers within the muscle (Burke and Edgerton, 1975). The terminal portion of a motor axon branches in order to innervate nonadjacent fibers in the motor unit. This partially explains why contractions of individual fibers are asynchronous even though the motor unit follows the "all-or-none" rule. This permits summation of activity to produce a stronger, smoother muscle contraction than would be seen if all fibers contracted together.

Motor units have biochemical and physiological differences as well as structural ones. This is true even in muscles with uniform fiber type composition, such as the soleus muscle of the leg. Some of the physiological characteristics of motor units are given in Table 2-2.

From the characteristics described in Table 2-2, we can see that motor units have specialized functions. Control mechanisms exist that allow a coordinated and effective use of appropriate motor units in muscle action to accomplish work most efficiently. This control mechanism centers on the anterior motor neuron (AMN), which lies in the anterior horn of the gray matter of the spinal cord. It determines the functional characteristics of a motor unit by acting directly on the muscle fibers of the motor unit.

Larger motor neurons have larger motor units. An important control system for movement depends upon the frequency of depolarization (activation or inhibition) of these AMNs. Smaller AMNs, generally, depolarize and conduct

Table 2-2: Characteristics of Motor Units The physiological and biochemical characteristics (column 1) of three motor unit types (slow twitch [S], fast twitch fatigue resistant [FR], and fast twitch fatiguable [FF] are shown in columns 2-4. The levels of each characteristic are given as either low (L), intermediate (I), or high (H), or between high and intermediate (H-I). The S motor units are recruited first in a muscle work situation, and the FF units come in last. (*Adapted from Burke and Edgerton, 1975.*)

Characteristics	Motor Unit Types		
	Slow Twitch (S)	Fast Twitch, Fatigue Resistant (FR)	Fast Twitch, Fatiguable (FF)
Muscle Fiber Type	"I"	"IIA"	"IIB"
Maximum Twitch Tension	L	I	H
Fatigue Resistance	H	H-I	L
Oxidative Capacity	H	H-I	L
Glycolytic Capacity	L	H-I	H
Frequency of Use	H	I	L

L = Low
I = Intermediate
H = High

impulses to their motor units more easily than larger ones do. This relationship, known as the *size principle*, states that there is an inverse relationship between AMN cell size and firing frequency (Henneman, Somjen, and Carpenter, 1965). The logic of this control is clear when the physiological characteristics of these motor units are considered. For example, prolonged standing does not require large muscle forces, yet continuous muscle activity is required. The small AMN motor units would be stimulated. These motor units are well suited to the physiological needs since they are fatigue-resistant, even though they have relatively low force output. If a sudden movement, like jumping aside, is required, the larger AMNs with larger and stronger motor units, are recruited. Repetitive movements like this are difficult since the large motor units fatigue rapidly.

Although, in general, the size principle has been shown to be operative, there are variations from this pattern: the AMN's activity and function can be altered by inhibitory or excitatory impulses coming from higher, or supraseg-mental, motor control centers.

The examples of motor unit involvement in a single, strong movement and in repetitive movements, such as jumping, are extremes. Movement can be controlled more precisely. There are two mechanisms that allow graded force output by the muscles through control of motor unit activity. These are temporal and spatial summation. Temporal summation occurs when the firing rate of a specific motor unit increases, resulting in higher muscle fiber activity within a given time period. Spatial summation occurs when additional active motor units are recruited. A finely controlled gradation of force within the functional range of muscle tension can occur in response to movement requirements by using these control mechanisms together or independently. This control permits a person to adjust muscle tension as the external load changes. For example, as an empty container is filled while being held, the muscles supporting it must increase their resistive forces. Increased activity of the AMNs that are already active and increased recruitment of other AMNs ensure that the appropriate muscle force is generated. The muscle tension "feedback" systems, or proprioceptors, that monitor the external load and muscle fiber tension are discussed later in this chapter.

D. MUSCLE MECHANICS AND MOVEMENT

The forces developed by muscles in the body in order to perform industrial tasks are the result of the contributions of the muscles' contractile and connective tissue and of the arrangement of the muscles on the skeleton. These aspects of muscle function as related to the production of movement are discussed here as well as in greater detail under the biomechanical analysis of work in Part III.

1. MUSCLE MECHANICS

Tension development in individual myofibrils is a key component in the production of movement, but many factors interact with the process to determine the actual function of the unit in the performance of work. Among the most important factors in determining tension levels are the muscles' mechanical properties. These include muscle length and the amount of connective tissue present (which determines the amount of stiffness in the muscle). The contractile elements of the muscle have a clear length-tension relationship with less tension developed at lengths less than and greater than the resting length of the sarcomeres (see Figure 2-3). The change in a muscle's ability to produce tension as a function of length is probably due to a reduced number of attached cross-bridges between the actin and myosin proteins at greater sarcomere lengths or to a high degree of overlap at shorter muscle lengths (Winter, 1979).

The length-tension characteristics of the muscle are also influenced by the connective tissue that surrounds the contractile elements. This connective tissue, or parallel elastic component, acts like an elastic band. At resting length or less, the parallel elastic element is slack and the muscle's length-tension curve is similar to that for the contractile elements. As the muscle is lengthened, tension increases in the connective tissue in a nonlinear fashion (Inman and Ralston, 1954). Muscles with greater amounts of connective tissue tend to be stiffer and will have higher passive tensions. Postural muscles that continually act against gravity show this characteristic. Muscles with less connective tissue, such as the biceps of the arm, will have lower passive tensions.

The total tension in a muscle is the sum of the contractile and parallel elas-

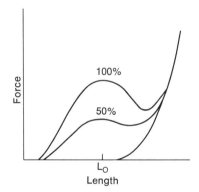

Figure 2-3: Length-Tension Curves as a Function of the Percentage of Maximum Strength Used   The tension, or force, measured on a muscle's tendon is shown on the vertical axis for two strengths of contraction: 50 percent and 100 percent of maximum voluntary contraction (MVC). Muscle length is on the horizontal axis, with L representing the resting length. At 50 percent MVC, the total tension is lower because the contractile tension is less. At lengths greater than resting length, the contractile tension is decreasing and passive tension from the elastic elements is increasing. The 50 percent and 100 percent curves merge at the point where contractile tension becomes negligible; the passive tension is constant for any given muscle length. (*Adapted from Winter, 1979.*)

tic tension. Since contractile tension depends on stimulus strength, the overall length-tension characteristics of a muscle are determined by the magnitude of motor neuron excitation (Figure 2-3). The amount of force measured on a muscle's tendon will increase as the percentage of maximum voluntary isometric contraction force (%MVC) is increased. This is true to the point where muscle length is so great that the contractile mechanism cannot operate. At this point, the only force generated is in the connective tissue elastic elements.

The length-tension characteristics of muscle are generally of concern during isometric, or static, contractions when muscle length remains relatively constant. Muscle strength measurements, including many of the ones given in Appendix A, are often made isometrically. Holding tasks and many postures require static effort of this kind where muscle length is constant during force development. Muscle length must change to produce movement, however. When this happens, the amount of tension that the muscle can develop depends also on how rapidly length changes, or the velocity of contraction. Figure 2-4 shows the force-velocity relationships for skeletal muscle at different muscle lengths. Velocity changes in the positive direction are shortening, or concentric, muscular contractions, such as those seen in the arm flexor muscles when an object is lifted upward. Negative velocity changes are lengthening, or eccentric, contractions. This is not a relaxation lengthening. The muscle's contractile machinery is actively developing tension but it is subjected to an external force greater than the force it can develop, and the muscle is forcibly lengthened. This can happen if the object being handled is heavier than expected and its weight straightens out the arms instead of allowing them to maintain a 90-degree elbow bend. The intersection of the force-velocity curve with the zero velocity axis represents the isometric force at a specific muscle length.

Much less information is available on the force-velocity characteristics of a muscle that is lengthened in an eccentric contraction. Eccentric contractions are as important to human movement as concentric ones are, and this relative lack of information is unfortunate. An example of eccentric contractions would be the lowering of fragile equipment from a shelf to a table top. The force-velocity curve for lengthening muscle does not follow the well-defined mathematical relationship for shortening muscle. Figure 2-4 shows that, at any velocity of lengthening, the force developed by the muscles is greater in lengthening than in either isometric or shortening (isotonic or concentric) contractions.

The mechanical properties of muscle are complex and are not independent of one another. Other factors that can modify muscle function are central nervous system activity, the chemical composition of muscle, fatigue, experience, and motivation. The relationships of some of these factors to movement and task completion are discussed later.

2. MOVEMENT

Muscles play several roles in the body. A muscle that provides the primary force for a specific movement is called the prime mover, or *agonist*. For example, the biceps brachii is the prime mover during elbow flexion. Depending on the com-

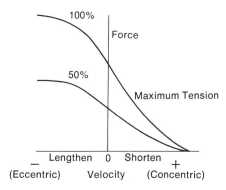

Figure 2-4: Force-Velocity Curves at Different Percentages of Muscle Strength—Concentric and Eccentric Contractions   The amount of force that can be generated by a muscle (vertical axis) as it is affected by the speed, or velocity, of shortening or lengthening (horizontal axis) is illustrated. Curves are given for two strengths of contraction: 50 percent and 100 percent of maximum voluntary contraction (MVC). At lower %MVCs, total force is reduced because the contractile tension is lower. Increasing positive velocity is a concentric, or shortening, contraction. The amount of force generated falls sharply as the speed of concentric contraction increases, reflecting a reduction in cross-bridge formation in the sarcomeres. A lengthening of the muscle during contraction is known as an eccentric contraction. As the speed of this contraction increases, more force can be developed by the muscle, reflecting a continued optimum condition for cross-bridge formation in the sarcomeres. Eccentric contractions can occur when a person picks up a heavy load and is unable to support it in the desired position. The muscles lengthen even though the signals to them are stimulating them to shorten. *(Adapted from Winter, 1979.)*

plexity of motion and force requirements, one or more muscles may act as agonists. A muscle can also be an *antagonist*; that is, its contraction produces motion opposite to that of the agonist. The triceps muscle of the upper arm is an antagonist for the biceps during elbow flexion, while the biceps is an antagonist for the triceps during elbow extension. Muscles can also act as stabilizers. They can fixate or support a bone or body part so that another active muscle has a firm attachment upon which to pull. Muscles can be synergists too, working together with other muscles to produce a desired action.

Individual muscle fibers and motor units by themselves do not give a person the ability to perform work. When an assembly-line worker lifts heavy product subassemblies onto and off a conveyor line, the complex interactions among the basic muscle function discussed previously, energy metabolism, and circulatory adaption are brought into play to develop and sustain the appropriate muscle tension. The muscle tension, or force, acts through the lever systems of the body, the bones, to create usable forces, or torques. A *torque* is a force exerted against a lever arm to counteract rotation and is expressed in newton-meters (or inch-pounds).

Figure 2-2 illustrates the various subelements of muscle and how they are bound together by connective tissue. These connective tissues merge at the ends of the muscle where they form tendons, which connect the muscle to the bone. The tendons provide the pathway for transmission of muscle force to the bones. It is this force applied against a bone that provides the torque required by the assembly-line worker to complete a task requiring the exertion of strength, such as tightening a part with a wrench, lifting the product and turning it over, or preparing the product for final packaging.

The tendon collects and transmits forces from many different muscle fibers. It connects to specific areas on the bones that are the origin and insertion points. When a muscle contracts strongly, both bones to which it is attached tend to move. The bone that moves the least is considered stationary, and the attachment of the tendon to this bone is called the origin. The tendon's attachment to the more freely moving bone is the insertion. In the development of torques, the muscle's insertion is where the force is applied. The force arm for the bony lever is the distance from the insertion point to the joint around which the lever rotates. Figure 2-5 illustrates these relationships and torques for a one-handed lift of a metal sphere.

The concept of torques is very important in the analysis of handling tasks and reaches. This topic is discussed in more detail in Part III as part of the biomechanical analysis of work. The relationship between human strength and torque will be considered here.

The questions "How much can a person lift?" or "How long or how often can a person lift?" imply a need for data on the maximum lifting strength capabilities of industrial populations. The answers to both questions are related to people's maximum strengths or some fraction of them. The term *maximum voluntary contraction* (MVC) is often used to describe the maximum force output of a muscle or a group of muscles. It is more correctly called a maximum torque. MVC is a generic term that can apply to an individual muscle or to a group of muscles. A single MVC does not exist, even for a single muscle. The maximum force will change as muscle length changes and as the biomechanics of the transmission of the muscle force through the lever system varies. For this reason, the joint angle and postures should be known whenever maximum strength data are used to determine biomechanical stress in evaluating a job or workplace design problem. Protocols for measuring isometric and dynamic strength are given in Appendix B: data from some of these tests are included in Appendix A.

The influence of muscle length and of the velocity of muscle shortening on the tension developed by a muscle or muscle group has been discussed earlier. Other factors contribute to skeletal muscle force development capabilities. Force or tension development is proportional to the cross-sectional area of a muscle. The enlargement of a muscle through weight training or through performing heavy effort on the job is associated mainly with an increase in the cross-sectional area of the muscle fibers (see Figure 2-2). Along with this size increase comes an increase in strength. Part of the training associated with jobs demanding heavy physical effort, localized to one limb or, generally, across many muscle groups, is to develop muscle strength by increasing fiber cross-

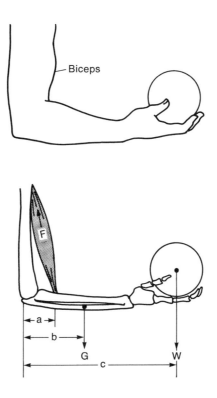

Figure 2-5: The Lever System of the Forearm   A person holds a metal sphere in the right hand with the elbow bent at 90 degrees to the upper arm. The lower drawing illustrates the forces acting on the forearm as measured from the axis of the elbow. Force W is the weight of the metal sphere; it acts on the lever arm c. Force G is at the center of mass of the forearm and represents the force due to gravity acting on the mass of the arm. Force G acts on the lever arm b. Force F is the muscle force needed to counteract the tendency of the forearm to fall away from the 90 degree angle due to the torques created by the mass of the forearm and the mass of the sphere acting on their respective lever arms. The muscle force creates torque by acting on lever arm a. F × a should equal (G × b) + (W × c) to keep the sphere supported as shown. Examples of work task evaluations using biomechanical analysis techniques are given in Part III and in Appendix C. (*Adapted from Williams and Lissner, 1962.*)

sectional area. The training is very specific to the muscle groups under load and will be maintained only as long as the load is present. With a larger cross-sectional area, the muscle fibers can generate more tension, so the job demands become a smaller fraction (%MVC) of the muscles' capacity. See the section on training in Part IV for more information.

Maximum torque development also depends on the biomechanical relationship between the muscle and bone. The hypothetical example in Figure 2-6 shows the importance of this relationship. Even though the force developed by the muscle is the same in both examples, the value displayed on the force indicator is different. This is important because it is the force on the gauge, not the actual muscle force, that counts when it comes to lifting objects or moving them during work. Further discussion can be found in Part III.

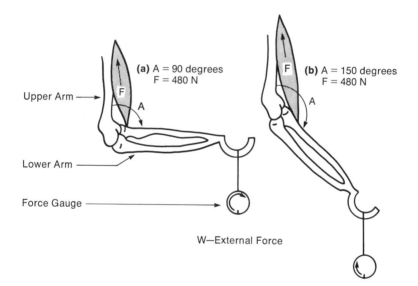

Figure 2-6: The Effect of Joint Angle on Measured Force for Equal Muscle Forces   The importance of elbow angle (A) in developing an external force (W) with the biceps muscle is shown. The elbow angle is formed by the upper and lower arm positions. It is about 90 degrees in *a* and about 150 degrees in *b*. In both examples the force exerted by the muscle is constant. The impact of elbow angle on external force development, as measured at the gauge on the hand, is shown. Increasing the elbow angle reduces the amount of force exerted on the gauge by changing the biomechanical lever system. The measured force is the one of interest when guidelines for force application and manual handling are set. Graphs showing the effect of elbow angle on maximum isometric forearm flexion and extension strengths can be found in Appendix A.

The concepts of MVC and maximum strength are important in the study of work physiology and ergonomics. These terms are sometimes used in a general sense, but they are very specific and apply only to the conditions under which they are measured. The difficulty in answering questions about how much people can lift is apparent when you consider that each work task can have its own set of postural requirements and can involve different muscles. No single MVC can be applied routinely to solve all handling design problems in the plant. The guidelines for manual handling and force exertions in Part VI of this volume are to be used as rough guides only. Each job situation should be evaluated in terms of the factors that will determine how much strength can be developed by people in the postures and under the other job conditions present, such as the frequency and duration of performing the task. Because the guidelines in Chapter VI are chosen to accommodate a large majority of the industrial workforce, however, following them is not likely to result in overestimates of the lifting and force exertion capabilities of most workers.

E. MOTOR CONTROL

A complex and versatile machine like the human body requires an elaborate control system. Our motor control system continually processes sensory information relating to our movement and force requirements and initiates the commands necessary for successful task completion. This sensory information is sent to all levels of the central nervous system (CNS), the cerebral cortex, cerebellum, brain stem, and spinal cord. The integrated response of the system determines what information the anterior motor neurons (AMN) receive and translate into movement. Figure 2-7 shows a block diagram of the major motor subdivisions of the central nervous system from which motor control information is obtained and where the processing of sensory information and the initiation of motor activity take place. A very brief discussion of each of these motor control centers follows. As motor control is a complex process, the interested reader should refer to standard texts on neurophysiology, motor behavior, and neuroanatomy for detailed information (cf. Chusid, 1979; Eyzaguirre and Fidone, 1975; Sage, 1977).

1. PROPRIOCEPTORS

There are several specialized receptors in the muscles and tendons of the human body that provide information critical to the control of movement. These receptors, called proprioceptors, are responsible for our kinesthetic sense, the knowledge of where our body parts are located in space. The execution of efficient movement patterns requires this kinesthetic information. There are three proprioceptors to be considered: muscle spindles, Golgi tendon organs, and joint receptors.

The muscle spindles are collections of specialized muscle fibers that provide information about muscle length to the central nervous system. These re-

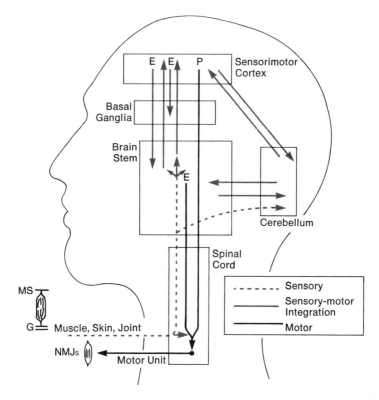

Figure 2-7: Nervous System Centers Involved in Movement Control
The diagram summarizes the major motor centers in the central nervous system. The sensory apparatus in muscles, joints, and skin (dashed lines) sends information to the cerebellum and the brain stem, as well as to motor neurons in the spinal cord. G represents the Golgi tendon organs in the muscle tendons and MS indicates the muscle spindles that lie parallel to the muscle fibers. NMJ indicates the neuromuscular junctions. The sensory information is relayed to other parts of the brain stem and to the cerebral, or sensorimotor, cortex. It is integrated with other activity and the response is sent out through two major pathways. The pyramidal system (P) conveys impulses from the motor cortex directly to the spinal motor neurons of the motor units that effect movement. The extrapyramidal system (E) sends information out to the basal ganglia and the brain stem. The basal ganglia are nerve centers that integrate activity from many parts of the nervous system. Motor activity is initiated from the brain stem and influences the activity of the motor unit neurons primarily by affecting the activity of interneurons around them. The cerebellum is also involved in integrating activity of the sensorimotor cortex so that smooth and coordinated muscle actions can occur. The integrating pathways are shown as dot-dashes, and the motor pathways of the pyramidal and extrapyramidal pathways as solid lines. (*Modified from Mountcastle, V. B., Medical Physiology, 12th Edition, St. Louis: The C. V. Mosby Company, 1968 and from Eyzaguirre, C. and Fidone, S. J., Physiology of the Nervous System, 2nd Edition. Copyright © 1975 by Year Book Medical Publishers, Inc.*)

ceptors are important in determining how many muscle fibers should be recruited to perform a task. If too little or too much effort is applied, the muscle spindles relay this information to the motor control centers.

Golgi tendon organs (GTOs) are important proprioceptors that play a role in preventing injuries. The GTO is located in the tendon of the muscle and is sensitive to stretch. If the GTO is stretched sufficiently, its sensory information will result in relaxation of the muscle and, thereby, a reduction in tension. This helps prevent overexertion injuries.

The joint receptors include a number of specialized sense organs that supply information to the central nervous system concerning joint angle, acceleration, and pressure in the tissues. These receptors are located in joints, muscles, tendons, and ligaments. They continually provide information used in the control of movement to the central nervous system.

2. SUPRASPINAL CONTROL

a. Motor Control Areas of the Brain

Our ability to control movement voluntarily permits us to accomplish useful work. While the information provided by the proprioceptors is important for motor control, complex voluntary movements require direction from motor control areas in the brain. There are several areas in the brain that are responsible for the control of movement; these include the motor area of the cortex, the basal ganglia, and the cerebellum. Other areas of the brain play supporting roles in the control of movement.

The motor area of the cortex influences most neurons associated with motor control. It integrates information from other centers, leading to rapid, discrete movements.

The main function of the basal ganglia appears to be in controlling coordinated movement, especially the generation of slow movements (Evarts, 1973).

The cerebellum is also associated with the control of complex and coordinated movements, and, therefore, receives and integrates information from the proprioceptors and other motor control areas. The cerebellum may also play an important role in the initiation of fast, ballistic movements.

b. Ascending and Descending Pathways for Motor Control

The control of movement requires the integration of information from the proprioceptors by the central motor control centers. Following this integration, the appropriate motor control commands must be sent to the anterior motor neurons, which have been called the final common pathway of motor control. This communication link among proprioceptors, motor control centers, and effectors is comprised of bundles or tracts of nerve fibers. These tracts conduct information in one direction only. There are independent tracts for proprioceptive information and motor control signals. For additional information, see a standard text on neurophysiology (cf. Sage, 1977).

F. MOTOR LEARNING AND SKILL ACQUISITION

Motor learning is a very important, yet often overlooked, factor in the training of new employees on jobs requiring motor skills. When new movements are performed, the ease of task completion depends on familiarity with the task difficulty. Inefficient task performance may result in the following:

- Increased energy demands on the body.
- Greater stress on muscles and joints, leading to increased susceptibility to injury.
- Increased task completion time.
- Low quality work.
- High error rate.
- Increased perceived exertion.

Recent research suggests that muscle contractions to produce movement are controlled by a "motor program." A motor program is a set of neural units that are structured before a movement sequence begins (Sage, 1977). When the program runs, a movement sequence is completed that is determined by previously established routes through the motor control system. The motor program selected during movement depends on the task to be performed. With repetition the program becomes refined and the desired movement is accomplished. Presumably, motor programs can be modified by new or accumulated information. For instance, as one continually repeats a fine assembly task, skill levels increase. During the initial stages of learning, feedback to the worker provides the necessary information for the establishment of a motor program. This information might come from proprioceptors in the muscles of the hands, from the visual system as one looks at the end product, or from the auditory system when techniques are explained. Eventually, the task becomes routine and is performed almost automatically. In other words, the motor program has been "debugged," and when it is run the desired task is performed.

Many motor programs can be reused to perform new tasks, with only slight modifications. Tasks that are similar can usually be learned quickly since entire programs do not have to be established. For more complex tasks, using more muscle groups and requiring more coordination of effort, multiple motor programs can be linked together. Subtasks controlled by specific motor programs are combined for the completion of an entire task.

Feedback is an important aspect in the learning of complex motor tasks. The development of an accurate motor program requires appropriate feedback on performance levels. Feedback from the completion of the task, or terminal feedback, is necessary to show that the motor program was run as intended. More importantly, this feedback is needed to correct the motor programs if errors occur. There are two types of terminal feedback that are important to an individual learning a new skill (Sage, 1977):

- Knowledge of results (KR), which provides information about the out-come of the movements.

- Knowledge of performance (KP), which provides information about the movements themselves.

Knowledge of results appears to be the strongest variable affecting performance and learning; it must be given in order to see performance improvements. This might be done for a new assembly worker through an on-line quality audit sampling of the product being assembled. The audit information can give the assembler immediate feedback on the appropriateness of the muscle movements and can accelerate skill development. In another application of feedback in motor training programs, computer-controlled word processing equipment can be programmed to alert the inexperienced operator when a keying error is made as part of a self-education program.

Knowledge of performance is concerned with information such as the timing, positioning, sequencing, and force-exerting aspects of movement. It may be provided by demonstrations, descriptions of movement patterns, or by using filmed or videotaped replays. When a worker is new to a job that requires the acquisition of new motor skills, his or her learning can be accelerated by providing knowledge of both results and performance. This can be in the form of figures and drawings, film or videotapes, discussions with experienced operators, or critiques of completed work. Although manual work routines and productivity may suffer initially, extra attention and feedback to a new worker can increase the rate of learning of the new task. The section on training in Part IV discusses training techniques and the development of increased muscle strength and coordinated movements in more detail.

G. FATIGUE AND PERFORMANCE IMPAIRMENT

Fatigue is a complex concept that continues to challenge the scientists who study its mechanisms. Part of the complexity is due to lack of agreement as to the site and cause of the fatigue process. Two general types of fatigue have been termed central and peripheral fatigue, depending on the predominant site of impairment. These often occur together, although pure central fatigue has been noted during prolonged observational tasks (Stegeman, 1981). The following discussion touches upon physiological fatigue factors that may influence task performance in an industrial setting.

The development of fatigue within a muscle is probably related to the adequacy of its blood supply. A contracting muscle requires high-energy phosphates for work performance, and the majority of these are supplied by oxidative metabolism (see the first section of this chapter). During the active period of muscle contraction many metabolites may be produced, including carbon dioxide, water, pyruvic acid, and lactic acid. These metabolites effectively change the chemistry of the muscle fibers. Shifts in the concentration of ions, changes in nutrient levels, altered levels of hormones and catecholamines, and increases in muscle temperature also influence muscle function.

Adequate blood flow through a muscle is needed to carry away waste products and to maintain an appropriate muscle chemistry. During muscle contraction, the flow of blood through the muscle can be impeded at levels as low as 40 percent of the MVC for the muscle in the posture used to do the task (Lind and McNichol, 1967). During static muscle contractions the reduction in blood flow continues as long as the contraction is maintained. In dynamic contractions the reduction in blood flow is intermittent and occurs only during the contraction period. When the blood flow to an active muscle becomes impaired, metabolites accumulate, and the oxygen supply to the muscle can be quickly depleted. There may then be a switch to anaerobic metabolism and increased production of lactic acid. As the acidity of the muscle increases, the metabolic mechanisms that provide energy will become less efficient. The concentration of high-energy phosphates may thus become inadequate for continued work.

Probably the most important factor in the prevention of fatigue is the maintenance of adequate blood flow to the active muscles. Jobs should be designed to reduce the requirements for static muscle loading, such as gripping, extended reaches, and awkward postures. The lowest possible percentage of the muscle's maximum isometric, or static, effort should be designed for in situations where static effort is part of the job, such as in tool use. Since lower tension or effort in a muscle results in less impairment of blood flow, the muscle will take longer to fatigue. Dynamic, or rhythmic, muscle contractions are preferred because they allow blood to flow between contractions, and this "washes out" some of the metabolites that build up during the work period. However, if the intensity and/or rate of contraction is high enough, this intermittent muscle blood flow may still be inadequate. A recovery, or rest, period will be needed to replenish the energy stores, its duration depending on how much fatigue has developed. The section on work/rest cycles in Part IV addresses the design of work and recovery patterns to minimize fatigue.

Reductions in work capacity can also occur due to the development of central fatigue. Central fatigue is related to the psychophysical aspects of work capacity reduction. It is often accompanied by a feeling of tiredness or a perception of heightened exertion during work at the same level as earlier in the shift. Generally, a decline in central nervous system performance is thought to lead to central fatigue. Central fatigue can be associated with, and may be due to, the onset of peripheral fatigue (the accumulation of metabolites and reduction in energy stored in the muscles). However, central fatigue can occur independently of local muscle changes; for example, in jobs with high mental or perceptual demands, there may be central fatigue but little evidence of muscle fatigue. Individual motivation and interest complicate the measurement of central fatigue. Long periods of continuous work without scheduled recovery breaks are more likely to result in central fatigue than a work pattern that alternates frequently between work and recovery activities (Floyd and Welford, 1953). The ultradian rhythms of performance discussed in Chapter 4 (Lavie, 1982) may reflect the phenomenon of central fatigue. However, there is still little data on which to base appropriate work schedules for mentally and perceptually demanding tasks.

REFERENCES

Astrand, P. O., and K. Rodahl. 1977. *Textbook of Work Physiology*. 2nd ed. New York: McGraw-Hill, 681 pages.

Brown, D. E. S., and F. J. Sichel. 1936. "The Isometric Contraction of Isolated Muscle Fibers." *Journal of Cellular and Comparative Physiology, 8:* pp. 315-328

Burke, R. E., and V. R. Edgerton. 1975. "Motor Unit Properties and Selective Involvement in Movement." In *Exercise and Sport Sciences, Volume 3*. New York: Academic Press, pp. 31-81.

Chusid, J. G. 1979. *Correlative Neuroanatomy and Functional Neurology*. 17th ed. Los Altos, Calif.: Lange Medical Publications, 464 pages.

Cooper, S., D. E. Denny-Brown, and C. S. Sherrington. 1926. " 'Reflex' Fractionation of a Muscle." *Proceedings of the Royal Society of Medicine, B100:* pp. 448-462.

Evarts, E. V. 1973. "Brain Mechanisms in Movement." *Scientific American, 227:* pp. 96-103.

Eyzaguirre, C., and S. Fidone. 1975. *Physiology of the Nervous System*. 2nd ed. Chicago: Year Book Medical Publishers, 418 pages.

Floyd, W. F., and A. T. Welford, editors. 1953. *Fatigue*. Symposium held at a meeting of the Ergonomics Research Society, College of Aeronautics, Cranfield, England, March 1952. London: H. K. Lewis, 196 pages.

Gollnick, P. D., and L. Hermansen. 1972. "Biochemical Adaptions to Exercise." In *Exercise and Sport Sciences Reviews, Volume I*. New York: Academic Press, pp. 1-43.

Harper, H. A. 1959. *Review of Physiological Chemistry*. 7th ed. Los Altos, Calif.: Lange Medical Publications, pp. 112-122, 170-186.

Henneman, E., G. G. Somjen, and D. O. Carpenter. 1965. "Functional Significance of Cell Size in Spinal Motorneurons." *Journal of Neurophysiology, 28:* pp. 560-580.

Hunt, C. C., and S. W. Kuffler. 1954. "Motor Innervation of Skeletal Muscle: Multiple Innervation of Individual Muscle Fibers and Motor Unit Function." *Journal of Physiology, 126:* pp. 293-303.

Huxley, H. E. 1958. "The Contraction of Muscle." *Scientific American, 199:* p. 66.

Inman, V. T., and H. J. Ralston. 1954. "The Mechanism of Voluntary Muscle." Chapter 11 in *Human Limbs and Their Substitutes*, edited by P. E. Klopsty and P. D. Wilson. Washington, D.C.: National Research Council Advisory Committee on Artificial Limbs, Department of Medicine and Surgery, U.S. Veterans' Administration, Surgeon General, Department of Army, pp. 296-317.

Issekutz, B., and H. Miller. 1962. "Plasma-Free Fatty Acids during Exercise and the Effect of Lactic Acid." *Proceedings of the Society of Experimental Biology and Medicine, 110:* pp. 237-239.

Jones, N. L., E. J. Moran-Campbell, R. H. T. Edwards, and D. G. Robertson. 1975. *Clinical Exercise Testing*. Philadelphia: W. B. Saunders, 214 pages.

Lavie, P. 1982. "Ultradian Rhythms in Human Sleep and Wakefulness." Chapter 9 in *Biological Rhythms, Sleep, and Performance,* edited by W. B. Webb. New York: John Wiley and Sons, pp. 239-272.

Layzer, R. B., and L. P. Rowland. 1971. "Cramps." *New England Journal of Medicine, 285:* pp. 31-40.

Lind, A. R., and G. W. McNicol. 1967. "Circulatory Responses to Sustained Handgrip Contractions Performed During Exercise, Both Rhythmic and Static." *Journal of Physiology, 192:* pp. 595-607.

Margaria, R. 1972. "The Source of Muscular Energy." *Scientific American, 226:* pp. 84-91.

Mathews, D. K., and E. L. Fox. 1976. *The Physiological Basis of Physical Education and Athletics.* 2nd ed. Philadelphia: W. B. Saunders, 577 pages.

Mountcastle, V. B., editor. 1968. *Medical Physiology.* 12th ed. Saint Louis: C. V. Mosby, 2 volumes, 1858 pages.

Sage, G. H. 1977. *Introduction to Motor Behavior: A Neuro-Psychological Approach.* 2nd ed. Reading, Mass.: Addison-Wesley, 610 pages.

Sherrington, C. 1947. *The Integrative Action of the Nervous System.* New Haven: Yale University Press, 433 pages.

Stegeman, J. 1981. *Exercise Physiology.* Translated and edited by J. S. Skinner. Chicago: Year Book Medical Publishers, 345 pages.

Williams, M., and H. R. Lissner. 1962. *Biomechanics of Human Motion.* Philadelphia: W. B. Saunders, 147 pages.

Winter, D. A. 1979. *Biomechanics of Human Movement.* New York: John Wiley and Sons, 202 pages.

Circulatory and Respiratory Adjustments to Work

CHAPTER 3. CIRCULATORY AND RESPIRATORY ADJUSTMENTS TO WORK

T he muscle activity discussed in the previous chapter can only be sustained if the muscle fibers are continuously supplied with nutrients and oxygen and can get rid of the end products of metabolism, especially carbon dioxide. Respiration accomplishes the exchange of oxygen and carbon dioxide with the environment. Nutrients are prepared for transportation to the cells by being broken into smaller molecules in the digestive system and absorbed into the blood through the wall of the small intestine. Circulation of the blood provides the transportation system to move gases, nutrients, and waste products of metabolism between the cells and the environment or to other parts of the body such as the liver and kidneys. In addition, blood distribution may be modified to act as a heat exchanger when work is done in hot environments. This chapter describes the functional anatomy of the circulatory and respiratory systems and how these systems are adjusted during work. These descriptions serve as background to the workload guidelines present in Parts IV, V, and VI and to the discussion of aerobic capacity testing in Appendix B.

A. THE CIRCULATORY SYSTEM

The discussion of muscle contraction and movement presented in Chapter 2 identified the role of the circulatory system in delivering oxygen and nutrients to and removing waste materials and carbon dioxide from the working muscles. The role is accomplished through (1) an efficient oxygen carrying system in the blood, (2) a centrally controlled blood vessel system that distributes blood to the active muscles in proportion to their workload, and (3) an efficient pump, the heart, that matches its output to the demand for oxygen in the working muscles or to the need to eliminate excess heat during exposure to or work in hot environments. The following subsections look at each of these aspects of circulatory function.

1. THE ROLE OF THE BLOOD

The blood components of most interest in work physiology are the hemoglobin-rich red blood cells that carry oxygen to the cells and help in the removal of carbon dioxide from them.

a. Oxygen Transport

The unique properties of the hemoglobin molecule (Hb) permit it to carry 4 molecules of oxygen very efficiently. For healthy people at sea level, Hb is 98 percent saturated when it leaves the lung in the arterial blood after being exposed to an oxygen pressure (pO_2) of 95 to 100 mm Hg. When it reaches the tissues, especially the working muscles, the hemoglobin molecule no longer holds onto the oxygen as tightly and the cells are provided with the oxygen needed to oxidize nutrients and generate ATP for muscle contraction and other life processes. The more work being done by a muscle, the more ATP is needed, and the greater the need for oxidative metabolism to replenish the ATP supply in the muscle cells. The tissue pO_2 will be in the range of 15 to 25 mm Hg, making the movement of oxygen out of the arterial blood very rapid (Honig, 1981).

The difference between the amount of oxygen in the arterial blood going to the muscles and the amount in the venous blood returning to the lungs is a measure of the amount extracted by the working muscles. When the arterial and venous samples are taken from the general circulation, not from the vessels supplying and draining a specific working muscle group, this oxygen extraction times the total blood flow (or cardiac output, $\dot{Q}$) provides a measure of the aerobic workload. Part IV and Appendices A and B include more detailed discussions of oxygen consumption and its relationship to muscle work.

b. Carbon Dioxide Transport

A second critical role of the blood is to remove the carbon dioxide (CO_2) formed during aerobic metabolism in the working muscle cells. If the CO_2 is not removed from the working muscles, the local buildup of acidity will slow down aerobic metabolism and result in less efficient work. Increased ventilation of the lung's alveoli - that is, more exchange of gas per minute in the lung air sacs - increases the amount of carbon dioxide "blown off" from the venous blood to the environment, making room for more CO_2 to be picked up by the blood on its next pass through the working muscles. Inadequate ventilation will permit CO_2 to accumulate in the muscles and in the venous and arterial blood, and it will limit the amount and efficiency of muscular work performed.

In muscular work there are large increases in the amounts of oxygen consumed and carbon dioxide produced, often 6 to 15 times the resting levels. In heavy exercise lactic acid is produced, increasing the acidity of the active muscles and of the blood. This increased acidity helps the release of oxygen from hemoglobin as arterial blood passes through the muscle. The amount of oxygen extracted by the active muscle is increased, and the amount of carbon dioxide taken into the venous blood is also increased. Because blood with a high pCO_2 and high acidity from strenuous muscle work has a reduced oxygen tension compared to resting values, the gradient for diffusion of oxygen from the lungs into the blood is increased, helping to keep the arterial blood oxygen supply at normal levels during work. Since the amount of muscular work that can be done is dependent on the availability of oxygen and removal of CO_2, maximum work capacity will be reduced any time the amount of oxygen delivered to the muscle cells per minute is lowered.

The amount of oxygen and carbon dioxide carried by the blood will be affected if the red cell count is abnormal, if the barometric pressure is reduced (as it is at high altitudes), or if the environmental gas breathed is not 21 percent oxygen and 0.03 percent carbon dioxide, the balance being nitrogen. In anemias, the red cell count can be reduced significantly and the hemoglobin content may be less than half of the normal values. Internal bleeding associated with ulcers or heavy loss of blood with menses can also reduce the amount of hemoglobin available for oxygen carriage. When the number of red blood cells is reduced, the carbon dioxide storage capability of the blood is also reduced. Some chemicals selectively combine with hemoglobin and form methemoglobin, which cannot carry oxygen as hemoglobin does; this effectively reduces the hemoglobin content of the blood. Carbon monoxide can also make hemoglobin unavailable for O_2 carriage (Roughton, 1954).

All of these factors affect a person's work capacity by reducing the amount of O_2 delivered to and CO_2 removed from the active muscles per unit of blood flow. To make up for the lower amount of O_2 per unit of blood, the heart has to work harder to increase the amount of blood to the tissues per unit time (blood flow) so that the oxygen needed by the cells is delivered. Eventually the limits of the circulatory system's response will be reached. This is discussed in more detail in the next section.

The composition of gas breathed in from the environment will also affect the levels of oxygen and carbon dioxide in the blood. If one is working in an atmosphere of reduced oxygen supply, as may occur in work that is done in confined spaces or at altitudes above sea level, the amount of oxygen carried in the blood will also be reduced. The atmospheric pO_2 is reduced, and the alveolar pO_2 is correspondingly lower (Boothby, et al., 1954). Although alveolar gas exchange will be increased somewhat to compensate for the lower pO_2, work capacity can be severely limited. The extreme of this limit is seen in the final ascent of Mt. Everest where very fit mountaineers can barely proceed at one step a minute. The lower alveolar pO_2, even when breathing pure oxygen, cuts the oxygen carrying capacity of blood at least in half. Although this is unlikely to occur in industrial tasks, the tying up of hemoglobin with carbon monoxide in heavy smokers or in people exposed to gasoline motor exhaust, and the exposure to other chemicals that affect the ability of hemoglobin to carry oxygen or to assist in carbon dioxide transport can reduce aerobic work capacity by 5 to 15 percent (Roughton, 1954).

2. BLOOD FLOW

The transport of oxygenated blood to the tissues of the body, especially to working muscles, is accomplished by the circulatory system. The heart is a muscular pump that creates the pressure needed to move blood along the arteries, arterioles, capillaries, venules, and veins. Figure 3-1 illustrates the heart and vessels of the whole body, or systemic, circulation and of the lung, or pulmonary, circulation. Blood flows through these vessels and the heart at a rate that is adjusted to workload and to the need to eliminate the excess body heat that accumulates in muscular work.

Cardiac output ($\dot{Q}$) is the amount of blood pumped per minute from the left side of the heart. It is a primary determinant of a person's aerobic work capacity. It is influenced by:

- The total amount of muscle mass involved in the work.

- The environmental load, such as heat and humidity, that also demands blood flow.

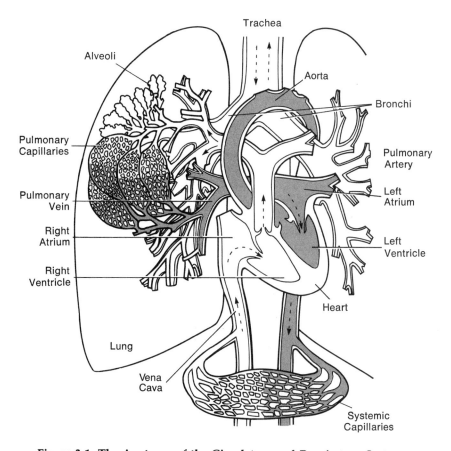

Figure 3-1: The Anatomy of the Circulatory and Respiratory Systems
The systemic, or general body, circulation is at the bottom of this illustration and the pulmonary, or lung, circulation is shown at the upper left. The trachea, bronchi, and alveoli of the respiratory system are
shown in relation to the heart and large vessels (aorta, vena cava, pulmonary artery, and pulmonary vein). Arterial blood is in color. Mixed
venous blood returns from the systemic capillaries via the vena cava
and enters the right atrium of the heart. It passes through the valve to
the right ventricle when the heart is relaxed between beats (diastole).
From the right ventricle it is pumped, during systole, into the pulmonary artery where it enters the pulmonary capillaries that surround
the lung alveoli. The blood emerges arterialized (less carbon dioxide,
more oxygen) and is carried by the pulmonary vein back to the left side
of the heart, first to the left atrium and then into the left ventricle.
When the left ventricle contracts (systole), the arterial blood is pumped
out the aorta to the rest of the body. From the aorta it goes through
arteries, arterioles, capillaries, venules, and veins before being returned
to the vena cava to complete the circulation. *(From "The Lung" by J. H.
Comroe. Copyright, February, 1966, by Scientific American, Inc. All rights
reserved.)*

- The psychological content of the job and how it influences the heart rate and the resistance of arterioles.
- The individual worker's cardiovascular fitness level, that is, how healthy the heart and blood vessels are.

a. Determinants

The sustained flow of blood to the muscles is produced by the pumping action of the heart, which generates a pressure to move the blood forward with each beat. This flow is inhibited somewhat by resistance in the vessels between the heart and muscle cells. The blood vessels' resistance is determined by their radii and length; small radii and longer vessels produce greater resistance to blood flow.

Very small changes in vessel size can produce large changes in blood flow. During physical work, the blood flow is matched to the muscles' demands for oxygen. If the blood vessels dilate in the arm muscles, however, there should be simultaneous constriction of arterioles in other parts of the body in order to maintain blood pressure. Other adjustments also are seen: more blood comes into the heart, for example, which helps the heart to beat more effectively and, in heavy work, the adrenal gland releases hormones that strengthen the heartbeat and increase the blood pressure to keep blood flowing to the working muscles (Folkow and Neil, 1971).

Increased blood pressure means more work by the heart muscle to pump against the resistance. Thus, the strain on the heart is greater. A person with chronically elevated blood pressure has less reserve work capacity available. In situations where it is necessary to increase the blood flow on a temporary basis, such as in short-duration heavy effort in an occasional handling task, such a person will be at a disadvantage. The heart muscle may be unable to respond to the demand for oxygen by the arms and legs and more anaerobic energy sources will be used, resulting in earlier fatigue.

b. Distribution

At rest, the skeletal muscles receive about 15 to 20 percent of the total cardiac output. During moderate work, about 45 percent of the cardiac output is needed by the muscles to deliver oxygen and remove carbon dioxide, and during very heavy work about 70 to 75 percent of the blood flow goes directly to the working muscles. The digestive system receives 20 to 25 percent of the blood flow at rest, but this fraction falls to 4 or 5 percent of the total flow during moderate to heavy work. Because the cardiac output is increasing, this fall in the percentage of blood flow to the internal organs does not mean that they are getting inadequate oxygen; the resistance of the vessels in them is increased in order to distribute more of the total flow to the muscles. If, however, a person eats a heavy meal or works in a very hot environment, the digestive system or skin will compete with the muscles for blood flow. Table 3-1 illustrates this redistribution of blood

Table 3-1: Blood Flow Distribution in Different Working Conditions   The major organs of the body are shown in column 1; one rest and two working conditions are shown in columns 2 through 4. The percentage of the cardiac output, or blood flow, that is distributed to each organ or system is shown in the table. At rest the cardiac output is about 4 to 5 liters/minute (L/min). In heavy work it is about 23 L/min, and in moderate work in the heat it is about 17 L/min. The majority of the resting blood flow is distributed to the digestive system, kidneys, muscles, and brain. In heavy work, without a heat load, it is distributed largely to the working muscles; other organs get proportionally less of the total increase in blood flow. During work in the heat, a large part of the blood flow goes to the skin so the body can maintain heat balance. The heart's coronary artery blood flow is maintained in all conditions. Brain blood flow is also quite constant; although it is proportionally less of the total cardiac output with increased work, the brain receives the same amount of blood per minute as at rest. The implications of these blood flow changes on work capacity are discussed in the text. (*Adapted from Astrand and Rodahl, 1977; Brouha, 1967.*)

Organs	Blood Flow Distribution (Percent of Cardiac Output)		
	Resting	Heavy Work, 21°C (70°F)	Moderate Work, 38°C (100°F)
Muscles	15 – 20	70 – 75	$\cong$ 45
Skin	5	10	$\cong$ 40
Brain	15	3 – 4	$\cong$ 4
Bone	3 – 5	0.5 – 1	
Kidneys	20	2 – 4	$\cong$ 6.5
Digestive System	20 – 25	3 – 5	
Heart	4 – 5	4 – 5	$\cong$ 4.5

for rest, very heavy work in a temperate climate, and for moderate work in a hot environment.

Guidelines for work in hot environments recognize the potential stress on the heart when it has to respond to both increased demands for oxygen and a need to distribute more blood to the skin in order to eliminate excess body heat. Table 11-4 in Part IV illustrates that work capacity is curtailed very significantly in hot and/or humid environments. This effect has also been illustrated in the section on temperature and humidity in Chapter V of Volume 1 of this series.

Increased blood flow to the muscles and skin during muscular work is accomplished by dilating (increasing the radii of) the vessels in the working muscles and skin. Control of smooth muscle "tone" in the arteriolar walls is done through both neural and humoral pathways. The neural pathways are from the vasomotor centers of the brain stem; the humoral pathway is through a direct effect on the muscle cells of a buildup of metabolic end products in the active skeletal muscle. These chemical substances encourage relaxation of the arteriolar smooth muscle cells, resulting in greater blood flow through adjacent capillaries (Johnson, 1964). This "autoregulation" of local blood flow supersedes the neural control, especially during heavy muscular work. It ensures that the muscle cells will receive maximum blood flow (and thus oxygen), even at the expense of regulation of body temperature. Thus, heat exhaustion and heat stroke are more possible during heavy work in the heat, and workloads must be carefully monitored.

In heavy work, circulating chemicals called *hormones* can also influence the radii of arterioles and, therefore, their resistance. Epinephrine, or adrenaline, produces vasodilatation in the skeletal muscle arterioles (Folkow and Neil, 1971), ensuring that blood flow through them can deliver oxygen rapidly to the active muscle fibers. The epinephrine effect usually lasts longer than the autonomic effects do and is thought to play an important role in maintaining muscle blood flow in sustained moderate or heavy work.

The amount of blood returning to the right heart every minute is called *venous return*. Since the cardiovascular system is circulatory, the flow of blood into the right side of the heart must be the same as the flow out of the left side into the aorta. When a person is standing upright, the blood has to be returned to the heart against the hydrostatic pressure created by gravity, for example, between the heart and the leg. During exercise, venous return is helped by the increased venal tone produced by autonomic nervous system activity, the pumping action of the active muscles (caused by dynamic work), and the pulling of the blood towards the heart by increased respiratory activity. Valves in the veins keep the blood moving towards the heart even during the resting period between heartbeats. Table 3-2 summarizes circulatory controls that permit blood flow to be increased in muscular work. The blood flow, by controlling how much oxygen is delivered per minute to the working muscles, is the primary determinant of a person's capacity for aerobic work.

If static, instead of dynamic, work is done, venous return is impaired, and the heart has to beat faster to make up for the reduced amount of blood it can empty into the aorta per beat (stroke volume). Extended work in the heat may produce some dehydration that results in reduced blood volume. This, too, results in higher heart rates because of the reduced stroke volume. For further information, see the discussions of static work and work in the heat in Part IV in this volume and in Chapters II and V of Volume 1, as well as the examples of heart rate responses to work in the heat in Chapter 9 of this volume.

Table 3-2: Circulatory Controls to Adjust Blood Flow During Work   Six circulatory control mechanisms are given in column 1, and their effect on blood flow is described in column 2. Muscle blood flow is given primary consideration since the regulation and distribution of blood flow during work determine a person's maximum aerobic work capacity. The cardiovascular controls are designed to maximize blood flow to the working muscles at the expense of blood flow to most other parts of the body, except for the heart muscle and the brain. *(Adapted from Barcroft, 1963.)*

Mechanism	Effect on Blood Flow
Local muscle temperature increase; increased CO_2 and metabolites in a muscle.	Increased blood flow within the muscle; opening up of additional capillaries to improve oxygen delivery to the muscle cells.
Increased sympathetic nervous system vasoconstrictor discharge, stimulated via circulatory reflexes or skin thermal receptors.	Increased resistance and decreased blood flow through the vessels of the digestive system organs, kidney, and resting muscles. More blood flow is delivered to the working muscles to allow them to work aerobically.
Increased sympathetic vasodilator activity, triggered through the midbrain, associated with emotional stress.	Increased dilatation of the vessels of the active skeletal muscles, resulting in increased blood flow to them. This prepares a person for sudden action, if needed.
Increased body temperature.	Increased blood flow to the skin; diverted from working muscles, if necessary, to get rid of excess body heat.
Release of adrenaline from the adrenal glands.	Increased blood flow in the working muscles due to vasodilatation. Very large effect initially; it is quickly reduced, and a slight vasodilatation is sustained for some time afterward. This may be related to local metabolites in the muscles.
Mechanical pressure on limb blood vessels; intermittent.	Increased venous return through better pumping of the blood back to the heart against gravity. This will increase cardiac output and deliver more blood to the working muscles as long as the pressure is intermittent.

c. Muscle Blood Flow

Skeletal muscle activity may not increase venous return if the contraction is isometric and is more than about 40 percent of the muscle's maximum voluntary contraction (MVC) (Lind and McNichol, 1967). In these conditions the muscle contraction can impede blood flow through the muscle capillaries running parallel to the sarcomeres (see Figure 3-2). This phenomenon was mentioned in the discussion on muscle fatigue in Chapter 2.

Because blood flow is impeded, the muscle fiber's oxygen supply is compromised. To sustain the effort, the muscle uses its local stores of ATP, CP, and myoglobin O_2, and anaerobic metabolism becomes the primary source of new high energy phosphates. Physical pressure on the vessels prevents them from dilating in response to their metabolic environment as long as muscle tension is kept high. Once the muscle is relaxed, even if for a very short time, blood rushes into the capillary bed at an accelerated rate, known as reactive hyperemia. The capillary bed is dilated due to the presence of products of anaerobic

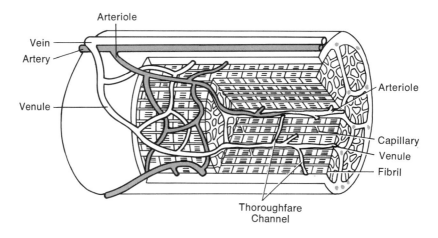

Figure 3-2: Muscle Capillaries   A longitudinal section of a muscle fiber is shown with its cross section on the right. An artery and vein are shown running parallel to the fiber and sending branches into the fibrils, which are made up of the banded sarcomeres (see Figure 2-2 for more information). The arterioles form many capillary beds that, in general, run parallel to the sarcomeres. The venules collect the capillary blood and return it to the vein. Two thoroughfare channels, vessels that connect arterioles to venules directly without going through a capillary bed, are indicated running across the sarcomeres. These are important in controlling the distribution of blood flow in the body; they are used to shunt blood away from resting muscles so that more blood flow is available for working muscles. In the cross section of the fiber at the right, the arterioles, capillaries, and venules can be seen as small circles. The largest circle is the venule, the second largest with thicker walls is the arteriole, and the smallest circles are the thin-walled capillaries. The muscle fibrils are distinguished by their outer membranes and are further divided into small groups by connective tissue. The arterioles and venules travel within the connective tissue. *(From ''The Microcirculation of the Blood'' by B. W. Zweifach. Copyright, January, 1959, by Scientific American, Inc. All rights reserved.)*

metabolism in the muscle. Limitations on muscle fiber circulation during isometric, or static, work explain why endurance times for work requiring more
than 40 percent MVC exertions are so short (see Figure 7-5 In Part III). Alternating contractions with relaxation periods, as in dynamic work, favors blood
flow in the muscle through the massaging effect of the contractions on the veins.

3. MEASURES OF CIRCULATORY AND
   RESPIRATORY ADJUSTMENTS TO WORK

Blood flow to the working muscles in physically demanding jobs or to the skin
in hot environments is not easy to measure directly in the workplace. One needs
a noninvasive measurement that allows the worker to perform the job without
interference and yet gives a good indication of the degree of stress on the heart
from performing the sustained and peak work activities.

   Since blood flow, or cardiac output, is determined by the amount of blood
pumped per beat of the heart (stroke volume) times the number of heartbeats
per minute, an estimate of the blood flow during work can be made by measuring heart rate. This estimate is very good at workloads above 50 percent of
maximum aerobic capacity ($\dot{V}_{O_2}$ max) (Astrand and Rodahl, 1977) because stroke
volume elevations with work are complete at about 40 percent of $\dot{V}_{O_2}$ max for
whole body work. At lower workloads, heart rate is still commonly used to
predict the effort level for physical work, but it should be used cautiously. This
is particularly true for mental and perceptual tasks and for emotionally demanding work situations. The pathways that increase heart rate in those conditions do not affect heart rate and stroke volume in the same way that physical
work does. Methods for measuring heart rate and interpreting job heart rate
data are given in Appendix B and Part III, respectively.

   Blood pressure changes with physical effort are also important, but measuring them may interfere with the way the job is done. Most work requires
that the worker use arm muscles. Since blood pressure is usually measured on
the arm, the worker's task performance must either be interrupted or done with
a temporary occlusion of forearm blood supply. Most blood pressure measurements are made as soon as possible after a peak load has occurred or at the
beginning of a recovery period after a sustained workload. Work with a large
static muscle effort component may be improperly evaluated if only a heart rate
measure is taken. Sustained holding or gripping tasks, such as operating a
grinding tool or working above shoulder level for extended periods, may show
minor changes in heart rate but very large elevations in mean blood pressure
(Lind and McNichol, 1967). Methods for measuring and interpreting blood pressure responses to work are given in Appendix B.

   The amount of physical work a healthy person can do is usually limited
by the cardiovascular system's ability to get adequate oxygen to the muscles
through increased blood flow. Table 3-3 summarizes average values for circulatory and respiratory adjustments to work in three conditions: resting, doing
moderately heavy work such as walking a mail route, and doing maximal vol-

Table 3-3: Circulatory and Respiratory Adjustments to Work
Normal resting (column 2), moderate work (column 3), and maximal work (column 4) values for several circulation, respiration, and blood parameters (column 1) are listed. These values are quite typical for an industrial population doing whole-body work. The values are ranges and will vary according to the age, sex, fitness level, and health status of the individual. The terms have been defined in the text except for the extraction of oxygen, which is the difference (in milliliters per 100 mL of blood) between the amounts of oxygen in the arterial and in the mixed venous blood. Increased workload is accompanied by increases in blood flow, blood pressure, ventilation, and extraction of oxygen from the blood. *(Developed from Asmussen 1965; Folkow and Neil, 1971; and Rodgers, Eastman Kodak Company, 1969-1973.)*

Measurement	Rest	Moderately Heavy Work	Maximal Work
Circulation			
Cardiac Output (L/min)	4 – 5	12 – 16	25 – 30
Heart Rate (beats/min)	65 – 80	120 – 150	(220 – age in years)
Stroke Volume (mL/beat)	50 – 80	80 – 120	110 – 150
Systolic Blood Pressure (mm Hg)	95 – 140	150 – 220	220 – 300
Diastolic Blood Pressure (mm Hg)	60 – 85	60 – 90	60 – 100
Double Product (Heart Rate times Systolic Blood Pressure divided by 100)	70 – 100	170 – 320	>330
Respiration			
Minute Ventilation (L/min, BTPS*)	6 – 10	25 – 45	60 – 150
Oxygen Consumption (mL/min, STPD**)	200 – 300	1500 – 3500	2500 – 4500
Carbon Dioxide Production (mL/min, STPD**)	160 – 250	1200 – 3000	2500 – 4500
Blood			
Arterial pO_2 (mm Hg)	90 – 100	90 – 100	95 – 105
Arterial pCO_2 (mm Hg)	39 – 41	38 – 42	35 – 42
Arterial pH	7.35 – 7.45	7.30 – 7.45	7.15 – 7.40
Venous pO_2 (mm Hg)	38 – 45	20 – 40	15 – 30
Venous pCO_2 (mm Hg)	45 – 48	45 – 52	47 – 55

Table 3–3 *(Continued)*

Measurement	Rest	Moderately Heavy Work	Maximal Work
Blood			
Venous pH	7.30 – 7.40	7.25 – 7.40	7.15 – 7.30
Extraction of Oxygen, Arterio-Venous Difference (mL/100 mL blood)	4 – 6	6 – 10	8 – 15

*BTPS = Body temperature and pressure, saturated.
**STPD = Standard temperature and pressure, dry.

untary work in a standard test of walking or running on a treadmill. These are meant to give a comparison of the degree of adjustment made in several physiological systems when physically demanding work is done. The differences between resting and maximal values are indications of the capacities of these systems to adapt to increased workload. Individual responses will vary. The effect of age on maximal heart rate levels, for instance, ensures that many older people will not be able to elevate their cardiac output as much as many younger people will. Thus, their capacity for physically demanding work will be, on the average, lower than that of a group of younger people. However, it does not mean that all older and no younger people will have difficulty doing a physically demanding task. Designing tasks to be within the capacities of most people reduces the impact of these age-related lower capacities of the cardiovascular and respiratory systems. Part IV gives guidelines for the design of sustained and peak work that recognize the varying work capacities of an industrial population.

a. Heart Rate

The most commonly studied circulatory parameter is the continuous recording of the electrical activity of the heart through the electrocardiogram (ECG). A discussion of how data on heart rate from such recordings can be interpreted to evaluate job demands is found in Chapter 9, Part III. Techniques for measuring the heart rate are given in Appendix B.

Heart rate and stroke volume together determine the amount of cardiac output, or blood flow per minute, going to the body. A person has a cardiac output of about 5 liters per minute (L/min) at rest. In moderate work the cardiac output will be about 15 liters per minute, and in very heavy work it may reach 25 liters per minute (see Table 3-3). These increases in output are accomplished through changes in the amount of blood per beat (stroke volume) and the number of beats per minute (heart rate) of the heart:

$$\frac{\dot{Q}}{\text{L/min}} = \frac{\text{HR}}{\text{beats/min}} \times \frac{\text{SV}}{\text{L/beat}}$$

Stroke volume is 0.050 to 0.060 liters per beat at rest and can more than double with work. Most of this increase is seen at workloads below 50 percent of maximum aerobic capacity (Astrand and Rodahl, 1977). Above this level, cardiac output is determined largely by heart rate increases.

Heart rate is used to estimate the changes in cardiac output even in the lighter workloads. For physical effort tasks the estimate is reasonably accurate, providing fairly large muscle groups are active. For emotional and mental stress, the relationships between heart rate and cardiac output are far less clear.

There are cardiac centers in the brain stem that affect heart rate by exerting a direct effect on the pacemaker cells of the sinoatrial node of the right atrium. One way that they control heart rate is by controlling the amount of vagus nerve activity to the heart's pacemaker. More activity slows the heart, and less activity increases its rate of beating.

Electrical activity from higher centers in the brain, from the respiratory centers, from receptors in muscles and joints, and from the skin and the midbrain temperature regulating centers also affect the activity of the cardiac centers. Stretch receptors in the heart and large vessels feed information to these centers about the amount of venous return. Heart rate is modified after all of the information received has been integrated by the centers (Folkow, Heymans, and Neil, 1965).

Heart rate reflects muscle work but it also responds to mental, emotional, and environmental loads. Heart rate will change with the cycle and depth of breathing (sinus arrhythmia), digestion, hormonal secretions (or the lack of them), and brain activity changes during sleep. It has a pattern of increases and decreases associated with the time of day (circadian rhythm), which is discussed in Chapter 4. Because it reflects so many body activities and changes, the heart rate changes measured on a job must be carefully interpreted, especially if they are being used to estimate the physical workload of a task. Where possible, a parallel measure of the oxygen demands of tasks should be made so that the other factors can be separated out and quantified.

A person's maximum heart rate is defined as the point when increasing workload no longer increases the heart rate. This value declines with age, falling roughly one beat per year. It can be estimated by subtracting a person's age from 220, but the error in this estimation is plus or minus 10 beats per minute for all age groups (Astrand and Christensen, 1964). Maximum heart rates are further discussed in Chapter 9 and in Appendix B.

b. Blood Pressure

Heart rate alone does not give a complete picture of the strain on the heart of increased workload or environmental and psychological stress. Blood pressure measurements, although very difficult to make during a task, can give information about the amount of static muscle work being done and the degree of psychological stress, especially. Where heart rate may be little changed because of low workloads, blood pressure may reflect the amount of tension caused by difficult mental or emotionally stressful job demands, including tasks that are done under strict time constraints or emergency conditions.

Blood pressure is measured at two times in the cardiac cycle: when the heart is contracting (systolic) and when it is relaxing (diastolic). Measurement techniques are found in Appendix B. Systolic pressure is directly related to heart muscle contractility and blood flow. Contractility can be improved by:

- Stretching the heart muscle fibers through increased venous return.

- Stimulation of the heart's pacemaker in the right atrium via inhibition of the parasympathetic, or vagus, branch of the autonomic nervous system or via excitation of the sympathetic branch.

- Direct action on the heart muscle of circulating epinephrine, or adrenaline, secreted by the adrenal glands in heavy work or in emergency response situations, and of nonepinephrine released in sympathetic nervous system stimulation.

Diastolic pressure is related to the total resistance of the vascular bed, especially the arterioles and veins. It is also partly determined by heart rate; higher heart rates leave less time for relaxation and cut off the fall in pressure that is measured during diastole (see Figure 27-12 in Part VII).

Mean, or average, arterial blood pressure (MAP) can be estimated for rest or light work using the following formula (Rushmer, 1972):

$$MAP = Diastolic\ BP + 0.33\ (Systolic\ BP - Diastolic\ BP)$$

This estimate may be inaccurate at higher heart rates or larger pulse pressures, which are the differences between systolic and diastolic blood pressures. Mean blood pressure is a determinant of the strain on the heart muscle of a given workload. It is correlated with heart muscle oxygen consumption, or work.

The *double product*, a derived measure, is also used to estimate the degree of heart strain (Froehlicher, 1972). It is the product of the heart rate and the systolic blood pressure, divided by 100, and it can be measured at rest or during a given task or condition such as walking or lifting:

$$Double\ Product = (HR \times Systolic\ BP)/100$$

At rest the double product is usually less than 100. During work it should not be permitted to exceed 350 for most industrial tasks except for very fit individuals.

Blood pressure is difficult to measure in many industrial jobs because diastolic pressure readings are difficult to detect in noisy surroundings and because the measurement technique (restricting blood flow to an arm for a short period) can interfere with the way a person normally does the job. If the task requires use of both arms, blood pressure readings are usually taken at the end of a cycle, rather than during it. This means that the worker gets an unscheduled recovery break each time his or her blood pressure is taken. These problems of measurement result in fewer studies of jobs including blood pressure as an indicator of job stress.

B. THE RESPIRATORY SYSTEM

The respiratory system includes the lungs, airways from the mouth and nose to the lungs, and the muscles of the chest wall and diaphragm. Through the action of these muscles, air is brought into the gas-exchanging part of the lungs called the alveoli. Blood flowing through the pulmonary capillaries picks up oxygen (O_2) from the alveoli and gives off carbon dioxide (CO_2). Figure 3-1 illustrates the pulmonary circulation and alveoli where gas exchange takes place.

When work is done, more carbon dioxide is formed in the muscles and more oxygen used. The respiratory system has to respond to this increased demand for O_2 and for CO_2 elimination by increasing the amount of gas exchanged in the alveoli per minute, known as the alveolar ventilation. Increasing the blood flow will not deliver enough oxygen to the muscles unless there is a constant supply of fresh O_2 for the blood to pick up as it passes through the lungs. Similarly, unless the carbon dioxide is breathed out, it will accumulate in the blood and, therefore, in the tissues, thus reducing the efficiency of muscle work.

1. LUNG CAPACITIES AND THE WORK OF BREATHING

The amount of alveolar ventilation a person can develop and maintain during work will depend on the capacity of his or her lungs and the amount of effort required to breathe deeply. Part of the lung capacity includes gas that remains in the lungs even after a maximum expiration; this gas is called the *residual volume*. It provides a store for O_2 and CO_2 and keeps their blood levels controlled within a fairly narrow range. This control is important in the regulation of breathing during work. The *vital capacity* is the usable volume of the lungs; it is the maximum amount of gas that can be breathed out after a maximum inspiration. Air that is brought down to the alveoli on each breath mixes with the residual volume gas; the resulting gas is the one with which the blood in the pulmonary capillaries exchanges CO_2 and O_2.

At rest, the volume breathed in per breath (tidal volume) is about 500 mL. A person with 5-liter vital capacity would have an expiratory reserve (ERV) of about 2.0 liters and an inspiratory reserve (IRV) of about 2.5 liters. During heavy exercise that same person might have a tidal volume of about 2.0 liters (Comroe, et al., 1955). This larger volume is developed by using some of the inspiratory and expiratory reserve capacity (Figure 3-3). Large tidal volumes ensure that more fresh O_2 reaches the alveoli per minute and that more CO_2 is removed from them. This exchange helps unload CO_2 and load O_2 onto the blood returning from the working muscles.

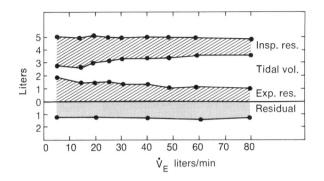

Figure 3-3: Functional Lung Volumes at Rest and During Exercise
During exercise, the tidal volume increases in order to ventilate the lung alveoli better. The figure illustrates the way this volume increase is accomplished for higher ventilation rates ($\dot{V}_E$), in liters per minute (L/min). The total lung volume in liters is shown on the vertical axis. This person has a residual volume of about 1.2 liters, indicated as the volume below zero because it cannot be expired from the lung voluntarily; it stays constant during work. The vital capacity for the subject is 5 liters, the total volume above the zero line on the vertical axis. The vital capacity (VC) is made up of the expiratory reserve volume (ERV), the tidal volume (V_T), and the inspiratory reserve volume (IRV). With increasing workload and increased tidal volume, the ERV and IRV are reduced. The muscle effort required to increase the V_T this way puts an additional load on the person. *(Asmussen and Christensen, 1939.)*

Air is brought into the lungs by muscle work that expands the chest and lowers the diaphragm. This increase in chest volume makes atmospheric pressure higher than lung pressure, so air comes in. When the chest and diaphragm muscles relax, the pressure in the lung becomes high relative to atmospheric pressure, so the gas moves out. The amount of work done by the respiratory muscles increases with larger tidal volumes. At high breathing frequencies, the passive relaxation of these muscles may not move air out of the lungs rapidly enough, and additional muscles of the chest, neck, shoulder, and abdomen are recruited to aid expiration (West, 1974). Heavy physical work, such as digging ditches, involves large increases in both the tidal volume and frequency of breathing. The work of breathing can be up to 10 percent of the person's demands for oxygen when near-maximum aerobic workloads are reached (Otis, 1964).

Resistance to breathing can be increased through internal narrowing of the conducting pathways, as in asthma, or through the addition of external resistance, as occurs when some types of respiratory protective equipment are worn. Increased resistance to flow means that more muscle work has to be done to get the same volume of air into the alveoli per minute. At very heavy workloads, the work of breathing may limit a person's capacity for work if the resistance is high. Strenuous labor requiring operators to wear respirators without positive pressure assists can fall in this category. In addition, working in a confined space or in a posture that makes it difficult to expand the chest fully (for example, when a person is stooped over) can also limit the amount of air brought into the lungs per breath.

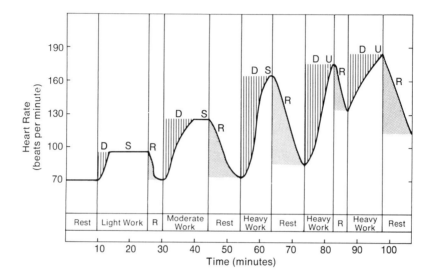

Figure 3-4: Heart Rate Changes in Intermittent Work Steady versus Nonsteady State A theoretical example of the changes in a person's heart rate (vertical axis) during a series of tasks and rest periods (horizontal axis) is shown. The heart rate is in beats per minute and is recorded continuously. The three levels of tasks are light, moderate, and heavy. They require heart rates of about 95, 125, and 170 beats per minute, respectively, in the steady state of work (S), so that enough blood flows to the working muscles to allow the work to be done aerobically. "At very heavy workloads, the steady state may not be reached (U)." From the beginning of each work period to the point when the steady-state heart rate has been achieved, there is a period of oxygen debt (D) for the working muscles; it is designated by vertical stripes. This is an unsteady state of work when tissue oxygen needs are not met by respiration and circulation and the muscle uses its stores of ATP and anaerobic glycolysis to support contraction. At the end of work, the debt is repaid (R), indicated by the dotted areas. In well-designed jobs, the areas under D and R are equal and the heart rate returns to the resting level before the next task is started (as in the first two work cycles). If the recovery time is not adequate to restore the ATP supply, resting heart rates will not be achieved, and the next work period will require a greater elevation of heart rate to supply the needed oxygen; this situation is seen in the last two work cycles. Recovery from this pattern takes longer because more anaerobic metabolism was used.

2. THE CONTROL OF ALVEOLAR VENTILATION

Not all of the air breathed per minute (the *minute ventilation*), gets to the alveoli. Some remains in the conducting pathways, known as the anatomical dead space. This volume is relatively constant, so increasing the tidal volume will increase the amount of air getting to the alveoli more than increasing the frequency of breathing will. The dead space is ventilated on each breath. The more shallow the breath, the higher the percentage of its volume that goes to dead space instead of to alveolar ventilation. In the control of breathing, the body balances changes in depth and frequency of breathing carefully in order to match alveolar ventilation to the workload. Table 3-3 shows the respiratory adjustment to moderately heavy and maximum work compared to resting.

If the alveolar ventilation is not increased in response to the increased metabolism, p_ACO_2 will rise and p_AO_2 will fall in the alveoli, and the arterial blood will reflect these changes (Rahn and Fenn, 1955). The amount of oxygen carried by hemoglobin will decrease as the oxygen tension is reduced, and this will limit the person's aerobic work capacity, making it difficult to sustain work. The higher arterial pCO_2 will make it more difficult to load the CO_2 from the tissues into the blood, resulting in a buildup of CO_2 and acid in the muscles, which will also limit the muscles' work capacity.

Fortunately, respiration, specifically alveolar ventilation, does increase to match the workload. It is adjusted by a complex control system, involving both neural and chemical components, to keep the arterial pCO_2 nearly constant. This ensures that the metabolic waste materials are removed from the active muscles and that the oxygen tension in the arterial blood also remains quite constant. Respiratory centers in the brain and major arteries near the heart monitor the CO_2 and O_2 pressures and the hydrogen ion concentration of the arterial blood to identify when to adjust the frequency and depth of breathing. Neural signals from the higher brain centers, the working muscles and moving joints, the stretch receptors in the lungs, and the temperature and circulatory control centers also affect the pattern and frequency of breathing (Dejours, 1959; 1966). The exact mechanism of this control in moderate to heavy exercise is still not known, but alveolar ventilation is closely matched to the muscles' need for oxygen in physical work. Most people still have excess breathing capacity when the maximum aerobic workload is reached, which indicates that blood flow, not alveolar ventilation, is usually the limiting factor in work capacity.

3. THE RELATIONSHIP BETWEEN GAS EXCHANGE IN THE LUNG AND TISSUE METABOLISM

Measurements of the amount of oxygen consumed ($\dot{V}_{O_2}$) and the amount of carbon dioxide produced ($\dot{V}_{CO_2}$) during work help to establish how heavy a job is. This information can be related to population work capacities to establish what percentage of the population can be expected to be able to do the job safely. (See Part IV for guidelines on workload design.) In field studies, the exchange of gases across the lungs is measured to estimate the metabolic work.

The amount of oxygen removed from the lung by the venous blood as it becomes arterialized is assumed to be in balance with tissue oxygen consumption. This is true in the steady state of muscular work, the point achieved when the heart rate and oxygen exchange in the lungs are constant for a given workload. It is not true in the initial stages of work when the heart rate and oxygen exchange are increasing to meet the new demands of the muscles for oxygen and nutrients. Figure 3-4 gives examples of steady and nonsteady states for heart rate, which are directly related to the tissue oxygen demand in repeated work and rest cycles. This work pattern is seen in industrial tasks such as loading trucks in a warehourse, shoveling, or in other construction activities.

Part IV shows patterns of work and gives design guidelines to keep physical workloads within the capacities of most people so that whole-body fatigue will be unlikely to occur. In well-designed jobs, the debt incurred at the beginning of each work period is repaid during the subsequent recovery period. When the recovery periods are too short for the work intensity, the heart rate will remain elevated, and each successive period will put more stress on the heart. More examples of unsteady state heart rates in work can be found in Chapter 9.

There are also transient increases in respiration without increased metabolism. These can be seen in tasks where psychic or emotional stress is present, as in highly paced machine monitoring or inspection jobs or in an emergency response to a machine jam or control failure. Because little additional oxygen is used, measurements of oxygen consumption will not correspond to the perceived stress on these jobs. Minute ventilation, however, will usually show increases as will blood pressure and heart rate. When physical work is superimposed on these stresses, it tends to overwhelm the psychic element, and ventilation corresponds more strictly to the oxygen consumption requirements of the task. Techniques for measuring oxygen consumption and carbon dioxide production from expired air samples are given in Appendix B.

REFERENCES

Asmussen, E. 1965. "Muscular Exercise." Chapter 36 in *Respiration: Volume II Handbook of Physiology,* edited by W. O. Fenn and H. Rahn. Washington, D.C.: American Physiological Society, pp. 939-978.

Asmussen, E., and E. H. Christensen. 1939. "Die Mittlekapazitat der Lungen bei Erhohten O_2 Bedarf. *Skandinavisches Archiv fur Physiologie, 82:* pp. 201-212.

Astrand, P. O., and E. H. Christensen. 1964. "Aerobic Work Capacity." In *Oxygen in the Animal Organism,* edited by F. Dickens and E. Neil. New York: Pergamon Press, pp. 295-314.

Astrand, P. O., and K. Rodahl. 1977. *Textbook of Work Physiology: Physiological Bases of Exercise.* 2nd ed. New York: McGraw-Hill, 681 pages.

Barcroft, H. 1963. "Circulation in Skeletal Muscle." Chapter 40 in *Circulation: Volume II Handbook of Physiology,* edited by W. F. Hamilton and P. Dow. Washington, D.C.: American Physiological Society, pp. 1353-1386.

Boothby, W. M., W. R. Lovelace II, O. O. Benson, Jr., and A. F. Strehler. 1954. "Volume and Partial Pressures of Respiratory Gases at Altitude." Chapter 4 in *Handbook of Respiratory Physiology: Respiratory Physiology in Aviation,* edited by W. M. Boothby Project No. 21-2301-0003. Randolph Field, Tex.: USAF School of Aviation Medicine, pp. 39-50.

Brouha, L. 1967. *Physiology in Industry.* 2nd ed. New York: Pergamon Press, p. 126.

Comroe, J. H., Jr. 1966. "The Lung." *Scientific American, 220:* pp. 56-68.

Comroe, J. H., Jr., R. E. Forster II, A. B. DuBois, W. A. Briscoe, and E. Carlson. 1955. *The Lung - Clinical Physiology and Pulmonary Function Tests.* 2nd ed. Chicago: Year Book Medical Publishers, 390 pages.

Dejours, P. 1959. "La regulation de la ventilation au cours de l'exercise musculaire chez l'homme." *Journal de Physiologie (Paris), 51:* pp. 163-261.

Dejours, P. 1966. *Respiration.* Translated by L. E. Fahri. New York: Oxford University Press, 244 pages.

Folkow, B., and E. Neil. 1971. *Circulation.* New York: Oxford University Press, 593 pages.

Folkow, B., C. Heymans, and E. Neil. 1965. "Integrated Aspects of Cardiovascular Regulation." Chapter 49 in *Circulation: Volume III Handbook of Physiology,* edited by W. F. Hamilton and P. Dow. Washington, D.C.: American Physiological Society, pp. 1787-1824.

Froehlicher, V. H., Jr. 1972. "Effects of Physical Conditioning in Healthy Young Individuals and in Coronary Heart Disease Patients." Chapter 5 in *Exercise Testing and Exercise Training in Coronary Heart Disease,* edited by J. Naughton and H. K. Hellerstein; I. C. Mohler, coordinating editor. New York: Academic Press, pp. 63-78.

Honig, C. R. 1981. *Modern Cardiovascular Physiology.* Boston: Little, Brown, 347 pages.

Johnson, P. C., ed. 1964. "Symposium on Autoregulation of Blood Flow." *Circulation Research, 15 (Supplement 1),* 290 pages.

Lind, A. R., and G. W. McNichol. 1967. "Circulatory Responses to Sustained Handgrip Contractions Performed During Exercise, Both Rhythmic and Static." *Journal of Physiology, 192:* pp. 595-607.

Otis, A. 1964. "The Work of Breathing." Chapter 17 in *Respiration: Volume I Handbook of Physiology,* edited by W. O. Fenn and H. Rahn. Washington, D.C.: American Physiological Society, pp. 463-476.

Rahn, H., and W. O. Fenn. 1955. *A Graphical Analysis of the Respiratory Gas Exchange: The O_2-CO_2 Diagram.* Washington, D.C.: American Physiological Society, 41 pages.

Roughton, F. J. W. 1954. "Respiratory Functions of Blood." Chapter 5 in *Handbook of Respiratory Physiology: Respiratory Physiology in Aviation,* edited by W. M. Boothby. Project No. 21-2301-0003. Randolph Field, Tex.: USAF School of Aviation Medicine, pp. 51-102.

Rushmer, R. F. 1972. *Structure and Function of the Cardiovascular System.* Philadelphia: W. B. Saunders, 296 pages.

West, J. B. 1974. *Respiratory Physiology - The Essentials.* Baltimore: Williams and Wilkins, 185 pages.

Zweifach, B. W. 1959. "The Microcirculation of the Blood." *Scientific American, 213:* pp. 54-60.

CHAPTER 4

Biological Rhythms

CHAPTER 4. BIOLOGICAL RHYTHMS

A *biological rhythm* is defined as any cyclic change in the level of a measure or chemical in the body. Examples are hormones, such as levels of circulating adrenal corticoids or thyroid hormone, and body temperature. The cycle time, or periodicity, of a rhythm can be very long, as in pregnancy; moderately long, as in the menstrual cycle; within a day (circadian rhythm) (Halberg, 1959); or within a few hours (ultradian rhythm) (Lavie, 1982).

The rhythms of most interest to ergonomists are the circadian and ultradian rhythms since there is evidence that human performance can vary with time of day or within a few hours of continued work on a task. These rhythms have implications for the design of hours of work, especially the choice of shift schedules, and also for the determination of appropriate work and rest schedules for sustained mental and perceptual work. This chapter includes a description of the biological rhythm of body temperature and its use in determining peoples' adjustment to altered activity cycles. A discussion of sleep, one part of the activity cycle, is also included with a short description of some of the effects of sleep loss on performance. Ultradian rhythms are discussed briefly, and a general description of the terminology used in research on biological rhythms is included to help the reader understand this complex subject.

A. BIOLOGICAL RHYTHMS—TERMINOLOGY AND DESCRIPTIONS

1. TERMINOLOGY

A biological rhythm is characterized by a periodic changing level of a substance or measure that repeats its pattern at regular intervals. Figure 4-1 illustrates a sine wave and shows its periodicity (or frequency), amplitude, phase, and acrophase. The amplitude of the rhythm is defined as the difference between the highest (maximum) and the lowest (minimum) points on the curve. The phase of the rhythm is defined as the relation of the curve to a fixed time, such as 12:00 noon. The maximum value of the curve is usually used to mark phase; this is also known as the acrophase (Monk, 1982).

2. DESCRIPTIONS

a. Internal versus External Generation

Biological rhythms appear to be determined both by internal, or endogenous, factors and by external, or exogenous, cues, called *zeitgebers* (Aschoff, 1978). The internal rhythm's source has not been identified although it is thought to be built into the genetic material of the body. The external zeitgebers range from natural phenomena, like the light/dark cycle of the day and night, outside thermal conditions, and seasonal changes, to social behaviors that affect the activity cycle, such as time of sleeping, level of activity, and diet (Aschoff, 1981). Most internal rhythms have a "free-running" period of about 25 hours. This has been determined through experiments with people deprived of external time cues. People who volunteered to live isolated in a cave for several months or who

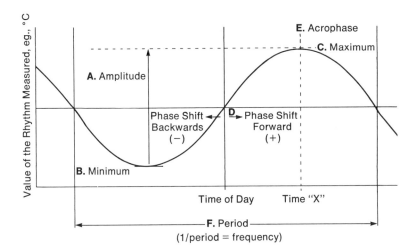

Figure 4-1: Terminology Used in Biorhythm Research   A theoretical sinusoidal rhythm is labeled to demonstrate its amplitude (A), minimum (B) and maximum (C) values, phase (D), acrophase (E), and periodicity or frequency (F). The horizontal axis shows time and the vertical axis is the measure of interest, often the percentage increase over an average value (see Figure 4-2), degrees Celsius or Fahrenheit (see Figures 4-3 and 4-4), or the concentration of a substance in the blood. Performance rhythms are similar to biological rhythm curves and vary according to the type of task. *(Colquhoun, 1982.)*

participated in several months of laboratory isolation-chamber studies were found to develop free-running body temperature rhythms of 25 hours (Aschoff, 1981; Weber, 1979) and to have sleep/wake activity cycles that shifted gradually to the right of the time scale over the months.

The link between biological rhythms, as shown by body temperature, and the activity cycle in a person isolated from external cues is known as synchronization. When the body temperature is at its lowest point, the person tends to be asleep. At its highest point, the person is most alert and tends to be active. Of concern in studies of people who are not isolated from other external influences is a desynchronization of the rhythms by changes in behavior or by scheduling of work outside of the daytime hours. Desynchronizing the rhythms is thought to have some effect on a person's health since biological rhythms are often disturbed in disease. It is not clear yet just what that health effect is or whether it is the result of or the cause of disturbances in the underlying biological rhythms (Akerstedt and Froberg, 1976). People who show the peak in body temperature early in the afternoon and begin to show the decline towards the minimum temperature in the mid-evening hours are known as "morning types" or "larks." Those who experience their peak later in the evening and their minimum point in the late morning are known as "evening" types or "owls." "Larks" have difficulty in adjusting to night shift work and are likely to experience more sleep and digestive disturbances than "owls" do (Froberg, 1981).

b. Entrainment of Rhythms

Another characteristic of biological rhythms is that they tend to become entrained to changes in the activity cycle over time; that is, they adjust their phase to try to match that of the activity cycle (Brown, 1972). In the early studies of Benedict (1904), an attempt was made to show that a person who works night shift for long periods will adjust his or her temperature curve to fit the reversed activity cycle. The nightwatchman studied by Benedict had spent eight years on a fully reversed activity schedule, living weekends as well as workdays in the reversed pattern; yet he had not fully reversed his body temperature curve. Rhythm reversal like this is thought likely to occur only in full isolation studies, where the reversal can be seen in about three and a half weeks in carefully controlled conditions (Wever, 1979). Despite the lack of evidence that the biological rhythms can be reversed to respond to activity pattern reversals, there is no question about the fact that external factors affect the phase and amplitude of the biological rhythms. This is discussed further in the section on body temperature rhythm and performance.

c. Amplitude Changes in the Rhythm

The amplitude of a biological rhythm will vary greatly from one person to the next. There is some evidence that the more amplitude a person has in his or her temperature rhythm, the more time it takes for changes in activity pattern to affect the rhythm. Younger workers tend to adjust their rhythms faster than older workers do (Reinberg, Andlauer, and Vieux, 1981). A person with a low-amplitude rhythm may take less time to adjust but be less able to sleep on a weekly rotating shift schedule. The need for sleep, however, appears to be greater for those with a higher-amplitude rhythm than for those with a relatively flat rhythm (Wever, 1981). There is evidence that women have lower-amplitude temperature rhythms, and their phase is different from the curves shown for both young and older men (Akerstedt and Froberg, 1976; Mellette, et al., 1951). Much more work needs to be done with women to determine whether this is a consistent finding and whether it bears any relationship to changes in body temperature due to the menstrual cycle. The variability in body temperature rhythms among men is so high that it is difficult to describe group responses as differing significantly except when they are used as their own controls, that is, before and after a change in activity cycle.

d. Hormonal Rhythms

Not all endogenous body rhythms follow the same pattern. The body temperature rhythm is usually used as the example of the effects of external factors because it is convenient to measure, does not change rapidly, and is well regulated through the circulatory and central nervous systems. The rhythm of adrenal gland cortisol released into the blood has different characteristics, the peak occurring at 8:00 a.m. in most people (Mills, 1966). Figure 4-2 illustrates this rhythm.

The blood level of epinephrine, or adrenaline, can be affected by a sudden change in the person's environment, such as a fright or sign of danger. This biological rhythm is of interest to researchers trying to identify factors that increase chronic stress in very demanding or highly paced work (Johansson, Aronson, and Lindstrom, 1978). When an external factor desynchronizes the body temperature rhythm, it may also desynchronize the adrenaline rhythm relative to it. The resychronization of these rhythms may take some hours to accomplish. Whether the person affected is at increased risk of illness or injury at this time, or whether his or her performance level will be affected, depends on the nature of the job, the performance level required, the individual's ability

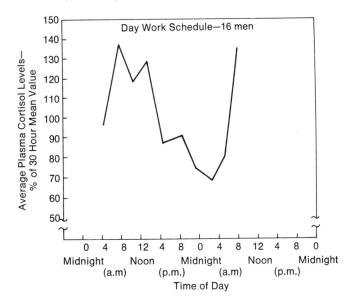

Figure 4-2: Daily Variation in Plasma Cortisol Levels   An example of a 24-hour rhythm in plasma cortisol during a day-work/night-sleep activity cycle is shown to illustrate a hormonal rhythm. Cortisol is one of the hormones released by the adrenal glands in response to stress. It is expressed as a percentage increase or decrease over its average value (vertical axis) for a 24-hour period (horizontal axis). The peak values are seen in the hours from 6 to 8 a.m. and the minimum values from midnight to 2 a.m. This is quite different from the normal oral temperature rhythm (see Figures 4-3 and 4-4). With changes in the activity cycle or amount of sleep, the cortisol rhythm tends to shift to the left and increase in amplitude. (*Adapted from Frank et al, "Circadian Periodicity, Adrenal Corticosteroids, and the EEG of Normal Man" and reprinted with permission from Journal of Psychiatric Research, Volume 4, Copyright © 1966, Pergamon Press, Ltd.*)

to adapt rhythms, and other factors (Aschoff, 1981; Murphy, 1979). More research is needed, preferably in the workplace, to answer these questions. The complex interactions of rhythms make predictions of individual or group responses hazardous at best.

c. Effects of Time Zone Changes on Body Temperature Rhythms

Another aspect of biological rhythm research has focused on the effect of crossing time zones on performance and on underlying physiological processes. With the vast increases in speed associated with supersonic transport compared to travel by ship, new concerns have arisen about the wisdom of people making high-level decisions after flying across several time zones and losing sleep. Flying from west to east—for example, from New York City to London—results in a 5-hour time shift towards evening. At the time most Londoners are ready for bed, the American is near the peak of his or her body temperature rhythm. The next morning at 9:00 a.m., for example, when decisions are required, the American's body temperature may be at its minimum; his or her performance may be affected on certain tasks. Some companies have policies that no decisions are to be made for at least 24 hours after arrival, but even this may be inadequate time for some travel distances, such as from Tokyo to New York (Cassuto, 1972). Travel from east to west, as when a Londoner comes to America, is somewhat easier as long as it is possible to go to sleep early in the evening and the person is not forced to stay up when the body temperature rhythm is at its low point. In longer travel across more time zones, such as from London to Singapore, the adjustment problem can be greater and more time will be needed to adapt to the time zone changes. This problem is recognized by international transport companies; layover times for crews are adjusted to permit some adaptation. The International Civil Aviation Organization has developed a travel-time formula to determine how much recovery time is needed to readjust the circadian rhythms (Cassuto, 1972):

Rest Period	=	(0.5) Travel Time	+	Time Zones	+	Departure
(in tenths		(in hours)		(in excess		Coefficient
of days)				of 4)		(local time)

				+	Arrival
					Coefficient
					(local time)

In this formula, the rest period is rounded off to the nearest half-day (0.4 becomes 0.5 day, for example) and travel time includes short breaks in the trip but not stopovers. If an overnight stopover is included in a long flight, the rest period calculation should be made only for the last part of the trip. Time zones are measured as every 15 degrees of longitude from the Greenwich meridian, and the departure and arrival coefficients are taken from Table 4-1.

Using the formula, one can calculate that a trip from Paris to New York,

Table 4-1: Departure and Arrival Time Coefficients   The local time for either arrival or departure is shown in column 1. The appropriate arrival and departure coefficients (in columns 2 and 3) are added to the International Civil Aviation Organization (ICAO) formula to calculate the recommended rest period after traveling across several time zones. *(Adapted from Cassuto, 1972).*

Local Time	Departure Time Coefficient	Arrival Time Coefficient
8 a.m. – 11:59 a.m.	0	4
12 noon – 5:59 p.m.	1	2
6 p.m. – 9:59 p.m.	3	0
10 p.m. – 12:59 a.m.	4	1
1 a.m. – 7:59 a.m.	3	3

which takes about 9 hours and crosses 5 time zones, will require 0.65, or one-half, day of rest if the passenger leaves at 2:00 p.m. and arrives in New York at 6:00 p.m., but it will require 0.95, or 1, day of rest if local departure and arrival times are 7:00 p.m. and 11:00 p.m., respectively. Traveling from New York to Paris takes an additional hour, so a person who left New York at 10:00 p.m. and arrived in Paris at 1:00 p.m. would need 1.2 rest periods or 1 day of rest. Travel by an SST reduces travel time about 5 hours, but if the flight left New York at 10:00 p.m. and arrived in Paris at 8:00 a.m., the required rest period would still be 1 day. A noon departure would require only 0.55, or one-half, day to recover.

f. Menstrual Cycle

The longer biological rhythm of the menstrual cycle is known to be associated with some changes in physiological and psychological well-being for many women. The basic rhythm is 28 days with large variations between individuals that appear to be related to external factors and to health and diet. The degree of physical incapacity associated with the menses will determine the suitability of some types of physical work during this time. The premenstrual syndrome can include very mild or very severe depressive symptoms, irritability, nausea, and difficulty in concentrating, and it may also affect performance on more demanding jobs (Redgrave, 1971). Because individuals respond very differently to the cyclic changes, it is not appropriate to anticipate that all women will have difficulty on a particularly demanding job during the premenstrual part of the cycle or during menses. Men may have a similar 28-day behavior cycle. Longer-cycle menopausal changes in performance and health are inadequately researched in the field of biological rhythms.

g. Physical, Emotional, and Intellectual Biorhythms

Much popular attention in the past 15 years has been given to another set of long-cycle rhythms known as *biorhythms*. These are characterized as 23-, 28-, and 33-day rhythms in physical, emotional, and intellectual functioning, respectively (Comella, 1976). The marketers of biorhythm charts and computers indicate that by giving your date of birth, you can find out what your rhythm chart is or was for any period of time in the future or past. By seeing where your physical, intellectual, and emotional rhythms are at a maximum or minimum or where they converge in crossing the base line, you can anticipate which days are good or which are critical ones when you may be at greater risk for illness or poor performance (Thommen, 1972).

The concept of using biorhythm charts to reduce accidents has been advanced (Drieske, 1972), but there have been few rigorous studies in the field. In one of the few such studies that have been done, the correlation between critical days and accidents was found to be insignificant, observed cases being explained by chance alone (Workman's Compensation Board of British Columbia, 1971; Soutar and Weaver, 1983). In a study in Japan where long-distance truck drivers were reminded of their critical days by a card placed on the dashboard at appropriate times, a reduction in accidents was seen (Drieske, 1972); but this could be explained by the attention given the drivers (a "Hawthorne effect") as easily as it could be by the biorhythm theory. A control was not used to rule out that possibility in the study.

Fourier and autocorrelation analyses of the rhythms have seldom been done to determine whether these rhythms are predictive of increased risk for accidents or poor performance (Kintz, Eastman Kodak Company, 1975; Monk, 1982). Because studies of these rhythms have not been carried out scientifically with careful controls, they have not found acceptance in the biological rhythms research field. At this time, they are considered more akin to astrology than to circadian rhythms, with most of the successful predictions being attributed to chance.

h. Ultradian Rhythms

Ultradian rhythms have periodicities of less than two hours and are best demonstrated in rhythms of sleep, hunger, and mental and perceptual functioning (Lavie, 1982). They are more apparent during performance on demanding and sustained tasks, such as visual inspection of moving products or trouble-shooting in a time-paced maintenance operation. The controller of these rhythms is not known, although it is clear that they can be modified substantially by external and internal factors like individual motivation, the task's interest level, physical activity, slight variety added to the task, and its difficulty (Lavie, 1982). They interest ergonomists primarily because they identify an underlying reason for organizing job demands to provide fairly frequent recovery or rest breaks in order to prevent fatigue or performance decrements on sustained work.

B. BODY TEMPERATURE RHYTHM AND PERFORMANCE

Many studies of the body temperature rhythms of students, naval recruits, coal miners, production workers, and military personnel have been carried out since Kleitman (1963) demonstrated that job or task performance and the temperature rhythm appear to be linked (Colquhoun, 1971; Colquhoun, 1982; Colquhoun and Edwards, 1970; Colquhoun, Blake, and Edwards, 1968a; van Loon, 1963). Because temperature curves are easy to measure, such studies have been acceptable to industrial workers and have been the primary physiological measure of their adjustment to shift work schedules. This section reviews the characteristics of these temperature rhythm changes and their association with performance on experimental tasks or on a job.

1. TWENTY-FOUR-HOUR TEMPERATURE RHYTHMS

The average oral temperature curve for a group of naval recruits is shown for a 24-hour period in Figure 4-3. Readings were taken every hour, except during the sleeping hours when they were taken every 2 hours. The minimum value is seen in the early morning hours and the maximum value in the later afternoon/early evening hours. The amplitude of the rhythm is 0.9°C (1.7°F). The acrophase is at 8:00 p.m., and the rhythm can be assumed to be in phase with the activity cycle, which is marked near the top of the graph.

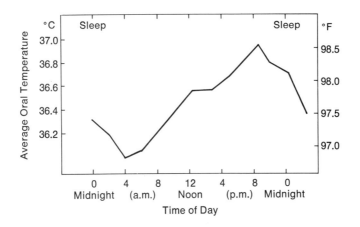

Figure 4-3: Average Oral Temperature Readings During a 24-Hour Period   The average oral temperature values over a 24-hour period are shown for 59 naval recruits. A clear minimum in temperature is seen at 4 a.m. with a maximum value at 8 p.m. This pattern repeats itself every 24 hours if the activity cycle remains the same. (*Adapted from Colquhoun, Blake, and Edwards, 1968a*).

2. LESS THAN 24-HOUR TEMPERATURE CURVES IN SHIFT WORKERS

In the workplace it is less easy to obtain 24-hour studies of body temperature, so temperature is often measured only during the work hours. If the workers are on a rotating shift schedule, one can fabricate a 24-hour curve from the measurements taken at different times of the week or month on each shift. Such a curve is shown in Figure 4-4 for 17 packaging operators working a weekly-rotating 5-day shift pattern. The oral temperature data were collected daily, six to seven readings per shift, over three weeks; the average responses for all 17 operators are shown for each shift and for the first (Monday) and last (Friday) days of each week. It is clear that the basic temperature rhythm is similar to that for the naval recruits shown in Figure 4-3. The adaptation of the rhythm from

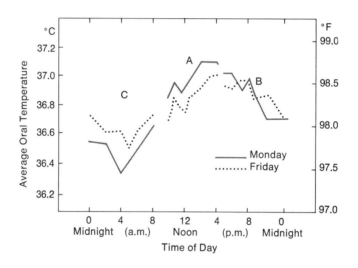

Figure 4-4: Average Oral Temperatures of 17 Packaging Operators
The average oral temperatures of 17 male packaging operators are shown over three shifts; they were taken over three weeks on a weekly-rotating 5-day shift schedule. The average age of the workers was 34 years, with a range of 21 to 45 years, and the average length of experience with this shift system was about eight years. The scales are the same as those in Figure 4-3. The shifts started at midnight (C), 8 a.m. (A), and 4 p.m. (B). The solid curve shows the results of measurements taken on the first day of each shift (Monday), and the dotted line indicates the measurements taken on the last day of the shift (Friday). The adaptation is seen as a reduction in amplitude of the composite temperature curve. The C-shift curve moves up and the A-shift curve moves down to accomplish the reduction in amplitude. There is also a slight shift to the right of the whole curve so that the minimum and maximum (acrophase) points occur later in the day. (*Rodgers, Eastman Kodak Company, 1972*).

the beginning to the end of the week is most obvious during the late-night, or C-shift, rotation. The adaptation consists of a reduction in amplitude of the curve from 0.7 to 0.45°C (1.3 to 0.8°F) and a phase shift to the right of one to two hours. The phase shifts are marked by the maximum, or acrophase, temperatures in these curves. No reversal of the rhythm occurred during the week, nor would it be likely to occur during a more extended rotation period (Knauth and Rutenfranz, 1976).

Similar changes in oral temperature rhythm amplitude have been demonstrated by other investigators. Table 4-2 summarizes several of these study results by shift system. The amount of change in amplitude appears to be related to the activity cycle and to the age and experience of the shift worker. It is also dependent on the type of work done, and individual variability on the same type of work is high. Caution should be exercised in extrapolating the results beyond the specific conditions of each study.

3. BODY TEMPERATURE RHYTHM ADAPTATION OVER LONGER SHIFT-ROTATION PERIODS

The temperature rhythm's phase shift observed over one week of night shift work has been used as a reason to extend the shift rotation cycle to a month. Proponents argue that a person needs time to adapt to the changed activity cycle, and that the natural adaptation time is from three to four weeks (Winget, Hughes, and LaDou, 1978). One study has shown that people who moved to a 21-day shift rotation schedule from a weekly rotation were less affected by sleep and digestion problems after nine months on the new schedule (Czeisler, Moore-Ede, and Coleman, 1982).

Although the physiological advantages of longer rotation schedules may be demonstrated, it is the psychosocial aspects of shift schedules that primarily determine their acceptability (Murphy, Eastman Kodak Company, 1972, 1973, 1976, 1978, 1980; Rutenfranz, Colquhoun, and Knauth, 1976; Wedderburn, 1967). Longer rotations are particularly dissatisfying on late-night and afternoon/evening shifts because of their interference with family activities (see Part V for discussions of different shift schedules and their advantages and disadvantages). Putting too much emphasis on the design of shift schedules to satisfy the physiological needs, then, ignores one of the most important factors in job satisfaction—how the schedule affects the worker's leisure time and interactions with family and friends.

A rapidly-rotating shift schedule where the workers never have more than three days in succession on the night shift has been used in Europe and in some American companies to address the biological rhythms desynchronization concern and also to improve the psychosocial aspects of shift work. Figure 4-5 shows oral temperature curves for chemical operators working 12-hour shifts, either day or night, and rotating after three days to rest days. The same reduction of amplitude and a small phase shift to the right are seen as were observed in Figure 4-4.

Table 4-2: Body Temperature Adaptations to Different Work Schedules The average oral temperature curve amplitudes for shift workers on three types of shift schedules (column 1) are shown in columns 3 and 4. The characteristics of the eight studies are given in column 1, and the variables for comparision are shown in column 2. N, D, M, and F refer to night shift, day shift, Monday, and Friday, respectively. A, B, and C refer to morning, afternoon/evening, and night shifts, respectively. A number after one of these letter indicates the number of days or weeks of work on the shift. Column 5 gives the percentage change in amplitude relative to the first value in each study, either night versus day, Friday versus Monday, B- and C-shift versus A-shift, and so on. This table indicates that there is a reduction in oral temperature curve amplitude of 20 to 72 percent when night work is done. The interested reader should consult the references (column 6) for further discussion of the specific experiments. *(Developed from data in the references cited in the table).*

| Schedules/ Conditions | Oral Temperature Curve | | | | |
| | | Amplitude | | | |
	Variable	°C	°F	% Change*	Reference
Day/Night Alternation					
3 young men, not experienced	D	0.7	1.3	—	van Loon, 1963
	N	0.4	0.75	− 42	
17 railway workers, average age 36 years, heavy work; 3-week rotation, 9-hour shifts	D	0.55	1.0	—	Akerstedt, 1977
	N (wk 1)	0.25	0.45	− 55	
	N (wk 3)	0.45	0.8	− 20	
10 – 11 young men; 12 days per shift, 8-hour shifts	D	0.8	1.4	—	Colquhoun, Blake, and Edwards, 1968b
	N6	0.5	0.9	− 36	
	N12	0.2	0.4	− 72	
10 – 11 young men; 12 days per shift, 12-hour shifts	D	0.6	1.15	—	Colquhoun, Blake, and Edwards, 1968b
	N6	0.47	0.85	−26	
	N12	0.45	0.80	−30	
Weekly Rotating, 5-Days, 8-Hours					
17 male packaging operators, experienced, average age 34 years; composite curve over 3 shifts, Monday/Friday comparison; 7 points per shift	M	0.8	1.4	—	Rodgers, Eastman Kodak Company, 1972
	F	0.45	0.8	− 43	

Table 4-2   (*Continued*)

Schedules/ Conditions	Oral Temperature Curve				
		Amplitude		%	
	Variable	°C	°F	Change*	Reference
Weekly Rotating, 5-Days, 8-Hours					
19 operators in a plas-tics factory, experi-enced males; composite curve over 3 shifts; Monday/Friday compari-son; 3 points per shift	M F	1.1 0.8	1.9 1.4	— − 26	Guberan, et al., 1969
Rapidly Rotating					
2 students, not experienced; 2-2-2, A-B-C shifts; rectal temperature	A1 A2 B1 B2 C1 C2 off off	1.3 0.9 1.2 1.3 1.1 0.8 1.3 1.3	2.3 1.6 2.2 2.3 2.0 1.4 2.3 2.3	— − 30 − 4 0 − 13 − 39 0 0	Knauth and Ruten-franz, 1976
8 submariners; 3-day duty cycle; 16–8–4 hours; 15 duty cycles total	Cycle 1 Cycle 5 Cycle 10 Cycle 15	0.30 0.28 0.24 0.15	0.55 0.50 0.43 0.27	— − 9 − 22 − 51	Colquhoun, Paine, and Fort, 1979

* % Change = Relative to first value in each of the studies shown.

4. BODY TEMPERATURE AND TASK PERFORMANCE

In reviewing the findings of studies of task performance at different times of the day and night and the relationship of performance to body temperature levels, one must recognize that the circadian rhythms of performance vary con-siderably (Colquhoun, 1982). Some performance rhythms change with body temperature; lower performance on monitoring tasks and on tests such as re-action time and a digit-span test of short-term memory have been associated with lower body temperatures in the early morning hours (Blake, 1967; Folkard, et al., 1976; Kleitman and Jackson, 1950). Other rhythms show increased per-

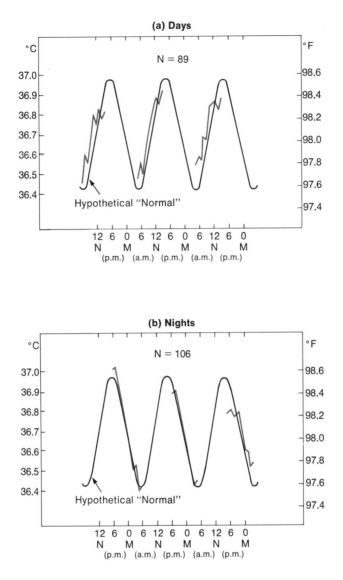

Figure 4-5: Average Oral Temperatures of Chemical Operators on a Rapidly Rotating 12-Hour Shift Schedule   The average oral temperature readings for chemical operators working three successive 12-hour days (a) and nights (b) are shown as solid curves superimposed on a hypothetical normal sinusoidal temperature curve. Temperatures are shown in degrees Celsius (°C) on the left vertical axis, and in degrees Fahrenheit (°F) on the right vertical axis. Time of day is shown on the horizontal axis. The day shift started at 6 a.m. and concluded at 6 p.m.; the night shift was from 6 p.m. to 6 a.m. As seen in the studies presented in Figure 4-3 and Table 4-2, a reduction in amplitude, or flattening out, of the temperature rhythm is associated with night work. This reduction is produced both by a reduced peak and a higher minimum value. This amplitude change was reversed a few days after day shift work was resumed. (*Adapted from Proceedings of the 25ʰ Human Factors Society, 1981, Rochester, N.Y.: pp 207–210 by G. D. Botzum and R. L. Lucas. Copyright © 1981 by the Human Factor Society, Inc. and reproduced by permission*).

formance at times when the temperature is decreasing and decreased performance when it is rising—for example, the postprandial (after-meal) dip in performance often observed in the early afternoon hours (Colquhoun, 1982).

External factors, or zeitgebers, can affect task performance in either direction, thereby desynchronizing it from the temperature rhythm. Incentive pay has been shown to reduce the amplitude of a performance curve in a repetitive task (Blake, 1971), resulting in more consistent performance over the shift. The amplitude of a performance curve, however, will depend largely on the difficulty of the task. More difficult tasks usually have larger performance curve amplitudes because the person is working close to capacity and may not be able to summon the extra effort needed as the shift proceeds. Incentives or motivation may not be able to keep performance high on demanding tasks, whether they are characterized by high physical, perceptual, or mental workloads. The results of some productivity, error, and accident studies at different times of the day are discussed in Part V. No physiological measures were made in these studies, so performance could not be linked to underlying biological rhythms.

Another confounding factor in studying job performance and body temperature changes in shift work is that the night shift worker may also be exhibiting performance problems associated with inadequate sleep. This is discussed in more detail later in this chapter.

The comments of night shift workers that physically demanding tasks seem harder in the early morning hours are difficult to reconcile with physiological tests of work capacity or heart rate responses on the job. In a study where heart rate was monitored on three shifts during light to heavy work in a hot environment, it was noted that heart rate changes during the early morning hours were similar to or a little lower than those seen on the day shift for the same activity done by the same person. The subjective response to the task, however, was that the work was far heavier and more difficult to sustain (Rodgers, Eastman Kodak Company, 1973). Studies of aerobic work capacity throughout the day do not indicate that capacity changes enough to explain the subjective feelings (Davies and Sargeant, 1975; Wahlberg and Astrand, 1973; Wojtczak-Jaroszowa, 1976). More studies need to be done to link physiological, performance, and psychophysical ratings of stress on industrial tasks in order to learn why the objective and subjective findings do not always correspond. A recent study using psychophysical and physiological measures showed that neuromuscular capacity may be reduced in the early morning hours (Ilmarinen, et al., 1980).

One additional reason for the inability to link body temperature levels and job performance is that it is often difficult to measure pure performance on a task in industry. The productivity of the worker can be confounded by a number of factors, such as:

- Whether the same work is done on all shifts, or whether special responsibilities are included in the day shift (for example, running special tests, training, or talking with visitors and supervisors).

- Production-limiting conditions that the workers do not have control over, such as equipment failure, inadequate supplies, quality control problems with parts, or the time needed to complete a production cycle in chemical making.
- The nature of the products being manufactured and the volume of each, since frequent changeovers of the manufacturing line will reduce total productivity per shift.
- The number and capabilities of the maintenance and service backup personnel, which can influence how long it takes to repair a problem or deliver the necessary materials to a manufacturing line.
- The supervisory quality, quantity, and style.

Because performance measures of workplace productivity are confounded by the above and by other differences between shifts, most studies assessing the relationship between performance and body temperature have been done in laboratory settings (Colquhoun, 1982). The results of laboratory studies, however, cannot be extrapolated to the workplace because of the multitude of external factors that can affect the relationship out in the field. Studies that compare very similar work on three shifts with rotating supervision and in isolation from other controlling factors will be needed to establish how predictable the relationship between performance and body temperature is in industry. For the time being, it is generally agreed that attention, or arousal, is lower in the early morning hours unless specific motivating factors are provided, and errors of omission and less clear thinking are possible at this time (Blake, 1971).

C. ACTIVITY CYCLES AND SLEEP

The relationship between inherent biological rhythms, such as body temperature or hormone levels, and a person's activity, or sleep/wake, cycle has been discussed earlier in this chapter. External social factors determine an individual's choice of hours of sleep and of work and recreational activity. The traditional breakdown of a 24-hour day is eight hours of sleep, eight hours of work, and eight hours of recreation. In reality, with changing work schedules, travel time, late-night entertainment on television, and other diversions, most people do not divide their time equally among these three pursuits. Table 4-3 summarizes data on the average hours of sleep for industrial workers working day shifts.

When the hours of work are changed from the daytime to the afternoon/evening shift, sleep is affected by being pushed about two to three hours further ahead in the day, usually starting at 1:00 or 2:00 a.m. instead of 10:00 or 11:00 p.m. Because the evening shift worker can usually sleep in the morning, this small shift in sleeping hours seldom results in sleep loss problems (Tepas, Walsh, and Armstrong, 1981). Some shift workers find this sleep the most satisfying of any they receive during a three-shift rotation (Murphy, Eastman Kodak Company, 1972).

Table 4-3: Average Hours of Sleep at Night Following Day-Shift Work   The average hours of night sleep recorded by shift workers during the weeks they worked the day shift are shown in column 2 for four different types of schedule (column 1). References for the data are given in column 3. The number of workers studied varies from less than ten to several hundred in these studies; all are male except for the nurses. The number of hours of night sleep obtained during the period of day shift work is less on weekly-rotating 5-day schedules or on rapid-rotation schedules than it is on permanent day work or on 3-shift rotating 7-day schedules. (*Developed from data in the references cited in the table*).

Shift Schedule	Average Hours of Sleep	Reference
Permanent Day	8	Mott, et al., 1965
	7	Tasto, et al., 1978
	7	Tepas, Walsh, and Armstrong, et al., 1981
	7.4	Tune, 1969
	7.8	Akerstedt, 1976
Weekly-Rotating 5-Day Schedule	5.6	Colquhoun and Edwards, 1970
	6.8	Smith, 1979
	7.1	Ostberg, 1973
	6.7	Murphy, Eastman Kodak Company, 1972
	7	Mott, et al., 1965
Rotating 7-Day Schedule	8	Maasen, Meers, and Verhaegen, et al., 1980
	8.5, 7.5 (Nurses)	Folkard, Monk, and Lobbam, et al., 1978
	6.0, 6.2	Akerstedt, 1976
Rapid-Rotation 7-Day Schedule	4.7 (Nurses)	Smith, 1979
	6.7	Webb and Agnew, 1978

Moving the hours of work from day to afternoon/evening shifts to late-night shifts will move the sleeping hours forward by about nine to eleven hours, to a time when a person would usually be active. In addition, noise from the household and street, interruptions from phone calls and delivery (or other day-schedule) people, and the need to do some personal business errands combine to make sustained sleep difficult during the morning and early afternoon hours. Guidelines for shift workers that should help them to overcome some of these interruptions are given in Part V. The result of these problems, though, is that most people get less, or less satisfying, sleep during periods when they are working the late-night shift.

Sleep satisfaction is associated with the amount of rapid eye movement

(REM) sleep obtained during the sleep cycle. REM sleep reflects brainwave activity thought to be needed for proper rest. Some laboratory studies of sleep at different times of the day show that the amount of REM sleep decreases substantially when the sleep cycle occurs during the day (Czeisler and Guilleminault, 1980; Webb, 1982b; Webb and Agnew, 1975). Whether long- or short-term health or performance problems are associated with this loss of REM sleep during night shift work is not known (Pearlman, 1982).

The nature of the shift schedule will determine how much sleep loss is accumulated during night shift work. There will be some adaptations in the activity cycle and in the temperature rhythm that will make REM sleep more prevalent as the night shift period proceeds (Webb, 1982). However, some workers never accomplish a full adjustment to the night-work/day-sleep pattern, and they accumulate sleep loss and fatigue with time. Table 4-4 illustrates the hours of sleep that shift workers reported for each shift of permanent (fixed) 5- and 7-day weekly rotating, and rapidly rotating shifts. Experienced shift workers on permanent night shifts get more sleep than those who rotate weekly; sleep appears to be better for people who work a 2-2-3 (A-B-C or C-B-A) rapidly rotating schedule (see Part V for more information about rapid-rotation schedules).

D. SLEEP LOSS AND PERFORMANCE

The relationship between the body temperature rhythm and performance on a task was addressed earlier in this chapter. The relationship between sleep loss and performance was mentioned as a confounding factor in these studies. Controlled studies of sleep in laboratory settings show a wide range of responses from no effect to profound effects, and some studies even show improved performance (Alluisi, 1972; Colquhoun, 1982; Folkard, et al., 1976). Two factors probably account for the disparity of these results: the intensity of the task used to determine the performance levels; and the length of time spent performing the task. It has been shown that a person with extensive sleep loss (24 to 70 hours) can still motivate him- or herself to perform short-duration tasks, ones that usually take less than two minutes to perform, but certainly under 15 minutes (Johnson, 1982; Wilkinson, 1961; Williams, Lubin, and Goodnow, 1959). The ability to sustain performance is considerably reduced, especially if the task is demanding. Studies of longer-duration performance indicate that sleep loss and its associated fatigue are contributors to reduced performance in detecting signals and with increased errors of omission (Blake, 1971; Colquhoun, 1982; Johnson, 1982). This appears to be a reduction in the central nervous system's arousal level, as shown by a decreased response in the brain's electrical activity (measured by the electroencephalogram, or EEG) to auditory signals (Bonnet, 1982). Decrements in task performance are greatest for very demanding tasks done over several hours. About 15 minutes of testing is required to overcome the motivation factor that keeps performance at normal levels on short tasks (Colquhoun, 1982; Wilkinson, 1961).

Table 4-4: Average Hours of Sleep on Several Shift Schedules The data from Table 4-3 are extended and summarized for the average hours of sleep for workers after they have been on the day, afternoon/evening, and night shifts, columns 2 through 4, respectively. Three types of shift systems are designated in column 1. The source of the data is found in the references in column 5. In addition to the average number of hours of sleep, the percentage of hours as compared to sleep after day shift work is shown (in parentheses) for the afternoon/evening and night shifts. Day shift hours start between 6 and 8 a.m. in most of these studies; afternoon/evening shifts start between 2 and 4 p.m.; and night shifts start between 10 p.m. and 12 midnight. A common finding is that sleep after working on the afternoon/evening shift is better than that obtained after the other two shift rotations. Sleep after working on the night shift is usually shorter than sleep after the other shifts and is also less satisfying. In one study (Murphy, Eastman Kodak Company, 1972), sleep after night shift work was rated 68 percent as satisfying as sleep after working on the day or afternoon/evening shifts. *(Developed from data in the references cited in the table.)*

Shift Schedule	Average Hours of Sleep (Percent of Day Shift Sleep)			
	Day	Afternoon/Evening	Night	Reference
Permanent	8	8 (100%)	7.7 (96%)	Mott, et al., 1965
	7	7.4 (106%)	6.3 (90%)	Tasto, et al., 1978
	7	7.3 (104%)	6.0 (86%)	Tepas, Walsh, and Armstrong, et al., 1981
	6.7	—	5.1 (76%)	Walsh, Tepas, and Moss, et al., 1981
Weekly-Rotating 5-Day	6.7	8.3 (124%)	7.5 (112%)	Murphy, Eastman Kodak Company, 1972
	5.6	8.9 (160%)	7.8 (139%)	Colquhoun and Edwards, 1970
	6.8	8.9 (131%)	7.0 (102%)	Smith, 1979
	6.2	8.8 (142%)	— —	Akerstedt, 1976
	7.1	7.6 (107%)	6.1 (86%)	Ostberg, 1973
	7	7.8 (111%)	6.7 (96%)	Mott, et al., 1965
Rotating 7-Day	8	7.8 (98%)	7.3 (91%)	Maasen, Meers, and Verhaegen, et al., 1980
	6.0	9.2 (153%)	5.8 (97%)	Akerstedt, 1976
	6.8	7.3 (107%)	5.4 (79%)	Tepas, Walsh, and Armstrong, et al., 1981
	7.0	7.2 (103%)	5.2 (74%)	Armstrong and Tepas, 1979
	8.5	— —	6.3 (75%)	Folkard, Monk, and Lobbam, et al., 1978
	7.5 (Nurses)	— —	5.0 (67%) (Nurses)	
Rapidly-Rotating 7-Day	4.7	8.1 (172%)	5.3 (118%)	Smith, 1979
	6.7	8.7 (130%)	7.5 (112%)	Webb and Agnew, 1978

Difficulties inherent in measuring performance on industrial jobs and the problems of attributing changes in performance to time-of-day effects from biological rhythms or to fatigue associated with accumulated sleep loss during the night shift period make application of this basic physiological and psychological data to shift system design imperfect. Recognition that human performance may be rhythmic and subject to both internal and external influences, however, is important in developing the job design guidelines included in this book. Because there is a large variation in the way individual workers respond to changes in their activity cycle produced by night work, jobs should be designed to be within the capacities of most workers. Critical tasks that are likely to occur in the early morning hours should be designed to make performance of them relatively automatic; this includes following population stereotypes for the movement of controls, providing aids to alert the worker and make trouble-shooting easier, and providing feedback about the results of the worker's action (see Chapter III of Volume 1 of this series for equipment design guidelines). Finally, because worker performance may vary with time of day, scheduling should be flexible and permit a person to choose the appropriate work and rest cycle pattern for optimum job performance. Applications of this information to work/rest cycle design and to shift work and overtime scheduling can be found in Parts IV and V.

REFERENCES

Akerstedt, T. 1976. "Interindividual Differences in Adjustment to Shiftwork." Adapted from the *Proceedings of the 6th Congress of the International Ergonomics Association,* July 1976, University of Maryland, Santa Monica, Calif.: Human Factors Society, pp. 510-514. Included in *Studies of Shiftwork,* edited by W. P. Colquhoun and J. Rutenfranz, 1980. London: Taylor & Francis, Ltd., pp. 121-130.

Akerstedt, T. 1977. "Inversion of the Sleep-Wakefulness Pattern: Effects on Circadian Variations in Psychological Activation." *Ergonomics, 20:* pp. 459-474. Included in *Studies of Shiftwork,* edited by W. P. Colquhoun and J. Rutenfranz, 1980. London: Taylor & Francis, Ltd., pp. 65-80.

Akerstedt, T., and J. E. Froberg. 1976. "Shift Work and Health—Interdisciplinary Aspects." In *Shift Work and Health: A Symposium,* edited by P. G. Rentos and R. D. Shepard. HEW Publication No. (NIOSH) 76-203. Washington, D.C.: U.S. Government Printing Office, pp. 179-197.

Alluisi, E. A. 1972. "Influence of Work-Rest Scheduling and Sleep Loss on Sustained Performance." In *Aspects of Human Efficiency,* edited by W. P. Colquhoun. London: English Universities Press. Cited in Tepas, 1982.

Armstrong, D., and D. I. Tepas. 1979. Personal communication. Cited in Tepas, Walsh, and Armstrong, 1981.

Aschoff, J. 1978. "Features of Circadian Rhythms Relevant for the Design of Shift Schedules." *Ergonomics, 21:* pp. 739-754. Included in *Studies of Shiftwork,* edited by W. P. Colquhoun and J. Rutenfranz, 1980. London: Taylor & Francis, Ltd. pp. 19-34.

Aschoff, J. 1981. "Circadian Rhythms: Interference with and Dependence on Work-Rest Schedules." In *Biological Rhythms, Sleep and Shift Work,* edited by L. C. Johnson,

D. I. Tepas, W. P. Colquhoun, and M. J. Colligan. New York: S.P. Medical & Scientific Books, pp. 11-34.

Benedict, F. G. 1904. "Studies in Body Temperature. I. Influence of the Inversion of the Daily Routine; the Temperature of Nightworkers." *American Journal of Physiology, 11:* pp. 145-169.

Blake, M. J. F. 1967. "Time of Day Effects on Performance in a Range of Tasks." *Psychonomic Science, 9:* pp. 349-350.

Blake, M. J. F. 1971. "Temperament and Time of Day." Chapter 3 in *Biological Rhythms and Human Performance,* edited by W. P. Colquhoun. New York: Academic Press, pp. 109-148.

Bonnet, M. 1982. "Performance During Sleep." Chapter 8 in *Biological Rhythms, Sleep, and Performance,* edited by W. B. Webb. New York: Wiley, pp. 205-237.

Botzum, G. D., and R. L. Lucas. 1981. "9-3 Slide Shift Evaluation—A Practical Look at Rapid Rotation Theory." In *Proceedings of the 25th Human Factors Society,* October 12-16, 1981, Rochester, NY. Santa Monica, Calif.: Human Factors Society, pp. 207-210.

Brown, F. A., Jr. 1972. "The 'Clocks' Timing Biological Rhythms." *American Scientist, 60:* pp. 756-766.

Cassuto, J. 1972. "Health Education Newsline: Circadian Rhythm." *Journal of Occupational Medicine, 14 (9):* pp. 716-717.

Colquhoun, W. P. 1971. "Circadian Variations in Mental Performance." Chapter 2 in *Biological Rhythms and Human Performance,* edited by W. P. Colquhoun. New York: Academic Press, pp. 39-107.

Colquhoun, W. P. 1982. "Biological Rhythms and Performance." Chapter 3 in *Biological Rhythms, Sleep, and Performance,* edited by W. B. Webb. New York: Wiley, pp. 59-86.

Colquhoun, W. P., and R. S. Edwards. 1970. "Circadian Rhythms of Body Temperature in Shift-Workers at a Coalface." *British Journal of Industrial Medicine, 27:* pp. 266-272.

Colquhoun, W. P., M. J. F. Blake, and R. S. Edwards. 1968a. "Experimental Studies of Shift-Work I. A Comparison of 'Rotating' and 'Stabilized' 4-hour Shift Systems." *Ergonomics, 11 (5):* pp. 437-453.

Colquhoun, W. P., M. J. F. Blake, and R. S. Edwards. 1968b. "Experimental Studies of Shift-Work II. Stabilized 8-hour Shift System." *Ergonomics, 11 (5):* pp. 527-546.

Colquhoun W. P., M. W. P. H. Paine, and A. Fort. 1979. "Changes in the Temperature Rhythm of Submariners Following a Rapidly Rotating Watchkeeping System for a Prolonged Period." *International Archives of Occupational and Environmental Health, 42:* pp. 185-190.

Comella, T. M. 1976. "Biorhythm. Personal Science or Parlor Game?" *Machine Design, 48 (5):* pp. 104-108.

Czeisler, C. A., and C. Guilleminault. 1980. "REM Sleep: Its Temporal Distribution." *Sleep, 198 (2):* pp. 285-346, 377-463. Cited in Webb, 1982, pp. 98, 108.

Czeisler, C. A., M. C. Moore-Ede, and R. M. Coleman. 1982. "Rotating Shift Work Schedules That Disrupt Sleep Are Improved by Applying Circadian Principles." *Science, 217:* pp. 460-463.

Davies, C. T. M., and A. J. Sargeant. 1975. "Circadian Variation in Physiological Responses to Exercise on a Stationary Bicycle Ergometer." *British Journal of Industrial Medicine, 32:* pp. 110-114.

Drieske, P. 1972. "Biorhythms." *Family Safety, 31 (2):* pp. 15-16, 23.

Folkard, S., P. Knauth, T. H. Monk, and J. Rutenfranz. 1976. "The Effect of Memory Load on the Circadian Variation in Performance Efficiency Under a Rapidly Rotating Shift System." *Ergonomics, 19:* pp. 479-488.

Folkard, S., T. H. Monk, and M. C. Lobban. 1978. "Short- and Long-Term Adjustment of Circadian Rhythms in 'Permanent' Night Nurses." *Ergonomics, 21:* pp. 785-799.

Frank, G., F. Halberg, R. Harner, J. Matthews, E. Johnson, H. Gravim, and V. Andrus. 1966. "Circadian Periodicity, Adrenal Corticosteroids, and the EEG of Normal Man." *Journal of Psychiatric Research, 4:* pp. 73-86.

Froberg, J. E. 1981. "Shift Work and Irregular Working Hours in Sweden: Research Issues and Methodological Problems." In *Biological Rhythms, Sleep and Shift Work,* edited by L. C. Johnson, D. I. Tepas, W. P. Colquhoun, and M. J. Colligan. New York: S.P. Medical & Scientific Books, pp. 225-240.

Guberan, E., M. K. Williams, J. Walford, and M. M. Smith. 1969. "Circadian Variation of F.E.V. in Shift Workers." *British Journal of Industrial Medicine, 26:* pp. 121-125.

Halberg, F. 1959. "Physiologic 24-hour Periodicity: General and Procedural Considerations with Reference to the Adrenal Cycle." *Zeitschrift fur Vitamin- , Hormon- und Fermentforschung, 10:* pp. 225-296. Cited in Harker, 1964, pp. 3, 105.

Harker, J. E. 1964. *The Physiology of Diurnal Rhythms.* Cambridge, England: University Press, 114 pages.

Ilmarinen, J., R. Ilmarinen, O. Korhonen, and M. Nurminen. 1980. "Circadian Variation of Physiological Functions Related to Physical Work Capacity." *Scandinavian Journal of Work, Environment, and Health, 6:* pp. 112-122.

Johansson, G., G. Aronsson, and B. O. Lindstrom. 1978. "Social, Psychological and Neuroendocrine Stress Reactions in Highly Mechanized Work." *Ergonomics, 21:* pp. 583-600.

Johnson, L. C. 1982. "Sleep Deprivation and Performance." Chapter 5 in *Biological Rhythms, Sleep, and Performance,* edited by W. B. Webb. New York: Wiley, pp. 111-141.

Kleitman, N. 1963. *Sleep and Wakefulness.* Chicago: University of Chicago Press, 638 pages.

Kleitman, N., and D. P. Jackson. 1950. "Body Temperature and Performance under Different Routines." *Journal of Applied Physiology, 3:* pp. 309-328.

Knauth, P., and J. Rutenfranz. 1976. "Experimental Shift Work Studies of Permanent Night, and Rapidly Rotating Shift Systems. 1. Circadian Rhythm of Body Temperature and Re-Entrainment at Shift Change." *International Archives of Occupational and Environmental Health, 37:* pp. 125-137.

Lavie, P. 1982. "Ultradian Rhythms in Human Sleep and Wakefulness." Chapter 9 in *Biological Rhythms, Sleep, and Performance,* edited by W. B. Webb. New York: Wiley, pp. 239-272.

Maasen, A., A. Meers, and P. Verhaegen. 1980. "Quantitative and Qualitative Aspects of Sleep in Young Self-Selected Four-Shift Workers." *International Archives of Occupational and Environmental Health, 45:* pp. 81-86.

Mellette, H. C., B. K. Hutt, S. E. Askovitz, and S. M. Horvath. 1951. "Diurnal Variation in Body Temperatures." *Journal of Applied Physiology, 3:* pp. 665-675. Cited in Tepas, Walsh, and Armstrong, 1981.

Mills, J. N. 1966. "Human Circadian Rhythms." *Physiological Reviews, 46 (1):* pp. 128-171.

Monk, T. H. 1982. "Research Methods of Chronobiology." Chapter 2 in *Biological Rhythms, Sleep, and Performance,* edited by W. B. Webb. New York: Wiley, pp. 27-58.

Mott, P. E., F. C. Mann, Q. McLoughlin, and D. P. Warwick. 1965. *Shift Work: Its Social, Psychological, and Physical Consequences.* Ann Arbor: University of Michigan Press, 351 pages.

Murphy, T. J. 1979. "Letters to the Editor: Fixed vs Rapid Rotation Shift Work." *Journal of Occupational Medicine, 21:* pp. 319-322.

Ostberg, O. 1973. "International Differences in Circadian Fatigue Patterns of Shift Workers." *British Journal of Industrial Medicine, 30:* pp. 341-351. Included in *Studies of Shiftwork,* edited by W. P. Colquhoun and J. Rutenfranz, 1980. London: Taylor & Francis, Ltd., pp. 131-141.

Pearlman, C. A. 1982. "Sleep Structure and Performance." Chapter 6 in *Biological Rhythms, Sleep, and Performance,* edited by W. B. Webb. New York: Wiley, pp. 143-173.

Redgrave, J. A. 1971. "Menstrual Cycles." Chapter 6 in *Biological Rhythms and Human Performance,* edited by W. P. Colquhoun. New York: Academic Press, pp. 211-240.

Reinberg, A., P. Andlauer, and N. Vieux. 1981. "Circadian Temperature Rhythm Amplitude and Long Term Tolerance of Shiftworking." In *Biological Rhythms, Sleep and Shift Work,* edited by L. C. Johnson, D. I. Tepas, W. P. Colquhoun, and M. J. Colligan. New York: S.P. Medical & Scientific Books, pp. 61-74.

Rutenfranz, J., W. P. Colquhoun, and P. Knauth. 1976. "Hours of Work and Shiftwork." In *Proceedings of the 6th Congress of the International Ergonomics Association,* July 11-16, 1976, University of Maryland. Santa Monica, Calif.: Human Factors Society, pp. XLV-LII.

Smith, P. 1979. "A Study of Weekly and Rapidly Rotating Shiftworkers." *International Archives of Occupational and Environmental Health, 43:* pp. 211-220.

Soutar, G. N., and J. Weave. 1983. "Biorhythms and the Incidence of Industrial Accidents." *Journal of Safety Research, 14 (4):* pp. 167-172.

Tasto, D. L. , M. J. Colligan, E. W. Skjei, and S. J. Polly. 1978. *Health Consequences of Shift Work.* HEW Report No. (NOISH) 78-154. Washington, D.C.: U.S. Government Printing Office, 137 pages.

Tepas, D. I., J. K. Walsh, and D. R. Armstrong. 1981. "Comprehensive Study of the Sleep of Shift Workers." In *Biological Rhythms, Sleep and Shift Work,* edited by L. C. Johnson, D. I. Tepas, W. P. Colquhoun, and M. J. Colligan. New York: S.P. Medical & Scientific Books, pp. 347-356.

Thommen, G. 1972. *Is This Your Day? How Biorhythm Helps You Determine Your Life Cycle.* New York: Crown, 160 pages.

Tune, G. S. 1969. "Sleep and Wakefulness in a Group of Shiftworkers." *British Journal of Industrial Medicine, 26:* pp. 54-58.

van Loon, J. H. 1963. "Diurnal Body Temperature Curves in Shift Workers." *Ergonomics,* *6:* pp. 267-273.

Wahlberg, I., and I. Astrand. 1973. "Physical Work Capacity during the Day and at Night." *Scandinavian Journal of Work, Environment and Health, 10:* pp. 65-68.

Walsh, J. K., D. I. Tepas, and P. D. Moss. 1981. "The EEG Sleep of Night and Rotating Shift Workers." In *Biological Rhythms, Sleep and Shift Work,* edited by L. C. Johnson, D. I. Tepas, W. P. Colquhoun, and M. J. Colligan. New York: S.P. Medical & Scientific Books, pp. 371-381.

Webb, W. B. 1982. "Sleep and Biological Rhythms." Chapter 4 in *Biological Rhythms, Sleep, and Performance,* edited by W. B. Webb. New York: Wiley, pp. 87-110.

Webb, W. B., ed. 1982. *Biological Rhythms, Sleep, and Performance.* New York: Wiley, 248 pages.

Webb, W. B., and H. W. Agnew, Jr. 1975. "Sleep Efficiency for Sleep-Wake Cycles of Varied Length." *Psychophysiology, 12(6):* pp. 637-641.

Webb, W. B., and H. W. Agnew, Jr. 1978. "Effects of Rapidly Rotating Shifts on Sleep Patterns and Sleep Structure." *Aviation, Space and Environment, 49:* pp. 384-389.

Weber, R. A. 1979. *The Circadian System of Man.* New York: Springer-Verlag. Cited in Aschoff, 1981.

Wedderburn, A. A. I. 1967. "Social Factors in Satisfaction with Swiftly Rotating Shifts." *Occupational Psychology, 41:* pp. 85-107. Included in *Studies in Shiftwork,* edited by W. P. Colquhoun and J. Rutenfranz, 1980. London: Taylor and Francis, pp. 275-297.

Wever, R. A. 1981. "On Varying Work-Sleep Schedules: The Biological Rhythm Perspective." In *Biological Rhythms, Sleep, and Shift Work,* edited by L. C. Johnson, D. I. Tepas, W. P. Colquhoun, and M. J. Colligan. New York: S.P. Medical & Scientific Books, pp. 35-60.

Wilkinson, R. T. 1961. "Interaction of Lack of Sleep with Knowledge of Results, Repeated Testing and Individual Differences." *Journal of Experimental Psychology, 62:* pp. 263-271.

Williams, H. L., A. Lubin, and J. J. Goodnow. 1959. "Impaired Performance with Acute Sleep Loss." *Psychological Monographs, 73 (14):* 26 pages. Whole No. 484. Cited in Johnson, 1982.

Winget, C. M., L. Hughes, and J. LaDou. 1978. "Physiological Effects of Rotational Work Shifting: A Review." *Journal of Occupational Medicine, 20:* pp. 204-210.

Wojtczak-Jaroszowa, J. 1976. "Health and Work Shifts: Discussion II." In *Shift Work and Health: A Symposium,* edited by P. G. Rentos and R. D. Shepard. HEW Publication No. (NIOSH) 76-203. Washington, D.C.: U.S. Government Printing Office, pp. 72-86.

Workman's Compensation Board of British Columbia. 1971. *An Investigation on the Biorhythm Theory: Report from the Statistical Research Section.* Victoria, B. C.: Workman's Compensation Board of British Columbia, 39 pages.

PART **III**

EVALUATION OF JOB DEMANDS

CONTRIBUTING AUTHORS

David M. Kiser Ph.D., Physiology

Suzanne H. Rodgers Ph.D., Physiology

The physical, mental, and perceptual demands of industrial jobs are of most interest to practitioners of human factors and ergonomics. Physical work can be quantified by measuring what the worker is doing, such as:

- Weights handled
- Work pattern
- Reaches
- Heights
- Repetition frequency
- Postures

The physiological responses of the worker can also be monitored. These include heart rate, blood pressure, breathing, metabolism, body temperature, and muscle activity.

Mental and perceptual work includes decision making and using the senses to see, hear, and touch. These can be evaluated by monitoring performance. For example, one can measure:

- The time taken to accomplish the task.
- The ability to detect defects.
- The number of errors made.

In addition, psychophysical techniques can be used to gather information about the difficulty of the decision making, monitoring, or inspection task. These techniques have also been used to determine the acceptability of a given load in physically demanding tasks (Borg, 1970; Snook, Irvine, and Bass, 1970—see References at the end of Chapter 5) and for assessing local muscle and joint stress (Pandolf, 1978; Cain, 1973). This methodology has been discussed in Appendix B of Volume 1.

Some techniques for the evaluation of job demands are presented in detail in this section. The methods are ones that have been used in our industrial studies. They do not, in general, require substantial expenditures for equipment.

Table 5-1: Uses of Job Evaluation Techniques   Six job evaluation techniques are listed in column 1. Use of each method for quantification of activity on the job and/or the worker's resulting physiological responses is indicated with an X in columns 2 and 3. The (X) found under ''Physiological Responses for Timed Activity Analysis'' refers to the use of this method to quantify work/rest cycles. These are important in evaluating the physiological load on the worker, although they do not directly measure a physiological response. Columns 4 and 5 indicate the appropriateness of using each method for evaluating existing or new jobs. The (X)'s found under ''New Jobs'' for both ''Survey Methods'' and ''Timed Activity Analysis'' refer to the use of these techniques to analyze similar jobs or to simulate the expected work pattern. Each of the techniques in column 1 is discussed in detail in this part.

Evaluation Technique	Quantification of		Use for	
	Activity on Job	Physiological Responses	Existing Jobs	New Jobs
Survey Methods	X		X	(X)
Timed Activity Analysis	X	(X)	X	(X)
Motion Analysis	X		X	
Biomechanical Analysis		X	X	X
Heart Rate Interpretation		X	X	
Estimation of Job Energy Demands		X	X	X

They are most often used in the evaluation of existing jobs to determine whether a workload problem exists. However, the survey methods, the energy demands estimation, and the biomechanical analyses can be used to predict the suitability of a given design before a job is initiated. The timed activity analysis can be used to estimate labor needs for new production workplaces. Using the techniques to study similar jobs before a new production area is opened can also prevent perpetuation of a design that makes the work difficult for a large part of the work force. Table 5-1 summarizes uses for each of the methods presented in this part.

CHAPTER 5

Survey Methods

CHAPTER 5. SURVEY METHODS

This chapter describes four steps to take in collecting information about job demands and assessing the need for change. The first step is to perform a primary analysis of the problem to determine whether the workplace, job, individual worker, or working situation is the major contributor to the stress. Once the problem is defined, methods to measure the workload are needed, and then the worker's performance on the job and the physiological cost of that performance should be quantified. Finally, a survey should recommend actions to be taken to reduce the problem identified.

Strategies for choosing an appropriate solution in accordance with the problem's frequency and duration of occurrence are also included in this section.

A. PRIMARY ANALYSIS OF THE PROBLEM

In the design of new jobs and workplaces, the designer can anticipate problems by relating his or her plans to the guidelines given in this book. If the discrepancy is large, changes should be made in the initial design. Once the workplace,

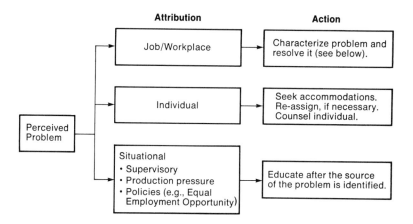

Figure 5-1: Primary Analysis of a Perceived Problem in an Existing Job   This illustration outlines the first steps to take in responding to a request from a production or office area to help resolve a perceived job problem. Before measurements of the job are made, you must determine whether the problem is attributed to a job or workplace characteristic, to the capabilities of an individual worker, or to a situation related to management policies or actions. Each of these attribution categories will require a specific approach or action, ranging from measurement to job accommodations to education. Early identification of the appropriate attribution and the action to be taken increases the efficiency with which ergonomics problems can be resolved in the workplace.

equipment, or job has been established, changes can be expensive and have to be weighed against the seriousness of the problem. Factors that decrease productivity but do not produce safety or health hazards may not be obvious enough to persuade others of the need for redesign. For instance, a handling job that requires a person to lift cases from a pallet on the floor to a table or conveyor can be physiologically stressful because the weight of the upper body also has to be lifted with each case. Raising the pallet so that the upper body is lifted less far each time reduces the physical effort and allows the handler to work for longer periods before having to rest. This increases the number of units that can be lifted per shift if the operation is paced by the worker, not by a machine. Often it is not possible to determine how much productivity will be gained by redesign until the redesign is done.

When existing jobs are being evaluated, it is important to establish first whether the perceived problem is related to the job and workplace, to the individual, or to a situation such as supervisory style, production pressures, or other psychosocial factors. Figure 5-1 indicates an approach for breaking down the problem and gives ways of dealing with each. If the problem appears to be an individual one, accommodations for the worker in the workplace or in another job are often possible. If the person is representative of a larger group such as women or older people, however, the issue should be dealt with as a workplace or job problem. If it is a situational problem, a brief evaluation of job demands may be useful in educating people on how to reduce the stress. The remainder of this chapter addresses how the job and workplace can be analyzed to characterize the stress of and to resolve the problems on existing jobs.

B. MEASURING WHAT THE PERSON IS DOING ON THE JOB

Characterization of job requirements is important in order to identify the types of physical, environmental, mental, and perceptual stresses present. Three types of information prove useful in characterizing physical and perceptual work: biomechanical, motion, and timed activity analyses. Biomechanical analysis of the forces required and the stresses on joints and muscles is useful in assessing manual handling stresses and the impact of workplace and equipment design on people. Motion analysis, a subset of biomechanical analysis called kinematics, is useful in the characterization of highly repetitive work and in describing light physical effort tasks. Timed activity analysis can be used in the assessment of external pacing, work pattern, and the impact of job organization on people. The measurements associated with these three types of analyses are shown in Table 5-2. These methods are discussed in detail in Chapters 6 through 10.

The timed activity analysis is useful for characterizing any job whether physical activity is a significant part of the job or not.

In addition to physical effort analysis, the job can be described in terms of its environmental, mental, and perceptual stresses. These are shown in Table 5-3.

Table 5-2: Factors Analyzed to Assess the Physical Work Load on a Job Examples of measurements that can be made to quantify the biomechanical, motion, and timing stresses of a job are summarized.

Biomechanical

- Reaches (including height and distance in front of the body)
- Weights
- Forces and torques
- Object dimensions

Motions

- Frequency
- Degree of rotation
- Duration
- Dexterity/coordination requirements (complexity)

Timed Activity Analysis

- Pattern of activities over the shift
- Distribution of heavy and light physical effort activities
- Time to do a task (used to assess time pressure)
- Frequency of occurrence of more demanding activities

C. PERFORMANCE MEASURES AND PHYSIOLOGICAL AND PSYCHOPHYSICAL RESPONSES IN THE EVALUATION OF JOB DEMANDS

The impact of job demands on people in the workplace can be measured by quantifying performance on the job and by monitoring the physiological responses of workers. Performance is often conditioned by production goals or pay incentives and gives only part of the picture about job demands. Measuring the physiological responses of people on the job allows one to identify at what personal cost those performance measures are achieved and maintained.

1. PERFORMANCE MEASURES

The ease of measuring performance will vary according to job type. Highly repetitive assembly, packing, or handling tasks are well suited for performance measures (see Table 5-4). Maintenance and monitoring tasks are harder to quantify.

Table 5-3: Factors Measured to Assess Environmental, Mental and Perceptual Job Demands Examples are given of measurements that can be made to determine the amount of stress on the worker from environmental factors and from task requirements for mental and perceptual work. Discussions of most of these factors can be found in the environment and information transfer chapters of Volume 1 of this series. Pacing is discussed in Part IV of this volume as are work and rest cycles for mental and perceptual work.

Environmental

- Temperature/humidity/air movement
- Noise
- Illumination
- Shift schedule
- Extended hours of work
- Other physical or chemical factors (such as odors, floor surfaces)
- Use of protective clothing or equipment

Mental and Perceptual

- Visual requirements
- Auditory requirements
- Complexity
- Information-handling requirements
- Decision-making requirements
- External pacing

More demanding jobs may show higher error rates and less productivity per hour as the shift progresses; both changes reflect fatigue development in people who have lower work capacities.

2. PHYSIOLOGICAL AND PSYCHOPHYSICAL RESPONSES TO WORK

An individual's physiological response to job demands depends on his or her capacity for work and tolerance for working conditions—physical, environmental, perceptual, and mental. Physically demanding work can be assessed at two levels: systemically (across the whole body) and locally (confined to one part of the body). Heavy physical work, for instance, produces an increase in heart rate, oxygen consumption, blood pressure, breathing, body temperature and hormone levels such as catecholamines (epinephrine and norepinephrine). At the

Table 5-4: Performance Measures for Assessing Job Demands Eight measures are listed that quantify the productivity and quality of performance by the worker. A job being evaluated can be compared to others if a common performance measure can be used. These measures show what the worker is doing in terms of moving product "out the door." The amount of waste and the types of errors made show how he or she is coping with job demands by organizing work and rest cycles through arbitrary and secondary work and by making decisions about how the work is done.

Productivity over the Shift

- Total units per shift at different levels and durations of effort and/or exposure
- Units per hour, compared to a standard
- Amount of time on arbitrary work breaks or secondary work
- Amount of waste
- Work interruptions - distractions, accidents

Quality of Output · Errors

- Missed defects/communications
- Improper actions
- Incomplete work

---

same time, local muscle fatigue may develop from sustained effort of hand, arm, or shoulder muscles. In light work, the systemic responses will be less apparent, but the local effects may be significant. They may be identified by asking people on the job to rate the effort or comfort levels, using a psychophysical scale such as the Rating of Perceived Exertion (RPE) (Borg, 1970; Pandolf, 1978).

Environmental stressors and conditions that increase anxiety or other emotions on the job, such as jobs with high accountability, may be apparent from systemic physiological responses. Elevations of heart rate, blood pressure, hormone levels, and breathing levels, or a gradual increase in their values over the shift, may indicate the presence of environmental or psychic job elements of concern.

Table 5-5 indicates some of the physiological and psychophysical measurements used to evaluate job demands in industry. Other methods are in development or are used in laboratory simulations; the ones listed in the table are those found useful in the workplace for the following reasons:

- They are noninvasive, that is, they do not penetrate the skin.
- Their measurement does not interfere significantly with the job.

Table 5-5: Physiological and Psychophysical Measures Used in the Assessment of Job Demands   Examples are given of six physiological and three psychophysical measures that can be used to assess physical, environmental, mental, and perceptual job demands. Table 5-6 indicates which aspects of job demands each of these measures best characterizes.

Physical Effort

- Heart rate
- Blood pressure
- Oxygen consumption
- Minute ventilation (volume of air exchanged per minute)
- Surface electromyography (muscle signals)
- Psychophysical scaling—Rating of Perceived Exertion scale

Environmental Factors

- Heart rate
- Minute ventilation
- Body temperature
- Psychophysical scaling—Comfort Rating scale
- Visual acuity/fatigue

Mental and Perceptual Factors

- Heart rate
- Blood pressure
- Minute ventilation
- Visual acuity/fatigue
- Psychophysical scaling—Stress Measurement scale

- They do not require investments in equipment of more than $10,000.
- They provide comprehensive data about job pattern and demands.
- They can be done by one or two investigators in the field.

Of these measures, the heart rate is the most versatile one for evaluating whole-body stress; psychophysical scales are probably most versatile for the assessment of local stress. Both of these measures reflect a number of factors that may influence the results: individual capability, job demands, and emotional responses to the job. Therefore, some independent measures may also be needed. For physical work, oxygen consumption is quite similar from person to person on the same job, even when heart rates vary greatly. Body temperature elevation is a reliable indicator of environmental temperature and humidity loads. The responses of many people can be measured in order to reduce the impact of any one person.

Table 5-6: Physiological and Psychophysical Measures Used in Characterizing Job Demands. The six physiological and three psychophysical measures given in Table 5-5 are shown in column 1. Six job characteristics are shown across the top of the table. An X is used to indicate which measures are suitable for measuring each job characteristic. The best measure for each is circled. Heart rate and psychophysical measures, such as Rating of Perceived Exertion, are the best universal measures of job demands.

Measurement	Job Characteristic					
	Total Work Load	Peak Work Load	Heat and Humidity Load	Specific Muscle Load	Work/ Rest Pattern	Work Pace/ Mental Load
Heart Rate	(X)	(X)	(X)	X	(X)	X
Oxygen Consumption	X	X				
Blood Pressure		X	X	X		X
Body Temperature			X			
Minute Ventilation		X	X			X
Electro-myography				X	X	
Visual Acuity/ Fatigue	X				X	X
Rating of Perceived Exertion	X	X		(X)		
Other Psychophysical Measures (e.g., Comfort Ratings			X		X	(X)

Table 5-6 summarizes some physiological and psychophysical measures and the job characteristics they best evaluate. Multiple measures are useful to get the best characterization of job demands. The preferred measure depends on the amount of information it provides and the ease of collecting the data.

Table 5-7: Strategies for Solving Job Problems. Four strategies for solving problems of job demands are shown across the top (columns 2 through 5). The frequency and duration of the problem is categorized as percentage of work time (column 1), extending from constant to very rare. The recommended strategies for each frequency and duration category are shown by an X in the appropriate column. Primary tasks are those that are done for more than 15 percent of the work period and occur daily. Secondary tasks occur for less time per shift or on an occasional basis during the work week. Redesign may not always be feasible from a technological or economic perspective, so (a) is used to mark the entries where this should be considered and where another strategy may be preferable. The more infrequently a task or job problem occurs, the more possible it is to use administrative actions, such as personnel selection or additional help, to deal with it. Redesign is the recommended approach if the problem occurs more than 25 percent of the time.

Percent of Time the Problem Occurs	Strategies for Problem Solving			
	Redesign, Eliminate Problem	Provide Aids (such as Handling Aids, Visual Aids)	Provide Help (Personnel); Alter Job Structure	Be More Selective in Placing People on the Job
Total Job, Constant: >50% of the time	X			
Primary Tasks: >15–50% of the time	X (a)	X		
Secondary Tasks: 5–15% of the time	X (a)	X	X	
Seasonal or Infrequent Tasks: <5% of the time		X	X	X
Contingency, Rare			X	X

(a) = If feasible from a technological or economic standpoint.

D. STRATEGIES FOR RESOLVING PROBLEMS WITH JOB DEMANDS

When a job is identified as being difficult for many people, two factors should be specified: the number of people affected, or what percent of the potential work force will have difficulty doing the job or task; and the frequency of the difficult work during a shift. A physically demanding task that occurs infrequently can often be dealt with by providing a mechanical aid or an additional person to help. For frequent activities, if more than 75 percent of the potential work force is unable to do the job, redesign is the preferred approach. If less than 25 percent of the potential work force is affected, the provision of aids is preferred; if this is not feasible, a validated screening test for the job may be needed. Table 5-7 summarizes the preferred strategies for resolving job problems as a function of the percentage of time that the problem occurs.

Selection of people for problem jobs is used only when the problem activity occurs infrequently or rarely, and when other strategies are not feasible. Redesign or the provision of aids are the preferred strategies for resolving most job problems.

REFERENCES

Borg, G. A. V. 1970. ''Perceived Exertion as an Indicator of Somatic Stress.'' *Scandinavian Journal of Rehabilitation Medicine, 2:* pp. 92-98

Cain, W. S. 1973. ''Nature of Perceived Effort and Fatigue: Roles of Strength and Blood Flow in Muscle Contractions.'' *Journal of Motor Behavior, 5 (1):* pp. 33-47.

Pandolf, K. B. 1978. ''Influence of Local and Central Factors in Determining Rated Perceived Exertion During Physical Work.'' *Perceptual and Motor Skills, 46:* pp. 683-698.

Snook, S. H., C. H. Irvine, and S. F. Bass. 1970. ''Maximum Weights and Workloads Acceptable to Male Industrial Workers.'' *American Industrial Hygiene Association Journal, 331:* pp. 579-586.

CHAPTER 6

Timed Activity Analyses

CHAPTER 6. TIMED ACTIVITY ANALYSES

A. Job Activity Analyses

B. Work Task Analyses

C. Work Cycle Analyses

D. Methods-Time Measurement (MTM)

References for Chapter 6

Once a survey or problem identifies the need to study a job in more detail, there are several techniques available for quantifying job stress. These include: various time and motion analyses, biomechanical analyses, physiological monitoring, and effort estimation techniques. In each of these it is essential to make observations of what the worker is doing and how it is being done in time, or the *work pattern*. Timed activity analysis is a general category of techniques that range from recording very broad activities, like loading a truck, to timing very small movements of the hands in a repetitive assembly task with methods-time measurement (MTM). Between these extremes there is a range of timed activity analyses that quantify job tasks and cycles and permit job demands to be defined more accurately in terms of the physiological and psychological stresses on a

Figure 6-1: Examples of Timed Activity Analyses Jobs are divided into activities, tasks, cycles, and motions. The times to the right of each division are average values for each bar. Job activities are measured over the full shift and can be as simple as defining work and break periods or can include the amount of time during a shift that a particular activity is done (see Table 6-1). Each work period can be divided into tasks (A, B, C, D). These are discrete activities that take place sequentially (Table 6-2) or simultaneously (Table 6-3) within the work period. Each task can usually be broken down into cycles; these may be less than a minute to several minutes in length. The work (W) cycle is defined as the time to complete an action on the product or equipment (see Table 6-4). A recovery (R) cycle is the time between work cycles and may include light work. Within each cycle it is possible to measure motions, such as moving (M), placing (P), or fastening (F) a part. The motions are measured in seconds using a stopwatch (see Table 6-4 and 6-5). Analyses of the distribution in time of work and recovery tasks and of the motions used to accomplish the work assist the ergonomist in quantifying job demands that are not clearly defined by physiological measures alone.

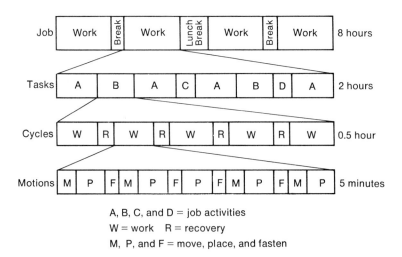

A, B, C, and D = job activities
W = work   R = recovery
M, P, and F = move, place, and fasten

112

worker. Figure 6-1 summarizes the timed activity analysis categories discussed in this chapter. Examples of each are given in the sections that follow with descriptions of job situations that they help define.

A. JOB ACTIVITY ANALYSES

The amount of energy expended, or oxygen used, during work is not usually measured throughout the shift. It has to be estimated from measurements of the workload during specific tasks, which are then weighted by the amount of time a person does them during the shift. An example of this type of job activity analysis is shown in Table 6-1. Four activities are shown with their measured

Table 6-1: Job Activity Analysis for a Handler   The overall job activities for a handler are given. Column 1 gives the four activities measured during the shift, three requiring moderate to heavy effort and the fourth being a recovery activity. Column 2 includes the measured oxygen demands of the activities in milliliters of oxygen per kilogram of body weight per minute (mL O_2 kg^{-1}min^{-1}). The percent of time spent per activity during the shift is shown in column 3. Column 4 gives the time-weighted oxygen demands for each activity and is calculated by multiplying the actual demands by the fraction of shift time from columns 2 and 3. The sum of the values in column 4 is the average oxygen demand for the full shift, in mL O_2 kg^{-1}min^{-1}. To calculate the demands of the work periods alone, we must subtract the standby activity from the total, and the percent of shift time must be recalculated for 100 percent work time. If the regularly scheduled breaks are included in these calculations, the total shift oxygen demands will be substantially lower. Basal oxygen consumption is included in all of the values shown above. This type of activity analysis is used to estimate the workload for a shift from oxygen measurements made by sampling in the work and rest periods. It also allows the ergonomist to predict how much recovery time will be needed to make a job suitable for more people.

Activity	Oxygen Demands (mL. kg^{-1}min^{1})	Percent of Shift Time	Total Time-Weighted Demands
Load and seal cases	14	85	11.9
Handle pallet trucks	16	7	1.1
Get supplies	12	5	0.6
Standby	6	3	0.2
			$\Sigma = 13.8$

Total 8-Hour Job Demands = 13.8 mL O_2 kg^{-1}min^{-1}.
Weighted Working Average = 14.5 mL O_2 kg^{-1}min^{-1}.

oxygen demands and the percent of shift time they take. The weighted demands are calculated, and a summation of the shift's workload is given both with and without the standby activity included. This provides the weighted working average, which is a measure of the actual physical demands of the work without including rest breaks or recovery periods between major activities.

The job activity analysis describes what is required of the worker in very general terms, somewhat like a job description. Within each of the major work periods there can be standby time or lighter tasks that also qualify as rest periods. These are recognized at the next level of timed activity analysis, where tasks are measured. The major advantage of measuring the whole job is that it defines the overall physical workload demands, thus permitting one to establish the need for recovery time that will bring the job demands within the capacities of many workers. These work/rest guidelines are discussed in Part IV.

B. WORK TASK ANALYSES

A detailed description of activities during a shift is shown in Table 6-2 for a shipping dock handler. The amount of time spent on different activities is recorded sequentially. The distribution of light and heavy work tasks can be determined by using the effort equivalency information in Appendix A to define task energy demands. This analysis is particularly helpful in studies of physically demanding jobs. The pattern of work and rest tasks can make a large difference in determining who can do the job. Short-duration heavy activities followed by standby time or light work to permit recovery constitute better job design than continuously heavy or moderately heavy work for more than an hour at a time. The influence of work patterns on the percent of the population that can perform a heavy job is discussed in Part IV.

In this example, the ratio between the physically demanding and the rest or light work tasks is 2:1 for the total time period shown. However, the ratio is 4:1 for the 1.5 hour period from 10:02 to 11:33 and 3:1 for the 1.8 hour period from 12:45 to 2:30. The higher the work/rest ratio, the more opportunity there is for fatigue to occur on a job; recovery periods are likely to be too short for the heart rate to return to resting levels (see Chapter 9).

The characterization of job task demands using timed analysis permits one to identify the tasks that are most likely to exceed worker capacities. These tasks can then become the focus of equipment, workplace, or job design changes. Figure 6-2 diagrams the approach used to relate work task analysis to some options for redesign or job restructuring that can reduce the difficulty of problem jobs.

A second type of task analysis is where there are several discrete tasks in the job that must be "time-shared"; that is, the worker must do the tasks simultaneously or intermittently. For example, running a production machine can require an operator to do several tasks in the same time period, such as:

- Loading supplies.
- Removing the product and inspecting it.

Table 6-2: Timed Activity Analysis of Job Tasks and Effort Levels A chronological diary of job tasks and recovery periods for a five-hour segment of a shipping dock handler's day is given. Column 1 gives the actual clock time in hours and minutes, and column 2 shows the elapsed time in minutes. The tasks are shown in column 3. Below the diary is a summation of the total minutes spent on physically demanding work, such as loading pallets on the trailer truck, moving boxes, walking, and sweeping. The total amount of time spent on rest, recovery, or light work activities is also shown. The ratio between work and rest activity times is about 2:1. The ratio is an indication of how well the handler is able to recover from the working periods; the lower the ratio, the less the opportunity for fatigue to develop.

Time on Clock	Elapsed Time (min)	Task
10:02	3	Walking
10:05	5	Sweeping truck
10:10	32	Loading pallets on truck
10:42	45	Moving boxes in truck
11:27	23	Paper work
11:33	17	Loading pallets on truck
11:50	55	Lunch
12:45	105	Loading pallets on truck
2:30	33	Rest break
3:03	—	—

Work Time Shown = 207 minutes

Rest Break/Lunch/Light Work Time = 111 minutes

Ratio = 2:1

- Preparing the product for packing.
- Making machine adjustments.
- Clearing machine jams.
- Keeping records.

Each of these activities can be considered a task within the job of running the machine, but they are not done separately and sequentially. An analysis of this pattern of time-sharing helps define job stresses that are sometimes difficult to identify from physiological measures such as heart rate. Time pressure stress, particularly, may only be revealed by a detailed task analysis that can indicate

Figure 6-2: Identification of Problem Tasks in a Job This flow chart shows the use of timed activity analysis to help identify problem tasks within a job. The time analyses are shown in the blue boxes. Initial indication of a problem job may come from the accident record or when difficulties are experienced in replacing a person on vacation. To determine why a job is difficult, one has to define the tasks and characterize their oxygen demands (see Appendix A) and durations. Using the guidelines presented in Part IV, you can identify problem tasks and initiate redesign. This approach is shown on the left side of the flow chart. It is possible that no one task is problematic but that the organization of tasks within the shift results in fatigue. The right side of the flow chart shows how the tasks can be grouped into "heavy" and "light" ones so the time sequence of continuous effort in each category can be determined. The calculation for the overall shift work/rest ratio, similar to the one in Table 6-2, is shown on the far right. The other calculations are for ratios within the shift. High ratios are 4:1 and above. Once these are identified, job task restructuring can be initiated.

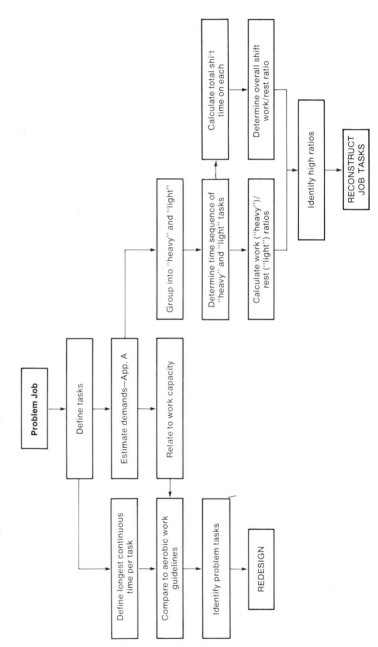

116

how much control the worker has over the way the job is done. This is discussed under paced work in Part IV.

Table 6-3 shows 13 minutes of a timed activity analysis of a production machine operator's job. The machine speed determines when each task must be done and how much time there is available to do it. In this example, most of the tasks had to be done in less than one minute because some other task required the operator's attention. A less detailed analysis of the tasks per shift

> **Table 6-3: Analysis of Time-Shared Work Tasks in a Machine Operator's Job** A detailed diary of a production machine operator's activities during a 13-minute period is shown. The activities, or tasks, that are time-shared are listed in column 3. Column 1 gives the time (in minutes) on the stopwatch, and column 2 gives the elapsed time per activity in minutes. Very few of the tasks were sustained for more than one or two minutes at a time since the operator had to move from task to task to keep the machine running. This need to accomplish several tasks simultaneously requires concentrated mental effort from the operator and often results in the machine controlling the operator. This stress and the time pressure stress are significant contributors to overall job demands. They can be best defined through a detailed timed activity analysis of the operator's activities, as shown here. *(Rodgers and Mui, 1981).*

Stopwatch Time (min)	Elapsed Time (min)	Activity
20.6	0.2	Start
20.8	1.1	Load vinyl
21.9	0.9	Inspect
22.8	0.7	Adjust lead foil
23.5	1.2	Load film
24.7	0.7	Inspect
25.4	0.9	Load end boards
26.3	1.1	Inspect
27.4	0.3	Change tray
27.7	2.1	Load end boards
29.8	0.5	Inspect
30.3	1.1	Load film
31.4	0.3	Inspect
31.7	0.3	Check film supply
32.0	0.6	Ticket work
32.6	1.3	Leave room
33.9	0.6	Correct die toggle jam

would show the percentage of time spent on each but would not explain why time pressure made this job difficult for many workers. This job is described in more detail in Part IV.

C. WORK CYCLE ANALYSES

A cycle can often be identified within a task. The cycle may be defined in several ways, for example:

- Time to complete an assembly.
- Time to inspect one unit of the product, such as a lot.
- Time to type one page.
- Time to make an equipment repair.
- Time to unload one pallet of the product.

Most cycles are expressed in minutes, with highly repetitive task cycles often being less than one minute. Work cycle times are generally expected to be fairly uniform over a shift unless there is some unpredictability built into the job, such as a repair operation. If cycle time increases during a shift, it is often assumed that fatigue is occurring. If it decreases in the first few days on a new job, training or skill acquisition is indicated.

Table 6-4 shows a more detailed analysis of pallet loading in a trailer by the shipping dock handler whose task analysis was given in Table 6-2. Here the cycle is defined as the time from the point when the pallet is moved into the trailer with a forklift truck until the next pallet is ready to be moved in. Each cycle in this example takes an average of 1.65 minutes, although there is sub-stantial variability in the times. Theoretically, more than 120 pallets could be handled during the working period sampled in Table 6-2. In this time period the handler actually handled about 75 pallets, the extra time being taken up by communications, personal business, paper work, and secondary tasks related to the main activity. It appears that the worker was regulating his own workload to a level that was acceptable to him for the total shift and would not result in excessive fatigue.

The information gained from a work cycle timed activity analysis is useful in establishing the variability in performance of a repetitive task and the de-velopment of fatigue over a shift if the task cycle times are quite predictable. The analysis helps to show how individual workers adjust their workload by taking "mini-rest breaks." These breaks can be accomplished by adding sup-plementary light tasks to the primary work task or by using communication or personal needs (such as getting a drink of water) to provide more recovery time. The more intense the effort during the work task, the more frequently these recovery tasks will be added by the operators.

D. METHODS-TIME MEASUREMENT (MTM)

The field of time and motion studies has long been associated with industrial engineering for setting job standards and estimating wages. The literature should

Table 6-4: Work Cycle Analysis for a Shipping Dock Handler's Job A more detailed analysis of the activities performed by a shipping dock handler (see Table 6-2) is given for an 8.5-minute period during the loading of pallets into a trailer truck. Clock time in hours and minutes is shown in column 1. Column 2 shows elapsed time in minutes, and column 3 gives total cycle time in minutes. The cycle is defined as the time from the point when a loaded pallet is taken into the trailer truck on a forklift to the time when another loaded pallet is on the forklift and is ready to go into the trailer. Each task activity is shown in column 4. "In" and "Out" refer to the movement of the forklift truck in and out of the trailer truck; "Unload" is the removal of shipping cases from the pallet and stacking them in the trailer; and "On forklift" is when a new pallet-load of cases is placed on the forklift in preparation for entering the trailer truck again. Cycle time varies as a function of work interruptions, quality problems with the product, inadequate paper work, and other factors beyond the handler's control. It also can vary as a function of the operator's ability to keep up the effort levels required. On the average, this cycle time is about 1.65 minutes. Cycle variability is useful in measuring job demands, the amount of control a person has over the task, and in detecting fatigue that may affect performance.

Time on Clock	Elapsed Time (min)	Cycle Time (min)	Task
10:10	0.25		– In
10:10 ¼	0.50		– Unload
10:10 ¾	0.25		– Out
10:11	0.50	1.5	– On forklift
10:11 ½	0.25		– In
10:11 ¾	0.25		– Unload
10:12	0.50		– Out
10:12 ½	0.25	1.25	– On forklift
10:12 ¾	0.25		– In
10:13	0.75		– Unload
10:13 ¾	0.50		– Out
10:14 ¼	0.25	1.75	– On forklift
10:14 ½	0.75		– In
10:15 ¼	0.25		– Unload
10:15 ½	0.25		– Out
10:15 ¾	0.75	2.00	– On forklift
10:16 ½	0.25		– In
10:16 ¾	1.00		– Unload
10:17 ¾	0.25		– Out
10:18	0.25	1.75	– On forklift
10:18 ¼	0.25		– In
10:18 ½	—	—	—

be consulted for a detailed discussion of MTM techniques (Barnes, 1968). One such technique is to record the motions of the hands and body during a task or work cycle. By analyzing these motions, you may be able to define more efficient motion patterns and sometimes to identify motions that may contribute to worker discomfort on a job (Gilbreth and Gilbreth, 1917).

A simplified approach to motion analysis on a job is shown in Table 6-5 for a graphic artist. This table extends the cycle analysis but is not as complex as a true MTM. Some of the differences between cycle and motion analyses are:

- Cycle time includes both motion and no-motion periods.

- Motion time is usually measured in seconds and fractions of seconds, while cycle time is more frequently in minutes or fractions of minutes.

Table 6-5: Motion Analysis of a Graphic Artist   A description of most of the motions made by a graphic artist during the preparation of art work for a chart is shown to illustrate more detailed motion analysis. Column 1 shows the time, in seconds, on a stopwatch, and column 2 shows the elapsed time per motion. The motions are described in column 3. This is not a true MTM analysis, but it describes what the artist is doing in a way that helps the ergonomist evaluate the stress on the small muscle groups of the hand and forearm. Modifications of this technique can be used to determine the amount of static loading of muscles that a job task requires. Inefficient motion patterns or awkward hand and wrist motions that may contribute to joint soreness can also be identified using this technique.

Time on Stopwatch (sec)	Elapsed Time (sec)	Motion
0	1.5	Lift label tape
1.5	2.0	Press tape on paper
3.5	2.0	Scrape tape edge
5.5	0.5	Move knife to right
6.0	2.5	Scrape ink on paper
8.5	0.5	Reach for paint brush
9.0	0.25	Grasp brush
9.25	0.5	Move brush to left
9.75	5.75	Apply paint
15.5	0.5	Move brush back to right
16.0	3.0	Pause
19.0	1.0	Blow on paint
20.0	3.0	Pause

- Motion analysis helps identify how long a highly repetitive task should take or how it can be modified to reduce the time required.

- Cycle analysis is more useful for identifying the variability in performance of a task and the potential fatigue factors.

In this example the task is to prepare art work for a book illustration. Each cycle is the completion of a discrete element of the illustration. Label tape is applied, cleaned up, and touched up with white paint. The motions are indicated by the verbs in each entry—lift, press, scrape, move, reach, and so on.

This type of analysis is less useful than task or cycle analysis to characterize job demands because it is too finely divided to make an accurate estimate of the energy demands and physiological stresses on the worker. Such analyses are often made in less than 15 minutes for highly repetitive assembly jobs. The standard times for completion of a part or cycle are derived from the sum of the times for the individual motions.

While MTM is important for detailed analysis of job demands on some very repetitive tasks, its applications in ergonomics are more limited. It is more useful in studying repetitive motion stress and can be extended to include a description of the location of the hand in space. Wrist flexion and extension and hand ulnar and radial deviations can be quantified if this modified motion analysis technique is used. See Part IV for additional discussion about the design of repetitive motion tasks.

In summary, timed activity analysis techniques can be used to describe job demands. They can range from roughly defined work periods and breaks to very detailed descriptions of every work motion used to accomplish a specific task. Task analyses are useful in assessing the total and peak physical work requirements of jobs and in determining the suitability of the pattern of light and heavy tasks throughout the shift. When tasks are "time-shared," an activity analysis of the tasks in time is an important way to define the time pressures on the operator. Use of these techniques to arrive at a recommendation for job redesign is discussed further in Part IV.

REFERENCES

Barnes, R. M. 1968. *Motion and Time Study: Design and Measurement of Work.* 6th ed. New York: Wiley, 799 pages.

Blum, M. L., and J C. Naylor. 1968. *Industrial Psychology: Its Theoretical and Social Foundations.* New York: Harper and Row, 633 pages.

Gilbreth, F. B., and L. M. Gilbreth. 1917. *Applied Motion Study.* New York: Macmillan. Cited in Blum and Naylor, 1968.

Rodgers, S. H., and M. Mui. 1981. "The Effect of Machine Speed on Operator Performance in a Monitoring, Supplying, and Inspecting Task." In *Machine Pacing and Occupational Stress,* edited by G. Salvendy and M. J. Smith. London: Taylor & Francis, pp. 269-276.

CHAPTER 7

Biomechanical Analyses of Work

CHAPTER 7. BIOMECHANICAL ANALYSES OF WORK

A. Force and Torque Relationships in the Body

   1. Lever Systems
   2. Torque and Moment
      a. Calculation of Simple Torque
      b. Calculation of Torque for Nonperpendicular Force Applications
      c. Torque Calculations Including Body-Segment Contributions

B. Biomechanics of the Back

   1. Lifting and Back Stress
   2. Postures and Back Stress

C. Shoulder Biomechanics

D. Biomechanics of Grasp

E. Dynamic Biomechanical Relationships

   1. Force-Mass Relationships
   2. Lifting Dynamics

F. Steps in Identifying and Solving Biomechanical Problems

References for Chapter 7

Biomechanics is an interdisciplinary science that integrates the factors influencing human movement. Knowledge and techniques for biomechanical analysis are drawn from such basic sciences as physics, mathematics, chemistry, physiology, and anatomy. Because of the complexities of human movement during work, biomechanics has played, and will continue to play, a major role in studying and optimizing human performance on the job, particularly in evaluating manual handling tasks and work postures.

Although many in-depth biomechanical analyses utilize sophisticated force and movement data acquisition systems and advanced computer techniques for data reduction and analysis, these techniques are often not easy to apply in the factory, warehouse, or office where specific work-related problems arise. On the other hand, there are a number of basic biomechanical principles that can be applied in the general sense to human movement. These simplified approaches to biomechanical analyses, while less precise, provide valuable information for the evaluation and design of tasks and workplaces. This chapter includes basic biomechanical concepts and techniques that can be used to study the motions of people at work. The physiological concepts presented in Part II should be reviewed as background to this discussion. Data on muscle strength, torque, and cross-sectional area; limb lengths and centers of gravity; and methods of measuring strength are found in Appendices A and B.

A. FORCE AND TORQUE RELATIONSHIPS IN THE BODY

Forces developed by the muscle-tendon complex act on bones at their points of insertion and cause rotation, or torque, around a joint. The muscles and bones act as a series of levers. When a worker lifts a box with both arms, the box acts as a load or resistance on a lever, the forearm. By understanding the mechanics of a task, the job designer can identify problems and correct or avoid them. Thus, a familiarity with the lever systems of the body is necessary.

1. LEVER SYSTEMS

All three classes of levers (first, second, and third) can be found in the body, although they are far from being equally represented. The first-class lever and an example of one in the human body are presented in Figure 7-1. Its fulcrum, or center of rotation (joint), is located between the resistance (load) and force (muscle) on either end of the lever. The length of the lever from the fulcrum to the resistance is called the resistance arm (RA), and the lever between the fulcrum and force is the force arm (FA). If FA is greater than RA, a mechanical advantage exists.

Figure 7-1 shows an example of this lever system using the worker who works at a microscope. The fulcrum is the atlanto-occipital joint connecting the head and spinal column. The resistance is the mass of the head, and the force

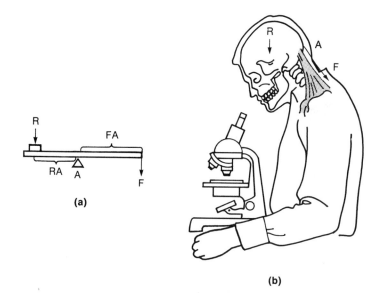

Figure 7–1: A First-Class Lever A schematic diagram of a first-class lever is shown in *a*. The lever rotates around the fulcrum (A). The force arm (FA) is the length of the lever from the applied force (F) to the fulcrum. The resistance arm (RA) is the length of the lever arm from the load offering resistance (R) to the fulcrum. An example of a first-class lever is shown in *b*. The mass of the head creates a force (R) that tends to rotate it downward. This tendency is resisted by the action of muscles on the back of the neck and upper back (F). The head rotates around the top cervical vertebrae (A), which act as a fulcrum. The distances between the fulcrum and each of the lines of force are the resistance (RA) and force (FA) arms. The location and orientation of the microscope eyepieces determine how far forward the head is bent and, therefore, how much work the neck and back muscles have to do to counteract the head's tendency to rotate forward. *(Developed from information prepared by Kamon and Terrell, Eastman Kodak Company, 1978).*

to counteract the resistance comes from the muscles on the back of the neck. Complaints of neck pain by microscope users can be traced to prolonged static contraction of the neck muscles used to maintain this working posture. Similar problems can be found in typists and video display terminal (VDT) operators when their workplaces force them to take awkward postures to see the screen.

Second-class levers are not prevalent in the human body. Figure 7-2 shows the second-class lever, in which a mechanical advantage always exists. There are few examples of second-class levers that are relevant to work performance. One example in the body is opening the mouth against resistance (Rasch and Burke, 1978).

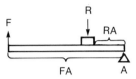

Figure 7-2: A Second-Class Lever   A schematic diagram of a second-class lever system is shown, with the same components as were identified in Figure 7-1. In the second-class lever system, the force arm (FA) is always larger than the resistance arm (RA), the fulcrum being at one end of the lever. There are few second-class levers in the body.*(Adapted from Tichauer, 1973).*

There are many examples of the third-class lever shown in Figure 7-3. Lever systems of this type are always at a mechanical disadvantage since RA is always larger than FA. This means that a proportionally higher force (F) is required to offset the load at (R). However, this lever system is important during movement since speed and range of movement are optimized by it. The example

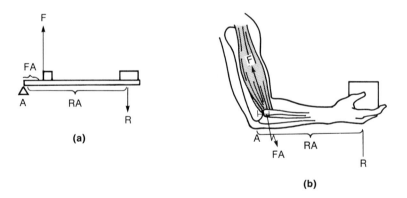

Figure 7-3: A Third-Class Lever   A schematic diagram of a third-class lever system is shown in *a*. RA is always larger than FA in this system, resulting in a mechanical disadvantage for generation of force. An example of a third-class lever in the body in shown in *b*, where a weight is held in the hand. The forearm and part of the hand constitute the lever rotating around the elbow, which acts as the fulcrum (A). The resistance arm (RA) is the distance from the elbow to the center of the load in the hand. The force arm (FA) is the distance from the elbow to the point of insertion of the upper arm's flexor muscles, which provide the force (F) to resist the load (R). This system is common in many handling and assembly task activities. *(Developed from information prepared by Kamon and Terrell, Eastman Kodak Company, 1978).*

of the third-class lever shown in Figure 7-3 is common in many jobs. Here the biceps muscle develops force to counteract the resistance of the load in the hand. FA is the distance from the elbow (fulcrum) to the insertion point of the biceps muscle, and RA is the distance from the elbow to the center of gravity of the load.

2. TORQUE AND MOMENT

The rotational tendency caused by application of a perpendicular force on a lever arm at some distance from its axis of rotation is called *moment* or *torque*. The torques generated by the human body translate muscle contraction into mechanical work on the job. The movement of an arm or leg is a torque; the axis of rotation is at the center of a joint, such as the elbow, shoulder, knee, or hip. The force initiating the torque is a balance between respective lever arms and forces generated by muscles and the force required to overcome some resistance or load. The resistance may come from a box to be lifted, a hand truck to be pushed, or a lever to be pulled.

Torque can be dynamic or static. If the product of muscle force and the force arm (F • FA) is greater than the product of force required to overcome the resistance and the resistance arm (R • RA), movement (dynamic torque) will occur. This is known as positive work. Dynamic torque can also occur if a load or resistance requires a force greater than that which is being developed by a muscle, and the muscle is forcibly lengthened. This is known as negative work (Winter, 1979). Negative work often occurs when a box or bag is lowered in a controlled manner. Static torque occurs when the products of opposing forces and lever arms balance.

Information about torque can be useful in evaluating the stress of a specific task on the musculoskeletal system of workers. Torque estimates for specific work tasks can be used in the design and evaluation of new workplaces by comparing them with information on human capacities for torque development (see Appendix A).

Although many forces can act on specific joints, estimates of torque are derived by relatively simple means. The following subsections describe torque calculations.

a. Calculation of Simple Torque

Torque is equal to the product of force (F) and the perpendicular ($\perp$) distance (D) from the axis of rotation, or:

$$T = F \times D\perp$$

This is illustrated by Figure 7-4a and the following calculations:

$$T = 0.5 \text{ meters} \times 10 \text{ newtons}$$
$$= 5 \text{ newton-meters (Nm)}$$

In this example, the system will rotate in a counterclockwise direction, and the torque is defined as positive (+). A clockwise rotation would be identified as a

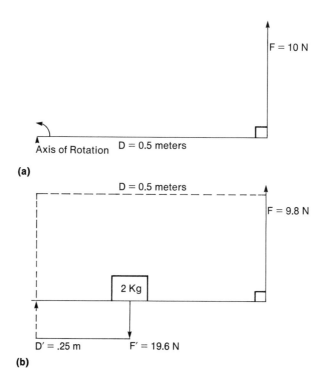

Figure 7-4: Torque Calculations. Information for two simple torque calculations is presented. In *a*, all movement will occur around the axis of rotation. The lever arm (D) is 0.5 m long, and the force acting on the lever system is equal to 10 N. In *b*, a static torque calculation is shown where equal and opposite torques exist. A force of 9.8 N (F) is applied against a lever arm of 0.5 m (D). The resultant torque is countered by a torque created by a force of 19.6 N (F') applied against a lever 0.25 m (D').

negative (−) torque. This designation of directional rotation is useful when more than one torque is identified. In the human body this is normally the case. By summing (Σ) the torques with their respective signs, one can determine the net torque with direction and magnitude. Under static conditions, the sum of all torques must equal zero (Σ T = 0). An example (Figure 7-4*b*) demonstrating a net torque calculation for more than one torque in opposite directions is:

$$\Sigma T = (0.5 \text{ m} \times 10 \text{ N}) + (-0.25 \text{ m} \times 20 \text{ N})$$
$$= 5 \text{ Nm} - 5 \text{ Nm}$$
$$= 0$$

This example represents a static condition in which no movement occurs, as indicated by a zero value for the sum of all torques. Figure 7-3 can serve as an example of how torque calculations can be used to study work performance. The following information is known about the third-class lever of the arm:

R = the perpendicular ($\perp$) force tending to create negative torque due to a 5 kg mass in the hand. The force is equal to the mass times acceleration due to gravity, or 5 kg × 9.8 m/sec^2 = 50 N.

RA = the resistance arm, or the distance from the elbow to the center of the hand. In this example, it is 0.33 m.

FA = the force arm, or the distance from the elbow to the point of insertion of the elbow flexor tendon. In this example, it is 0.03 m.

F = the force developed by the elbow flexors perpendicular to the forearm while holding the mass.

Since F • FA = R • RA, then:

$$F = \frac{R \cdot RA}{FA} \text{ and } F = \frac{50 \text{ N} \times 0.33 \text{ m}}{0.03 \text{ m}}$$

Therefore, F = 550 N. This means that in this situation, the flexor muscles of the elbow have to generate a force of 550 N in order to hold the 5 kg mass. This example also demonstrates the great mechanical disadvantage of the third-class lever.

Uses for these torque calculations in workplace analyses can be found later in this chapter; a brief example is presented here. The maximum elbow flexor strength of a given group of workers can be measured or estimated from previously published studies (Kamon and Goldfuss, 1978). The forces required to perform a specific task, as calculated in the example above, can be compared to these estimated maximum capacities and expressed as a percentage of maximum strength capacity for any person. The frequency and duration of acceptable task performance depends on the percentage of maximum capacity required by the task. The general relationships are described by the curves in Figure 7-5. At workloads that require a high percentage of muscle strength or torque capacities, performance time is limited and recovery time must be increased substantially. This relationship is discussed in more detail in the section on work/rest cycles in Part IV.

b. Calculation of Torque for Nonperpendicular Force Applications

Unfortunately, the example presented in Figure 7-3 is idealized; usually the analysis is more complicated. In this first example, the forearm was positioned so that all the forces acted perpendicular to it. The mass of the forearm was also ignored, which means that the force that must be developed by the elbow flexors was underestimated.

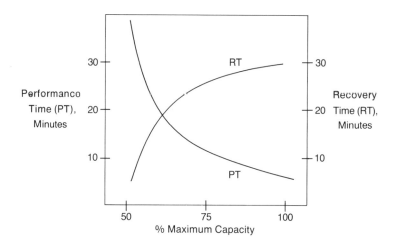

Figure 7-5: Performance and Recovery Times as a Function of Percentage of Maximum Strength Capacity   The general relationship between the intensity of work (percent of maximum capacity, horizontal axis) and the duration of work performance (performance time, left vertical axis) is shown by the curve labeled PT. The relationship of work intensity and the time required for recovery after the work (recovery time, right vertical axis) is shown by the curve labeled RT. As work intensity increases, moving from left to right on the horizontal scale, performance time decreases and recovery time increases. People work more efficiently if high percentages of their work capacities are not routinely required on the job. (*Adapted from Clarke, 1971; Kamon, 1981; Stull and Clarke, 1971*).

From the definition of torque, we know that only the force acting perpendicular to a lever arm contributes to its rotational tendency. If a force acts at an angle other than 90 degrees to the lever arm, only a fraction of the force can be used in the calculation of torque. An example of a system in static equilibrium is given in Figure 7-6a. It shows the effects of angles of force application other than 90 degrees.

In this calculation, the negative, or clockwise, torque is 98 N × 0.33 m, or 32 Nm. The positive, or counterclockwise, torque is F' (the perpendicular force) × 0.03 m. Since the system is static, the sum of the torques equals zero.

Since $\Sigma T = 0$, the positive torque must balance the negative torque already calculated:

$$F' \times 0.03 \text{ m} = 32 \text{ Nm}$$

Therefore, F' = 1067 N.

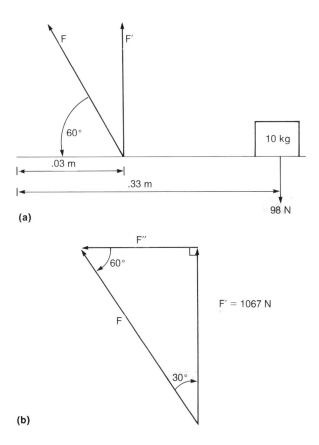

(a)

(b)

Figure 7-6: Calculating Muscle Force in a Static Equilibrium Condition The force exerted by a muscle (F) that is not pulling at 90 degrees to the lever arm (D) is illustrated in *a*. The negative torque is calculated as 98 N × 0.33 m, the resistance arm. The positive torque is the perpendicular force (F') × 0.03, the force arm. Because the muscle is pulling at 60, not 90, degrees, only part of its force (F) contributes to F'. The triangle of forces is illustrated in *b*. The perpendicular force (F') divided by the cosine of 30 degrees determines the muscle force needed to support the 10 kg weight shown in *a*. See the text and Appendix B for more information about calculating torques in nonperpendicular force applications.

The developed force (F) acts at an angle of 60 degrees to the lever arm, so only a portion of the force contributes to the F' of 1067 N. F is clearly greater than F'. F can be determined by using the trigonometric relationships of a right triangle in which F is the hypotenuse. An example of the triangle forces is shown in Figure 7-6*b*. Since:

$$\cos 30° = \frac{F'}{F}$$

$$F = \frac{F'}{\cos 30°} = \frac{1067}{0.87}$$

then F = 1226 N.
  To calculate F":

$$\sin 30° = \frac{F''}{F} \qquad\qquad \text{or} \qquad \tan 30° = \frac{F''}{F'}$$

$$F'' = \sin 30° \times F = 613 \text{ N} \qquad\qquad F'' = \tan 30° \times F' = 616 \text{ N}$$

The important effect of angle of pull on muscle force requirements can be seen by these calculations. A force of 1226 N (276 lbf) has to be developed in the biceps to lift a 10 kg (22 lbm) load in the hand. Force F", which acts parallel to the lever arm, is important in the analysis of work performance, too. It acts to stabilize or destabilize the elbow joint depending on the magnitude and direction of the force.

c. Torque Calculations Including Body-Segment Contributions

The lines of force (vectors) due to the mass of an object that is resisting a muscle must be analyzed in the same manner as the muscle force. Only the force component perpendicular to the lever arm contributes to torque around the joint. The parallel component, in the case of the elbow, tends to destabilize the joint. Figure 7-7 shows an example of a heavy object picked up with one hand, a common situation in industrial and everyday tasks. By measuring or estimating the angle of the elbow, one can estimate the force required of the elbow flexors to hold the load.

For the static case:

$$F_\perp \cdot FA = R_\perp \cdot RA$$

Since the angle and the weight of the object R are known, $R_\perp$ can be calculated using trigonometric relationships, as have been illustrated in the previous subsections. FA and RA can be estimated from anthropometric measurements. $F_\perp$ can then be calculated from the static torque relationship, and F can be determined by trigonometric calculations. By comparing the calculated muscle force, F, with information on the maximum strengths of specific muscle groups (Appendix A), one can determine how many people will be able to generate the necessary force. These calculations can be used in conjunction with muscle endurance information to develop guidelines for task frequency and duration.

In Figure 7-3, torque around the elbow joint was calculated by considering the mass held in the hand as the only resistance to muscle effort. In reality, the forearm and hand also have inherent mass that requires muscular effort for support and movement. These masses must also be included in elbow torque calculations. Figure 7-8 gives an example in which the masses and centers of gravity

Figure 7-7: A One-Handed Lifting Task.   An example of a one-handed lifting task is lifting a bucket onto a shelf. As the bucket is lifted from the worker's side to the 75-cm (30-in.) high shelf, the effective moment arm increases, and therefore the required torque increases, up to an angle of 90 degrees. The biomechanics of this task are discussed in the text and illustrated in Figure 7-8.

for the forearm, hand, and load have been used to calculate the total static torque around the elbow. The resistance arms for torque calculations are assumed to be the distances from the joint to the center of the mass. This information can be obtained from tables of anthropometric data found in Appendix A.

In Figure 7-8, not only does the object offer resistance but the hand and forearm also have mass that tends to create negative torques (clockwise rotation). Muscle force is necessary to generate positive torque (counterclockwise rotation) and to flex the elbow even when no weight is in the hand. In Figure 7-8b, the vertical lines of force due to the mass of the body segments and load are indicated. These forces act vertically downward from their centers of gravity. Figure 7-8c shows the resolution of the forces into those that contribute to torque ($F\perp$) and the lever arm on which they act. By analyzing for static torque, as previously described, we can calculate the net negative (clockwise) torque that must be resisted by muscular effort:

$$- \text{Torque} = (F_F \cdot \sin\beta \cdot r_F) + (F_H \cdot \sin\beta \cdot r_H) + (F_R \cdot \sin\beta \cdot r_R)$$

There are many movements in industrial work that can be studied by using the simplified biomechanical techniques just discussed. The remainder of this chapter provides examples that illustrate some of these biomechanical analyses

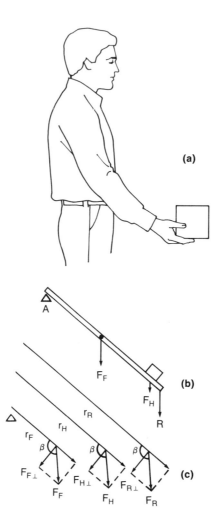

Figure 7-8: Static Torque at the Elbow Joint in One-Handed Lifting
The common industrial task of holding an object in a hand located in
front of the body is shown in *a*. The three forces creating torque on the
elbow that must be countered by forearm flexor muscle effort are shown
in *b*. These include forces due to the mass of the forearm (F_F), hand
(F_H), and load (F_R). This example is a third-class lever system with the
fulcrum at the elbow. Information for static torque calculations of
the lifting task demands is given in *c*. The moment arms for each of
the forces in *b* are identified and correspond to the distances from the
elbow to the centers of gravity of each mass unit—forearm, hand, and
load. They include r_F for the forearm, r_H for the hand, and r_R for the
load. The perpendicular forces that contribute to torque are also identified. These forces, calculated from trigonometric relationships (see
Appendix B), are $F_F\perp$, $F_H\perp$, and $F_R\perp$. β is the angle of force application
of the forearm, hand, or load that must be resisted by the flexor mus-
cles. *(Developed from information prepared by Kamon and Terrell, Eastman
Kodak Company, 1978).*

in the study of work. These examples include the biomechanics of the back, shoulder, and grasp, and discussions of dynamic lifting as well as how to analyze a job or workplace for biomechanical problems.

B. BIOMECHANICS OF THE BACK

There are two primary situations in which the biomechanics of the back should be analyzed: during manual lifting activities, and to evaluate the effects of different body postures, such as slouching or bending over from the waist, during seated or standing work. Work tasks having these types of activities are discussed in the following subsections.

1. LIFTING AND BACK STRESS

Lifting and carrying objects are common to many industrial tasks. Several factors influence the load placed on the spine during the performance of these tasks (Lindh, 1980):

- The position of the load relative to the center of the spine.
- The degree of flexion or rotation of the spine.
- The characteristics of the object: size, shape, weight, and density.

Most of the sources of stress on the spine can be identified by applying basic torque equations. The same analysis often leads to strategies for stress reduction. Figure 7-9 shows a box-handling task that can be evaluated through simple biomechanical analyses.

In this example, each of the boxes is held in front of the body. They weigh the same but, because they have different dimensions, the stresses on the spine and back muscles are not the same. From our knowledge of torques, we know that two factors are important: the torque moment arm and the force perpendicular to it. The force is vertical and downward due to gravity, acting through the center of gravity of the box. For a large box, the effective moment arm, or distance from the axis of rotation (spine) to the center of force increases. Even though the two boxes have the same mass, handling the larger box results in a 33 percent increase in forward bending moment, which must be countered by increased muscular effort. In this example, a packaging change could significantly reduce stress on the back.

Figures 7-10 through 7-13 illustrate lifting situations where stress on the back is increased by increasing the moment arm.

Unlike carrying, lifting often requires some forward rotation of the trunk. The high incidence of back problems during lifting has been associated with the stresses placed on back muscles, ligaments, and intervertebral discs. By applying the basic biomechanical relationships previously developed, we can identify clearly the origin of these stresses. During lifting and carrying there are two load factors of concern. Certainly the load carried is of prime importance, but in most cases the mass of the upper body is greater than the mass of the load to be moved. In any movement, such as bending, stooping, squatting, or car-

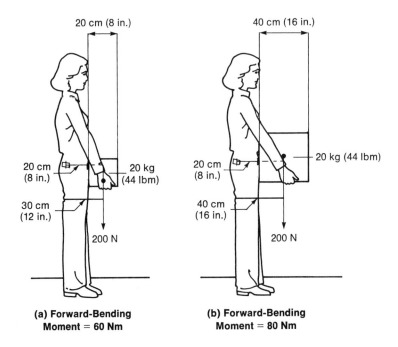

20 cm (8 in.)   40 cm (16 in.)

20 cm   20 kg   20 cm   20 kg (44 lbm)
(8 in.)   (44 lbm)   (8 in.)

30 cm   40 cm
(12 in.)   (16 in.)

200 N   200 N

(a) Forward-Bending   **(b) Forward-Bending**
Moment = 60 Nm   **Moment = 80 Nm**

Figure 7-9: Biomechanical Analysis of Box Handling   The effect of object size on the level of back stress is demonstrated here for a handling task. Both boxes are of equal weight, but the box in *a* is smaller. Thus, the center of gravity of the box is closer to the body. Since the distance from the handler's vertebrae to the front of her body (RA) is the same in both situations (about 20 cm or 8 in.), the forward-bending moment is a function of box width. This moment is calculated as the force due to the box (R) times the moment arm (RA). For the smaller box in *a*, this is 200 N × 0.3 m = 60 Nm; for the larger box in *b*, the forward-bending moment is 200 N × 0.4 m = 80 Nm. The greater box width adds 20 Nm to the stress on the back, increasing it by 33 percent compared to the smaller box. (*Adapted from Lindh, 1980*).

rying, upper body mass must be considered as a major source of load stress. Figure 7-14 shows how the total stress of lifting can be determined by including the mass of the body in the analysis.

The 20 kg (44 lbm) load in the hands results in a force of 200 N. The weight of this individual's upper body results in a force of 450 N (100 lbf) on the back. To calculate the torque, or moment, we must also identify the lever arms. For the load in the hands, the lever arm (L_p) is 0.3 m (12 in.). Thus, the moment due to the load in the hands is calculated as:

$$M_L = 200 \text{ N} \times 0.3 \text{ m}$$
$$M_L = 60 \text{ Nm}$$

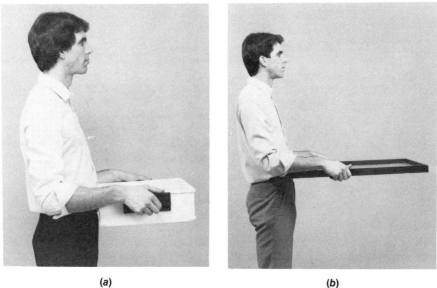

(a)   (b)

Figure 7-10: Tray Lifting   Two types of trays are shown as they are lifted to and above the handler's waist level. The tray on the left (*a*) is 50 × 33 × 15 cm (19.5 × 13 × 6 in.) in length, width, and height. The tray on the right (*b*) is 76 × 61 × 5 cm (30 × 24 × 2 in.). The distance from the handler's spinal column (vertebrae) to the tray's center of mass, which is approximately the middle of the tray's width, is increased by about 15 cm (6 in.) with the wider tray. This puts more stress on the back for the same tray weight. Wide trays can be found in manufacturing operations where the product is large or where the product, such as a food or chemical, is spread out to permit more rapid drying. If such a tray cannot be redesigned easily to reduce its width, methods should be developed that allow the handler to slide the tray instead of having to lift it.

However, the force due to the mass of the upper body also creates a moment in the same direction. The lever arm for the moment calculation of the upper body (L_w) is estimated as 0.02 m (0.8 in.). The moment due to the upper body mass is:

$$M_B = 0.02 \text{ m} \times 450 \text{ N}$$
$$M_B = 9 \text{ Nm}$$

The total forward-bending moment is the sum of the load moment and the body mass moment:

$$M_T = M_L + M_B$$
$$M_T = 60 \text{ Nm} + 9 \text{ Nm}$$
$$M_T = 69 \text{ Nm}$$

Figure 7-11: Lifting over a Vertical Obstruction. Here, the handler has to lift an object over a vertical barrier. Examples of barriers are the sides of packing cases and the sides of water-filled settling tanks. The barrier prevents the handler from placing a foot next to the load and makes it difficult to bend the knees in order to lift. This results in awkward lifting away from the body, as can be seen. The long resistance arm compared to the force arm results in a significant mechanical disadvantage for the back and leg muscles that are active during the lift. This type of lift also puts additional strain on the shoulder muscles.

It is this total forward-bending moment that stresses the muscles, connective tissues, and structural components of the back. In the upright position, the mass of the body contributes only 15 percent of the total forward-bending moment. Under actual lifting conditions, however, the situation can change significantly. Figure 7-15 demonstrates what happens when a person bends at the waist.

This situation is entirely different since the lever arm from the vertebrae to the center of gravity of the load (L_p) has increased substantially. The magnitude of change is clear when static biomechanical calculations are used. The forward-bending moment due to the load becomes:

$$M_L = 200 \text{ N} \times 0.4 \text{ m}$$
$$M_L = 80 \text{ Nm}$$

Forward bending increases the distance from the vertebrae to the center of gravity of the upper body (L_w) from 0.02 to 0.25 m (0.8 to 10 in.). Thus, the forward-bending moment due to the mass of the upper body is:

Figure 7-12: Lifting with an Extended Reach.   This worker is placing a product on a conveyor that runs along the outer margin of her assembly work station. She has to make this lift with her arms fully outstretched. This increases the length of the resistance lever arm and that puts additional stress on the muscles of the back and shoulders, which are countering the tendency of the product to rotate downward. Such workplaces can often be modified to permit the worker to slide the product onto the conveyor instead of lifting it.

$$M_B = 450 \text{ N} \times 0.25 \text{ m}$$
$$M_B = 112 \text{ Nm}$$

The total forward-bending moment now becomes:

$$M_T = M_L + M_B$$
$$M_T = 80 \text{ Nm} + 112 \text{ Nm}$$
$$M_T = 192 \text{ Nm}$$

By bending forward to pick up or put down the 20 kg load, the worker increased the forward-bending moment by 278 percent over the upright position. The moment due to the mass of the upper body contributes 58 percent of the total forward-bending moment in the trunk's flexed position.

The above calculations show the importance of analyzing the postures during lifting and carrying even very light weights. The body mass may, in itself, create proportionately high stresses. Often a person is forced into an undesirable

Figure 7-13: Low Lifting of Sheet Materials.   The handler is lifting a pack of large-sized (90 × 90 cm, or 36 × 36 in.) paper sheets off the floor. The sheets are too large to fit between his legs as they are lifted, so he raises them without the benefit of his leg muscles. If he were to take a full knee bend, he would have to support the sheet material much farther in front of his body than its location in this example. This would result in even higher compressive forces on his lumbar discs. Devices that raise materials up to at least 50 cm (20 in.) above the ground reduce the need for handlers to lift large-sized products in awkward postures like this.

body position due to the size of objects to be lifted (Figure 7-13) or because of workplace obstacles over which the object must be lifted (Figure 7-11).

Specific lifting styles and techniques are often suggested to help minimize stress on the back. Biomechanical calculations reveal some interesting facts about lifting techniques. Figure 7-16 provides comparison of three back moments where different techniques are used to lift a mass of 20 kg.

In Figure 7-16a, the person is lifting with the knees bent and back straight (squat position). Part of the load passes between the knees and thus minimizes the lever arm for the M_L calculation. The L_w distance is less than in Figure 7-16 b and c since the body is maintained in an upright position. The calculation of forward-bending moment is the same as for Figure 7-15 except for the lever arm changes:

$$M_T = M_L + M_B$$
$$M_T = (200 \text{ N} \times 0.35 \text{ m}) + (450 \text{ N} \times 0.18 \text{ m})$$
$$M_T = 151 \text{ Nm}$$

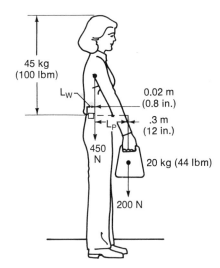

45 kg
(100 lbm)

L_w

0.02 m
(0.8 in.)

.3 m
(12 in.)

450
N

20 kg (44 lbm)

200 N

L_p

Total Forward-Bending Moment = 69 Nm

Figure 7-14: Biomechanical Analysis of Two-Handed Load Carrying
Two torques are tending to rotate the body forward. These are due to
the resistance of the load carried (200 N) and the mass of the upper
body (450 N). The total forward-bending moment is the sum of the two
forces times their resistance arms; L_p is the distance from the vertebrae
to the center of the load carried and L_w is the distance from the center
of the upper body mass to the vertebrae. The optimal posture for load
carriage in both hands is shown; the body is erect and the load is close
to it. As a person walks, leg movement will hit the load and cause it
to be moved farther from the body. This increases the stress on both
the back and shoulder muscles. The equations for calculating the torque
in this example are found in the text. *(Adapted from Lindh, 1980).*

Figure 7-16*b* shows the lifting situation described in Figure 7-15. The body
is flexed forward and the calculated forward-bending moment is 192 Nm. Figure
7-16*c* shows an important, yet often overlooked, point. Lifting with bent knees
does not always result in the lowest forward-bending moment. In this example,
the object is too large to pass between the knees. This results in an increase in
L_p and L_w, the latter to 0.5 m (20 in). Calculating the forward-bending moments
shows that the force on the back increases relative to that seen with bending
from the waist. The lever arm of the force due to upper body mass also increases
when this lifting technique is used. The total forward-bending moment, then,
can be calculated as follows:

$$M_T = (200 \text{ N} \times 0.5 \text{ m}) + (450 \text{ N} \times 0.25 \text{ m})$$
$$M_T = 212 \text{ Nm}$$

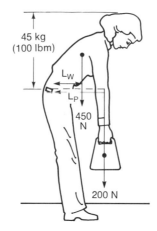

45 kg
(100 lbm)

L_W

L_P

450
N

200 N

Total Forward-Bending Moment = 192 Nm

Figure 7-15: Biomechanical Analysis of Bending at the Waist   In this illustration the handler is bending over with a load in both hands, as might be observed if the object is being lifted over a vertical barrier (see Figure 7-11). Bending forward increases the resistance lever arms for both the load (L_p) and the upper body mass (L_w). Although upper body mass (45 kg) and load mass (20 kg) are the same as they were in Figure 7-14, the total forward-bending moment is increased almost threefold. See the text for the torque calculations. *(Adapted from Lindh, 1980).*

2. POSTURES AND BACK STRESS

Many jobs are performed with the operator seated (Figure 7-17). Because the weight is supported on the chair, these postures would seem to reduce stress on the body, yet back pain and fatigue are common complaints of seated operators such as typists and VDT (video display terminal) operators. A science of chair design has evolved to address problems. Many factors must be considered when designing seated workplaces (see Chapter II in Volume 1). Biomechanical analyses can provide insight into the factors that contribute to back fatigue and discomfort with prolonged sitting and can help identify ways to reduce them.

Slight changes in posture can lead to increases in forward-bending moments (see Figure 7-15). If the body is to remain in a stationary position, these forward-bending moments must be resisted by backward-bending movements, developed by the back's extensor muscles. Increases in back extensor muscle force also lead to increases in compressive force on the intervertebral discs. Figure 7-18 shows relative compressive forces on the third lumbar disc in various body postures compared to the force when a person is standing erect.

Disc compressive forces are higher during sitting than during standing

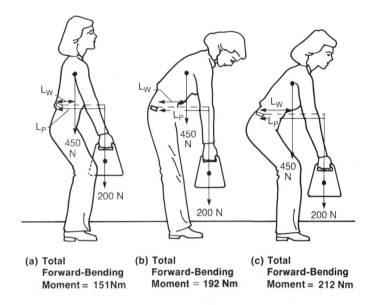

(a) Total
Forward-Bending
Moment = 151Nm

(b) Total
Forward-Bending
Moment = 192 Nm

(c) Total
Forward-Bending
Moment = 212 Nm

Figure 7-16: Biomechanical Analysis of Three Lifting Techniques
Three lifting techniques show the total forward-bending moments on
the lumbar spine, with the variations in resistance lever arms for the
load (L_p) and the upper body mass (L_w) indicated. Moving the object
closer to the center of gravity of the body reduces the forward-bending
moment and results in less stress on the back and shoulder muscles. *a*
shows the recommended posture for lifting, but it assumes that the
load can fit between the legs during the lift, which is often not possible.
b is similar to lifting over a vertical obstruction (see Figure 7-11). *c* is
analogous to the low lifting of sheet material (see Figure 7-13). *(Adapted
from Lindh, 1980).*

erect. If the worker's seated posture is poor, these forces can be greater than
those measured during standing with a flexed trunk. Figure 7-18 provides some
insight into the differences between sitting and standing. It is the mass of the
torso that tends to create the forward-bending moment around the fifth lumbar/
first sacral (L_5/S_1) vertebrae. Support of this weight below the axis of rotation,
such as when a person sits on the buttocks, does not reduce the forward-bending moment. In Figure 7-19 three postures are shown that give the expected
changes in horizontal distance between the vertebrae and the vertical force vector for upper body mass. It is this horizontal distance that determines the lever
arm (L_w) for the forward-bending moment calculations.

With relaxed sitting, the L_w shifts forward more than it does during standing. With erect sitting, the forward shift of L_w is reduced, but it is still greater
than that for standing. Since the mass of the torso is constant, the forward-

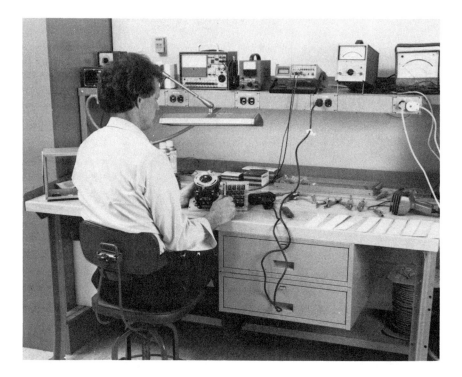

Figure 7-17: Sitting Posture at an Electronics Workbench The technician is seated on a stool that has been adjusted so he is working at elbow height on the workbench surface. The work requires fine visual attention, resulting in his having to lean forward and down to be able to see clearly. This forward bend of the head and trunk puts stress on the lower spine and the neck muscles, causing them to fatigue if the posture has to be maintained for several minutes at a time. If the technician slumps, or sits in a chair without lumbar spine support, the compressive forces on the lower spine will be even greater, as is demonstrated in Figure 7-18.

bending moment is a function of L_w. Therefore, in the sitting position the forward-bending moment goes up, and back extensor muscle activity and disc compressive forces increase proportionately.

Biomechanical considerations can also be used to evaluate the effects of chair design on back stress. Figure 7-20 indicates the relative changes in disc pressure that can be expected when the inclination of a back rest is changed (A and C) or when lumbar support is added (B and D). The undesirable effect of back support in the thoracic region is also demonstrated (E). The changes in disc pressure with different chair designs are related to the L_w values. Backward

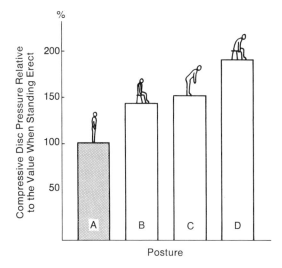

Figure 7-18: Compressive Forces on a Lumbar Disc in Different Postures   The effects of body posture (illustrated above each bar on the horizontal axis) on the relative compressive forces measured at the third intervertebral lumbar disc (L_3) are shown. The compressive force on L_3 in the standing erect posture (a) is about 686 newtons for a 70 kg person. The vertical axis shows compressive force as a percentage of the standing erect value. Seated postures, whether erect (b) or slouched (d), result in higher disc forces than standing erect. Leaning forward from the waist while standing (c) puts less compressive force on the third lumbar disc than does sitting in a slouched posture (d). *(Adapted from Nachemson, 1975).*

inclination of the seat and/or lumbar support reduce L_w and, therefore, reduce forward-bending moment and disc pressure. Likewise, the addition of support in the thoracic spine leads to increased disc pressure because the body is pushed forward, thus increasing L_w.

C. SHOULDER BIOMECHANICS

The shoulder joint and associated muscles are often called upon to perform industrial tasks. Any task that requires an outstretched arm will require activity of the muscles of the shoulder joint. This muscle-joint complex is particularly important when lifting objects above waist level or away from the body. By making a number of assumptions, we can make simplified static moment calculations for the shoulder. These calculations provide useful information for evaluating existing tasks, and they also provide a method of recognizing and avoiding problems associated with new workplaces and work tasks.

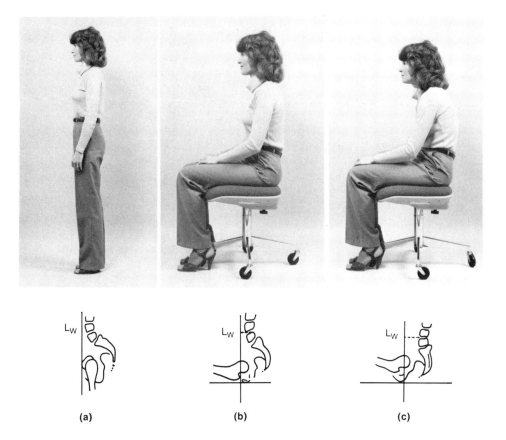

Figure 7-19: The Lever Arm of the Spine in Sitting and Standing
Three postures—erect standing (a), erect sitting (b), and slouched sitting (c)—are illustrated with schematic diagrams of the corresponding resistance arm of the upper body mass (L_w) below the photographs. The larger the L_w, the more the compressive force on the lumbar discs, and the more potential to aggravate a lower back problem. Backrests that give lumbar support help to push the spine forward to shorten the L_w (see Figure 7-20). (*Adapted from Lindh, 1980*).

The reasoning behind the common recommendation of designing a workplace so the worker can work at or below elbow height, and not have to abduct the upper arms, can be illustrated through the use of a simplified biomechanical approach. Shoulder abduction may be difficult to avoid, however, in some maintenance work when a large piece of equipment is being repaired at a workbench (Figure 7-21).

Figure 7-22 presents a simplified scheme for calculating the muscle force needed to maintain the arm in an abducted, or elevated, position.

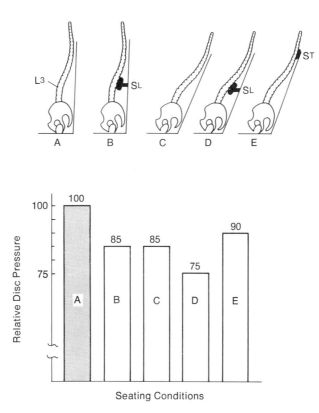

Figure 7-20: Effects of Chair Design on Back Stress   The relative compressive forces on the third lumbar disc (L_3) are shown in the lower graph for the five seating conditions illustrated in the upper part of the figure. A and B are upright backrests with and without lumbar support (S_L). C, D, and E are angled backrests (about 110 degrees) with and without lumbar support and with thoracic support (S_T), respectively. An angled backrest with lumbar support (D) is the recommended seating condition for minimal compressive force on the lumbar discs, this being about 75 percent of the force measured at L_3 when the person sits on a chair without lumbar support and with the backrest at 90 degrees (A). *(Adapted from Andersson et al., 1974; Lindh, 1980; Nachemson, 1975).*

Certain assumptions have been made in this example, as in other biomechanical problems. Although more than one muscle contributes force, the resultant muscle force can be represented by force M (F_M). The moment arm for F_M is assumed to be 0.03 m (1.2 in.). By using the static equilibrium relationship that the sum of all moments equals zero, F_M can be estimated. In this example, only the moment due to the mass of the arm acting at a perpendicular distance

Figure 7-21: Maintenance Work with Elbow Elevation   An example
of an occupational task that results in elbow elevation, or shoulder abduction, is shown. The maintenance worker is repairing a component
that is located 150 cm (59 in.) above the floor. To repair it, the maintenance worker has to have both elbows elevated. This puts additional
stress on the shoulder muscles to hold the elbow away from its resting
position alongside the trunk. A schematic biomechanical analysis of
elbow elevation is presented in Figure 7-22.

from the shoulder will be used to determine the force required by muscle M.
The mass of the arm (3 kg, or about 7 lbm) and the distance between the shoulder joint and the center of gravity of the arm (0.3 m, or 12 in.) are given. However, they could also be estimated from known relationships among anthropometric measures, segmental masses, and centers of gravity. Tables presenting
these relationships can be found in Appendix A, and an example of their use
is given in Chapter 28.

Using the static equilibrium relationships for the shoulder joint, the required muscle force (F_M) can be calculated as follows:

$$F_M = \frac{29 \text{ N} \times 0.3 \text{ m}}{0.03 \text{ m}}$$
$$= 290 \text{ N}$$

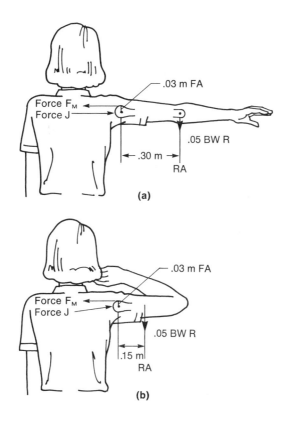

Figure 7-22: The Biomechanics of Shoulder Abduction   These drawings show the biomechanics of abducting the shoulder (or elevating the elbow) with the arm fully extended to the side (*a*) and with the upper arm extended and the elbow flexed (*b*). The shoulder muscle force needed to keep the elbow elevated (F_M) can be calculated using the information about the force arm (FA), the mass of the arm (0.05 BW, or 0.05 times the worker's body weight), and the resistance arm (RA). The latter is 0.3 m (12 in.) in *a* and 0.15 m (6 in.) in *b*, the difference caused by the position of the forearm relative to the trunk. Force J is the reaction force at the shoulder joint; it tends to stabilize the joint during abduction. When the upper arm is elevated to 90 degrees from the trunk, as it is in this illustration, force J is parallel to force F_M and approximately equal to it. (*Adapted from Matsen, 1980*).

Thus, a force of 290 N is required by the shoulder abductor muscles to hold the arm in position A. Force J is also shown in Figure 7-22. This is the joint reaction force that provides stability to the joint. In this instance, it is approximately parallel and equal to F_M. Figure 7-22*b* shows the effect of seemingly minor body position changes on the force requirements of the muscle. Flexing the elbow results in a shift in the center of gravity towards the shoulder. Although the mass of the arm does not change, the moment arm is one-half of the previous value. This reduction in the length of the moment arm reduces F_M to one-half of the value required for the extended arm. If the arm is abducted and the elbow is pulled in closer to the body, a similar effect will occur (Figure 7-23). The vertical line through the center of gravity of the arm will move closer to the vertical line through the axis of shoulder rotation. Thus, the moment arm is reduced and the force needed to be developed by the muscle is less.

It should be remembered that the functional capabilities of a muscle also vary with position changes because of changes in muscle length and insertion angles. The important ratio is:

$$F_{required} / F_{maximum}$$

This is the fraction of maximum voluntary contraction strength required by the task. If the ratio is multiplied by 100, it becomes the percent maximum

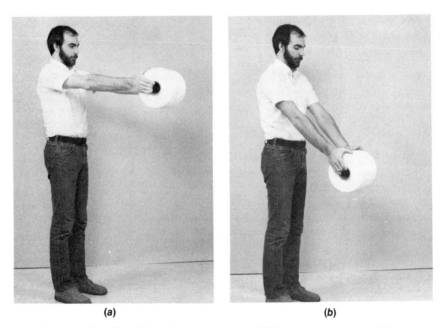

(a)   *(b)*

Figure 7-23: Shoulder Moment Arm Differences in a Handling Task   The effect of changing the angle of a lift requiring arm extension and forward reach is shown for a roll handling task. In *a* the roll is held at full arm extension close to shoulder height, resulting in a near-maximal resistance arm for lifting. In *b* the roll is still held with the arm extended, but it is lower and much closer to the body. Shoulder torque increases as the roll's weight increases or as its distance from the shoulder joint becomes greater, thereby increasing RA (see Figures 7-3 and 7-12).

strength, or %MVC. If a change in the position of a work task results in a change in the muscle's required force, consideration must also be given to changes in the maximal force capacity of the muscle in the new position. Tables of muscle strength measured at different body positions provide information that can be used for these analyses (see Appendix A).

D. BIOMECHANICS OF GRASP

Many work tasks depend on the ability of the hands to develop forces. The hands provide the link between the worker and the objects to be lifted, carried, pushed, pulled, or manipulated. A weak or inefficient grip can make work more difficult or even impossible to perform. The size of an object influences how it is grasped and how much force can be developed to do the task (see Appendix A). The following biomechanical analysis of a grasping task demonstrates these relationships. The forces generated by the hand when closing the lid on a can are estimated in Figure 7-24.

The hand in Figure 7-24a is closed halfway with the fingers encircling the can quite comfortably. The tendon (T) that pulls on the bone at the second joint of the finger is connected to a forearm muscle. The direction of pull, F, is such that the component F_y is almost parallel to it. The angle α between F and F_y is small. Consequently, F_y, the component that is pushing on the lid, is relatively large, while F_x, the component that stabilizes the joints between the two middle links of the finger, is small.

If the force needed to snap the lid, F_y, is 50 N (11 lbf), the muscle force (F) pulling on the tendon is:

$$F = F_y/\cos \alpha = 50 \text{ N}/\cos 15° = 52 \text{ N (11.6 lbf)}$$

In Figure 7-24 b, the hand is opened more widely and uncomfortably. The last link of the finger exerts much of the pressure on the lid. The pull of F is closer to the component F_x, which acts to stabilize the joint. In contrast, the component F_y acting to press on the lid is at an angle of 75 degrees. For the same 52 N of force on the tendon, the component F_y is now:

$$F_y = F \times \cos \alpha = 52 \text{ N} \times \cos 75° = 13 \text{ N (3 lbf)}$$

F has to be larger in b than in a to be able to close the lid on the can. For $F_y = 50$ N, $F = 50/\cos 75° = 193$ N (43 lbf). This substantial increase in the muscle force, F, caused by changes in the angles between finger links, makes snapping the lid on the can much harder, even impossible, for people with small hand spans.

E. DYNAMIC BIOMECHANICAL RELATIONSHIPS

In earlier discussions, static equilibrium relationships were used to estimate the static muscle forces needed to perform various tasks. While static postures and efforts are commonly used during work, movement is also very important. From

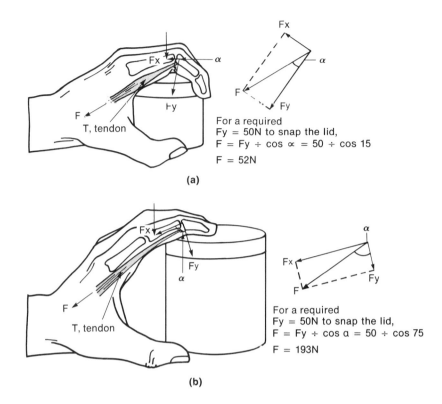

For a required
Fy = 50N to snap the lid,
F = Fy ÷ cos α = 50 ÷ cos 15
F = 52N

(a)

For a required
Fy = 50N to snap the lid,
F = Fy ÷ cos α = 50 ÷ cos 75
F = 193N

(b)

Figure 7-24: The Biomechanics of a Grasping Task   The muscle work required (F) to snap a lid on a can with a force (F_y) or 50 N (5 kg or 11 lbf) is illustrated for two sizes of can. The smaller can (*a*) permits the fingers to curl around it and to transmit the muscle force quite directly to the lid. A diagram of the forces is to the right of each drawing. It shows the tendon force (F), the force applied to the lid (F_y), the angle between the applied and tendon forces (α), and the force component acting perpendicular to the force required to close the lid (F_x). As the hand is spread farther apart with the larger can (*b*), the angle α gets larger, and less of the force developed by the muscles is utilized to snap the lid on (F_y). Thus, a larger muscle force is required in *b* than in *a* to do the task. (*Developed from information prepared by Kamon and Terrell, Eastman Kodak Company, 1978*).

the biomechanical perspective, dynamics is the study of man in motion. Dynamic relationships of work can be evaluated on three levels (Miller and Nelson, 1973). The first is a temporal analysis of work, which deals with the timing, or rate, of work performance. The second is a kinematic analysis, which is concerned with the geometric aspects of motion. Displacement, velocity, and acceleration information are part of a kinematic analysis. The third level is kinetics, the study of the forces that influence movement, and it is perhaps the most detailed of all dynamic analyses.

Many biomechanical problem-solving techniques have been developed for the study of dynamic relationships. Some of the more useful approaches are discussed later in this chapter. In the following subsection, a brief discussion is presented of how movement influences forces acting on and developed by the human body during work. Advances in dynamic biomechanical analysis using computer and electronic technology make dynamic analyses of work more feasible now. Techniques for measuring motion are discussed in Chapter 8.

1. FORCE-MASS RELATIONSHIPS

Newton's Second Law can be simply expressed by the formula:

$$F = ma$$

where F equals force, m equals mass, and a equals acceleration.

This relationship has important implications to the worker who is required to lift, lower, push, or pull an object. Such a broad work requirement encompasses the entire industrial population. Newton's Second Law states that a force is needed to move an object or one's own body from one position to another. To move the same object more quickly, you need a greater force. Acceleration, or the rate of change of velocity, has a direct effect on force. This effect and its influence on work performance may be seen more clearly by the following relationships. Starting with $F = ma$ and substituting the following for a:

$$\frac{\text{Final Velocity} - \text{Initial Velocity}}{\text{Change in Time}}$$

or

$$(V_f - V_i)/\Delta t$$

one gets:

$$F = \frac{m(V_f - V_i)}{t}$$

Multiplying through by m gives:

$$F = \frac{mV_f - mV_i}{t}$$

The factors mV_f and mV_i are referred to as *momentum*. For example, consider the task of lifting a box from a box from a pallet. The box is initially at zero velocity so $mV_i = 0$. The force required to lift the box, then, is described by the following:

$$F = \frac{mV_f}{t}$$

This relationship says that two factors can influence the force necessary to lift the box: the final velocity of the box during lifting, and the length of time during which force is applied. The relationship becomes much more complicated since the capacity for force generation by the human body changes during the lift due to neuromuscular and biomechanical factors. Object mass alone does not determine force requirements.

Another example of this relationship helps to explain some occupational injuries. Many injuries can be related to temporary losses of control while lifting or carrying in which a falling object is "caught." During the carrying phase, adequate forces are developed to support the mass of the object. If it is dropped, the object accelerates downward due to the effects of gravity. To stop the object from falling to the ground, the worker must apply a force to the object. Looking again at the formula:

$$F = \frac{mV_f - mV_i}{t}$$

V_f must become zero in order to stop the fall. Therefore, the force required to stop the falling object is described by the following equation:

$$F = \frac{-mV_i}{t}$$

where V_i is the velocity of the object just prior to force application, and the negative sign indicates that force and velocity are in opposite directions. To stop the fall, F will have to be greater than that measured during the carry. Depending on the magnitude of V_i and t, the required F may be high enough to cause damage to muscle, joint, or connective tissues.

2. LIFTING DYNAMICS

A biomechanical model has been developed that uses expected dynamic relationships to calculate forces during sagittal-plane lifting tasks (Ayoub and El-Bassousi, 1978). Figure 7-25 shows the calculated compressive and shear forces on the L_5/S_1 disc while performing two lifts, one with the back and the other with the legs. The object being lifted weighs 4.5 kg (10 lbm). These drawings show the dynamic characteristics of lifting. The peak compressive load on the back develops early in each lift and declines, with a smaller secondary peak near the end of the lift. Any change in the time required to perform the lift would be expected to affect the magnitude of these forces.

The effects of acceleration of a load during lifting can be seen in Figure 7-26, obtained from strain gauges placed in the handles of a box (Ekholm, Arboarelius, and Nemeth, 1982). In this example, the weight of the box was 12.8 kg (28 lbm); it is represented by the 126 N force required to support the box statically. Initiation of the lift results in a force on the handles that actually exceeds the weight of the load by approximately 20 percent. This peak is followed by force levels below the weight of the load, yet the load is still moving upwards.

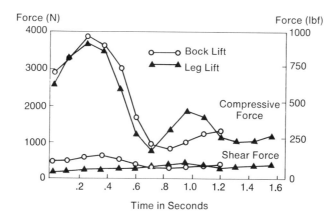

Figure 7-25: Compressive and Shear Forces on the Low Back During Lifting   A biomechanical model of dynamic lifting was used to generate these estimates of the compressive and shear forces on the low back (L_5/S_1 disc) during a two-handed low lift of a 4.5 kg (10 lbm) object. Force is shown in newtons (N) on the left vertical axis and in pounds (lbf) on the right vertical axis. The time for a single lift is shown on the horizontal axis. Estimates of these forces were made for both a back lift with the person bending over from the waist to lift, and a leg lift, bending the knees and using the legs to lift the body and object to the erect posture. The shear forces, shown at the bottom of the graph, were estimated as fairly steady throughout the lift period. They were somewhat higher for the back lift technique at the beginning of the lift. The compressive forces on the disc show peak forces at the beginning of each lift and have a second, lower, peak near the end of the lift, especially in the leg lift technique. These predicted compressive force changes indicate that the back is most vulnerable at the beginning of a lift. Faster lifting times can be expected to result in higher peak compressive forces; slower lifts should reduce the peak values. (*Adapted from Ayoub and El-Bassousi, 1978*).

This fluctuation in required force during lifting is a common finding in manual materials handling. A more practical example is lifting a heavy object onto a high shelf. The weight may be handled easily close to the ground where lifting capabilities are the highest. But strength and lifting ability decrease with increased vertical distance. A load that is too heavy to be held in front of the body at shoulder height can often be lifted from the floor onto a shelf at shoulder level. This is accomplished by using a ballistic technique in which very large forces are generated early in the lift when lifting capacity is high. This accelerates the load and effectively reduces the force required by the weaker shoulder muscles to complete the final lift at shoulder height.

Dynamic changes in abdominal pressure during a lifting task provide an-

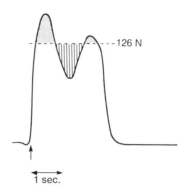

1 sec.

Figure 7-26: Force and Acceleration Measurements During Lifting
The electrical signal recorded from strain gauges mounted in the handles of a 13 kg (29 lbm) box is shown during a two-handed lift of the box. The static force required to support the box (126 N) is shown as a dotted horizontal line across the force trace. The lift was initiated at the arrow on the horizontal time axis. The initial force developed is greater than the required static force of 126 N (shaded area); a force lower than the required one follows (cross-hatched area); and finally there is another slight elevation near the end of the lift. The acceleration component of the lift accounts for the measured force fluctuations. If an object is too heavy to hold above waist level, a weaker person often uses this "ballistic" lifting approach to accelerate it rapidly when it is nearer the ground where the larger muscle groups of the legs and back can do the work. *(Ekholm, Arborelius, and Nemeth, 1982).*

other component of the complex systems available to protect the lumbar discs from excessive forces. This pressure is less easy to generate when the upper trunk is bent over (Bartelink, 1957).

F. STEPS IN IDENTIFYING AND SOLVING BIOMECHANICAL PROBLEMS

An approach to evaluating the biomechanical stresses in a work activity is summarized below (developed from information prepared by Kamon and Terrell, Eastman Kodak Company, 1978):

1. Look for the parts of the body that are needed to perform the task. These could be the fingers, hands, arms, shoulder, back, a foot, or a leg. Observe the links and joints involved in the task's performance.

2. Follow the motion of the limb during task performance. Identify the most active group of muscles. Such groups of muscles could be around the shoulder, the wrist (in flexion and extension), the ankle (pushing the foot down), or the knee (pushing with the knee bent).

3. Define the resistance force, such as the weight for a lift or a carry or the forces needed to push a pedal. Determine the direction of the force. The direction of the resistance should be defined in relation to the body part that is used for the task.

4. Determine the exact position of the joint around which the muscles are active. The position should be defined for the highest resistance introduced by the job. This refers to the angle of the joint and the position of the limb involved in performance of the task. The limb's position should be related to its resting, or natural, position—that is, to what extent its angle or level is different from that in the relaxed position.

5. Determine if there are postural changes. Note whether the trunk is in a natural, relaxed position either in standing or sitting.

6. If the posture is different from the natural position, define the difference with respect to gravitational forces and lever arms. Find the torques resulting from these postures.

7. Using information in Appendix A, identify the maximum torque of the most active muscle group (step 1). Do the same for muscle groups associated with awkward postures, if they are present in the task.

8. Quantify the forces involved in providing resistance to the main task (step 3) in terms of percent maximum voluntary contraction (%MVC), using the MVCs shown in the tables in Appendix A.

9. Find the endurance time for the quantity established in step 8 by referring to Figure 7-5. This endurance time is used to calculate recovery time allowance (see step 11).

10. Establish the time of contraction (t) for each group of muscles with regard either to the joint involved or to the postural demands, if present.

11. Determine the recovery time using information about percent maximum voluntary contraction strength (%MVC) and duration of continuous exertion from Figure 11-3.

12. Finally, design the speed of the work according to the total time of contraction (t) and the derived recovery time.

REFERENCES

Andersson, B. J. G., R. Ortengren, A. Nachemson, and G. Elfstrom. 1974. "Lumbar Disc Pressure and Myoelectric Back Muscle Activity During Sitting. I. Studies on an Experimental Chair." *Scandinavian Journal of Rehabilitation Medicine, 3:* p. 104.

Ayoub, M., and El-Bassousi, M. 1978. "Dynamic Biomechanical Model for Sagittal-Plane Lifting Activities." In *Safety in Manual Materials Handling,* edited by C. G. Drury, DHEW (NIOSH) Publications No. 78-185. Proceedings of a symposium held in Buffalo, New York, July 1976. Cincinnati: DHEW/NIOSH (Department of Health, Education, and Welfare/National Institute of Occupational Safety and Health), pp. 88-95.

Bartelink, D. L. 1957. "The Role of Abdominal Pressure in Relieving the Pressure on the Lumbar Intervertebral Discs." *Journal of Bone and Joint Surgery, 39 (B):* pp. 718-725.

Clarke, D. H. 1971 "The Influence on Muscle Fatigue Patterns of the Intercontraction Rest Interval." *Medicine and Science in Sports, 3:* pp. 83-88.

Ekholm, J., U. P. Arborelius, and G. Nemeth. 1982. "The Load on the Lumbo-Sacral Joint and Trunk Muscle Activity During Lifting." *Ergonomics, 25 (2):* pp. 145-161.

Kamon, E. 1981. "Aspects of Physiological Factors in Paced Physical Work." In *Machine Pacing and Occupational Stress,* edited by G. Salvendy and M. J. Smith. London: Taylor & Francis, pp. 108-115.

Kamon, E., and A. J. Goldfuss. 1978. "In-Plant Evaluation of the Strength of Workers." *American Industrial Hygiene Association Journal, 39 (10):* pp. 802-807.

Lindh, M. 1980. "Biomechanics of the Lumbar Spine." In *Basic Biomechanics of the Skeletal System,* edited by V. H. Frankel and M. Nordin. Philadelphia: Lea and Febiger, pp. 255-290.

Matsen, F. 1980. "Biomechanics of the Elbow." In *Basic Biomechanics of the Skeletal System,* edited by V. H. Frankel and M. Nordin. Philadelphia: Lea and Febiger, pp. 221-242.

Miller, D. I., and R. C. Nelson. 1973. *Biomechanics of Sport.* Philadelphia: Lea and Febiger, 248 pages.

Nachemson, A. 1975. "Towards a Better Understanding of Back Pain: A Review of the Mechanics of the Lumbar Disc." *Rheumatology and Rehabilitation, 14 (3):* pp. 129-143.

Rasch, P. J., and R. K. Burke. 1978. *Kinesiology and Applied Anatomy.* 6th ed. Philadelphia: Lea and Febiger, 496 pages.

Stull, G. A., and D. H. Clarke. 1971. "Patterns of Recovery Following Isometric and Isotonic Strength Decrement." *Medicine and Science in Sports, 3:* pp. 135-139.

Tichauer, E. 1973. "Ergonomic Aspects of Biomechanics." Chapter 32 in *The Industrial Environment - Its Evaluation and Control.* Washington, D.C.: U.S. Department of Health, Education, and Welfare, National Institute for Occupational Safety and Health, pp. 431-492.

Winter, D. A. 1979. *Biomechanics of Human Movement.* New York: Wiley, 199 pages.

CHAPTER **8**

Techniques for Analyzing Human Motion

CHAPTER 8. TECHNIQUES FOR ANALYZING HUMAN MOTION

Many techniques have been used to study the kinematics, or dynamic biomechanics, of the human body. The technique chosen depends on the information desired. All techniques have advantages as well as disadvantages or limitations. This chapter discusses some of these factors for eight of the simpler techniques used to analyze motion.

A. DIRECT MEASUREMENT TECHNIQUES

Direct measurement of the angles of joints, of the forces applied to objects in the workplace, or of the accelerations of a limb permits one to calculate the muscle forces required for occupational tasks. Four direct recording devices are discussed briefly in the subsections that follow.

1. ELECTROGONIOMETERS

The electrogoniometer, or elgon, is an electrical potentiometer that can be used to measure joint angle. Details of the construction of an elgon can be found elsewhere (Hutinger, 1971; Karpovich, Herden, and Asa, 1960). Figure 8-1 shows an elgon positioned for continuous measurement of knee joint angle changes, as would occur in walking or crouching. It must be placed over the axis of rotation of the joint with its arms attached to adjacent limb segments.

A constant voltage is applied to the elgon; the resistance of the potentiometer changes as a function of knee joint angle. The output of the elgon can be calibrated so that the measured voltage can be read as joint angle. The voltage signal can be recorded on a strip chart or can be run directly into a microprocessor to read out joint angle values during a work task.

There are several limitations to the use of an elgon:

- Its wires and attachment across the joint may interfere with normal movement patterns on the job.
- Its wires can also limit how far the person being studied can move around the workplace. It is best used for studies where the worker is seated or standing in one location.
- If several joints are to be studied, an elgon is needed for each, making the cost of this technique escalate if several workers are studied on a job that involves multiple joint activity.
- The technique assumes that a joint has a simple, fixed center of rotation, which is rarely true in human motion.

On the other hand, the elgon is relatively simple to construct and operate, and its output signal is easy to calibrate and record. It is also inexpensive when compared to other methods of motion analysis.

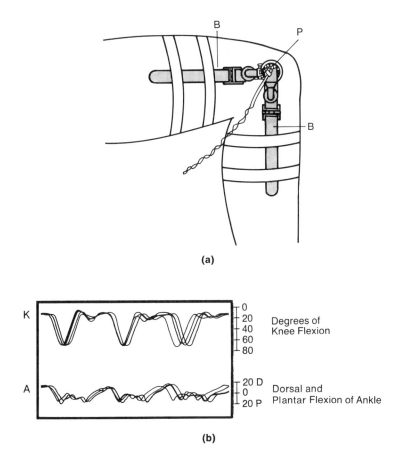

(a)

(b)

Figure 8-1: Measurement of Motion with an Electrogoniometer   An electrogoniometer (elgon) is positioned to measure knee joint motion in *a*. A potentiometer (P) is centered over the axis of rotation of the joint, and its blades (B) are attached to adjacent limbs. The type of information gathered by an elgon (*b*) is a continuous electrical signal that shows changes in the angle of the knee (K) and ankle (A) during a repetitive task. The scales for the degrees of movement for both joints are shown on the right side of the trace. Cautions about using elgons on the leg and arm joints are given in the text. (*Adapted from Miller and Nelson, 1973; Karpovich and Sinning, 1971.*)

2. FORCE TRANSDUCERS

The measurement of force is of prime importance in the study of the biomechanics of work. There are many approaches and levels of sophistication for measuring forces during the performance of work. Mechanical force transducers

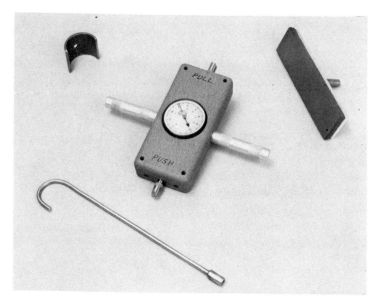

(a)

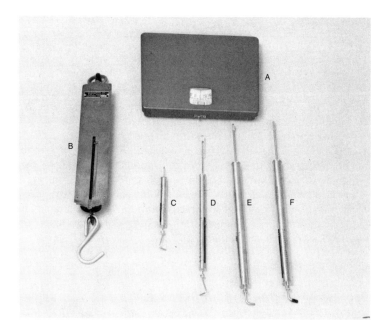

(b)

include spring dynamometers and cable tensiometers. While the accuracy and resolution of these transducers may be questioned, they are among the most valuable tools available for ergonomic field measurements.

Figure 8-2 shows some of the mechanical force transducers available. The force transducer with a continuously moving force indicator and a peak indicator is recommended (Figure 8-2a). This transducer can be used to measure the starting and sustaining forces during dynamic movements, as in pushing a box on a table or operating a chain hoist (Figure 8-3).

Figure 8-4 shows the use of this mechanical force transducer in measuring the forces needed to push or pull a hand truck. During starting or stopping of the truck, higher forces are required because of the truck's inertia. These forces are recorded by the peak indicator. As the truck assumes a steady velocity, the sustaining force will be relatively constant and can be read from the indicator dial.

Mechanical force transducers can also be used to measure the isometric strength of human muscles. This information can be very useful in a biomechanical analysis of work capacity (see Appendices A and B).

If increased accuracy is needed, or when dynamic force measurements are made, electronic force measurement techniques should be used. There are several electronic force transducers available; their characteristics and cost will help to determine the best one for a specific application. The most common types include the linear variable differential transformer (LVDT), the strain gauge, and the piezoelectric force transducer. All transducers work on the principle that an applied force causes a strain in the transducer. This strain results in a change in its electrical characteristics so that its output voltage is proportional to the measured force (Miller and Nelson, 1973). Figure 8-5 shows how electronic force transducers can be used to study industrial work tasks. The interested reader can learn more about the use and selection of these transducers from the references provided at the end of this chapter.

Figure 8-2: Examples of Mechanical Force Transducers. Commonly used mechanical force transducers are pictured. The transducer in *a* is a push-pull gauge with attachments to permit easier interfacing of the gauge with the objects being moved and measured. This gauge has a peak force indicator as well as a continuous force indicator needle. It can be used to measure peak and sustaining forces developed during the performance of work. *b* shows several other types of mechanical force transducers that depend on tension or compression of springs to measure force. These include: A, a platform scale such as a bathroom scale; B, a spring tension scale, often referred to as a "fish scale"; and C, D, E, and F, which are compression scales having force ranges from less than 0.45 kg (1 lb) to 11 kg (25 lb).

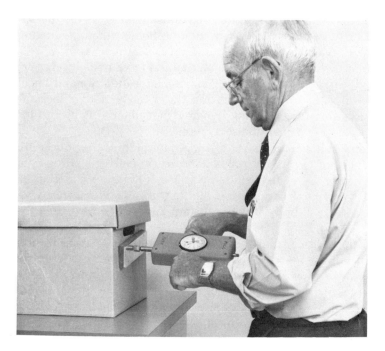

(a)

(b)

3. ACCELEROMETERS

In a more sophisticated biomechanical analysis, one may need to measure acceleration. Acceleration can be measured with an accelerometer, which is a specialized force transducer of known mass (m) attached to the force-sensing component of the transducer. If acceleration (a) of the transducer occurs, then the force exerted by the mass is F = ma. If force is measured by the force transducer and m is constant and known, a is proportional to the output of the accelerometer (Winter, 1979).

Although the concept is relatively simple, there are limitations to the technique. Only acceleration occurring at right angles to the accelerometer is measured. Many work patterns would not meet this limitation. Triaxial accelerometers are available that have three accelerometers mounted in a cube at right angles to one another (Figure 8-6). With these, acceleration in the vertical, horizontal, and lateral planes can be measured, but their use for industrial tasks is still limited.

4. FORCE PLATES

Multiple force transducers can be mounted on the undersurface of a metal plate to form a force platform. Force platforms can be used to study the ground reaction force that resists the feet during standing, walking, or running. They can be used in studies of the dynamic aspects of industrial tasks when a worker is standing, walking, or lifting.

Four types of information can be derived from force platform studies: three force measurements, including vertical force and two shear components along the force plate, and the instantaneous point of force application and moment about the vertical axis. The shear components represent forces in anterior—pos-

---

Figure 8-3: Using a Mechanical Force Transducer to Measure Pushing and Pulling Forces   Two uses of the push-pull gauge (Figure 8-2a) are shown. In *a*, the gauge is being used to measure the forces required to push a box across a work table. A rubber-backed attachment is used on the gauge to prevent damage to the box. Force is exerted on the handles of the gauge until the box begins to move. The peak-force indicator marks the initiating force when the box's inertia has been overcome. The sustaining force can be read from the indicator needle as the box is moved across the table; it is somewhat lower than the initiating force. The push-pull gauge can also measure the forces required to use a manual chain hoist (*b.*) A hook on the end of the gauge can be placed in a link of the chain hoist, allowing the operator to measure the peak and sustaining forces by pulling the chain down with the gauge handles. The intensity of the task can be evaluated by comparing the measured forces to the capacities of the population for that type of muscle work.

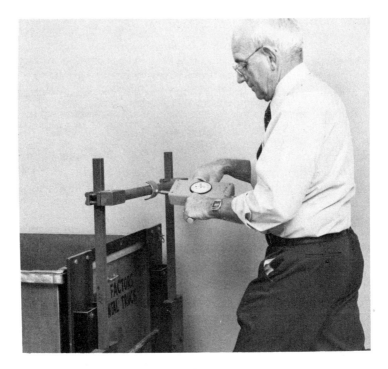

(a)

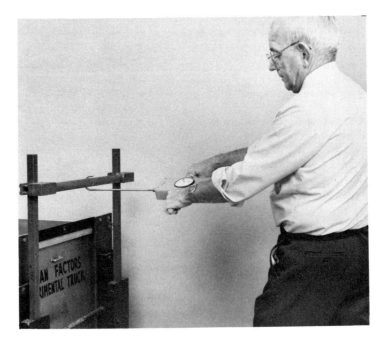

(b)

terior and medial—lateral directions, sometimes referred to as the horizontal and lateral forces.

These measures are sensitive indicators of changes in performance technique. For example, differences in lifting capacity and style can be studied using a force platform, since it is a good tool for the study of ground reaction forces. Its use, however, is not recommended for routine task analyses, as it is a relatively expensive tool, designed primarily for more sophisticated biomechanical analyses.

B. INDIRECT MEASUREMENT USING IMAGING TECHNIQUES

Due to the complexity of human movements, direct measurement techniques are of limited use in the study of many industrial tasks. When large amounts of data are to be generated in the analysis of complex tasks, imaging techniques may be the only practical approach to motion analysis. Many new imaging techniques have recently been developed, including electronic imaging and laser optical recordings. Only the most commonly used techniques are discussed in this section.

1. CINEMATOGRAPHY

Cine, or motion picture, cameras are classified by film type: 8 mm, Super 8 mm, 16 mm, 35 mm, and 70 mm. The two latter types provide large images that are ideal for analytical purposes but are also far more expensive to operate than the first three. The small film size of the 8 mm and Super 8 mm systems makes quantitative analysis of the images less accurate, but they are excellent for qualitative, or descriptive, studies of industrial tasks. The 16 mm systems are the preferred ones for motion analysis because the larger image permits improved quantification of the motion compared to the 8 mm formats (Miller and Nelson, 1973).

Figure 8-4: Measuring the Forces Required to Push or Pull a Hand Truck   A push-pull gauge can be used to measure the starting, stopping, and moving forces in handling a hand truck. In *a*, the gauge is used to push the truck, with a wide, curved attachment forming a stable interface for force development. In *b*, a large hook is used on the gauge to pull the truck by its handle while the forces are read off the dial. Use of a push-pull gauge in this application is helpful in evaluating truck design, including the best handle height and wheel or caster type. The effects of floor surface on the forces required for truck movement can also be measured.

Figure 8-5: Using an Electronic Force Transducer to Measure the Forces Required in an Industrial Task. An electronic force transducer is used to measure the force required to lift a tray. It is located on the cable that connects the bottom of the tray to the floor platform. The voltage output of the transducer is proportional to the force developed on the tray as it is lifted. The signal can be recorded on a chart recorder or measured by a computer. Electronic transducers provide a more precise description of force requirements than can be achieved with their mechanical counterparts.

There are many considerations besides the film and camera types involved in the use of this measurement technique. These include the following factors:

- exposure capability
- film sensitivity
- film speed capabilities

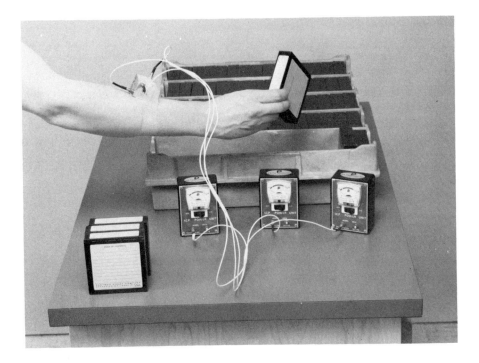

Figure 8-6: Using a Triaxial Accelerometer to Measure Movement During an Assembly Task   A triaxial accelerometer can be used to study arm motion in a repetitive assembly task. The accelerometer is placed over the center of gravity of the forearm. Most of the job stress on the arm is due to its repetitive motion and the forces required to accelerate and decelerate it. Since the mass of the forearm is constant, a relative indication of the forces required during movement can be obtained by measuring the acceleration of that segment. The acceleration signal will indicate high-force and, possibly, high-stress components of the motion. These recordings can be used to identify where new movement patterns should be designed to reduce the frequency and magnitude of acceleration in a limb.

- equipment calibration capability
- timing function
- data reduction and analysis needs

Cinematographic analysis to quantify motion requires careful planning, proper equipment, and rigid procedures. In the hands of experienced researchers, it is a powerful motion analysis technique (Miller and Nelson, 1973).

2. VIDEO RECORDERS

The use of video recorders has increased significantly in recent years and parallels the use of film in simple motion analysis studies. Video recording has such desirable features as allowing you to erase and use the tape again, if desired, and having an "instant replay" capability. Costs are reasonable, especially if only qualitative studies are desired and there is no need to keep a permanent record for further biomechanical analysis. Traditional videotape is not generally acceptable for quantitative analyses of motion because of its relatively low effective frame rate (Miller and Nelson, 1973), which does not provide sufficient detail for the study of rapid and complex tasks.

3. MULTIPLE EXPOSURE TECHNIQUES

A very simple and economical approach to motion analysis is the multiple exposure technique. This method utilizes a still camera in a darkened room; the camera shutter remains open during the entire task. A strobe light is used to illuminate the individual at preset intervals. Reflective markers or tapes can be placed on body segments, objects handled, or other items of interest. With every flash of light, a new exposure is made on the film. With a sequential task, the motion can be captured on one frame of film, the amount of information recorded being dependent on the strobe frequency and time of the task cycle. If reflective tape strips are placed along joint segments, the resulting print shows a stick figure of the subject performing the task.

Although this technique is quite simple and inexpensive, it has the following disadvantages:

- Few industrial tasks are performed in the dark, and those that are could not be done with the strobe light present. Therefore, the technique is more suitable for workplace or job simulations than for field studies in a manufacturing area.

- Although the motion is well described qualitatively, quantification and detailed analysis are crude.

4. OPTOELECTRIC TECHNIQUES

In recent years, new techniques for the analysis of human motion have been developed through optoelectronics applications. One system, called Selspot, uses infrared light sources, light-emitting diodes (LEDs), or laser light sources that are attached to the object or body part to be measured. For body motion analysis, the light sources would be attached to the joints and they would be seen as point sources of light by one or more Selspot cameras. The light spot is registered on the camera's light detector as a grid, with x and y coordinates. The output of the camera gives precise position information that can be fed into a computer or microprocessor for further data analysis.

Calculations of the speed, acceleration, and rotation of a limb or joint can be made easily with this system. Its advantages over traditional cinematography relate to the improved data collection and analysis capabilities of the SELSPOT system, but the cost is many times greater.

An example of an industrial application of this technique would be the analysis of a repetitive motion, such as tool use in an assembly task. Light sources could be located on the hand tool, hand, wrist, elbow, and the shoulder. The relative motions of the segments between each light source could be evaluated to study the efficiency of motion or the effectiveness of training or to investigate movements that might contribute to joint irritation.

C. SUMMARY OF MOTION ANALYSIS TECHNIQUES

This chapter has presented a very brief discussion of some common motion analysis techniques. Many factors must be considered when selecting a method to study motion in an industrial setting. The advantages and disadvantages of each technique should be evaluated with respect to the goals and requirements for data analysis in each specific application. In many instances, simple cinematography or videotaping is adequate to document the activity so that the motion can be studied in more detail, with additional people participating at a later time. In other situations, it is important to be able to measure the angles of joints and lengths of travel of body segments, so more complex equipment is desirable.

Force measurement can often be a "ballpark" figure, in which case simple mechanical transducers can be employed. This is the case when pushing and pulling forces are determined for the handling of hand carts and trucks. When limb accelerations and more dynamic motions are considered, the measuring equipment and the data collected become much more complex. The practitioner has to learn to balance the amount of analysis needed with the time required to provide an answer and with the importance of that information in developing a solution to the perceived problem.

REFERENCES

Hutinger, P. W. 1971. "Construction and Utilization of a Simple Electrogoniometer." In *Biomechanics*, edited by J. M. Cooper. Chicago: Athletic Institute.

Karpovich, P. V., and W. E. Sinning. 1971 *Physiology of Muscular Activity*. Philadelphia: Saunders, 374 pages.

Karpovich, P. V., E. L. Herden, and M. M. Asa. 1960. "Electrogoniometric Study of Joints." *U. S. Armed Forces Medical Journal, 11:* pp. 424-450.

Miller, D. I., and R. C. Nelson. 1973. *Biomechanics of Sport*. Philadelphia: Lea and Febiger, 248 pages.

Winter, D. A. 1979. *Biomechanics of Human Movement*. New York: Wiley, 199 pages.

CHAPTER 9

Heart Rate Interpretation Methodology

CHAPTER 9. HEART RATE INTERPRETATION METHODOLOGY

A. Methods for Collecting Heart Rate Information on the Job

B. Information Available from Heart Rate Recordings
   1. Heart Rate Level, Elevation, and Percent of Range
   2. Upward-Sloping Heart Rate Patterns
      a. Fatigue or Heat Accumulation
      b. Frustration
   3. Peak Loads

C. Use of Recovery Heart Rate Information to Evaluate Work Stress and Individual Fitness Levels
   1. Recovery from Physical Work—Indication of Fitness Level
   2. Recovery from Emergency Stress—Duration Effect
   3. Recovery Heart Rates with Light Activity
   4. Recovery from Work in the Heat
   5. Whole-Body Fatigue from Sleep Loss

D. Use of Individual Heart Rate Data to Assess Group Responses to Job Demands

References for Chapter 9

The heart rate, usually expressed in beats per minute, is the most convenient physiological measure of job stress. Increased rates can reflect the stress of the following types of job conditions (Brouha, 1970; Lehmann, 1962; Sternbach, 1966):

- Physical effort.
- Environmental heat and/or humidity.
- Psychic stress and/or time pressure.
- Some types of decision making and perceptual work.
- Other environmental factors (such as some chemicals, noise).
- Combinations of the above factors.

In addition, the heart rate reflects an individual's capacity for the work being done. This chapter describes the information that can be collected from monitoring the heart rates of people in the workplace. Some of the methods for measuring heart rate and their limitations in the workplace are given here and in Appendix B.

A. METHODS FOR COLLECTING HEART RATE INFORMATION ON THE JOB

The heart rate of a person doing a job can be measured continuously or discontinuously. The former involves investment in telemetry or tape-recording equipment; the latter can be done by a trained observer taking the worker's pulse rate at the wrist, using only a watch that registers seconds. Discontinuous measures of heart rate in a job are subject to sampling errors. Pulse rates are taken during recovery from a work period, resulting in a loss of important information about peak loads, durations of elevated heart rate, and patterns of recovery from heavier effort. The rate may also be affected by the way the pulse is taken, especially if the carotid artery in the neck is palpated, not the wrist. A careful activity analysis (discussed in Chapter 6) is needed to assess the workload.

Telemetry requires that the work be done in an area relatively free from electrical interference. An investigator has to be on site to record data and log the activities, and he or she may have to separate the noise that can accompany muscular work from the true heart rate in the signal. Tape-recording the continuous electrocardiogram, accompanied by a timed signal to permit accurate recording of rates, is the preferred method and does not require an investigator's continuous presence. The volunteer can keep a diary of activities to aid in interpreting the heart rate data. Although the cost of continuous monitoring equipment and data analyzers can be considerable, the heart rate information gathered makes the investment worthwhile.

B. INFORMATION AVAILABLE FROM HEART RATE RECORDINGS

Figure 9-1 shows a typical heart rate trace obtained by monitoring a person doing a physically demanding job. The level and pattern of the heart rate demonstrate

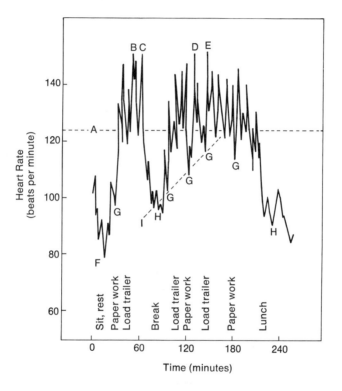

Figure 9-1: Measuring Job Demands by Electrocardiogram   The heart rate, in beats per minute, is shown on the vertical axis. It is taken from continuous recordings of a shipping dock worker's electrocardiogram (ECG) during a four-hour period (horizontal axis). Each of the major tasks and recovery periods is recorded just above the time scale. The horizontal dotted line (A) at 124 beats per minute represents the average heart rate over the four-hour period. Peak heart rates at B, C, D, and E are associated with manual handling of shipping cases of the product for three to five minutes at a time. The resting heart rate at the beginning of the shift is shown at the left of the graph (F). Several recovery heart rates during paper work (G) and breaks (H) are also indicated. A line sloping upward to the right (I) near the center of the graph indicates incomplete recovery with a gradually rising recovery heart rate level as the intermittent heavy lifting work continues. Each of these measures can be used to assess the demands of jobs on workers.

the large amount of information such a trace provides. The following measurements can be made from this trace:

- Resting heart rate (F).
- Average heart rate (A).
- Peak loads (B, C, D, E)
- Recovery heart rates, including rates of recovery (G, H).
- Work patterns (the shape of the curve).

These are discussed in the following subsections with additional examples.

1. HEART RATE LEVEL, ELEVATION, AND PERCENT OF RANGE

The absolute level of heart rate (HR—or fc, which stands for cardiac frequency) is only useful if the person studied on the job is typical of most other people. In general, it is better to use heart rate elevations to express the demands of jobs for the population. Individual responses can also be shown in relation to predicted maximum heart rate (HR max). The latter can be roughly estimated by subtracting a person's age from 220 (Astrand and Rodahl, 1977). To estimate the percent of maximum HR range required by a job or job activity, you can apply the following formula:

$$100 \times \frac{\text{(Average HR on Job - Resting HR)}}{\text{(Predicted HR max - Resting HR)}} = \begin{array}{c} \text{Percent Maximum HR Range} \\ \text{Required by the Job} \end{array}$$

The numerator (average HR - resting HR) is the heart rate elevation, and the denominator (HR max - resting HR) is the individual's heart rate range.

For many tasks, the percent of maximum heart rate range is closely related to the percent of maximum aerobic capacity, or maximum oxygen consumption (VO_2 max), required (see Appendices A and B). This is especially true if moderate-to-heavy whole-body work, such as lifting boxes onto pallets, is done in temperate workplaces. If, on the average, more than 33 percent of maximum HR range for whole-body work is required during the shift, the average worker is likely to fatigue. Jobs that may exceed 33 percent of maximum HR range during a shift include manual materials handler, stockkeeper, heavy-equipment repair operator, carpenter, painter, plasterer, and bricklayer. In most instances, people will structure their work to include lighter activities that reduce the average effort level to 33 percent of maximum HR range or less. When a worker does not have control over work pace, as can happen during work on a machine-paced task, workloads averaging more than 33 percent of maximum heart rate range are more likely to occur.

When other environmental or mental stressors, such as heat and time pressure, are present, the percent of maximum HR range helps to indicate the level of those stresses. The percent of maximum aerobic work capacity can be used to assess how much of the heart rate response is associated with the physical workload. The difference between the percent of maximum HR range and the percent of maximum aerobic capacity will indicate the nonphysical stress level. Studies of the same activity (driving a fire truck) under training and emergency conditions, for example, show percent of maximum HR range values of 30 and 50, respectively. The emergency stress, therefore, was calculated to account for (50 - 30)/50, or about 40 percent of the total heart rate elevation (Rodgers and Jones, Eastman Kodak Company, 1976).

This approach to distinguishing different job stresses is useful in defining the most effective intervention for reducing job stress. For example, a job in which a person must lift cases in a hot environment can be improved either by reducing the lifting requirements or by cooling the environment. If the lifting task is relatively heavy and difficult in any environment, simply reducing the heat level may not be the most effective intervention. Reducing the workload through redesign of the handling task could result in increased productivity and permit the hotter environment to be more easily tolerated, especially if heat is only a factor in the summer months.

2. UPWARD-SLOPING HEART RATE PATTERNS

a. Fatigue or Heat Accumulation

An increase in resting heart rate over time is often an indication of a fatiguing work pattern indicating incomplete recovery from work (Brouha, 1970). Figure 9-1 shows this, as does Figure 9-2 with a gradually increasing heart rate level on a job where heat was present and the workload was heavy and intermittent.

The increasing level of resting heart rate (between activities) as the shift proceeds can be attributed to a number of factors:

- Accumulating body heat, requiring higher skin blood flow to get rid of the heat. This increases the heart rate above that required simply to supply the demands for oxygen by the working muscles.

- Less efficient muscle work associated with increasing fatigue and the buildup of lactic acid in the muscles. This results in more muscle mass being involved and more demands being placed on the heart to deliver oxygen; therefore, a higher heart rate is seen.

- An increased requirement for blood flow to the digestive organs to aid in digestion after a meal. This explains the increased heart rate seen an hour or two after the mid-shift meal break. This effect is usually complete within two hours of eating a meal. By getting a good resting heart rate sample after the meal and by using this in place of the lowest HR of the shift to calculate the HR elevation, you can factor out the digestion effect when evaluating the stress levels of specific activities.

- Compensation for hemoconcentration, or decreased plasma volume, due to sweat-loss dehydration.

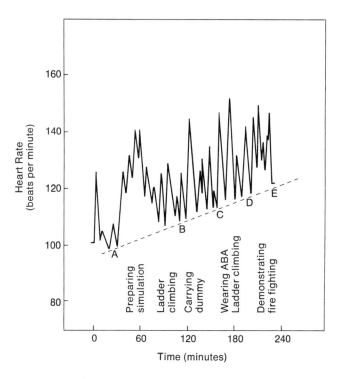

Figure 9-2: Example of a Fatigue Heart Rate Pattern   A continuous heart rate recording in beats per minute (vertical axis) is plotted against time (horizontal axis) for a firefighter in a simulated building fire. The average temperature was over 40°C (104°F) inside the building and about 30°C (86°F) outside of the building. The increased heart rate seen on the left side of the graph represents the setting up of the fire simulation. Successive peak loads include ladder climbing, carrying a 70 kg (154 lb) dummy, and wearing a self-contained breathing apparatus. Resting and recovery heart rates (A, B, C, D, and E) indicate the accumulation of body heat caused by working in a hot environment. The upward slope is characteristic of a fatiguing job; full recovery does not occur between work cycles. Appropriate adjustment of work and recovery cycles according to guidelines given in Part IV can reduce the fatigue illustrated here.

b. Frustration

Figure 9-3 illustrates an upward-sloping heart rate measured as a worker experienced trouble on a production machine. The frustration produced by the machine malfunctions was enough to elevate the heart rate 25 beats per minute. As soon as the problem was solved, however, the heart rate dropped to its

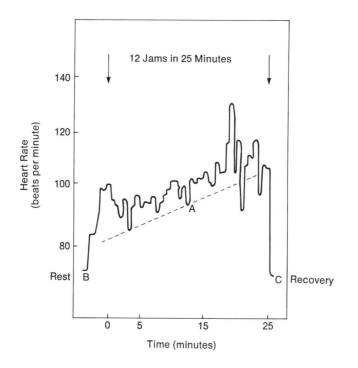

Figure 9-3: Heart Rate Elevation with Machine Jam Stress   Heart rate is shown (vertical axis) for a 30-minute period (time, horizontal axis). The person whose heart rate is shown was having difficulty with an assembly part that was outside the tolerance levels required by the machine. This resulted in 12 machine jams in a 25-minute period. The operator's frustration with the situation resulted in an increasing heart rate (A) until the problem was solved, at which time the heart rate returned rapidly to the resting value (B and C). Most frustration stress results in small elevations of heart rate. However, this response adds to the heart rate elevation associated with the physical and/or environmental stressors and makes the job more difficult for many people. *(Rodgers and Mui, 1981).*

previous level. From the heart rate trace, it is possible to determine the duration of elevation of the heart rate above the usual levels as well as the magnitude of the increase.

3. PEAK LOADS

To assess the peak loads in a job, one has to look at both the intensity and the duration of the load. A one-minute heart rate of 150 beats per minute, for instance, may be less stressful than a five-minute heart rate of 130 beats per minute. On the other hand, a one-minute heart rate of 180 beats per minute for a person over 40 years of age would be undesirable because it could represent a maximum level of work for the heart. See Appendix B for more information of the effects of age on maximum heart rate values. The heart rate response of an industrial firefighter is shown in light, moderate, and heavy effort, and in emergency stress in Figure 9-4. Peak loads are marked and can be quantified in terms of intensity and time.

Using the heart rate range (predicted maximum HR - resting HR) as a measure of the heart's work capacity, and the elevation of heart rate above the rest-

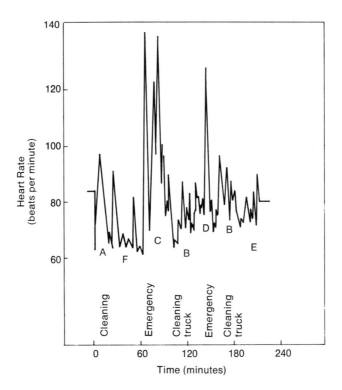

Figure 9-4: Peak Heart Rate Responses in Firefighting   The heart rate response (vertical axis) of an industrial firefighter is shown for a 3.5-hour period on the day shift (horizontal axis). The activities are indicated just above the time scale. They include moderately heavy cleaning, such as mopping floors (A) and cleaning fire trucks (B), and responding to alarms, both real (C) and false (D). In addition, there are moderate work tasks (E), such as working on the fire truck's engine and light work at a desk (F), which are associated with less heart rate elevation. Peak elevations in heart rate are seldom sustained for more than a few minutes at a time. Their duration and the subsequent recovery times determine whether fatigue, as reflected by the heart rate pattern, develops (see Figure 9-2).

ing value as a measure of the task load, one can estimate the percent of maximum capacity at which the heart is working. The higher this value, the more the potential for fatigue. There is also proportionately higher risk for heart problems in susceptible people at high, near-maximum heart rates (American Heart Association, 1972).

C. USE OF RECOVERY HEART RATE INFORMATION TO EVALUATE WORK STRESS AND INDIVIDUAL FITNESS LEVELS

After a period of heavy effort, heat exposure, or an emergency, the elevated heart rate will drop towards its resting level. The rate of fall of the heart rate is a function of:

- The individual's cardiovascular fitness.
- The duration of the previous stress.
- The nature of the activity done and the environmental conditions during the recovery period.

1. RECOVERY FROM PHYSICAL WORK— INDICATION OF FITNESS LEVEL

Figure 9-5 illustrates recovery heart rates for two people of the same age after they have walked on a treadmill for 13 minutes. The curve on the left is from a man who reached a heart rate of 132 beats per minute during the exercise and who took ten minutes to reestablish his recovery heart rate. The curve on the right is from a man whose heart rate reached 152 beats per minute during a higher level of exercise but who took only five minutes to recover his resting level. The man represented by the curve on the right is more fit, and this is reflected by his more rapid recovery to his resting heart rate.

In field job studies where it is not possible to measure the aerobic capacities of volunteers prior to monitoring their heart rates on the job, the heart rate recovery slope can be used to estimate an individual's fitness level (Figure 9-1). If there is no significant environmental or psychic stress on the job, a fast recovery rate after a physically demanding task indicates a fit person, whereas a sustained, or slowly falling, heart rate indicates poor fitness. From this information, one can estimate how others would respond to the same job demands.

2. RECOVERY FROM EMERGENCY STRESS— DURATION EFFECT

The curves in Figure 9-6 illustrate heart rate responses and recovery rates of industrial firefighters in emergency fire calls. As the emergency duration increases, the time for recovery of the heart rate to resting levels increases.

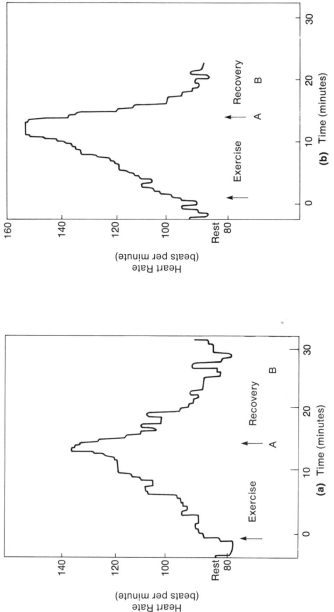

Figure 9-5: Heart Rate Recovery from Physical Effort Two graphs of heart rate (vertical axis) versus time (horizontal axis) are shown. Each person did a staged capacity test on a motor-driven treadmill (see Appendix B). At the point marked by arrow A in each graph, the exercise was terminated and the heart rate recovery time was measured. When heart rate returned to resting levels (B), the person was considered recovered from the work stress. The person having the heart rate response seen in *a* took ten minutes to reduce the 54-beat increase back to resting values. The person represented by *b* had been tested to a higher level of work and had a heart rate elevation of 63 beats, but he was fully recovered within five minutes. A more rapid recovery from work is desirable and indicates a higher level of cardiovascular fitness and physical work capacity.

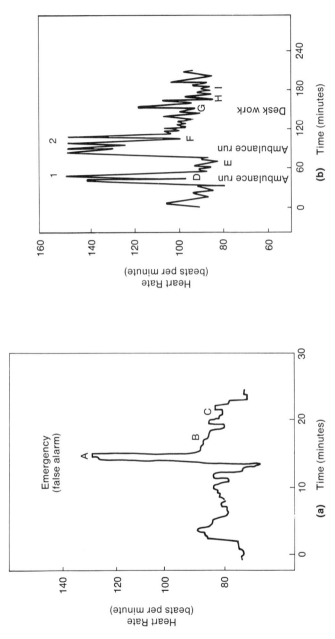

Figure 9-6: Heart Rate Recovery from Emergencies Two heart rate curves for industrial firefighters are shown with the heart rate on the vertical axis and time on the horizontal axis. The time axis is marked in 10-minute blocks in *a* and in 60 minute blocks in *b*. *a* illustrates how the heart rate changed during a false alarm. The sudden elevation of heart rate of more than 50 beats per minute lasted only about 1.5 minutes (A). Within 0.5 minutes it had returned close to the resting level (B), and it was fully back to resting within 5 minutes (C). *b* illustrates two ambulance calls of different durations. The first call (I) lasted about 20 minutes and the second call (II) took about 35 minutes to complete, although there was a recovery period of about 3 minutes after the patient reached medical attention and before the ambulance returned to the fire station. It took only about 3 minutes for the heart rate to fall to a few beats above resting levels after the first call was complete (D) and a total of 20 minutes to fully recover (E). After the second call, the heart rate recovered to a level 16 beats per minute above resting within 10 minutes (F) but had not recovered much further when a new activity was initiated 20 minutes after the call completion (G). After another 30 minutes of quite sedentary desk work (H) the resting level was reestablished (I). Longer emergencies result in longer recovery times. If another emergency or physically demanding activity occurs before recovery is complete, the worker's heart rate will be correspondingly higher.

185

3. RECOVERY HEART RATES WITH LIGHT ACTIVITY

The curves in Figure 9-7 illustrate recovery heart rates when activity during the recovery period is either complete rest (upper curve) or light activity (lower curve). The heart rate is reestablished at a higher level when light work is done during recovery but the slope of the recovery heart rate is not necessarily affected.

4. RECOVERY FROM WORK IN THE HEAT

Other factors also influence the rate and level of recovery of the heart rate on the job. Moderate or heavy work done in a hot and/or humid environment usually produces an elevation in body temperature. The increased blood flow to the skin to dissipate the heat results in higher heart rates in recovery than were seen prior to that activity. The environmental temperature and workload together determine how much skin blood flow is needed. In addition, the decreased volume of blood due to hemoglobin concentration causes an increased heart rate. Figure 9-8 illustrates heart rate during and after work in the heat; it demonstrates the increasing recovery heart rate as the shift progresses. The pattern of work and recovery periods determines how high the heart rate will rise in each successive work period.

5. WHOLE-BODY FATIGUE FROM SLEEP LOSS

Slower rates of recovery from effort and higher resting heart rate levels can also be associated with fatigue caused by sleep loss. Figure 9-9 illustrates the heart rate trace of a cafeteria attendant near the end of the day shift after she had had less than five hours of sleep the previous night. She expressed her feelings of being "tired" at the beginning of the shift, and she indicated that the feelings became stronger as the shift progressed.

The heaviest activity in the attendant's job was unloading an automatic dishwashing machine. This occurred just after lunch, in a 30°C (86°F) environment, and was under time pressure. The individual's heart rate on this task is a function of the physical effort (moderately heavy), the heat, and her digestive load from lunch. After this task was completed, her heart rate slowly decreased. It never reached its earlier resting level, despite several breaks and some lighter activity, during the remainder of the shift. Since the body heat accumulation from the unloading task was not enough to account for this elevation, it may be related to whole-body fatigue from sleep loss the night before.

D. USE OF INDIVIDUAL HEART RATE DATA TO ASSESS GROUP RESPONSES TO JOB DEMANDS

People vary in their physiological responses to job stress. One approach to ensuring that the job demands are being properly assessed is to measure several people on the same job. Figure 9-10 includes curves of the heart rate responses of four people, 25 to 32 years of age, on the same packing job. The work pattern

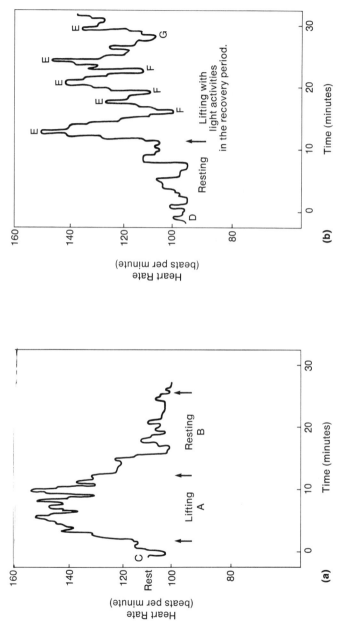

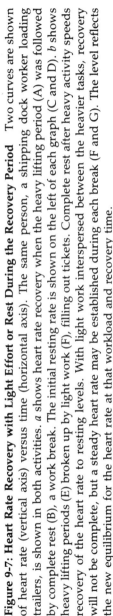

Figure 9-7: Heart Rate Recovery with Light Effort or Rest During the Recovery Period   Two curves are shown of heart rate (vertical axis) versus time (horizontal axis). The same person, a shipping dock worker loading trailers, is shown in both activities. *a* shows heart rate recovery when the heavy lifting period (A) was followed by complete rest (B), a work break. The initial resting rate is shown on the left of each graph (C and D). *b* shows heavy lifting periods (E) broken up by light work (F), filling out tickets. Complete rest after heavy activity speeds recovery of the heart rate to resting levels. With light work interspersed between the heavier tasks, recovery will not be complete, but a steady heart rate may be established during each break (F and G). The level reflects the new equilibrium for the heart rate at that workload and recovery time.

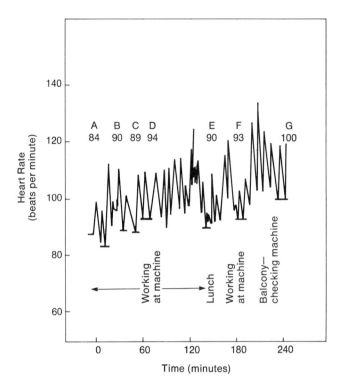

Figure 9-8: Heart Rate Patterns During Work in the Heat   The heart rate (vertical axis) of a person working on a two-story production machine having temperatures ranging from 32°C to 39°C (92°F to 104°F) is shown. Recovery heart rates over four hours (horizontal axis) are given just above the heart rate curve. The values are marked by horizontal lines on the heart rate curve and designated A through G. With continued moderate work in the heat, the resting heart rate increased from 84 to 100 beats per minute over four hours. A 30-minute work break (E) in a 24°C (75°F) cafeteria helped slow the upward rise of heart rate after the first 2.3 hours of work. The increase in resting heart rate can be related directly to the body's circulatory response to heat load. More blood is directed to the skin to get rid of excess body heat built up by work. The length of the recovery periods between work periods will determine how low the heart rate falls and, therefore, how high it will go during the next work period.

is determined by each operator, by the packing requirement for the product, and by the department's production goals. In each case, the individual worker met the production goals. How it was done was influenced by his or her capacity for the work and decisions about the appropriate work and recovery patterns to prevent local muscle and whole-body fatigue.

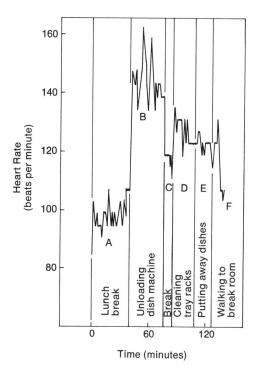

Figure 9-9: Heart Rate Response Due to Sleep Loss   The heart rate (vertical axis) of a cafeteria attendant is shown for the last part of the day shift (horizontal axis). Vertical lines indicate times when the activities, shown above the time scale, changed. The attendant had had less sleep than usual the night before this study and felt very tired. Her heart rate during rest breaks shows inadequate recovery from the heavier tasks, including unloading the dishwashing machine (B) and doing cleanup operations (D and E). Her subjective fatigue was objectively demonstrated by the elevated recovery heart rates (C and F) compared to the early morning resting rate of 80 beats per minute and the lunch break rate of 90 beats per minute (A).

Since job stress can only be quantified by monitoring the responses of a skilled worker who is doing the job on a typical shift, the volunteers who are studied must be "calibrated" on a standard task to determine their fitness level compared to the rest of the population. The standard task may be a measure of whole-body aerobic capacity, measured on a treadmill or bicycle ergometer, or a measure of upper body capacity, usually studied using a lifting or a two-arm cranking task. Protocols for these tests are included in Appendix B.

A person who has a high fitness level for aerobic work will demonstrate a lower heart rate on the job than one who is less fit. Data from the person with

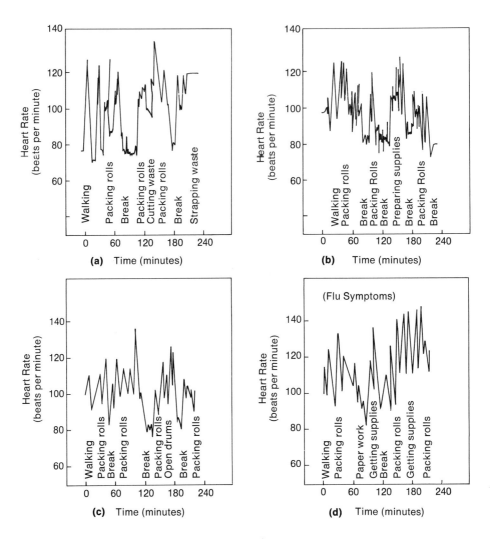

Figure 9-10: Heart Rate Responses of Several People Doing the Same Job. Four graphs of heart rates (vertical axes) are shown over the first four hours of the morning shift (horizontal scale). Each person performed the same job but on different days and with slightly different amounts of time on tasks other than roll packing. All were 25 to 32 years of age. In *a*, *b*, and *c*, the roll packers all show average heart rates over the four hours of about 100 beats per minute, although the heart rate patterns were quite different. In *a*, a woman newly placed on the job showed heart rate elevations of 40 beats per minute during roll packing. The heart rates of more experienced packers, *b* and *c*, were elevated about 25 to 35 beats per minute. The subject in *d*, who had flu symptoms and went home before the end of the shift, showed heart rate elevations during roll packing of about 40 beats per minute. Each person performed up to standard on the job, but the inexperienced worker and the one with flu symptoms did so at increased physiological cost. Heart rate measurements to quantify job demands are best made on several people doing the same job so the individual differences in work pattern or skill level are clear and so that an average response can be obtained.

low fitness will make the job appear to be heavier than it is for most people. By expressing the heart rate responses of an individual on the job in relation to his or her fitness level, you can determine the percent of the potential work force that may find the job difficult based on its total and peak demands. For instance, a handling task that requires 40 percent of aerobic capacity for a very fit person will be too difficult for most people. If it takes, on the average, 30 percent of the capacity of a person who has below-average fitness, most people should be able to do it without excessive fatigue. More discussion on the use of capacity data can be found in Appendix B. Guidelines for acceptable workloads can be found in the section on work/rest cycles in Part IV.

REFERENCES

American Heart Association. 1972. *Exercise Testing and Training of Apparently Healthy Individuals: A Handbook for Physicians.* Committee on Exercise. New York: American Heart Association.

Astrand, P. O., and K. Rodahl. 1977. *Textbook of Work Physiology.* 2nd ed. New York: McGraw-Hill, 633 pages.

Brouha, L. 1970. *Physiology in Industry.* London: Pergamon Press, 164 pages.

Lehman, G. 1962. *Praktische Arbeitsphysiologie.* 2nd ed. Stuttgart: George Thieme Verlag, 409 pages.

Rodgers, S. H., and M. Mui. 1981. "The Effect of Machine Speed on Operator Performance in a Monitoring, Supplying, and Inspecting Task." In *Machine Pacing and Occupational Stress,* edited by G. Salvendy and M. J. Smith. London: Taylor & Francis, pp. 269-276.

Sternbach, R. A. 1966. *Principles of Psychophysiology.* New York: Academic Press, 297 pages.

CHAPTER **10**

Estimation of the Energy Demands of Jobs

CHAPTER 10. ESTIMATION OF THE ENERGY DEMANDS OF JOBS

A. Primary Physical Effort Requirements

B. Supplementary Physical Effort Requirements

C. Estimate of Total Job Energy Demands

References for Chapter 10

F or a number of environmental conditions, such as heat, cold, and the presence of some chemicals, job designers need to estimate the physical workload of a job in order to assess the appropriateness of the exposure. Data on the oxygen demands of work and recreational tasks (Appendix A) may be used with caution to estimate workload, but oxygen consumption cannot always be measured and does not describe all categories of physical job demands. Over the years, methods have been developed to assess the physical effort for handling tasks and other job tasks (Garg, Chaffin, and Herrin, 1978; Passmore and Durnin, 1955); these are often based on elemental analyses of job tasks. Similarly, the method of workload estimation given in this chapter was developed to help quantify the physical effort levels of jobs for an evaluation of total job demands. When the results from a work physiology study on 21 jobs were compared to estimated effort levels (using the method presented here), the correlation between total points and average oxygen consumption on the job was 0.83 (Rodgers, Caplan, and Nielsen, 1976).

Based on the results of this study, the researchers identified five discrete categories of effort that emerge from the combinations of intensity and frequency of effort. For example, constant light (CL), frequent moderate (FM), and occasional heavy (OH) effort are substantially the same total effort level. The primary determinant of the effort level is the oxygen consumption, or the energy demands of the job studied. Figure 10-1 shows the five total effort level categories.

In Appendix C there are tables of effort equivalencies for different occupational and recreational tasks. These are grouped by five effort categories that are parallel to the ones described in Figure 10-1.

Physical effort stress can be assessed by identification of primary and supplementary job requirements. The analyst finds the intensity of a given task, such as lifting or pushing, by choosing the effort level according to the weights lifted or forces exerted. Each job task is similarly analyzed and the total time for each level of effort is calculated. The points for primary effort are determined from these two factors. Supplementary effort is recognized via additional points for specific job activities not covered under primary effort. The rest of this chapter describes this effort estimation technique.

A. PRIMARY PHYSICAL EFFORT REQUIREMENTS

Primary effort includes manual handling tasks and climbing (see Table 10-1). The amount of effort can be described in terms of an intensity (degree of effort) and a duration (percent of total shift time) for each degree of effort. First, one identifies each type of effort in a job. The amount of time each degree of effort takes can then be determined. The balance of the shift includes all other types of activities (total residual). Residual time can be calculated according to the breakdown of activities in Table 10-2.

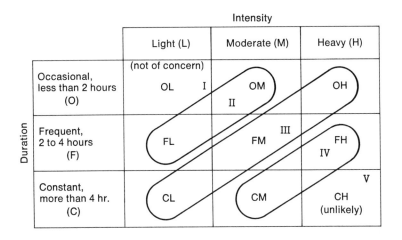

Figure 10-1: Categories of Physical Effort—Intensity and Duration
Nine combinations of effort intensity (across the top) and duration (column 1) are shown in columns 2 through 4. The equivalent effort levels are circled, such as FL and OM, for example. These combinations result in five discrete effort levels, which are indicated by I through V. Of these effort levels, jobs with only occasional light effort are not of concern for their physical effort demands. Jobs that require constant heavy effort are rare, and very few people can sustain them for the time required. Most industrial jobs fall in effort categories II, III, and IV. *(Adapted from Rodgers, 1978).*

An example illustrates the use of the primary requirements factor in analyzing a specific job. A chemical bagging job involves the following activities:

- Placing empty bags (1 kg or 2.2 lbm) on loading chutes, 60 times per hour.
- Pulling the filled bags (23 kg or 50 lbm) down the conveyor line, 60 per hour, forces of 90 newtons or 20 lbf.
- Lifting full bags off of the conveyor and onto a pallet, 60 times per hour.
- Procuring supplies (sheaves of empty bags, 25 kg or 55 lbm), eight to ten times per shift.
- Dragging empty pallets to the conveyor area (forces of 180 newtons or 40 lbf), 12 times per shift.

From the primary requirements factor in Table 10-1, you can see that the frequent handling of empty bags, except in a sheaf, is not included since bag weight is less than 1.8 kg (4 lbm). The 25 kg (55 lbm) sheaf of bags is relatively easy to handle, so it falls into the moderate effort category. Dragging the pallet

Table 10-1: Primary Physical Effort Requirements Three types of effort (lifting/carrying, application of forces, and climbing) are shown in column 1. Three degrees of effort (light, moderate, and heavy) are shown across the top of the table. Under each degree of effort, two parameters are given: the range of weight handled or force exerted in kilograms (kg) or newtons (N), with the pounds equivalent, and the ease of handling or exerting force, defined as easy or difficult. Examples of difficult handling are lifting or carrying a container of liquid, applying force or supporting a weight on a thin edge instead of on a broad surface, or carrying a bulky object when climbing up a ladder. Easy handling usually suggests that there are well-designed handholds on the object and that it is compact and balanced.

Degree of Effort

Type of Effort	Light		Moderate		Heavy	
	Weight or Force	Ease of Handling	Weight or Force	Ease of Handling	Weight or Force	Ease of Handling
Lift/carry (weight)	1.8–4.5 kg (4–10 lbm)	Easy/Difficult	5–34 kg (11–75 lbm)	Easy	>34 kg (75 lbm)	Easy
			5–18 kg (11–40 lbm)	Difficult	>18 kg (>40 lbm)	Difficult
Application of Forces (force)	18–180 N (4–40 lbf)	Easy	181–335 N (>40–75 lbf)	Easy	>335 N (>75 lbf)	Easy
Climbing (weight)	18–110 N (4–25 lbf)	Difficult	111–180 N (>25–40 lbf)	Difficult	>180 N (>40 lbf)	Difficult
	0–4.5 kg (0–10 lbm)	Easy/Difficult	5–18 kg (11–40 lbm)	Easy	>18 kg (>40 lbm)	Easy
			5–11 kg (11–25 lbm)	Difficult	>11 kg (>25 lbm)	Difficult

Table 10-2: Time Analyses—Percent of Time at Different Effort Levels per Shift. The calculation to account for 100 percent of shift time is shown in (a). The job analyst identifies what percent of total shift time is spent in light, moderate, or heavy activities (see Table 10-1). Any balance will be residual activity, which can be further broken into specific activities, as shown in (b). Heavy effort is counted in the primary requirements points (see Table 10-3) only if it occurs for 5 percent or more of shift time. If it occurs for less than 5 percent of the time, the effort is recognized through the supplementary requirements points (see Table 10-5). The percent of shift time spent on a specific activity is determined by dividing the total time per shift (often 480 minutes) into the total time spent on each degree of effort category.

(a) Total Light % _____ + Total Moderate % _____

   + Total Heavy %* _____ + Total Residual % _____ = 100% of Shift Time

*If less than 5 percent, use Table 10-4.

(b) Residual Time

Other physical activities	_____ %
Base/nonphysical activities	_____ %
Standby	_____ %
Paid lunch	_____ %
Breaks	_____ %
Total Residual	_____ %

---

is a moderate effort. Pulling the bag along the conveyor is a light effort. Lifting the bag onto the pallet is a heavy effort because the bag's contents will shift, making them difficult to handle, and they have to be turned from vertical to horizontal.

The percent of time in each effort category has to be determined from an activity analysis. In this instance, the large majority of the shift was spent in loading, pulling, and handling the bags; about two hours were spent on each activity each shift, on the average. The auxiliary-supplies handling tasks (pallets and bags) each took about 20 minutes per shift. In summary, then:

- Light: pulling bags for two hours.
- Moderate: dragging pallets for 20 minutes; carrying sheaves of bags for 20 minutes.
- Heavy: lifting bags for two hours.
- No effort category: loading empty bags for two hours.

If total shift time is 480 minutes, this analysis shows the following activity breakdown in time:

- 25 percent of time, light.
- 8 percent of time, moderate.
- 25 percent of time, heavy.

This leaves 42 percent of the shift in work activities other than handling. These residual activities can be accounted for using the analysis shown in Table 10-2:

- 25 percent, other physical effort.
- 12 percent, breaks.
- 5 percent, standby or nonphysical activities.

The points for primary requirements are found in Table 10-3:

$$25 \text{ percent light} = 10 \text{ points}$$
$$8 \text{ percent moderate} = 16 \text{ points}$$
$$25 \text{ percent heavy} = 26 \text{ points}$$

52 points, total Primary Requirements points

Table 10-3: Points—Primary Requirements. Three degrees of effort (light, moderate, and heavy) are shown across the top of the table. Three durations with their percentages of shift time are shown in column 1. These refer to the total time spent in that degree of effort, as calculated from Table 10-2. In the body of the table are the points assigned to each combination of effort level and duration for the primary requirements analysis. Some combinations of effort intensity and duration require that the points for other types of effort not be added to the final value. These limitations are given below the points table.

Duration	Degree of Effort		
	Light*	Moderate**	Heavy
Occasional 5–25%	10	16	26
Frequent 26–50%	19	38	57
Constant > 50%	38	76	115

*Omit points for Light Effort if any of the following occur:
  - Constant Heavy
  - Constant Moderate
  - Frequent Moderate and Constant Heavy
**Omit points for Moderate Effort if Constant Heavy Effort occurs.

B. SUPPLEMENTARY PHYSICAL EFFORT REQUIREMENTS

Table 10-4 shows supplementary requirements of jobs. These physical effort components increase the job stress but are significant only at a given intensity (such as the visual attention required) or after a given duration, such as the external pacing. Points are given for each component that exists in a job. Included are points for short duration heavy effort that occurs for less than 5 percent of the shift and receives no points under primary requirements.

Continuing with the chemical bagging job analysis, we can assign points for the following supplementary requirements (from Table 10-5):

Table 10-4: Supplementary Physical Effort Requirements.   Three levels of effort (low, medium, and high) are shown across the top of the table. They are defined for six types of effort (column 1) that are not covered under primary requirements. These include static muscle effort (a, b, and c), time pressure (d), repetitive small muscle work (e), and short duration heavy effort that takes less than 5 percent of shift time (f). The levels of effort in columns 2 through 4 are defined either by duration (a, b, d, and e) or by intensity (f). Visual and auditory requirements combine intensity (the difficulty of detecting a signal) with duration (more than 50 percent of the shift). Static muscle loading through restricted movement of the head or neck in a critical inspection task (RHN) is also a determinant of the task's degree of effort. All of these types of effort increase the job's physical effort demands, but many do not contribute greatly to whole-body oxygen consumption in comparison with the primary requirements factors

Type of Effort	Level of Effort		
	Low	Medium	High
a. Standing/walking	—	25–50% of time	>50% of time
b. Restrained posture (except neck and head)	Sit >75% of time	Awkward posture >5% of time	—
c. Visual or auditory requirements >50% of time; restricted head and neck posture (RHN)	Easily detected, no RHN	Easily detected, with RHN; Hard to detect, no RHN	Hard to detect, with RHN
d. Fixed external pace	—	>50% of time	—
e. Use of small muscle groups (fingers, hands, forearms, feet), up to 1.8 kg (4 lbm)	—	25–50% of time	>50% of time
f. Short-duration heavy effort (<5% of time)	—	Up to 23 kg (50 lbm)	>23 kg (>50 lbm)

Table 10-5: Points—Supplementary Requirements. The three levels of effort, defined for types of effort in Table 10-4, are shown in column 1. The points for each one are given in column 2. These points are additive. Each of the six types of effort from Table 10-4 having one of the designated effort levels adds points to the sum of the primary requirements points calculated in Table 10-3.

Level of Effort	Points
Low	6
Medium	13
High	22

Standing/walking, more than 50 percent of the time = 22 points
Repetitive use of small muscles, 25 percent of the time. = 13 points
External pacing, more than 50 percent of the time = 13 points

48 points

C. ESTIMATE OF TOTAL JOB ENERGY DEMANDS

Once the points for the primary and supplementary requirements have been determined, they are added to find the total effort level of the job over an eight-hour shift. Figure 10-2 shows the relationship between job oxygen consumption measurements and the number of points obtained from effort estimation for 21 jobs. The predicted energy demands can be related to population aerobic capacity data (Appendix A) to determine how many people will have difficulty with the job (see Part IV).

The total points for the bagging job come to 52 + 48, or 100. Using the relationship between oxygen consumption and points shown in Figure 10-2, we can estimate the metabolic demands of the bagging job at about 1.0 liters of oxygen per minute. This estimate is consistent with physiological measurements made on two people doing the job.

This technique of job effort analysis should not be used to estimate work demands in situations where individual suitability for work is being assessed, but it does allow a person to estimate reasonably accurately a level of effort in a job. This information can be used to evaluate the interaction of environmental factors, such as chemical exposures or heat and humidity, with workload in assessing total job stress levels.

REFERENCES

Garg, A., D. B. Chaffin, and G. Herrin. 1978. "Prediction of Metabolic Rates for Manual Materials Handling Jobs." *American Industrial Hygiene Association Journal*, 39 (8): pp. 661-674.

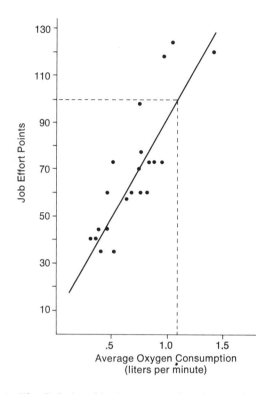

Figure 10-2: The Relationship Between Job Points and Average Oxygen Consumption Requirements   The average oxygen consumption demands of 21 jobs (•) are shown on the horizontal axis in liters per minute. They are plotted against the estimated physical effort points developed from Tables 10-3 and 10-5 and discussed in the text. The regression line drawn through the points can be used as a rough estimate of the energy demands of jobs for a given number of points (dotted lines). Points for jobs that are farthest from the regression line often represent jobs with proportionately more supplementary than primary job demands. Oxygen does not directly reflect the total effort level for supplementary types of effort.

Passmore, R., and J. V. G. A. Durnin. 1955. "Human Energy Expenditure." *Physiological Reviews, 35:* p. 801-840.

Rodgers, S. H. 1978. "Metabolic Indices in Materials Handling Tasks." In *Safety in Manual Materials Handling,* edited by C. G. Drury. DHEW/NIOSH Publications No. 78-185. Proceedings of a symposium, July 1976, in Buffalo, N.Y. Cincinnati: Department of Health, Education, and Welfare/National Institute Occupational Safety and Health, pp. 52-56.

Rodgers, S. H., S. H. Caplan, and W. J. Nielsen. 1976. "A Method for Estimating Energy Expediture Requirements of Industrial Jobs." Paper presented at the International Ergonomics Association meeting, July 1976, in College Park, Maryland. Abstracted in *Ergonomics, 19* (3): p. 349.

PART **IV**

Patterns of Work

CONTRIBUTING AUTHORS

David M. Kiser, Ph.D., Physiology

Suzanne H. Rodgers, Ph.D., Physiology

Regulating the way a job is done can make it possible for people of lower work capacity to perform competently on tasks that may appear to be excessively demanding. Workload can be described by two dimensions: the intensity of the effort, whether whole-body work, local muscle exertion, or visual requirements; and the duration of continuous work at that level. The more intense an effort, the less time it can be sustained before fatigue occurs or performance deteriorates. A skilled worker learns how to select a work pattern that permits more demanding work to be done in appropriate time periods so that fatigue does not become a problem.

Factors that influence how job tasks are done and what strengths, endurances, and perceptual demands are required of the workers may all affect productivity and safety in the workplace. Job satisfaction is higher when a person feels that he or she has control over the way a given task is accomplished (Sen, Pruzansky, and Carroll, 1981—see References at the end of Chapter 11). Frustration will occur if the job design forces a person to work in an inefficient manner. A job pattern that appears efficient on paper, such as dividing work into discrete, highly repetitive tasks with little variety, may, in the long run, limit productivity through development of local muscle fatigue or loss of attention in the worker.

Ergonomic principles of job design identify potentially fatiguing activities and provide work patterns that minimize them. Since individuals vary greatly in their tolerance to work, these principles emphasize self-regulation of effort.

Other principles of job design relating to job satisfaction and enrichment are not included here, but they complement the ergonomic design approach (Friedlander, 1964; Herzberg, Mausner, and Snyderman, 1959; Maslow, 1943).

Guidelines for the design of work patterns are given in Chapter 11 through a discussion of work/rest cycles for static and dynamic physical effort and for perceptual work, such as visual inspection. The impact of external pacing of work by machines or fellow workers is included in Chapter 12 to indicate how this can affect the adjustment of work patterns for individuals. Patterns for highly repetitive light work, such as small assembly and packing jobs, are also discussed (Chapter 13) with guidelines to help reduce the potential for repetitive-motion disorders of the shoulder, elbow, wrist, and hand. Finally, the subject of training and skill acquisition is included (Chapter 14) to illustrate how learning on the job helps each worker in the selection of an appropriate work pattern.

CHAPTER **11**

Work/Rest Cycles

CHAPTER 11. WORK/REST CYCLES

$\mathbf{T}$he amount of physiological or psychological stress a job activity produces is related to the ratio of its work and rest cycles. These cycles have been discussed in Part III for several types of jobs. The more intense a work period, the longer the rest period needed to recover from it (see Figure 7-5). The ratio of work to rest time in a job is also a way to assess whether an individual can regulate the cycles to avoid fatigue. Jobs where long-duration physically or perceptually demanding tasks are done without breaks, such as unloading product from a moving conveyor, usually have a high work/rest ratio. Jobs with many tasks in a variety of effort levels, such as cafeteria attendants, can be patterned to have low work/rest ratios by alternating between tasks of different effort levels.

Rest or recovery in job design is not synonymous with full cessation of activity or a scheduled work break. A rest phase may be any period of light activity, such as record keeping, that alternates with more demanding activities, whether they involve manual handling, cleaning, packing, or different inspection tasks. Rest can also refer to an activity that uses different muscle groups from those most active in the main task. In most self-paced operations, people provide their own mini-rest breaks to balance the work demands and lessen fatigue. These breaks may be very short (under one minute) or quite long (more than ten minutes), depending on the intensity of the work directly preceding them (see Part III).

Information about the design of work and rest patterns for physically and perceptually demanding jobs is included in this chapter in the context of designing jobs to reduce fatiguing activities. The concept of fatigue is confounded by observations that a person who feels fatigued at work may exhibit great energy with a change of activity (the Friday afternoon syndrome). Personal motivation and other social factors are strong components of the feeling of fatigue that is used as an end point when gathering information on attitudes about some types of occupational tasks (Yoshitake, 1971). Physiological fatigue, in contrast, may be one component in the feeling of fatigue, but it is a series of phenomena that cannot be altered by mental state, even though the symptoms can be suppressed (Floyd and Welford, 1953).

In the industrial environment there are a number of behavioral clues that allow the ergonomist to detect physiologically fatiguing activities:

- Shaking a limb or hand between work cycles usually points to a muscle group that is fatiguing.

- Taking frequent, short breaks from the work (for example, rearranging tools, going to check on supplies, getting a drink of water) often indicates a need by the operator to "recover" from the work demands. This behavior is frequently observed in areas where paced or highly repetitive work is done.

- Declining productivity throughout the shift usually indicates a fatiguing workload and inadequate work/rest cycle design. The more demanding the task, the more likely it is that performance will decline with time.

- High absenteeism, excessive personnel turnover, and more accidents near the end of the shift often indicate a job that fatigues many people.

A common industrial approach to designing rest breaks is to provide 15 minutes of rest after each two hours of work. During this rest period, the worker can get away from the workplace and attend to personal needs. This provides some recovery from the buildup of fatigue in the preceding two hours. However, the time given to a person doing heavy handling is usually the same as that given to a person doing light assembly work. To determine how much additional break time is needed for the heavier job, job analysts have established formulas based on information about how physiological fatigue develops. These formulas are presented later in this chapter. Fatigue allowances for static muscle loading in some working postures and for demanding perceptual work are also discussed in the sections that follow.

A. PHYSICALLY DEMANDING JOBS

The methods for evaluating job demands that are presented in Part III should be used to help identify the limiting activities in a physically demanding job. If the job is not yet being done, it can often be simulated, or similar jobs can be studied. The best source of information is the worker, who can identify whether the most difficult part of the job is a sustained effort, a specific strength, or an awkward posture. If the limiting factor is the energy demands of a specific task (as in short-duration heavy effort) or of the full shift (the eight-hour workload), the worker will indicate feelings of extreme fatigue or difficulty in sustaining a specific task for more than a few minutes continuously. If the limiting factor is a static muscle load required for exerting strength or maintaining an awkward posture, the worker will often mention sore muscles, a "sore back," or an inability to sustain a hold or carry as long as desired.

If a muscle group or muscle-joint complex is identified as the limiter of work capacity, the intensity of the effort and the required continuous duration of its application can be determined and compared to recommended guidelines for dynamic and static work. Dynamic work demands, as discussed in Part III and Appendix B, can be studied using physiological and psychophysical measures. These include heart rate and oxygen consumption measurements and ratings of perceived exertion (RPE). Static work requirements can be identified through biomechanical analyses of work postures and strength exertions, timed activity analyses, and motion analyses, all of which are discussed in Part III. The data collected on the job or in a job simulation can then be evaluated against the guidelines given here for dynamic and static work. Although most industrial activities include both static and dynamic work activities, such as holding an object (static) while walking (dynamic), guidelines are given for each type of work separately. (The manual handling guidelines given in Part VI were developed by combining the information about acceptable static and dynamic workloads for specific tasks.)

1. DYNAMIC WORK

Dynamic, or isotonic, muscle work is characterized by muscle contraction alternating with relaxation, and it results in the movement of a body part. For example, the leg and trunk muscles are dynamically active in walking, the finger and arm muscles in typing, and the arm and shoulder muscles in repetitive lifting. Each muscle contraction takes energy and that energy is supplied by way of oxidation of foodstuffs such as glucose (see Part II). To deliver the oxygen to the working muscles, the heart has to increase the blood flow, and this is done by increasing the heart rate and the beating efficiency of the heart. The amount of oxygen delivered to the blood in the lungs is also increased by deeper and faster breathing (see Part II). Consequently, the best way to evaluate the level of dynamic muscle work is to measure the oxygen usage and the heart rate elevations during that work. The methods for doing this are given in Appendix B. Examples of the oxygen consumption demands of several occupational activities are given in Appendix A, and illustrations of how the heart rate data are used to interpret job demands are given in Part III.

a. Duration and Intensity Relationships

The intensity of dynamic work that is acceptable in a job or task varies with the length of time it must be sustained. A workload that must be sustained for a full eight-hour shift, for example, should average not more than 33 percent of the worker's capacity for that type of work (Astrand and Rodahl, 1970). If the hours of work are extended to 10 to 12 hours, the acceptable average oxygen usage requirements must be reduced further (see Figure 15-1). The percent of maximum aerobic work capacity ($\dot{V}_{O_2}$ max) that can be used for much shorter-duration tasks is higher. For one hour of continuous work, the demands can average 50 percent of maximum oxygen usage for the task; for 20 minutes, it can average 70 percent; and for 10 minutes, it can average 85 percent. This relationship between intensity and duration of dynamic work is shown in Figure 11-1.

b. Estimating the Percentage of Maximum Aerobic Capacity Used

To develop guidelines for acceptable dynamic workloads, one has to be able to measure or estimate the type of work being done. Appendix B includes a discussion of methods for measuring these capacities. It indicates that work capacity is directly associated with the number or size of the muscles available to do the work. Whole-body work includes use of most of the leg, trunk, arm, and shoulder muscles, and the $\dot{V}_{O_2}$ max value is high. Work with primarily the upper body muscles, as occurs in many standing, wrapping, or packaging jobs, results in a maximum aerobic capacity of about 70 percent of the whole-body value

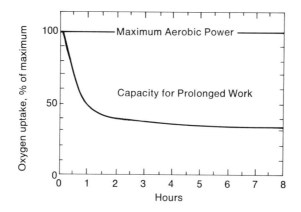

Figure 11-1: Effect of Duration on the Percent of Maximum Aerobic Work Capacity That Can Be Used for Dynamic Work   The percentage of maximum aerobic power (% $\dot{V}_{O_2}$ max) that can be used on the average over different continuous working times (horizontal axis) is shown. The oxygen consumption for the work is expressed as a percentage of maximum values (vertical axis) because absolute levels of aerobic fitness vary greatly within the industrial population. This relationship between work intensity and duration, however, remains quite constant for most industrial workers. A very fit individual can sustain higher workloads, averaging about 50 percent of maximum aerobic work capacity on an eight-hour shift. Few industrial workers meet these levels of fitness, however; a value of 33 percent is found to be a more appropriate upper limit for them. Upper body work and lifting capacities show similar relationships between % $\dot{V}_{O_2}$ max and time of continuous work, but the maximum aerobic capacity will vary with the number of active muscles. *(Adapted from Astrand and Rodahl, 1970).*

(Asmussen ad Hemmingsen, 1958). Because the frequency of lifting, the weight of the object being lifted, and its height above the floor at both the initiation and the completion of a lift all vary substantially, there is a continuum of maximum aerobic capacities for repetitive lifting tasks (Lind and Petrofsky, 1978; Petrofsky and Lind, 1978).

It is impractical to measure the actual maximum aerobic capacities of an industrial work force for all variations of lifting tasks, but one can estimate how closely a given task approaches its maximum aerobic level by looking at heart rate elevations on the job compared to predicted maximum heart rate levels. This is only appropriate for work with large muscle groups that is done in temperate conditions and with little psychic stress. Heart rate is affected by heat, humidity, and emotional and psychic stresses; if these are present in significant amounts, the relationship between heart rate and oxygen consumption will no longer be as linear (Astrand and Rodahl, 1977). Where these other factors are minimally present, however, one can equate the percent of the maximum aerobic work capacity required by a task to the percent of the available heart rate range, defined as the difference between the maximum predicted heart rate (HR max) and the resting heart rate (HR rest) (see Part III), or:

$$\frac{HR_{task} - HR_{rest}}{HR_{max} - HR_{rest}} \cong \frac{\dot{V}_{O2\ task} - \dot{V}_{O2\ rest}}{\dot{V}_{O2\ max} - \dot{V}_{O2\ rest}}$$

(Percent Heart Rate Range)   (Percent Maximum Aerobic Capacity)

The predicted maximum heart rate is roughly calculated as 220 − age in years (Astrand and Christensen, 1964; Cooper, et al., 1975). The relationship between percent of heart rate range and percent of maximum aerobic capacity holds best for workloads of 40 percent of maximum aerobic capacity or more, when virtually all of the increase in blood flow to the muscles is being accomplished by changes in the heart rate (see Part II). It can be used cautiously at workloads between 25 and 40 percent of capacity to estimate a person's approximate maximum aerobic capacity for the work being performed (Rodgers, Eastman Kodak Company, 1972). An example of how oxygen usage and heart rate measures on the job can be related to a predicted maximum oxygen consumption for the dynamic task of interest is provided in the following paragraphs. See Part III as well as the problems in Appendix C for further examples of calculating dynamic work levels and acceptable loads.

A 50-year-old person is working on a lifting task where boxes are taken off of a pallet on the floor and placed on a 75-cm (30-in.) high conveyor. The person's resting heart rate is 70 beats per minute and resting oxygen usage is 3 mL O_2 kg^{-1}min^{-1}. During the lifting period, a heart rate of 120 beats per minute and a $\dot{V}_{O2}$ level of 15 mL O_2 kg^{-1}min^{-1} are measured. Assuming that there are no other environmental factors of significance that will affect heart rate but not oxygen consumption, one can predict the person's maximum aerobic capacity for this lifting task by using the following formula for percent of the maximum heart rate range:

$$\frac{120 - 70}{[(220 - 50) - 70]} = \frac{50}{100}$$

and the following formula for percent of the maximum aerobic capacity:

$$\frac{15 - 3}{(\dot{V}_{O2}\ max - 3)} \ or \ \frac{12}{(\dot{V}_{O2}\ max - 3)}$$

Equating %HR max and %$\dot{V}_{O2}$ max values, one gets:

$$0.5 = \frac{12}{(\dot{V}_{O2}\ max - 3)}$$

or

$$\dot{V}_{O2}\ max = \frac{12 + 3}{.5} = 27 \ mL \ O_2/min/kg \ body \ weight$$

This value can be compared to the industrial population's whole body capacity data (Appendix A) to determine the difficulty of finding people to do the task for given time periods. If the work is primarily done with the upper body, the $\dot{V}o_2$ max value will be about 70 percent of the whole body value. Both the heart rate and oxygen consumption during work will also be lower. This estimation of maximum aerobic capacity using heart rate range should only be used in situations where measurement of oxygen intake on the job or in a capacity test is not possible. Environmental factors that add to job stress may make the heart rate changes greater than those related to the physical work alone.

c. Guidelines for Acceptable Workloads in Dynamic Tasks

It is possible to combine the information about the intensity and duration relationships in dynamic work (shown in Figure 11-1) with information about the maximum aerobic work capacities of industrial employees to define acceptable workloads for dynamic, whole-body tasks. Table 11-1 summarizes these recommendations using two physiological measurements, heart rate and oxygen consumption, as indices of acceptable stress levels. The oxygen consumption values chosen are ones that accommodate at least 50 percent of the female population and most males. This also includes most older workers in the design.

The values given in Table 11-1 are the highest ones for each duration. Although no activity should exceed any of the levels indicated, longer work periods can be kept within the guidelines by interspersing light tasks with the heavier work. Fatigue or rest allowances are also used in industry to reduce the total workload for the shift (Lehmann, 1962). Methods for determining the weighted working average and total workloads for any time period are included in the method for estimating the energy demands of jobs in Part III. They have also been demonstrated in some of the examples of timed activity analyses and in the work and recovery heart rate traces in the same part. The ergonomist must be as concerned with short-duration intense tasks as with the sustained workloads required by the job.

d. Determining Rest or Recovery Times Needed in Dynamic Tasks

There are two primary concerns in evaluating the need for additional rest periods or the introduction of light work tasks in a physically demanding job. The first is the need to curtail the continuous work time on heavy tasks so that lactic acid does not accumulate in the muscles (see Part II). This concern is discussed in more detail later in this section in relation to short-duration heavy work. The second concern is to keep the average workload for the shift within a level that can be replenished through caloric intake by the worker. Using the latter rationale, researchers have suggested that an average caloric expenditure of 15 mL O_2 kg^{-1}min^{-1} (5.2 kcal per minute) is the upper limit for a healthy, 35-year-old male (Bink, 1962; Bonjer, 1962; Lehmann, 1962). To accommodate more women and older workers, a value of 10 mL O_2 kg^{-1}min^{-1} (3.5 kcal per minute) should be used (National Institute for Occupational Safety and Health, 1981).

Table 11-1: Maximum Workloads for Whole-Body Dynamic Work The highest percentage of the maximum aerobic capacity (column 2) recommended for tasks using most of the muscles of the body is shown for four continuous work durations (column 1). This is the same information as is shown graphically in Figure 11-1. In addition, values for the average oxygen consumption are shown in column 3, based on a design guideline to try to accommodate the aerobic capacities of at least half of the female work force (see Appendix A). Assuming that no significant environmental stressors, such as heat, are present, column 4 gives an approximation of the steady state heart rate for the workload indicated, expressed as beats per minute elevation above the resting heart rate. Column 5 relates each level of effort to the classification scheme used in Appendix A. These heart rate and oxygen consumption values are levels that should not be exceeded during work if you wish to reduce accumulated fatigue. *(Developed from information prepared by Rodgers, Eastman Kodak Company, 1976; Rohmert, 1973b).*

| | | Upper Limit Value | | |
Duration	Percent Maximum Aerobic Capacity	Equivalent Oxygen Consumption $(mLO_2 \, kg^{-1} min^{-1})$	Approximate Heart Rate Elevation (beats/min above resting level)*	Equivalent Effort Level (see Appendix A)
8 hours	33	9	+35	Moderate
1 hour	50	13	+55	Heavy
20 min	70	18	+75	Very heavy
5 min	85	22	+90	Extremely heavy

*These elevations are based on a 40-year-old person. Older workers may exhibit lower elevations or may need somewhat reduced workloads.

Figure 11-2 illustrates the relationship between intensity and duration of effort and the rest pauses (in percent of work time) needed to recover from dynamic work. The steps of strain are the equivalent of the light, moderate, heavy, very heavy, and extremely heavy categories presented in the effort equivalency tables in Appendix A.

A person doing a 30-minute handling job that falls into the heavy category (step 3), for instance, would require a 40 percent rest allowance, or 12 minutes, on a light activity to recover from the effort. This effort is obtained by tracing from 3 on the horizontal scale up to the 10-to-30-minute curve; a line parallel to the horizontal axis is drawn to the vertical axis from the latter point to find the percent rest allowance.

A number of industrial tasks require short-duration (less than 15 minutes)

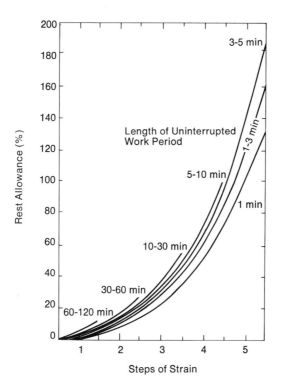

Figure 11-2: Rest Allowance in Dynamic Work   Five levels of physical effort, measured by the amount of oxygen used or the heart rate elevation, are indicated as levels of strain (horizontal axis). They are equivalent to the five effort levels presented in Appendix A. The percentage of the continuous work time that should be designated as compensatory recovery time (the rest allowance) is indicated on the vertical axis. Seven curves, describing the relationships between the rest allowance requirements and the intensity of the work, represent different continuous work periods, from one minute to two hours. These curves are based on the guidelines for acceptable workloads in dynamic work. Additional rest time is very large whenever the oxygen consumption requirements begin to exceed the maximum limits given in Figure 11-1. Rest allowances for one- and three-minute work periods are less than those for three to five minutes of work because less anaerobic metabolism is required for extremely heavy work of less than a few minutes duration (see Figure 2-1). *(Reichsauschuss fur Arbeitstudien, 1972; this material was taken from papers by W. Rohmert, Applied Ergonomics, vol 4, pages 91–95 and 158–162, 1973 and published by Butterworth Scientific Limited, Guildford, Surrey, UK.)*

heavy, very heavy, or extremely heavy effort. These tasks include the charging of smelting furnaces, machine start-ups, emergency recovery, and chemical mixing operations. In many instances, it is possible for the person doing these tasks to pace him- or herself. The guidelines given for work above 50 percent of maximum capacity should be followed, but the recovery time can be shortened considerably if short pauses are taken during the task. The physiological concern in this type of work is that the sudden strain on the heart and the buildup of lactic acid from inadequate blood supply to intensely active muscles will result in rapid fatigue (Hill, Long, and Lupton, 1923/1924; Margaria, Edwards, and Dill, 1933). Proper design of rest pauses can permit recovery of the muscle through oxidation of the lactate or its storage again as glycogen (DeCoster, 1971). Data in Table 11-2 show how introducing short rest breaks into very heavy work results in considerably less lactic acid accumulation, and thus a longer endurance time, than constant work at the same level.

An approach to rest break design that has gained wide acceptance recently is the use of "micropauses" in dynamic work tasks, also called *intermittent work* (Rohmert, 1965; Simonson, 1971b). A micropause lasts 0.5 to 3 seconds and probably helps prevent the development of muscle fatigue by assisting blood flow between powerful contractions and helping to restore the myoglobin and high-energy compound levels in the muscle cells so that more work can be done.

Table 11-2: Intermittent versus Continuous Work at an Extremely Heavy Workload Five patterns of work/rest cycles are shown in columns 1 and 2. The person studied (a marathon bicycler) was pedaling a bicycle ergometer and showed a very high work capacity for the short work period (column 3). His heart rate, in beats per minute (column 4), shows this to be a near-maximal test. Blood lactic acid levels, measured after each pattern of work, are shown in column 5 and are expressed as milligrams per 100 mL of blood. The length of time that the person was willing to sustain the effort is shown in column 6. The ratios of work to rest time are the same for all four intermittent work patterns, but oxygen consumption, heart rate, and lactic acid level varied considerably. The acceptable duration of the work was closely related to the physiological changes, especially the lactate levels. Very high effort levels should be confined to very short durations to avoid high lactate levels. (*Adapted from Astrand, et al., 1960; from E. Simonson, Ch. 18 in* Physiology of Work Capacity and Fatigue, *1971, courtesy of Charles C. Thomas, Publisher, Springfield, Illinois.*)

Minutes		$\dot{V}o_2$ (L/min)	Heart Rate (beats/min)	Oxygen Debt (lactate, mg%)	Subjective Acceptable Duration
Work	Rest				
Continuous	—	4.60	204	150	9 min
3	3	4.66	188	120	15 min
2	2	4.40	178	95	30 min
1	1	2.93	167	45	1 hour
0.5	0.5	2.90	150	20	1 hour

The duration of the work cycle in heavy effort is the primary determinant of the rate of fatigue since interruption of the effort after 10 to 30 seconds often permits restoration of the myoglobin oxygen levels in preparation for the next muscle work. Continuous work by the muscles, even if it does not take place at a very high force, may compromise their blood flow and result in earlier fatigue (Astrand and Rodahl, 1977).

Another type of rest break that is especially recommended for dynamic work tasks is an "active" break, where the worker continues working with muscles other than the ones that have been most heavily loaded (Simonson, 1971a). The work appears to increase the rate of recovery of the fatigued muscles, and the elevated blood flow helps to move the lactate out of the muscle either to be oxidized or to be stored again as glycogen. Use of light work tasks like record keeping alternating with the heavier work should increase personal productivity by reducing the amount of muscle fatigue.

In industrial situations, these data suggest that short, intense work activities should be paced by the individual doing them, not by a machine or by a time standard. Interrupting the work period with short-duration rest breaks will result in more efficient work and reduce the time needed to recover from heavy muscular effort.

2. STATIC WORK

Although dynamic work may lead to fatigue, especially if the person cannot control work/rest cycles, static work frequently results in local muscle fatigue even for short-duration activities. Where minutes and hours were the acceptable time periods for dynamic work, static work durations are measured in seconds and minutes. Table 11-3 illustrates how the duration of the effort is affected as one increases the percentage of maximum strength required to perform a task.

The values given in Table 11-3 are maximum voluntary holding times. The person performing this test experiences muscle pain and discomfort long before stopping, so jobs should not be designed to these limits. As was found for dynamic work, the longer the effort is sustained, the longer it takes to recover. Consequently, making short, intermittent exertions and shifting the load between several muscle groups are strategies used to reduce rapid fatigue in statically loaded postures.

A wide range of muscles of the arms, trunk, and legs are described by these intensity-duration relationships. Because the values are expressed as percentages of maximum strength, sex and age effects are minimal. Table 11-3 indicates the maximum holding or effort times; recovery from these efforts would take many minutes of rest after a short work period. Figure 11-3 presents curves of rest allowances expressed as a percent of the original work time for different percentages of strength and holding times. These data have been used in conjunction with the strength data in Appendix A to develop the guidelines for lifting, control design, and workplace design found in both volumes of this book.

In the design of jobs, reducing the static component of any task can pre-

Table 11-3: Static Work Duration as a Function of Intensity. The maximum amount of time that a muscle can sustain a given percentage of its maximum strength (column 1) is shown in column 2. This describes a percentage of maximum isometric or static force developed and relates it to changes in the amount of blood flowing in and out of the active muscle. There is less blood flow at higher forces and fatigue occurs very rapidly. (Figure 7-5 illustrates this relationship graphically.) These values represent upper limits; job design should ensure that static muscle loading will be minimal and will be of very short duration so that accumulating fatigue will not make muscular work more difficult as the shift continues. *(Adapted from Rohmert, 1973b; Scherrer and Monod, 1960).*

Percent of Maximum Static Strength	Maximum Endurance Time
100	6 seconds
75	21 seconds
50	1 minute
25	3.4 minutes
15	>4 minutes

vent local muscle fatigue from limiting productivity. The following guidelines for workplace and job design have the goal of reducing static effort:

- Avoid placement of displays so that rigid location of the head or eyes is required to monitor the output.
- Avoid reaches or lifts above 127 cm (50 in.).
- Avoid forward reaches more than 50 cm (20 in.) in front of the body when standing or 38 cm (15 in.) in front of the body when sitting.
- Design standing workplaces to avoid stretching or stooping.
- Provide seating or supports for leaning for people who must work on their feet much of the day. Provide adequate foot support at seated workplaces.
- Reduce force requirements on controls that have to be operated rapidly (>10 times per minute) or held for periods in excess of 30 seconds.
- Design foot pedals to reduce the need to keep a constant pressure on the pedal during an assembly operation.
- Provide rest breaks within highly repetitive tasks.
- Provide aids such as carrier bags or carts for carrying tasks taking more than one minute and involving objects weighing more than 7 kg (15 lbm).

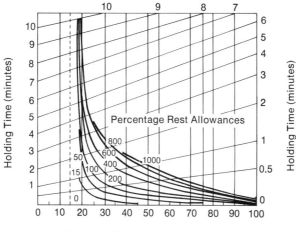

Force Developed (percentage of maximum force)

Figure 11-3: Rest Allowances for Static Effort Activities   The relationship between the exertion of a force for a given length of time (vertical axis) and the percentage of maximum force required by the task (horizontal axis) is shown in terms of the rest allowance needed after isometric work is done. (The rest allowance is expressed as a percentage of the holding time.) The nine curves show the percentage rest allowances needed for several combinations of force and holding time. Actual holding times may be very short (less than three seconds) in handling tasks. Rest allowances that exceed holding times are required only when exertions of more than 33 percent of maximum strength are held for more than six seconds. The lower the percentage of muscle strength used, the less rest allowance is needed. (*This material was adapted from a paper by W. Rohmert, Applied Ergonomics, volume 4, pages 91–95, 1973, and published by Butterworth Scientific Limited, Guildford, Surrey, UK.*)

- Use jigs and fixtures to reduce the requirement for holding in assembly tasks.
- Whenever possible, provide handles or handholds on objects to be lifted or carried.

Further information regarding these guidelines can be found in Chapters II and III, Volume 1, and Part VI of this volume.

Most industrial tasks involve both static and dynamic work. Since static work can limit productivity far more than can dynamic work, it is a good general practice to reduce the static component of the work whenever possible. An example of how job stress can be reduced in this way is a skimming operation in a chemical plant (Brouha, 1967). The operators were skimming the tanks at shoulder level. Their heart rates were at least 150 beats per minute during the task. By installing a work platform to raise them above the tank, job designers were able to reduce the static load on their shoulder muscles. The operators could now skim the tank at waist level, and their heart rates dropped to about 110 beats per minute for the same activity. The task was thereby reduced from a very heavy task to a moderately heavy one.

Similar effects of static effort limiting dynamic work performance can be seen in materials handling tasks. As the weight of an object increases, a greater percentage of maximum strength is needed to handle it. This increases the static component of the lift. At weights above 18 kg (40 lbf), the static component becomes very limiting for a large part of the potential work force. Without adequate handholds, pinching and grasping the corners of cases or boxes may become important limits on the amount of product that can be handled (see Part VI).

3. ENVIRONMENTAL EFFECTS—HEAT AND HUMIDITY

The guidelines for static work design presented in the previous sections assume that the person is working in a temperate environment that is not more than 23°C (75°F) with a relative humidity below 60 percent. Heat and humidity affect acceptable workload levels by demanding blood flow to the skin to get rid of accumulated body heat. The skin blood flow is no longer available to the muscles, so the work capacity is effectively reduced. For further discussion of temperature and humidity effects, see Chapter V, Volume 1.

When heat and humidity are present, heart rate elevations give a better indication of the stress on a worker than energy expenditure values do. The heart rate elevation values given in Table 11-1 can be used to estimate the acceptable duration of a given effort. At heavy workloads, these levels can increase very rapidly with time. In choosing rest cycles, one should try to minimize the upward trend of the heart rate during the activity. The guidelines in Table 11-4 for heat exposure have been derived from laboratory studies and from observations of work in hot environments. Humidity interacts synergistically with increasing temperature; if relative humidities are above 60 percent, the durations of acceptable continuous exposure will be even less than those shown in Table 11-4.

Since the individual doing the task is most aware of the combined heat and humidity stresses, he or she should be encouraged to pace the work to keep it within acceptable levels. A rough guideline to assist in planning the amount of rest is that each rise of one degree Celsius (1.8 degrees Fahrenheit) in ambient temperature above 25°C (77°F) is equivalent to a 1 percent increase in the percent of maximum effort on the task (Kamon, 1980). Thus, a task requiring 50 percent of maximum aerobic capacity performed at 46°C (111°F) would be roughly equivalent to a 70 percent maximum aerobic capacity task performed at 25°C (77°F).

Table 11-4: Effect of Temperature on Time for Work at Different Effort Levels. The duration of continuous exposure to six different heat conditions, in °C and °F (columns 1 and 2), and at a relative humidity at or below 50 percent, is shown for five levels of effort (columns 3 through 7). These are suggested upper limits of continuous exposure. They are approximations because the air velocity, clothing worn, and options for intermittent versus sustained work will all influence the interaction between temperature and workload. This interaction argues against doing heavy work at high temperatures because of the effect on body temperature levels. See Chapter V, Volume 1, for more discussion of temperature and humidity stress during work. (*Developed from information in Kamon, 1975; Lind, 1963*).

°C	°F	Duration of Continuous Exposure per Shift				
		Light	Moderate	Heavy	Very Heavy	Extremely Heavy
<24	<75	8 hrs	8 hrs	≈ 1 hr	≈ 20 min	≈ 20 min for especially selected people
24–29.5	75–85	8 hrs	6 hrs	≈ 1 hr	≈ 10 min	≈ 10 min for especially selected people
30–35	86–95	≈ 8 hrs	≈ 2 hrs	≈ 45 min	≈ 5 min	≈ 5 min for especially selected people
35.5–40.5	96–105	4 hrs	≈ 1 hr	≈ 20 min	Not recommended	Not recommended
41–46	106–115	≈ 1 hr	≈ 40 min	≈ 10 min	Not recommended	Not recommended
46.5–51.5	116–125	≈ 20 min	≈ 10 min	Not recommended	Not recommended	Not recommended

B. LIGHT EFFORT AND PERCEPTUAL WORK

Unlike the physiological fatigue of muscular work, feelings of fatigue in light effort and perceptual work are difficult to quantify. Jobs like writing and problem solving in maintenance and engineering work and some assembly tasks involve light physical work and a degree of mental effort. Extended monitoring of machines, product inspection, and fine visual control of assembly tasks are examples of perceptually demanding jobs. Subjective tiredness varies considerably among individuals doing the same task and within an individual depending on other psychosocial circumstances. In addition, it is highly task dependent (Murrell, 1965). Some of the signs of improperly designed light effort and perceptual work are discussed below with guidelines for job design to minimize them. Highly repetitive jobs are covered in Chapter 13.

1. PERFORMANCE EFFECTS

Performance on mentally or perceptually demanding tasks can improve or deteriorate over time (Floyd and Welford, 1953; Murrell, 1965). The former situation is termed a "warm-up" effect; the latter has been called "fatigue," but it is better defined as a decrement in performance. Improved performance is usually seen in the first hour of doing a task. The operator is fully skilled to the job but takes a little time to establish a working rhythm or reestablish a coordinated pattern of movement at the start of a shift. When a person is fully trained, these "warm-up" effects are usually apparent only after he or she has been away from the job for more than a few hours.

There are three types of performance decrement of concern in perceptually and mentally demanding work:

- Simple performance decrement.
- Disorganization of performance.
- Deterioration of performance.

Simple decrements in performance occur in very repetitive tasks where one aspect of performance can be altered without affecting others. The decrement may be in the rate of doing the work, the accuracy with which it is done, or the amount of time spent in activities that do not contribute to the productivity measure. It is unusual for the amount of time per cycle to be increased; the productivity reduction per hour is more often the result of increased time spent on auxiliary tasks (Graf, 1959).

Skilled performance may become disorganized in less repetitive tasks where complex work is done. The operator's control of performance on such tasks may be affected over time and result in errors or less coordinated actions. Effects on short-term memory can produce inefficient work patterns, for example, in information retrieval tasks such as invoicing or order filling.

Deterioration of performance can occur with time when continuous work is done in the presence of suboptimal external factors. The performance may be reduced by excessive noise, heat, or uncomfortable working postures. Some of these concerns have been addressed in Chapters II and V, Volume 1. In these situations, performance errors will often be detected by the worker and will increase the stress. A common behavior is to change the method of doing the task, particularly if one of the external factors is postural or small muscle discomfort. This produces a temporary improvement in performance, but it does not reduce the basic problem of job or workplace design contributing to reduced productivity over the shift.

The difficulty of identifying the cause of an observed performance decrement is illustrated by the symptom of "visual fatigue." People working in mail sorting, at CRT terminals, and in inspection tasks, for instance, may complain of visual fatigue and indicate that their eyes hurt, their vision is blurred, or that they find it difficult to focus on their work near the end of the shift (Smith, Hurrell, and Murphy, 1981; Weston, 1953). As a result of this discomfort, productivity per hour may fall as the shift progresses. Among the factors contributing to this performance decrement in a study of postal workers (Hurrell and Smith, 1981) were:

- The visual characteristics of the worker—that is, near- or farsightedness, aging effects on the eye, astigmatism, and eye muscle imbalance.

- The worker's constitution and health status, which could include poor nutrition, inadequate sleep, or medications.

- Awkward postural requirements of the visual task.

- Poor quality or inadequate quantity of ambient lighting, such as glare and shadows.

- Small size of objects being viewed.

- Object movement.

- Poor legibility or definition of objects to be read or seen (for example, carbon copies).

Attention to many of these factors may be necessary to reverse a performance decrement in a visually demanding task. Further discussion of visual demands and ambient illumination can be found in Chapters III, IV, and V, Volume 1.

2. GUIDELINES TO REDUCE PERFORMANCE DECREMENTS IN LIGHT EFFORT AND PERCEPTUAL WORK

Performance decrements may be associated with sensory system "fatigue," as in visual fatigue, or with motivational problems, like boredom. Assembly cycle-time consistency helps to indicate which factors are contributing to the performance problem. Fatigue usually shows decreased output per hour over time and exhibits fairly low variability in cycle time; boredom usually exhibits highly variable cycle time and more variable output per hour (see Figure 12-1). External incentives or pacing may improve the performance of a person who is "bored" but may be unable to stimulate the fatigued individual for more than very short periods (Murrell, 1965; Wyatt and Fraser, 1929). The principles of job design given here should be suitable for situations in which either fatigue or boredom is a problem. The two primary principles of job design for perceptual and light effort tasks are:

- Vary the tasks.
- Make the task easier to do.

a. Varying the Tasks

This principle is important in any job design but particularly so in perceptually demanding, light work tasks. Job enrichment and job enlargement are two job design techniques that employ this principle. A study of task variation is summarized in Table 11-5. Three subjects were given three tasks to perform either in continuous 2.5-hour periods or alternating for 50 minutes each. Two involved mental arithmetic and reasoning; the other was a simulation of muscular work. Alternating the tasks every 50 minutes resulted in improved performance on each task compared to the continuous sessions.

Introducing short rest pauses into the workday may actually increase the output per shift on light assembly tasks. Figure 11-4 shows the average seconds per cycle for a coil-winding job in an electrical assembly plant over eight hours. The assembler was given two work patterns: continuous work with only a lunch break, and five minutes of rest for each 55 minutes of work. The average time to complete one coil-winding cycle was measured throughout the shift; the higher the value, the less productivity per hour. Time spent on secondary work, such as procuring supplies and inspecting parts, and on arbitrary work breaks (ones not scheduled but devised by the assembler to reduce the task monotony) was also monitored.

The curves show that increasing the scheduled rest periods resulted in more consistent performance over the shift. The major changes were in the time spent on secondary tasks and arbitrary work breaks, both of which decreased when breaks were scheduled each hour. This resulted in a small increase in productivity over the shift, showing that increased rest is not necessarily associated with lost productivity in light assembly tasks. Earlier in this chapter it was mentioned that, in physically demanding jobs, rest breaks can be periods of lighter activity, not just full cessation of work. In light assembly tasks, rest breaks can be periods of heavier activity, such as procuring supplies or disposing of the product. Any change from the activity of the primary task qualifies as a rest break. It is important to identify tasks that are different enough to provide variety. The salutary effect of a totally different activity, even of short duration, is more desirable than rotating between tasks that are basically similar in their perceptual demands.

Table 11-5: Work Improvement with Task Variation The effects of task variation on the amount of work done and the number of errors made are shown for two people on two perceptually and one physically demanding tasks. The work done (column 3) is expressed as the percent improvement when the tasks are varied compared to the performance on a single task. The errors (column 4) are expressed as the percent reduction in the numbers of errors compared to the rates on the single task alone. *(Adapted from Murrell, 1965; Wyatt, 1924).*

| Task | Subject | Amount of Improvement Compared to Continuous Work Schedule | |
		Work Done, Percent Improvement	Errors, Percent Reduction
Adding	A	+24.2	26.1
	B	+ 8.8	55.1
	C	+10.6	25.1
Mechanical	A	+12.7	43.4
calculator	B	+ 4.2	9.2
	C	+18.2	27.9
Muscular	A	+ 7.8	—
	B	+ 2.4	—
	C	+ 5.1	—

The following points should be considered when organizing the tasks within a job:

- Alternate physically-demanding tasks with perceptually demanding tasks where possible; for example, allow handlers to spend time on lighter tasks like order checking.

- Alternate highly demanding perceptual tasks with ones having lower demands.

- Alternate long-cycle tasks with shorter cycle ones, even for short periods.

- Provide for frequent changes of posture, at least once per hour. Do not lock the worker into the workplace by having someone else move all parts and product into and out of the workplace; allow the seated operator to stand while getting supplies and doing paper work.

- Provide a few minutes of relief every half-hour for people doing continuous monitoring tasks such as inspecting a moving product. Performance on light tasks can probably be influenced by the phases of ultradian cycles, function rhythms with a cycle time of about 90 minutes (see Part II).

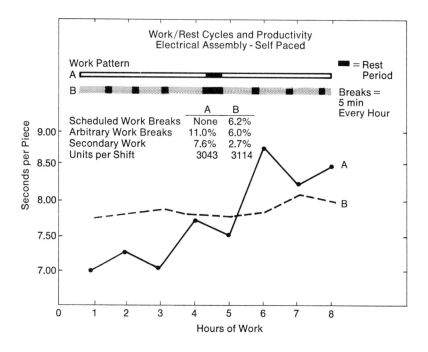

Figure 11-4: Effect of Work/Rest Cycle Patterns on Productivity in a Light Assembly Task A study of the average time, in seconds, to wind a coil in an electrical assembly operation (vertical axis) over an eight-hour shift (horizontal axis) is shown. Curve A represents a work pattern that has no scheduled work breaks except for a midday meal. Curve B represents the same operator's performance when a five-minute break was scheduled every hour. The work patterns are depicted near the top of the graph. An analysis of the productivity and distribution of official and unofficial work breaks for each work pattern also appears and shows that the number of units produced is higher when the extra work breaks are given. The change can be explained by less arbitrary break and secondary work time when the multiple small breaks are present. The unofficial breaks represent ways for a worker to vary the tasks in a perceptually demanding task, but they do not contribute to productivity ratings. (*Adapted from Graf, 1959; Lehmann, 1962.*)

b. Making the Tasks Easier to Do

For perceptual work, such as inspection and machine monitoring, the effort required is related to the ease of detecting a defect or changed machine status relative to other activity in the operator's environment. Information about ways to improve the visualization of defects or the identification of an auditory signal is found in Chapter V, Volume 1. Ways to reduce the demands of these jobs through coding and display are given in Chapter IV, Volume 1. The following general principles should be considered when organizing work patterns and designing equipment and workplaces:

- Increase the intensity of the defect; use color, shape, or special marking codes to set it off from the background (Murrell, 1965).
- Provide redundancy in alarms so that more than one sense is involved.
- When feasible, provide rapid feedback to the operator about performance. If a defect is detected later in the system, the operator who missed it should be informed.
- Where practical, provide the operators with visual aids for comparative judgments in visual tasks.
- Keep the number of different defects to be detected low.
- Reduce environmental fatigue enhancers, such as:
  - machine pacing
  - highly repetitive, monotonous tasks
  - isolation
  - awkward posture requirements
  - high heat and humidity
  - high noise levels
  - glare
  - nonadjustable workplaces

Making the task easier may compound the problem of monotony on some monitoring tasks. Improving the visualization of a defect may be effective on an inspection task where defects occur frequently, but it may not improve performance on a task where defects are rare and other factors make a larger contribution to the performance decrement (Murphy, Eastman Kodak Company, 1974). Effective designs should consider both of these problems and find compromises that provide the best working conditions.

REFERENCES

Asmussen, E., and I. Hemmingsen. 1958. "Determination of Maximum Working Capacity at Different Ages in Work with Legs or with the Arms." *Scandinavian Journal of Clinical Laboratory Investigation*, 10: pp. 67-71.

Astrand, I., P-O. Astrand, E. H. Christensen, and R. Hedman. 1960. "Intermittent Muscular Work." *Acta Physiologica Scandinavica*, 48: pp. 448-453.

Astrand, P-O., and E. H. Christensen. 1964. "Aerobic Work Capacity." In *Oxygen in the Animal Organism*, edited by F. Dickens and E. Neil. New York: Pergamon Press, pp. 295-314.

Astrand, P-O., and K. Rodahl. 1970. *Textbook of Work Physiology*. 1st ed. New York: McGraw-Hill, p. 292.

Astrand, P-O., and K. Rodahl. 1977. *Textbook of Work Physiology.* 2nd ed. New York: McGraw-Hill, 681 pages.

Bink, B. 1962. "The Physical Working Capacity in Relation to Working Time and Age." *Ergonomics, 5 (1):* pp. 25-28.

Blum, M. L., and J. C. Naylor. 1968. *Industrial Psychology—Its Theoretical and Social Foundations.* Rev. ed. New York: Harper and Row, pp. 543-546.

Bonjer, F. H. 1962. "Actual Energy Expenditure in Relation to the Physical Working Capacity." *Ergonomics, 5 (1):* pp. 29-31.

Brouha, L. 1967. *Physiology in Industry.* 2nd ed. New York: Pergamon Press, p. 126.

Cooper, K. H., M. L. Pollock, R. P. Martin, S. R. White, A. C. Linnerud, and A. Jackson. 1975. "Physical Fitness Levels versus Selected Coronary Risk Factors." *Journal of the American Medical Association, 236 (2):* pp. 166-169.

DeCoster, A. 1971. "Present Concepts of the Relationship Between Lactate and Oxygen Debt." Chapter 9 in *Frontiers of Fitness,* compiled and edited by R. J. Shephard. Springfield, Ill.: Thomas, pp. 174-191.

Floyd, W. F., and A. T. Welford, ed. 1953. *Fatigue.* A symposium held at the College of Aeronautics at a meeting of the Ergonomics Research Society, March 1952, Cranfield, England. London: Lewis, 196 pages.

Friedlander, F. 1964. "Job Characteristics as Satisfiers and Dissatisfiers." *Journal of Applied Psychology, 48 (6):* pp. 388-392. Cited in Blum and Naylor, 1968.

Graf, O. 1959. *Arbeitszeit und Produktivitat. Band 2: Garztagige Arbeitsablauf Untersuchungen an 200 Arbeitzplatzen.* Berlin. Cited in Lehmann, 1962.

Herzberg, F., B. Mausner, and B. B. Snyderman. 1959. *The Motivation to Work.* 2nd ed. New York: Wiley, 157 pages. Cited in Blum and Naylor, 1968.

Hill, A. V., C. N. H. Long, and H. Lupton. 1923/1924. "Muscular Exercise, Lactic Acid, and the Supply and Utilization of Oxygen." *Proceedings of the Royal Society of London, 96:* pp. 438-475. *Proceedings of the Royal Society of London, 97:* pp. 84-138. Cited in Simonson, 1971a.

Hurrell, J. J., and M. J. Smith. 1981. *Source of Stress Among Machine-Paced Letter-Sorting Machine Operators.* Department of Health and Human Services/National Institutes for Occupational Safety and Health report. Cincinnati: NIOSH. Cited in Smith, Hurrell, and Murphy, 1981.

Kamon, E. 1975. "The Ergonomics of Heat and Cold." *Texas Reports on Biology and Medicine, 33 (1):* pp. 145-182.

Lehmann, G. 1962. *Praktische Arbeitsphysiologie.* 2nd ed. Stuttgart: George Thieme Verlag, pp. 51-72.

Lind, A. R. 1963. "A Physiological Criterion for Setting Thermal Environmental Limits for Everyday Work." *Journal of Applied Physiology, 18 (1):* pp. 51-56.

Lind, A. R., and J. S. Petrofsky. 1978. "Cardiovascular and Respiratory Limitations on Muscular Fatigue During Lifting Tasks." In *Safety in Manual Materials Handling,* edited by C. G. Drury. Report on the international symposium held by the Department of Health, Education, and Welfare/National Institute for Occupational Safety and Health, July 18-20, 1976, Buffalo, New York. Washington, D.C.: U.S. Government Printing Office, pp. 57-62.

Margaria, R., H. T. Edwards, and D. B. Dill. 1933. "The Possible Mechanism of Contracting and Paying the Oxygen Debt and the Role of Lactic Acid in Muscular Contraction." *American Journal of Physiology, 106 (6):* pp. 689-715.

Maslow, A. H. 1943. "A Theory of Human Motivation." *Psychological Review, 50:* pp. 370-396. Cited in Blum and Naylor, 1968.

Murrell, K. F. H. 1965. *Human Performance in Industry.* New York: Reinhold, pp. 381-408.

National Institute for Occupational Safety and Health (NIOSH). 1981. *Work Practices Guide for Manual Lifting.* Department of Health and Human Services/National Institute for Occupational Safety and Health, No. 81-122. Washington, D.C.: U.S. Government Printing Office, 183 pages.

Petrofsky, J. S., and A. R. Lind. 1978. "Metabolic, Cardiovascular, and Respiratory Factors in the Development of Fatigue in Lifting Tasks." *Journal of Applied Physiology (Respiration, Environmental, and Exercise Physiology), 45 (1):* pp. 64-68.

Reichsauschuss fur Arbeitstudien (REFA). 1972. *Methodenlehre des Arbeitstudiums. Teil 2: Daterermittlung.* Munich, Federal Republic of Germany: Carl Henser Verlag. Cited in Rohmert, 1973b.

Rohmert, W. 1965. "Physiologische Grundlager des Erholungszeitbestimmung." *Zeitblatt Arbeit Wissenschaft, 19:* p. 1. Cited in Simonson, 1971b.

Rohmert, W. 1973a. "Problems in Determining Rest Allowances. Part 1: Use of Modern Methods to Evaluate Stress and Strain in Static Muscular Work." *Applied Ergonomics, 4 (2):* pp. 91-95.

Rohmert, W. 1973b. "Problems in Determining Rest Allowances. Part 2: Determining Rest Allowance in Different Human Tasks." *Applied Ergonomics, 4 (2):* pp. 158-162.

Scherrer, J., and H. Monod. 1960. "Le travail musculaire local et la fatigue chez l'homme." *Journal de Physiologie (Paris), 52:* pp. 419-501.

Sen, T. K., S. Pruzansky, and J. D. Carroll. 1981. "Relationship of Perceived Stress to Job Satisfaction." In *Machine Pacing and Occupational Stress,* edited by G. Salvendy and M. J. Smith. London: Taylor & Francis, pp. 65-72.

Simonson, E., comp. and ed. 1971a. *Physiology of Work Capacity and Fatigue.* Springfield, Ill.: Thomas, 571 pages.

Simonson, E. 1971b. "Recovery and Fatigue. Significance of Recovery Processes for Work Performance." Chapter 18 in *Physiology of Work Capacity and Fatigue,* compiled and edited by E. Simonson. Springfield, Ill.: 571 pages.

Smith, M. J., J. J. Hurrell, and R. K. Murphy, Jr. 1981. "Stress and Health Effects in Paced and Unpaced Work." In *Machine Pacing and Occupational Health,* edited by G. Salvendy and M. J. Smith. London: Taylor & Francis, pp. 261-267.

Weston, H. C. 1953. "Visual Fatigue—with Special Reference to Lighting." Chapter 12 in *Fatigue,* edited by W. F. Floyd and A. T. Welford. London: Lewis, pp. 117-135.

Wyatt, S. 1924. *On the Extent and Effect of Variety in Repetitive Work. Part B: The Effect of Changes in Activity.* IFRB Report No. 26. London: His Majesty's Stationery Office. Cited in Murrell, 1965.

Wyatt, S., and J. A. Fraser. 1929. *The Effects of Monotony in Work.* IFRB Report No. 56. London: His Majesty's Stationery Office. Cited in Murrell, 1965.

Yoshitake, H. 1971. "Relations Between the Symptoms and Feeling of Fatigue." *Ergonomics, 14 (1):* pp. 175-186.

CHAPTER 12

Work Pace

CHAPTER 12. WORK PACE

Increased use of automation to perform highly repetitive assembly tasks has produced an increase in machine pacing for workers who perform the more complex tasks on an assembly line. Conveyors often link the work stations and move the product from machine to person to machine. Because machine rates are constant, the line rate is either tied to the slowest of them or is limited by the assembly worker's time requirements. This can result in limiting the productivity of faster workers by not delivering parts fast enough for their operation. It can also produce time stress on the slower workers who have parts coming in constantly and may not be able to keep the next station supplied fast enough to meet the line demands. Line balancing will probably require either selection of those people who can perform at higher levels or slowing down the faster operations. Either approach can be costly in terms of productivity and training time.

A. VARIABILITY IN WORK RATE

As traditional manual industrial tasks are taken over by machinery, work pace becomes a more critical part of job design. People who formerly paced their own work according to pay incentives or individual motivation may now be part of a line- or machine-paced operation. Differences between individuals in performing a task must be recognized when setting conveyor speed in a sequential assembly operation. In addition, an individual will vary the speed with which a specific task is done, so conveyor speed should be selected to allow for such variability (Belbin and Stammers, 1972; DeJong, 1962). Figure 12-1 illustrates the variability in two operator's times for performing a light assembly task. It illustrates that the amount of variability in performance is influenced by one's attitude towards it as well as by the duration of performing it.

If conveyor speed is set to accommodate the median performance of assembly workers, one-half of the people will be pressed to work above their average speeds and one-half will be bored or frustrated because the speed holds them back. If the speed is set to accommodate most people's capabilities, 80 to 90 percent of the operators will be working at less than their optimal level. If speed is set to get maximum productivity, most people will be working above their capabilities, and stress and quality problems will be more likely to occur. Designing conveyor systems so they permit some people to work ahead and some to work behind the conveyor's set pace is a way of recognizing individual performance differences as well as each individual's need to vary the pace.

Task criticality is an important modifier of pace. The more critical the task, the more control the operator should have over work pace (Conrad, 1954). Dissatisfaction with externally paced work, where a machine or someone other than the individual sets the speed, is often related to the operator's concern that he or she will not have time to do the work properly. The more critical the task, the more concern and frustration if the work pace is above the individual's optimal rate.

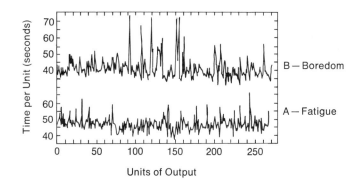

Figure 12-1: Variability in Time to Perform a Self-Paced Assembly Task   The time to complete a single unit in a small-parts assembly task is shown for 260 units during a working time of about three hours. Although the average time to complete an assembly is about 40 to 45 seconds in the upper trace (B) and 45 to 50 seconds in the lower trace (A), the range of times is from 32 to 72 seconds for the top trace and from 38 to 67 seconds for the lower trace. The upper trace is from data gathered on an operator who felt "bored"; it is characterized by wide fluctuations in assembly time and by frequent distractions. The lower trace is from data gathered on a person who felt "fatigued" on the job; it is characterized by more frequent, longer assembly cycles as the shift progresses, but the range of assembly times is not as wide as for the "bored" assembler. The amount of variability in assembly times explains why an operator can consider a fixed machine pace stressful. *(Adapted from Wyatt and Fraser, 1929).*

The ability to work ahead and then "coast" for a while to vary the task and make it more interesting becomes limited as conveyor speed is increased. In a "pulse conveyor" assembly line that was paced for the most skilled operators, there were complaints from operators that they "didn't have time to blow their noses" (Rodgers, Eastman Kodak Company, 1978). Although skilled assemblers were on the line and each person could work on up to three stations in the conveyor without moving from the workplace, there was not enough flexibility in work rate available to them to accommodate unforeseen events. In one three-minute observation period when the operators were not interrupted but two people stood two meters (six feet) from the line and watched them at work, the distraction of having people in the area resulted in all of the operators slipping behind in the assembly tasks. Most had been working ahead of themselves prior to the visit. When a line is paced with little tolerance of environmental change, quality problems are more likely to occur. In addition, this pace stress can make it difficult to find people who can sustain performance for a full shift.

B. MACHINE PACE AND PRODUCTIVITY—A CASE STUDY

One way to increase overall productivity in a high-volume operation is to reduce the amount of time required to produce a unit or to increase the number of units produced per hour. A logical way to do this is to increase the speed of a machine. When evaluating the impact of increasing machine speed, engineers usually direct attention to the ability of the machine to maintain product quality standards and to perform for long hours at the higher rate. The machine operator is expected to be able to adjust to the new situation and is often not thought of as a possible limiter of machine speed changes. A study that evaluated the potential impact of increased machine speed on operator productivity and job stress illustrates that increased speed does not necessarily result in increased productivity, especially when multiple short-cycle tasks are done simultaneously (Rodgers and Mui, 1981).

1. THE JOB

The operator's job was to load, unload, adjust, and monitor the operation of a machine producing small film packets. The object was to keep the machine running and turning out a high-quality product; the latter was determined by quality or "Go/No Go" inspection on the outcoming packs of 150 pieces each. Machine jams were cleared by the operator whenever possible.

Engineers had developed ways of increasing the machine speed from 580 pieces per minute to 720 pieces per minute as new machines were added to the production department. They had found ways to increase the speed to 1,000 pieces per minute, but the machine operators indicated that it would not be possible to perform their job at that rate. There had already been a tendency for less-skilled operators to find the machines runnings at 720 pieces per minute too difficult. This perceived stress could not be explained by increased physical workload.

2. THE ANALYSIS OF JOB REQUIREMENTS

Two types of measurements were made on four skilled machine operators, each of them being studied at the 580 and 720 pieces per minute speeds. Heart rate was used as a measure of the physiological stress and a detailed timed activity analysis measured the pace pressure and actual work demands (see Part III). The percentage of time spent on each of the major activities was identified for each speed and estimated from those data for the 1000 per minute speed. A small fraction of the timed activity analysis diary is shown in Table 6-3. The percent of shift time spent on each of the major activities is shown in Table 12-1. The "Other" category includes meetings, communications, paper work outside of that done for machine control purposes, and training programs in which each operator is expected to participate.

It is clear from the activity time distribution that increased speed requires increased machine loading and, proportionately, more jam clearance. The

Table 12-1: Percentage of Time Spent in Different Activities as a Function of Machine Speed. The percentage of total shift time spent on the machine operation tasks of clearing jams, loading and preparing supplies, and inspecting, as well as standing by and performing administrative and record-keeping tasks ("Other"), is shown for two measured machine speeds (columns 2 and 3) and one estimated speed (column 4). At increased machine speeds, the tasks of jam clearing and loading and preparing supplies increase in proportion to machine speed. All other activity times decrease in proportion to the speed change. The available inspection time does not keep pace with the numbers of units being produced, especially at the highest machine speed. The impact of these speed increases on time per pack for inspection is shown in Table 12-2. *(Rodgers and Mui, 1981).*

Activity	Percent of Time over a Shift		
	580 units/min	720 units/min	Estimated 1,000 units/min
Jam clearance	9.8	11.3	15.7
Load and prepare supplies	24.6	30.6	40.4
Standby	6.7	5.6	4.4
Inspection	21.6	23.4	15.0
Other	37.3	29.1	24.5

amount of product to be inspected increases as machine speed increases, but the amount of time available for inspection decreases compared to the time available at 580 pieces per minute, especially at the 1,000 per minute rate. Heart rates did not indicate significant stress levels. The only sign of increased stress that could be attributed directly to machine operation was an elevation in heart rate of 25 beats per minute associated with 12 machine jams in 25 minutes for one operator (see Figure 9–3). Since the final heart rate was 120 beats per minute, this did not indicate a serious problem; the rate had recovered to its previous level within a minute after the problems were resolved.

The inspection task was identified as the only activity over which the operator has some discretion or control. It was usually worked into the periods between loading and jam clearance and seldom could be sustained for more than one or two minutes at a time. The availability of an outgoing conveyor on which the packs could be temporarily accumulated gave the operator some options for when inspection had to be done. At the slower machine rates there was about five minutes of leeway to allow the conveyor to build up product before it had to be inspected. This dropped to about three minutes at the 1,000 per minute rate. The operator had to keep track of the outgoing conveyor's volume constantly because the production machine would stop if the conveyor filled up.

3. DETERMINATION OF THE TIME STRESS ON THE OPERATOR

Table 12-1 showed that the amount of time for quality inspection of the product was reduced in relation to the number of pieces produced as machine speed increased. When this factor was analyzed in terms of the amount of time available to inspect one pack (150 pieces), the true impact of the increasing machine speed on the operator's performance became clear (Table 12-2). Some people who had found 580 pieces per minute an acceptable work level felt stressed at 720 pieces per minute. The perceived stress was probably related to their concern about being able to do quality inspection in less time. At 1,000 pieces per minute it is clearly impossible to inspect the product properly, especially when a critical decision has to be made as to whether to reinspect it or to send it directly out to the customer.

An analysis of quality control data for the periods in which the ergonomics studies were done showed how the operator coped with increasing machine speed. It was noted that the percent of product rejected and sent back through a 100 percent reinspection process stayed fairly constant as machine speed increased, ranging from 10 to 12 percent of the machine's output. It was also noted that only about two percent of the product was ultimately rejected after the reinspection process, and only 0.2 percent of the pieces had ''critical'' defects,

Table 12-2: Impact of Increased Machine Speed on Time for Inspecting One Pack of Product. The time available for a machine operator to inspect one pack of 150 pieces of product (column 2) is shown as a function of machine production speed (column 1). At 580 units per minute, the operator had 7.3 seconds to do a quality inspection; as machine speed increased, the inspection time became far too short to do this task. At the highest speed, inspection time per pack was estimated from the time left after the loading, jam clearance, and other tasks were done. *(Rodgers and Mui, 1981).*

Machine Speed (units/min)	Time Available to Inspect Pack (sec)
580	7.3
720	4.8
1000	1.1*

*Estimated

ones that would influence product performance significantly. The machine operators, all of whom enjoyed the challenge of the job and performed at a high level, acknowledged that they chose to err on the side of rejecting too much rather than too little. If there was any question about product quality level during inspection of the pack, the reinspection route was chosen. With increased machine speed, they continued to err on the side of rejecting too many, and the amount of product rejected was slightly higher proportionately than at the slower rate.

4. AN ERGONOMIC APPROACH TO REDUCING THE MACHINE-PACING PRESSURE

In this example, the increased machine speed did not result in the productivity gains anticipated because the operator's ability to do the inspection task was affected by the increased amount of time spent keeping the machine supplied with rolls of plastic, film, and foil, and with cardboard pieces. Two approaches to reducing the machine pacing were identified: redesign the supply end of the machine to allow larger supply rolls to be used, thereby reducing the amount of time spent loading at the higher machine speeds; and provide for quality inspection at another station, asking the operator to do only that inspection necessary to ensure that the machine is operating properly. The former approach was adopted for design of future machines, as was a recommendation by the operators to reduce machine length and to put it in a horseshoe configuration. This configuration reduced the travel time between the supply and the outgoing conveyor ends of the machines, thereby freeing more time for inspection. When another station for quality inspection was provided, the amount of product that needed to be reinspected was cut markedly, increasing productivity for the department as a whole.

C. JOB PACING AND STRESS

In addition to machine-pacing pressure in occupational tasks, there can be external pacing pressure, from service demands, fellow workers, supervision, or an individual's need to earn the maximum pay allowance in a job that is on an incentive compensation system. Computerized control systems also pace an operator by making the retrieval of information more rapid and the opportunity for work breaks to get that information more limited.

The impact of these types of time pressure on an individual worker vary greatly and are related to the worker's capacity for information handling and/or physical work and his or her tolerance of sustained performance at these pace levels. Personality variables, such as one's ability to get along with others, degree of tolerance of change or unpredictability in the job situation, interest in the job's content, and motivation to perform quality work, will influence how an individual responds to a job situation where these types of pace pressure exist.

The degree of accountability the person has on the job, its complexity, the potential impacts of other people's work on his or her job performance and how much control can be exerted over that work, and the degree of interference by supervision or other people not directly associated with the line manufacturing activity will also influence the responses of operators to time pressure on the job. An ambulance or fire dispatcher, for example, has large accountability on the job because of the potential impacts of delayed response in these emergency situations. Time pressure can be very severe at times when multiple calls are coming into the dispatch board, yet only a fraction of them are of an emergency nature. A half-hour period of very heavy call activity may be followed by several minutes of quiet, but the dispatcher always has to be alert to the possibility that an emergency call will come in. In addition to answering the calls, the dispatcher must keep track of the ambulances or fire trucks and their crews so that they can respond to the emergency as rapidly as possible.

The need to act rapidly in identifying and responding to an emergency can also be seen in production operations where a machine or process appears to be moving out of its specified safe range of operation. If the process is computer-controlled, as it is in many manufacturing and utilities operations, the operator must respond quickly to large amounts of information made available on the video display terminal (VDT) and must take an action that will halt the undesired event. Making the proper decision under time pressure relies on thorough training of the operator and well-presented information on the VDT. The less clearly that information is presented, the more stress there will be on the operator in the emergency situation.

Although individuals vary greatly in their responses to time stress on the job, several studies have shown that long-term physiological changes can result from highly paced jobs. Figure 12-2 shows urine catecholamine levels for people in different occupations during their work periods (O'Hanlon, 1981). The degree of stress felt by an operator on a job is thought to be reflected by the levels of catecholamines in the blood and urine. Epinephrine levels appear to fluctuate more with time pressure, while norepinephrine levels are elevated in the presence of constraints in the job and increased worker frustration or irritation levels (Johansson, 1981; Johansson, Aronsson, and Lindstrom, 1978). The relationship of elevated catecholamine levels to long-term health is not always predictable for levels seen in most occupational tasks, but consistently high levels of these hormones are not desirable (Selye, 1976).

Time pressure can affect the measured anxiety levels for people who work at a job for extended periods (Broadbent and Gath, 1981). In a study of paced versus unpaced assembly workers, little difference in job satisfaction was noted, yet a twofold difference in anxiety rating was observed as measured by interviews with the workers. Those workers with more meticulous, less relaxed, personalities were most affected by paced work and exhibited the greatest anxiety. This anxiety was sufficient to increase the risk of some of the paced workers for anxiety neuroses and the need for subsequent psychiatric help.

Time pressure from externally paced work is also associated with increased

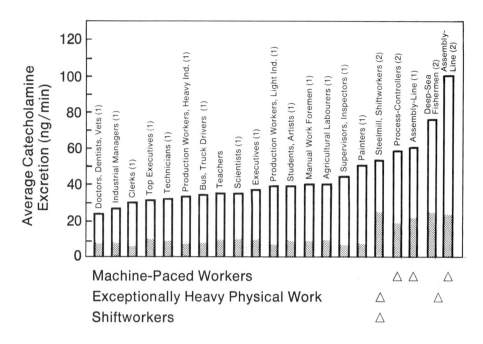

Figure 12-2: Catecholamine Excretion Rates in Different Occupations
The average nanograms per minute (ng/min) of epinephrine (dark bars) and norepinephrine (light bars) excreted in the urine are shown for 21 occupations. The five occupations to the right of the graph have the highest catecholamine levels, especially those jobs that are machine-paced or include exceptionally heavy physical work, indicated by the triangles at the bottom of the graph. These data are taken from studies summarized by Jenner, Reynolds, and Harrison (1980), indicated by (1) after the occupation, and by Astrand and Rodahl (1977), indicated by (2). An individual employee's values for these occupations may vary considerably from the average values, and as the work demands change from day to day; so may the absolute level of catecholamine excretion. However, machine pacing appears to be associated with elevated levels of both catecholamines; these could contribute to long-term health problems in some people. *(O'Hanlon, 1981).*

risk for some repetitive-motion disorders, such as tendonitis and carpal tunnel syndrome, in highly repetitive work (Arndt and Fabian, 1981; Hansen, 1982). These disorders are discussed in Chapter 13.

One other aspect of pace pressure or time stress is the effect it can have on the operator's ability to perform a mental task, such as problem solving. Requiring the person to keep track of time while also trying to sort out a problem that requires synthesis of many pieces of information can disrupt performance of the primary task. In one study, the performance of maintenance workers on paper-and-pencil trouble-shooting tasks was studied with and without time pressure (Rouse, 1976). Most people had been able to perform these tests in 40 seconds. One group was given the trouble-shooting diagrams and told to work as fast as they could and that errors would count against them. The other matched group was told that they had only 90 seconds to do each test, well over twice the time most people needed. The self-paced group performed similarly to previous groups on the tests, whereas many in the second, paced group had difficulty completing the tests in the required time of 90 seconds. The time pressure appeared to disrupt their problem-solving skills somewhat, resulting in worse performance, which might be reflected in decreased productivity in the plant. The source of the time pressure can be supervisory as well as line or machine pacing, so supervisors should be aware that too-close supervision with its implied time pressure may reduce productivity rather than increase it in some operations.

Some degree of pace pressure can be desirable in certain types of tasks (Kilbridge and Wister, 1970; McCormick, 1981; Salvendy, 1981), especially those where highly repetitive tasks are done. Much more research is needed to define the optimal level of pacing and to identify how much stress is appropriate in occupational tasks in relation to long-term health consequences.

D. DESIGN GUIDELINES FOR PACED JOBS

The following characteristics should be considered for operations where work pace is not fully under the operator's control. One should recognize that:

- Individuals vary in their capabilities for doing a task.

- Any one person will vary the way a task is done as a function of part variation, other distractions, or a desire to relieve monotony on a highly repetitive task (Welford, 1968).

- Any fixed speed chosen for a line operation will be optimal for only a small percentage of the people on the line.

- The longer people are asked to perform a paced task, the less percentage of their capability they can commit to it, on the average. Therefore, a work pace that is acceptable for an eight-hour shift may not be suitable on an overtime schedule (see Part V.)

- The more critical the task, the more desirable it is to let the operator set the pace.

- Tasks that require moderate muscular effort may overload the older or less fit worker if they are line- or machine-paced. The interaction of pace with other materials handling characteristics must be considered and is discussed in Part VI.

To reduce the stress of externally paced operations, job design should:

- Permit an operator to work at more than one conveyor station on a continuous line operation. Three stations should be accessible so an operator can work ahead of and behind the regular work station (Kellerman, van Wely, and Willems, 1963).

- Allow some self-pacing on a line-paced operation by providing in-process inventory space at each workplace (Murrell, 1965).

- Provide ample equipment for operators to accomplish the task within the time span given.

- Permit some control over line speed by key operators or supervision so that variations in parts or line staffing changes can be accommodated (Murrell, 1965).

- Allow operators to take short breaks (less than five minutes) from the line as needed for personal time or to get more supplies. A "floating" operator should be available to help balance out line problems when parts are uneven or individuals are not fully skilled in the task.

- Provide a well-designed workplace with good seating (see Chapter II, Volume 1). Avoid static loading of neck, shoulder, and arm muscles by extended reaches and locating parts too high.

REFERENCES

Arndt, R. H., and G. F. Fabian. 1981. "Report on Postal Worker Letter-Sorting Jobs." Paper given at the 1981 American Industrial Hygiene Association meeting in Portland, Oregon. Summarized in *Occupational Safety and Health Reporter*, May 28, 1981: p. 1572.

Astrand, P.-O., and K. Rodahl. 1977. *Textbook of Work Physiology*. 2nd ed. New York: McGraw-Hill, 681 pages.

Belbin, R. M., and D. Stammers. 1972. "Pacing Stress, Human Adaptation, and Training in Car Production." *Applied Ergonomics, 3 (3):* pp. 142-146.

Broadbent, D. E., and D. Gath. 1981. "Symptom Levels in Assembly-Line Workers." In *Machine Pacing and Occupational Stress*, edited by G. Salvendy and M. J. Smith. London: Taylor & Francis, pp. 243-252.

Conrad, R. 1954. "Speed Stress." Chapter 13 in *Human Factors in Equipment Design*, edited by W. F. Floyd and A. T. Welford. London: Lewis, pp. 95-102.

DeJong, J. R. 1962. "Speed of Work and Machines—Ergonomical Aspects." *Ergonomics, 5:* pp. 15-23.

Hansen, N. S. 1982. "Effects on Health of Monotonous, Forced-Pace Work in Slaughterhouses." *Journal of the Society of Occupational Medicine, 32:* pp. 180-184.

Jenner, D. A., V. Reynolds, and G. A. Harrison. 1980. "Catecholamine Excretion Rates and Occupation." *Ergonomics, 23:* pp. 237-246.

Johansson, G. 1981. "Psychoneuroendocrine Correlates of Unpaced and Paced Performance." In *Machine Pacing and Occupational Stress*, edited by G. Salvendy and M. J. Smith. London: Taylor & Francis, Ltd., pp. 277-286.

Johansson, G., G. Aronsson, and B. O. Lindstrom. 1978. "Social Psychological and Neuroendocrine Stress Reactions in Highly Mechanized Work." *Ergonomics, 21:* pp. 583-600.

Kellerman, F. T., P. A. van Wely, and P. J. Willems. 1963. *Vademecum—Ergonomics in Industry*. Eindhoven, The Netherlands: Phillips Technical Library, 102 pages.

Kilbridge, M. D., and L. Wister. 1970. "The Balance Delay Problem." In *Humanism and Technology in Assembly-Line Systems*, edited by T. O. Prenting and N. T. Thomopoulos. Rochelle Park, N.J.: Hayden, pp. 59-78.

McCormick, E. J. 1981. "Comments from the Sidelines." In *Machine Pacing and Occupational Stress*, edited by G. Salvendy and M. J. Smith. London: Taylor & Francis, pp. 1-4.

Murrell, K. F. H. 1965. *Human Performance in Industry*. New York: Reinhold, pp. 381-408.

O'Hanlon, J. F. 1981. "Stress in Short-Cycle Repetitive Work: General Theory and Empirical Test." In *Machine Pacing and Occupational Stress*, edited by G. Salvendy and M. J Smith. London: Taylor & Francis, pp. 213-222.

Rodgers, S. H., and M. Mui. 1981. "The Effect of Machine Speed on Operator Performance in a Monitoring, Supplying, and Inspection Task." In *Machine Pacing and Occupational Stress*, edited by G. Salvendy and M. J. Smith. London: Taylor & Francis, Ltd., pp. 269-276.

Rouse, R. 1976. Paper presented to the Western New York Human Factors Society and the Rochester Quantity Control Society, October, 1976.

Salvendy, G. 1981. "Classification and Characteristics of Paced Work." In *Machine Pacing and Occupational Stress*, edited by G. Salvendy and M. J. Smith. London: Taylor & Francis, pp. 5-12.

Selye, H. 1976. *The Stress of Life*. New York: McGraw-Hill, 515 pages.

Welford, A. 1968. *Fundamentals of Skill*. London: Methuen, 426 pages.

Wyatt, S., and J. A. Fraser. 1929. *The Effects of Monotony in Work*. IFRB Report No. 56. London: His Majesty's Stationery Office. Cited in Murrell, 1965.

CHAPTER 13

Repetitive Work

CHAPTER 13. REPETITIVE WORK

A. Repetitive-Motion Disorders

B. Individual and Task Factors Associated with Repetitive-Motion Disorders

C. Guidelines for the Design of Jobs with Repetitive Activity

References for Chapter 13

Ⅰn industrial jobs, the time to complete one unit of assembly or to inspect one item is defined as a cycle. This activity is usually considered repetitive if cycle time is two minutes or less and it is repeated throughout the shift. Highly repetitive tasks have cycle times of 30 seconds or less. Short-cycle work is often created in order to minimize training time on a job. A cycle time of 1.5 minutes is considered optimal for tasks with a fast working pace, such as television-chassis assembly (Kilbridge, 1962). These tasks are usually performed with the muscles of the fingers, arms, shoulders, or feet; many of the operations are done while seated.

Although the energy demands of these tasks are usually quite low, the repetitive use of small muscle groups and rotation around the wrist, elbow, and shoulder joints may be associated with symptoms of inflammation and soreness, collectively grouped as repetitive-motion disorders. Whether the job activity exacerbates previously existing illness or contributes to its development is not clear since, in most instances, only a small number of the people doing the job develop symptoms.

A. REPETITIVE-MOTION DISORDERS

The group of disorders collectively called repetitive-motion disorders ranges from joint inflammation to muscle soreness to nerve entrapment. They are also called "overuse" syndromes and "cumulative trauma" disorders. Although foot pedal operation can aggravate these symptoms in the lower leg and ankle, most repetitive-motion disorders are seen in the upper extremities and the neck. One analysis of the body parts affected in a group of 85 female electrical workers who reported symptoms during coil-winding, wiring, cable-forming, and spring-setting tasks is shown in Table 13-1. The heavy activity of the hands and wrists is reflected in the data, showing that more than one-half of the disorders were seen in those joints and muscles.

The most severe form of repetitive-motion disorder seen in tasks of this type is carpal tunnel syndrome. In this situation, the median nerve gets trapped in the wrist canal and becomes inflamed (Armstrong and Chaffin, 1979b; Tanzer, 1959). This inflammation leads to wrist pain and numbness or tingling of the fingers with a reduction in functional grip strength. An effective way to treat this kind of illness or irritation to the joint is to rest the hand and wrist and allow them to recover normal function.

Two other serious forms of repetitive-motion disorder are tendonitis and tenosynovitis. These are often overdiagnosed when a myositis is present or when muscles become sore when they are used heavily after not being very active for some time. True tendonitis and tenosynovitis are more specific (less diffuse) and follow the tendons of the affected muscles. Rest is the preferred treatment for these, too. Probably most symptoms are related to muscle tenderness and overuse, especially in workers who are learning a new job and are not yet fully skilled in it (Ferguson, 1971; Hymovich and Lindholm, 1966).

Table 13-1: Body Part Affected—Upper Extremity Disorders   The percent of female workers (n = 85) who reported symptoms due to repetitive operations at work is shown for five parts of the upper limb. A sixth classification for multiple parts affected and unclassified data is also given. Hands, fingers, wrists, and forearms are most heavily represented in these cases as the primary job activities were cable-forming, coil-winding, and wiring operations. The 85 women were seen over a 21-month period; 77 of them lost time from work because of their cumulative trauma disorders. *(Adapted from Ferguson, 1971).*

Body Part Affected	Percent of Electrical Workers with Upper Extremity Disorders
Hands and fingers	22%
Wrists and forearms	34%
Elbows and arms	16%
Shoulders and scapulae	14%
Neck	6%
Multiple or unclassified	8%

Bursitis and other inflammatory joint problems are also included when speaking of repetitive-motion disorders. These are common elbow and shoulder complaints where repetitive abduction and rotation of the joint is required throughout the shift, as in working above chest height or with extended forward reaches. Arthritis problems will also be seen in some individuals on highly repetitive tasks where force exertions are required. Whether these are causally associated with the work or are just brought out more in susceptible people is in need of further investigation (Hadler, 1977).

Other nerve entrapment syndromes in the neck, shoulder, elbow, leg, and hand are also associated with some occupational activities (Feldman, Goldman, and Keyserling, 1983). Table 13-2 summarizes some of the job task requirements and occupations where these motions are made regularly. The aggravating motion is identified for several types of work of the hands and wrists, arms and shoulder, and leg and foot.

B. INDIVIDUAL AND TASK FACTORS ASSOCIATED WITH REPETITIVE-MOTION DISORDERS

The causes of repetitive-motion disorders are very complex and no single factor can be identified. However, in order to further discuss ergonomic approaches to the prevention or reduction of these disorders, we will look here at some of

Table 13-2: Tasks and Occupations That May Aggravate Repetitive-Motion Disorders. Actions and motions that aggravate joints, tendons, or muscles and that are associated with the development of repetitive-motion disorders are shown for hand and wrist, arm and shoulder, and leg and foot joints (column 1). Some job requirements that include these types of motions are given in column 2, and examples of occupations using similar tasks are given in column 3. This table is not all-inclusive but is meant for use in identifying job characteristics that may predispose a susceptible person to such repetitive-motion disorders as tendonitis, bursitis, or carpal tunnel syndrome. *(Adapted from Feldman, Goldman, and Keyserling, 1983).*

Aggravating Actions and Motions	Job Requirements	Some of the Occupations Affected
A. Hand and Wrist		
Repeated forceful pronation of the hand in conjunction with forceful finger flexion.	Writing; manipulating controls and levers; scraping with a putty knife; using paint brush or roller; applying labels; sealing cartons.	Electronic wiring technician Assembly worker Painter Paint scraper Sheetrock installer Manual packaging worker Telecommunication repair worker
Repeated flexion-extension of the wrist; stress increased when accompanied by pinching and gripping; stress may be related to extent of flexion and extension, wringing action.	Pulling cloth; repeated handling of objects on conveyor belt or work table with a flexed wrist; use of ratchets and screwdrivers in awkward positions; use of paint roller and brushes; closing bags, wrappers, envelopes; placing brick stonework.	Chassis assembly worker Sewing machine operator Painter Cabinetmaker Fishing industry employee Manual packaging worker Musician (violinist) Mason
Repeated deviation of the wrist; stress increased with forceful grasp.	Hammering; shoveling; sweeping; using tin snips, side cutters, pliers, and cross-action tools.	Carpenter Cabinetmaker Janitor Electronic assembly worker
Repetitive pinching.	Grasping and pulling of fabrics, paper, or other materials; using of tweezers, forceps; inserting small parts with fingers.	Sewing machine operator Upholsterer Small-parts assembly worker Manual boxmaker
Repetitive pressure, pounding and compression into the palm.	Pressing tools into the palm; using the palm to apply pounding forces; using scrapers and wood gouges; shoveling.	Sailmaker Carpenter Cabinetmaker Painter Leatherworker Digging, earth-moving worker

Table 13-2 *(Continued)*

Aggravating Actions and Motions	Job Requirements	Some of the Occupations Affected
B. Arm and Shoulder		
Repeated forceful supination-pronation of the arm; forceful extension of the elbow.	Hammering with a straight elbow; lifting with extended arms; heavy packaging operations.	Carpenter Mason Handler, shipping dock worker Storeroom worker
Repetitive adduction and abduction movement of shoulder and arm.	Carrying heavy loads on shoulder; working overhead; signaling.	Carpenter Pipefitter Painter Sheetrock installer Electrician Traffic controller
C. Leg and Foot		
Repetitive crouching, squatting, kneeling.	Repair and maintenance; large-product assembly work; mining; floor scrubbing.	Equipment mechanic Janitor Equipment assembly worker Mining industries employee
Repetitive flexion and extension of foot.	Operating foot pedal; ladder climbing.	Heavy equipment operator Tractor driver Assembly worker Process control operator Press operator

the factors that have been suggested as increasing the likelihood of their development.

Table 13-3 lists some factors that are unique to individual workers as well as workplace factors that depend on specific aspects of the job. From the ergonomics perspective, the workplace factors are of primary interest. However, successful management of repetitive-motion disorders requires a thorough understanding of all factors.

1. INDIVIDUAL FACTORS

Many factors unique to individual workers have been identified as repetitive-motion disorders risk factors, in that they are present more often in people who develop the disorders. The ability to predict the occurrence of repetitive-motion disorders in a specific worker based on the presence of risk factors is not, however, always possible.

Table 13-3: Individual and Workplace Factors Associated with Repetitive-Motion Disorders. The upper half of the table lists nine factors that are associated with susceptibility for repetitive-motion disorders and that are sometimes observed in repetitive tasks involving high force exertions. The lower half of the table identifies 12 workplace and job factors that are thought to increase the risk for these disorders.

Individual Factors

1. Preexisting arthritis, bursitis, or other joint pain.
2. Peripheral circulatory disorders.
3. Preexisting neuropathy.
4. Reduced estrogen levels.
5. Small hand/wrist size.
6. New or inexperienced to the job.
7. Aggressive work methods.
8. Inefficient work methods requiring excess force application.
9. High personal stress level.

Job and Workplace Factors

1. General working postures required: elbows elevated, wrists deviated.
2. Machine pacing.
3. High-speed work.
4. High-frequency work.
5. Large number of movements, both regular and overtime work.
6. Incentive pay system, generating additional time pressure.
7. Off-specification assembly parts.
8. Direction of force application, requiring extra force to accomplish task.
9. Poor tool design.
10. Vibration from tools.
11. Cold or wet work environment for hands.
12. Poorly fitting gloves for hands, reducing grip strength.

Workers with preexisting medical problems are at higher risk of developing symptoms than healthy workers. Disorders such as arthritis, peripheral neuropathies, and circulatory disorders can be aggravated by the performance of repetitive tasks (Wells, 1961). Individuals with known disorders of this type should avoid regular assignment on jobs with the characteristics described in the second half of Table 13-3.

Alteration in female ovarian hormone levels, either related to surgery or to the use of oral contraceptives, has also been suggested as a factor that may increase the risk of repetitive-motion disorders (Cannon, Bernacki, and Walter, 1981). However, the exact relationship is not clear.

Small wrist or hand size has been suggested as a risk factor, particularly for the development of carpal tunnel syndrome. The force per unit of surface area on the median nerve during wrist deviations is higher for small wrists and hands (Armstrong and Chaffin, 1979b). The significance of wrist size for predicting who may be predisposed to carpal tunnel syndrome has not been proved (Armstrong and Chaffin, 1979a).

Symptoms of muscle, joint, and tendon soreness may be noticed by a new employee in the first several weeks on a new job. The new or inexperienced worker may be at greater risk for development of repetitive-motion disorders, either because of a higher individual susceptibility for these disorders, or because the untrained worker is less highly skilled at the tasks being performed. During the learning period, inefficient applications of force and overly aggressive work methods may be responsible for increased symptoms (Welch, 1972). New workers may be under additional stress due to their efforts to perform up to department standards, and this tension may contribute to their susceptibility for symptoms of repetitive-motion disorders. As the worker's muscles become accustomed to the work and as his or her skills are developed, the risk for repetitive-motion disorders appears to become less.

2. JOB AND WORKPLACE FACTORS

There are many workplace factors that have been associated with the risk of developing repetitive-motion disorders. Ergonomic interventions in the workplace begin with recognition of the contributions of the workplace, work methods, and work tools to the development of these problems.

Workplace design will influence body postures during the job, especially the amount of static muscle effort required to support the arm during an assembly or packing task. Work surfaces that are too high and make the worker abduct the elbows and shoulders, extended reaches that statically load the shoulder muscles, and orientation of the work piece so that large wrist deviations are required in the task all contribute to the risk for overexertion of hand, arm and shoulder muscles and joints (Tichauer, 1978). Visual task requirements may put additional stresses on the neck and shoulders if the viewing angles exceed the guidelines given in Chapter II, Volume 1. Static work is discussed in Chapter 11, and the effect of deviations of the wrist on grip strength is described in Appendix A.

Table 13-4 illustrates how hand and wrist positions differ in sewing machine operators who have or have not had a history of carpal tunnel syndrome. It is not clear if the differences in use of pinch grip and wrist flexion are causally related to the carpal tunnel symptoms, but both actions increase the stress on the median nerve and reduce functional grip strength (Armstrong and Chaffin, 1979a).

Table 13-4: Percent of Time in Different Hand and Wrist Positions and Carpal Tunnel Syndrome. *The wrist and hand positions of two groups of sewing machine operators (33 women in each) were compared at work. The members of one group had a history of carpal tunnel symptoms while the other group did not. The grip type and wrist flexion and extension positions were recorded by cinematography, and the data were gathered later through analyses of the film. The percent of time that each group used each of the recorded hand and wrist positions is shown in columns 2 and 3. Operators with a history of carpal tunnel syndrome were observed to use more pinching and wrist flexion than those without symptoms. (Adapted from Armstrong and Chaffin, 1978).*

	No History of Carpal Tunnel Syndrome (n = 33)	History of Carpal Tunnel Syndrome (n = 33)
Hand Position		
Pinch	39	56
Fingers opposing palm	39	34
Press	22	9
Wrist Position		
Extension	31	28
Neutral	56	42
Flexion	12	24
Not seen	1	6

Machine pacing and/or incentive pay can lead to work rates on repetitive tasks that do not allow appropriate recovery periods for heavily loaded local muscle groups. Forced-pace jobs paid on a piece-rate basis in a slaughterhouse, for example, were found to have a significantly higher ($p = 0.05$) number of people with shoulder, elbow, wrist, hand, back, and neck complaints than were jobs that were paced less rapidly or paid at an hourly rate (Hansen, 1982).

The speed of work will influence the forces developed on the tendons of the hand and arm muscles, and this also appears to be associated with increased risk for repetitive-motion disorders. At higher speeds, larger peak forces are generated, and repeated work at these levels may aggravate symptoms in susceptible people (Welch, 1972).

The number of repetitions of a movement within a given time period (frequency) and over the total work shift is associated with risk for developing repetitive-motion disorders. The larger the force required, or the more wrist deviation or pinch grip used, the higher percentage of work capacity of the active muscles is required to do the task, and the more opportunity there is for fatigue and inflammation to occur in the muscles and joints (Armstrong, 1983). If the repetition frequency is high, inadequate recovery time may increase the potential for disorders. If the total number of repetitions per work shift is very high, the wear and tear on the joint may be a significant factor as well (Welch, 1972). With overtime work or extended work weeks, there may be inadequate time for repair of the traumatized joints and muscles, and muscle and joint soreness may progress to more severe cumulative trauma disorders, such as tendonitis, carpal tunnel syndrome, or "frozen" shoulder (Bjelle, Hagberg, and Michaelsson, 1979).

Poorly designed, machined, or molded parts and components that do not assemble easily and require excessive forces from the hand and arm in order to be used, are associated with increased complaints of repetitive-motion disorders (Welch, 1972). Repetitive trauma from banging on a tool with the palm of the hand to dislodge a part or to clean it will increase the risk for tendonitis and carpal tunnel disorders (Tichauer, 1978). Hand tool design to reduce these symptoms is discussed in Chapter II, Volume 1. Some guidelines for tool design are summarized later in this chapter.

The use of vibrating tools is recognized as a factor that can lead to spasm of the small blood vessels of the hand, wrist, and arm. The impaired circulation has a direct effect on the function of these muscles, and continued work with them can lead to cumulative trauma disorders (CTD) (Armstrong, 1983). In a study of the physical stresses associated with the use of pneumatic screwdrivers, researchers found that a grip force of 110 newtons (24 lbf) was used to control the tool. This force, in conjunction with the vibration, increases the risk for CTD (Radwin and Armstrong, 1982). See Chapter V, Volume 1, for more information about vibration illness and guidelines to reduce the risk for it.

Hand function, which is an important factor in the risk of CTD of the hand and wrist, is also influenced by other workplace factors. Hands that are continually cold or wet may have impaired function due to reduced blood flow and altered neuromuscular function. The meat-cutting and fishing industries experience this problem, and both report a high incidence of CTD. The use of gloves to protect the hands can have undesirable effects if the gloves fit poorly or are the wrong type. Table 13-5 shows the observed loss in grip strength when gloves are worn. Wearing gloves may be an additional factor contributing to repetitive-motion disorders in people doing repetitive work with the hands requiring large force exertions.

C. GUIDELINES FOR THE DESIGN OF JOBS WITH REPETITIVE ACTIVITY

While there is no evidence that following the guidelines presented here will eliminate the development of cumulative trauma symptoms in susceptible peo-

Table 13-5: Grip Strength Reduction with Glove Use. The effects of several different types of gloves (column 1) on isometric grip strength are shown (column 2). Grip strength is measured with a grasp span between 5 and 6.25 cm (2 and 2.5 in.). The values in column 2 are expressed as a percent reduction in strength when wearing gloves compared to the bare-handed strength at the same span. The number of subjects by sex is indicated in column 3. The data sources are shown in column 4. This reduction in isometric grip strength should be considered when designing repetitive tasks where the hands must be protected. Because hand grip strength is reduced with the gloves, the same force requirement will be a larger fraction (percentage) of the capacity of the gloved hand than of the bare hand.

Glove Type	Average Percent Reduction in Maximum Isometric Grip Strength	N	Reference
Rubber kitchen gloves	19	5 F 5 M	Wang, Bishu, and Rodgers, 1982
Cotton gardening gloves	26	"	"
Asbestos furnace gloves	38	"	"
Unpressurized flight gloves	25	27 M	Garrett, 1968
Pressurized (3.5 psig) flight gloves	36	"	"
Air Force pilots' gloves	26	44 M	Hertzberg, 1955

ple, there are indications that the probability of their occurrence will be reduced. General and specific guidelines are given for the prevention and management of repetitive-motion disorders in the workplace.

1. GENERAL GUIDELINES

- Engineer products to allow machinery to do highly repetitive tasks; leave more variable tasks to human operators.
- Spread the load over as many muscle groups as possible to avoid overloading the smaller muscle groups.
- Design tasks to permit gripping with the fingers and palm instead of pinching.
- Avoid extreme flexion or extension of the wrist (see Appendix A). Design work surface heights, orientations, and reach length to permit the

joints to remain as close as possible to their neutral positions for maximum muscle strength.

- Keep forces low during rotation or flexion of the joint. Use power assists if forces are high. Avoid repetitive gripping actions.
- Provide fixtures to hold parts during assembly so that awkward holding postures can be minimized.
- Provide a variety of tasks over a work shift, if possible.
- Minimize time or pace pressures.
- Give people time to break into a new repetitive task.

2. SPECIFIC DESIGN GUIDELINES

- Keep the work surface height low enough to permit the operator to work with elbows to the side and wrists near their neutral position. Avoid sharp edges on workplace parts bins that may irritate the wrists when the parts are procured (Armstrong, 1983).
- Keep reaches within 50 cm (20 in.) of the front of the work surface so the elbow is not fully extended when the forces are applied (Armstrong, 1983).
- Keep motions within 20 to 30 degrees of the wrist's neutral point (see Figure 27-12) (Tichauer, 1978; Welch, 1972).
- Avoid operations that require more than 90 degrees of rotation around the wrist (Tichauer, 1978).
- Avoid gripping requirements in repetitive operations that spread the fingers and thumb apart more than 6.25 cm (2.5 in.) (Hertzberg, 1955). Cylindrical grips should not exceed 5 cm (2 in.) in diameter (Pheasant and O'Neill, 1975), with 3.75 cm (1.5 in.) as the preferable size (Ayoub and LoPresti, 1971). Hand tools that produce vibrations, require wide grip spans, or repetitively abrade the wrist area during use are of particular concern (Greenberg and Chaffin, 1977).
- For repetitive operations that require finger pinches, keep the forces below ten newtons (2.2 lbf). This represents 20 percent of the weaker operators' maximum pinch strength (Asmussen and Heebol-Nielsen, 1961). For gripping actions, keep the required forces to 21 newtons (4.8 lbf). This represents 20 percent of the isometric grip strength of the average woman when the hand is at its optimum span for force exertion (Kamon and Goldfuss, 1978).
- For continuous, highly repetitive operations, design a five-minute break for another activity into each hour.

- Select a glove with the least interference for gripping if hand protection is needed for a repetitive task. Provide a range of glove sizes to permit people to get the best fit for both large and small hands.

The guidelines in this section have been addressed more to the wrist and hands because many jobs require repetitive motions of the fingers and hands. The general guidelines also apply to other joints and muscles. Using 20 percent of maximum isometric muscle strength of the weaker worker (see Appendix A) as a guideline provides a safe limit for forces in repetitive work over a full shift.

3. HAND TOOL DESIGN FOR REPETITIVE TASKS

Many repetitive tasks require the use of hand tools. Ergonomic design of these tools can help reduce the potential for repetitive-motion disorders.

- Design handles that make use of the maximum strength capability of the hand by featuring a power or oblique grip involving the palm. Avoid pinch grip requirements. Make handle diameters as close as possible to 3.75 cm (1.5 in.) and the span on double-handled tools from 5 to 6.25 cm (2 to 2.5 in.).

- Make handles long enough (about 10 cm or 4 in.) to avoid applying repeated pressure to the base of the thumb, as when using a putty knife or a paint scraper.

- Orient the tool handle so it does not have to be used with the wrist deviated markedly in either the ulnar or radial directions.

- Design tools to reduce the need to exert a sustained force on a cold and hard surface. Properly textured handles increase the feeling of control on a powered tool; handle material with low thermal conductivity may also be desirable for some tasks.

- Reduce the vibration from a powered hand tool, such as an electric drill, as far as this is practical.

4. MANAGEMENT OF REPETITIVE-MOTION DISORDERS IN THE WORKPLACE

Some people will still experience repetitive-motion disorders on their jobs even if many of the preceding recommendations are implemented. Careful management of their workload and of their pattern of work on repetitive tasks requiring heavy force exertions may ensure that even these workers will lose little time from work because of their disorders. Some management approaches are:

- Rotate workers between jobs having different force requirements so no one person has to spend a full shift on the heaviest tasks. If rotation between jobs or tasks is not feasible, intersperse the primary task with several lighter tasks that provide a break for the muscles and joints most involved in the task.

- Train workers to recognize early signs of repetitive-motion disorders and to report them immediately so they can be reassigned to a less stressful job until the symptoms subside. Early detection of susceptibility can reduce the risk for more severe problems and decrease the time lost from work (Amoroso, Eastman Kodak Company, 1978).

- Identify the best ways to accomplish the more difficult repetitive tasks so that joint, tendon, and muscle strain are minimized. Teach these techniques to all new workers, and reinforce the training in the more experienced workers on a regular basis (Mitchell, Eastman Kodak Company, 1974).

- When people are starting a highly repetitive job with forceful exertions or are returning to work after more than two weeks' absence, rotate them between several activities until their muscles, tendons, and joints are accustomed to the work. A maximum of two hours of continuous work for a total of four hours per shift is recommended for the first few days on a highly repetitive job if musculoskeletal symptoms have been seen.

REFERENCES

Armstrong, T. J. 1983. *An Ergonomic Guide to Carpal Tunnel Syndrome*. Monograph prepared for the American Industrial Hygiene Association Ergonomics Guides Committee. 23 pages.

Armstrong, T. J., and D. B. Chaffin. 1978. *An Investigation of Occupational Wrist Injuries in Women*. Research proposal to NIOSH (No. 276-32-1563), obtained from the authors, 67 pages.

Armstrong, T. J., and D. B. Chaffin. 1979a. "Carpal Tunnel Syndrome and Selected Personal Attributes." *Journal of Occupational Medicine, 21 (7):* pp. 481-486.

Armstrong, T. J., and D. B. Chaffin. 1979b. "Some Biomechanical Aspects of the Carpal Tunnel." *Journal of Biomechanics, 12:* pp. 567-570.

Asmussen, E., and K. Heebol-Nielsen. 1961. "Isometric Muscle Strength of Adult Men and Women." *Communications from the Testing and Observation Institute of the Danish Association for Infantile Paralysis, 11:* pp. 1-44.

Ayoub, M. M., and P. LoPresti. 1971. "The Determination of an Optimum Size Cylindrical Handle by Use of Electromyography." *Ergonomics, 14 (4):* pp. 509-518.

Bjelle, A., M. Hagberg, and G. Michaelsson. 1979. "Clinical and Ergonomic Factors in Prolonged Shoulder Pain Among Industrial Workers." *Scandinavian Journal of Work Environment and Health, 5:* pp. 205-210.

Cannon, L. J., E. J. Bernacki, and S. D. Walter. 1981. "Personal and Occupational Factors Associated with Carpal Tunnel Syndrome." *Journal of Occupational Medicine, 23 (4):* pp. 255-258.

Feldman, R. G., R. Goldman, and W. M. Keyserling. 1983. "Peripheral Nerve Entrapment Syndromes and Ergonomic Factors." *American Journal of Industrial Medicine,* 4: pp. 661-681.

Ferguson, D. 1971. "Repetitive Injuries in Process Workers." *The Medical Journal of Australia, 2:* pp. 408-412.

Garrett, J. W. 1968. *Clearance and Performance Values for the Bare-Handed and the Pressure-Gloved Operator.* AMRL Technical Report 68-24. Wright-Patterson AFB, Ohio: Aeromedical Research Laboratory, pp. 82-93.

Greenberg, L., and D. B. Chaffin. 1977. *Workers and Their Tools. A Guide to the Ergonomic Design of Hand Tools and Small Presses.* Midland, Mich.: Pendell, 143 pages.

Hadler, N. M. 1977. "Industrial Rheumatology: Clinical Investigations into the Influence of the Pattern of Usage on the Pattern of Regional Musculo-Skeletal Disease." *Arthritis and Rheumatism, 20 (4):* pp. 1019-1025.

Hansen, N. S. 1982. "Effects on Health of Monotonous, Forced-Pace Work in Slaughterhouses." *Journal of the Society of Occupational Medicine,* 32: pp. 180-184.

Hertzberg, H. T. E. 1955. "Some Contributions of Applied Physical Anthropology to Human Engineering." *Annals of the New York Academy of Sciences,* 63: pp. 616-629.

Hymovich, L., and R. Lindholm. 1966. "Hand, Wrist and Forearm Injuries—The Result of Repetitive Motions." *Journal of Occupational Medicine,* 8: pp. 573-577.

Kamon, E., and A. Goldfuss. 1978. "In-Plant Evaluation of the Muscle Strength of Workers." *American Industrial Hygiene Association Journal, 39 (10):* pp. 801-807.

Kilbridge, M. D. 1974. "Non-Productive Work as a Factor in the Economic Division of Labor." In Chapter 6 of *Humanism and Technology in Assembly Line Systems,* by T. O. Prenting and N. T. Thomopoulos. Rochelle Park, N.J.: Hayden, pp. 59-78.

Pheasant, S., and D. O'Neill. 1975. "Performance in Gripping and Turning—A Study in Hand/Handle Effectiveness." *Applied Ergonomics, 6 (4):* pp. 205-208.

Radwin, R., and T. J. Armstrong. 1982. *A Study of the Physical Stresses Associated with Pneumatic Screwdrivers.* Ann Arbor: University of Michigan, College of Engineering and School of Public Health, 90 pages.

Tanzer, R. C. 1959. "The Carpal Tunnel Syndrome: A Clinical and Anatomical Study." *Journal of Bone and Joint Surgery, 41 (Al):* pp. 626-634.

Tichauer, E. R. 1978. *The Biomechanical Basis of Ergonomics: Anatomy Applied to the Design of Work Situations.* New York: Wiley Interscience, 98 pages.

Wang, J., R. Bishu, and S. H. Rodgers. 1982. Student project for Industrial Engineering 536, State University of New York at Buffalo. Paper submitted for publication.

Welch, R. 1972. "The Causes of Tenosynovitis in Industry." *Industrial Medicine and Surgery, 41 (10):* pp. 16-19. *Industrial Medicine and Surgery, 41 (12):* p. 34.

Wells, M. J. 1961. "Industrial Incidence of Soft Tissue Syndromes." *Physical Therapy Review,* 41: pp. 512-515.

CHAPTER 14

Training

CHAPTER 14. TRAINING

The time it takes an operator to become fully skilled in a task is a function of many factors:

- *Task:*
  - complexity
  - intensity
  - duration
  - repetitions and/or frequency
- *Operator:*
  - physical fitness
  - perceptual capabilities
  - decision-making capabilities
- *Environment:*
  - distractors
  - communications
  - stability of operations

If the task involves machine operation, the machine's design will determine the task complexity and demands. Among the factors influencing the worker are the design of controls and displays, force requirements, postural demands, and visual requirements.

Skill acquisition is usually measured by the time it takes an operator to reach the standard of performance expected on a job. Table 14-1 presents estimates of the number of repetitions required to establish a motor engram (a learned pattern of movement) for skilled performance in several occupational tasks. Less complex skills take fewer repetitions, so a job is often broken down into separate tasks, each of which can be learned more rapidly. As these skills are developed, they can be linked together to provide skilled performance on more complex tasks, the philosophy behind the design of many assembly lines.

Physiological training, which includes the development of muscular strength and endurance or tolerance to heat, is often forgotten in preparing people for jobs where moderate to heavy effort is required. Whereas skill training may only be partially lost if the motor patterns are not used for an extended period, physiological training may be lost if a person is away from the job for as little as three weeks, and it will start to decline within three days of stopping the work (Müller, 1970).

A. JOB DESIGN THAT MINIMIZES TRAINING TIME

When one or more of the following conditions exist, the cost of training people can be high, thereby affecting the productivity of the department:

- High production volume.
- Rapid turnover of new products.
- Entry-level job.
- High turnover of personnel (moving on to higher-level jobs).
- Assembly line—long-cycle tasks, especially.
- Multiple tasks—using the same skills

Among the individual capabilities often needed to perform these jobs are:

- Motor skills—dexterity, coordination.
- Visual skills—detection of shape, color, surface irregularities.
- Mental skills—three-dimensional spatial perception, memory, adaptability, time organization.

Since combinations of these skills are used in learning assembly tasks, and because personality variables are also important determinants of skill acquisition, it is very difficult to predict who will take a long time to master a specific task and who will learn quickly.

Table 14-1: Estimated Repetition Frequency Needed to Become Fully Skilled at a Task.   Five tasks are shown in column 1 that require skilled motor performance, especially manual dexterity. The number of repetitions of the basic motor pattern needed to acquire each skill has been estimated (column 3) either from performance records or from the number of units or area completed when a person is considered skilled (column 2). The people performing each task are designated in column 4. Repetitions establish a motor pattern, or engram, for each movement, and coordinated, skilled performance can emerge from a synthesis of several engrams. See Part II for further information on motor learning and skill acquisition. *(Adapted from Kottke et al., 1978).*

Task	Basis for Estimate	Repetitions	Performer
Hand knitting	20 sweaters at 75,000 stitches per sweater	1.5 million stitches	Women
Pearl handling	Performance records in a Japanese pearl industry	1.5 to 3.0 million	Women
Rug making	35 m² at 40,000 knots per m² (Ecuador)	1.4 million knots	Children
Cigar making	Crossman, 1959	3 million cigars	Young women
Marching	Army basic training	800,000 steps in six weeks	Men

Figure 14-1 illustrates a learning curve for an operator doing a relatively simple washer-and-plug assembly task. The amount of time needed to complete a bloc of 1,296 assemblies drops rapidly as the first 20 blocs are completed and then declines more slowly over the next 40 blocs. A financial incentive produced

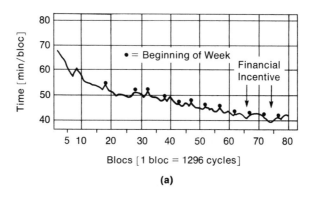

(a)

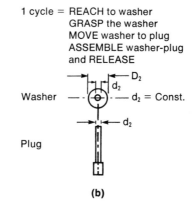

(b)

Figure 14-1: Training Time on a Repetitive Assembly Task   The number of minutes to complete a bloc of 1,296 washer-and-plug assemblies (vertical axis) is plotted against the total number of blocs completed (horizontal axis). The assembly task is diagrammed in (b) with the graph of the analysis of the movements required to complete each assembly in (a). The performance of this operator was monitored over approximately 12 work weeks; each week is indicated on the training curve by an open circle. The financial incentive offered in the last weeks of the study appeared to have a limited positive effect, improving productivity a few percent at the most. This fairly simple task was still being learned after about 70 hours. However, substantial improvement in operator performance could be seen in the first few days. More complex tasks should take correspondingly longer to learn. (*Adapted from Rohmert and Schlaich, 1966*).

some transient performance improvement after ten weeks on the job, but there was evidence that the training plateau was about to be reached. The first 20 blocs represent about 25,000 cycles; 80 blocs is about 104,000 cycles, and there was apparently still some performance improvement possible at that time.

The following guidelines for job design and initial training should improve the rate at which people are able to work up to the desired standard in these tasks.

- Keep cycle time in the range of 30 seconds to two minutes (Kilbridge, 1974) on assembly-line tasks. Keep the number of individual steps to be done down to about ten. If the cycle time is much shorter than 30 seconds, provide another activity with a longer cycle time, such as getting parts, for a few minutes each hour.

- Provide visual aids of the operation (graphics) at the workplace as a reference for the new operator.

- Provide aids for better visualization (such as magnifiers), fixtures for holding objects, and good workplaces to eliminate unnecessary effort in performing the task (see Chapters II and V, Volume 1).

- Provide feedback on the quality of the work being done, particularly during the first days on a new task (Blum and Naylor, 1968).

- If the product changes frequently, provide the operator with an instruction book so that he or she can refer to specific product needs during assembly.

- Train the person in discrete tasks; do not try to teach the whole job at once. Demonstrate the method and have the operator try it immediately (Blum and Naylor, 1968).

- Have the new operator practice the operation off-line initially, and then introduce her or him to the line operation for limited periods (one to two hours) before working up to a full shift. Provide frequent breaks and comfortable seating (Rutenfranz and Iskander, 1969).

- Review the training after the operator has been on the task for a few days. Keep communications open to anticipate where training may not have been adequate.

- When an operator is trained on one task and demonstrates capability for further training, start with tasks bearing some resemblance to the one already learned.

- Do not sacrifice job content and challenge in the interest of minimizing training time. Provide variety and design the job and workplace to remove unproductive effort, such as rehandling products (Davis and Canter, 1956).

B. PHYSICALLY DEMANDING JOBS— TRAINING GUIDELINES

For jobs that require moderate or heavy effort, whether measured as high strength or total effort over time, physical fitness specific to the job demands is necessary. Fitness training is needed not just when a new job is learned but also when a person has been away from the job for a few weeks. There is some evidence that highly repetitive tasks, which involve constant use of small muscle groups of the hand and arm and rotation around joints like the wrist and elbow, should only be done for two to four hours per shift initially instead of asking a person to work on them for a full eight hours from the start (Hymovich and Lindholm, 1966).

The following guidelines for training and work practices should reduce the potential for muscle, joint or cardiovascular overload for people who are new to the tasks or who are returning to work after an extended vacation or illness:

- For jobs that include heavy workloads or high muscle strength, allow the individual to pace the effort him- or herself. Avoid continuous work periods of more than 15 minutes and provide ample periods of recovery with light work activities. An untrained individual should not be subjected to maximum loads.

- When a person has been away from a highly repetitive job for more than two weeks or is starting a new job with these characteristics, rotate the operations every two hours, if possible. A maximum of four hours per shift on the operation of concern is recommended for the novice's first two weeks or the skilled operator's first week. Workers should be encouraged to report any muscle or joint soreness early so that appropriate adjustments to workload or pattern can be made.

- If a job requires occasional handling of heavy loads, such as the handling of 208-liter (55-gallon) drums, provide training for the operator that will allow frequent performance of this activity over several days. In this way the requisite skill can be obtained and the muscle strength requirements reduced.

- If a job requires occasional to frequent handling of heavy objects, such as 45 kg (100 lbm) bags, select a specially trained group of individuals to handle the work instead of shifting it around to everyone equally. Since the special team will be doing the work more frequently, they will be in better physical condition to sustain the effort than people who only perform the task once a week or month.

- If the person's job includes exposure to high levels of heat and humidity, he or she should gradually work into the area over three to four days, being allowed frequent breaks from the hot or humid area as needed (see Chapter V, Volume 1).

Because training is very specific to the job demands, a general fitness improvement program such as jogging or calisthenics may not be as effective as a controlled on-the-job training program (Astrand and Rodahl, 1977). However, a high general fitness level will reduce the time needed to train the specific muscles required for the job tasks. The absolute time period needed to train to maximum strength will depend on the person's initial strength and the training regimen. The less fit a person is, the longer it will take to reach maximum strength. In a laboratory experiment where the elbow flexors were contracted isometrically to their maximum for four to six seconds once each day for about three months, it took about 12 weeks to reach maximum strength values (Müller and Rohmert, 1963). If the frequency of exertions per day had been increased, the slope of the curve might have been steeper.

Because muscle training takes several weeks to complete, the new employee on a physically demanding job may be at increased risk for overexertion injuries. The guidelines for designing tasks requiring force exertions or lifting (Part VI) assume that men and women, older and younger workers, may be asked to do the task.

C. TRAINING PROGRAMS

Rapid changes in technologies and manufacturing techniques create an increased need for effective approaches to training new employees or those transferred from other jobs. Training and learning are very complex subjects supported by extensive psychological research; the following information is a brief overview of some of the factors that can influence the training for and learning of industrial tasks.

Training can be defined as the systematic development of the attitude, knowledge, or skill behavior patterns needed by an individual to perform adequately on a given task or job, such as an assembly task (Stammers and Patrick, 1975). New employees also have to be trained to perform tasks that rely on visual or auditory skills. New social skills may have to be developed by employees who interact with customers or others where jobs require effective interpersonal communication skills. The varying complexities of tasks require different types of training. Three approaches to dealing with a difficult task are:

- Design or redesign the job situation in order to reduce the task demands.
- Select people, by evaluating their aptitudes, abilities, or previous training, who have the capabilities to perform the tasks.
- Train people to deal with these task or job demands.

The ergonomics approach is illustrated in the first option; it has obvious benefits since more people will be available to do the task and training time should be reduced. Selection of new employees of high aptitude is desirable but not always possible; some training must be done in nearly all instances. The best approach is to combine all three of these approaches in order to develop a well-trained and efficient work force. The extent of supervisory commitment to these approaches is influenced by other considerations. These include:

- Available financial resources.
- Available labor force.
- Time constraints.
- Expected life of the manufacturing process.
- Current practices in the human resources field.
- Availability of specialists.
- Past tradition and precedents.

1. LEARNING AND SKILL ACQUISITION IN TRAINING PROGRAMS

The goal of any training program is to develop skilled performers, individuals who can produce high-volume, high-quality output. The achievement of the training goal depends on several factors, including:

- Task complexity.
- Abilities of the trainees.
- Abilities of the trainers.
- Type and amount of training.

These are discussed in the subsections that follow. The success and effectiveness of any training program depends on the integration of factors unique to each situation.

a. Task Complexity

The ability of an individual to perform successfully on any job depends on task difficulty. Thus, a task analysis should be performed prior to the implementation of a training program. The task analysis in this context is defined as ''a systematic analysis of the behavior required to carry out a task with a view to identifying areas of difficulty and the appropriate training techniques and learning aids necessary for successful instruction'' (Department of Employment, 1971). The design of the training program will emerge from the results of the task analysis, more time being spent on those job requirements that require greater skill or knowledge to perform.

b. Abilities of Trainees

A group of workers can have a wide distribution of types and levels of abilities, some of which are required to perform each job or task successfully. In an assembly task, for example, an individual with fine motor skills could be considered a skilled performer. The same individual might perform very poorly, however, in a task that required fine visual-discrimination skills, such as identifying subtle differences in color. Selection tests try to measure the skills required on a specific task in order to find people who are skilled in those particular areas. Because more than one skill is often required, the approaches of optimal design of tasks and training programs are preferred so that more people can be trained to perform most industrial jobs.

c. Ability of Trainers

Training itself is a skill, and the ability of individual trainers will vary. The success of a training program will depend to a large extent on the trainers. The same considerations should be given to training and evaluating trainers as to training new workers.

d. Type and Amount of Training

The type and amount of training will depend on the skills required to perform the task, which should be identified in the preliminary task analysis. A simple systems approach to training is shown in Figure 14-2.

The training objectives must first be defined. Once this has been accomplished, criterion measures to determine the success of the program can be developed. Failure of trainees to meet the criterion measures is a reflection on the

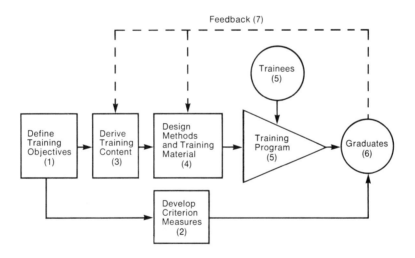

Figure 14-2: A Systems Approach to Training   A simplified training system is described by seven components. Development of the training program includes defining the objectives (1) and simultaneously developing the criterion measures to evaluate how well the objectives have been met (2). Specification of the content of the training program in order to meet the program objectives (3) will help define the training methods and equipment to be used (4). The trainees and materials then come together in the training program (5). After the trainees complete the training program and graduate (6), they are evaluated against the criterion measures, and any discrepancies are fed back to modify the content of the training program (7). (*Eckstrand, 1964; Stammers and Patrick, 1975*).

training program, not the trainees. The third stage is to specify what should go into the training program in order for the objectives to be met. Next, training methods and equipment must be developed to meet the needs of the training program. Then, trainees and training methods come together in the training program itself. Upon completion of the program, the graduates are evaluated against the criterion measures. Any discrepancy between the information learned and the criterion measures provides feedback that can be used to improve the training program.

Each job to be learned must have its own training system, the complexity of which will vary depending on the findings of the task analysis. One method of evaluating training programs is to develop performance curves. A performance curve is a method of displaying changes in performance due to an independent factor, such as time or a training program. Figure 14-3 shows the general form of performance curves that relate performance to the amount of practice.

Performance curves can be used to assess the effect of factors that influence the training of individuals, such as:

- Practice schedules

- Feedback on performance

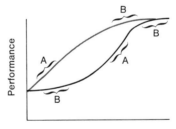

Amount of Practice

Figure 14-3: The Relationship of Task Performance to the Amount of Practice in Assembly Tasks   The relationship between the amount of practice (horizontal axis) and the performance of a skill (vertical axis) are presented for two types of job tasks. In general, increased performance can be expected as the amount of practice increases. Performance will change at different rates depending on the skill to be learned. Segment A on each curve indicates very rapid increases in performance while segment B indicates slow improvements. The upper curve is often seen for people working on tasks that require simple movements and are relatively easy to learn. The lower curve is typical of a person doing a task requiring unique movements that take some time to learn; once these are learned, performance increases rapidly.

- Individual motivation levels
- The environment
- Fatigue over time

By incorporating various combinations of these factors into a training program and by measuring performance curves, one can develop the best possible program. For example, the two theoretical performance curves for small assembly task workers in Figure 14-4 could result from two training programs, only one of which provided feedback to the trainees. From this example, you can see that practice plus feedback clearly results in more rapid improvement in performance.

2. TRAINING METHODS

Some common methods used for training are discussed below. These include: lectures, on-the-job training and simulation, job rotation, sound motion pictures and videotapes, and demonstrations and other visual aids.

a. Lectures

Lecturing is economical since many people can be presented with large amounts of information by one trainer. The efficiency of lectures from the standpoint of learning varies between individuals. They are, generally, of little value for developing job skills but are appropriate for the presentation of background information.

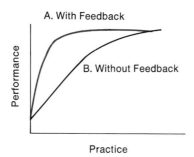

Figure 14-4: Performance Curve Variations—The Effects of Practice With and Without Feedback   The level of task performance is affected by practice and by feedback provided during practice. A typical performance curve for a hypothetical task shows that, with greater practice, performance increases up to a point, when a plateau is reached (B). The performance curve for the same task when practice is supplemented by appropriate feedback (A) shows that the rate of task learning, as reflected by increases in performance, is greater when feedback is presented. Note that the performance plateaus are identical.

b. On-the-Job Training and Simulation

When the acquisition of new motor skills is required to learn a job, the best training program is one that allows on-the-job training. Although training on the actual job is usually the best approach, there are many instances when off-the-job training or simulation is more effective. Some of the reasons to use simulations include:

- The costs and consequences of high error rates during training make the use of simulations preferable.

- Simulations provide an environment in which errors and poor performance are tolerated and where more intense feedback and guidance can be given.

- They can be used to remove distracting environmental stresses such as noise, heat, or light. This permits a worker to develop the basic skills before having to perform under less than optimal environmental conditions in the workplace.

- They can help make more efficient use of training time by simulating tasks that occur infrequently but must be learned, such as machine start-ups, shutdowns, or emergencies.

- They can isolate subtasks that require extra training time for practice prior to training on the overall task.

- It may cost less to train on a simulator than on the real equipment.

- They can be used to train people on tasks that do not yet exist. This occurs when new plants or processes are being developed or when tasks must be performed in hazardous environments.

The important factor in both on-the-job and simulation training is that the trainee is actually performing the task. Motor skill acquisition requires hands-on practice (see Part II). This type of training also provides opportunities for individualized instruction, feedback, and encouragement.

c. Job Rotation

Job rotation is a frequently used technique in industry that is a form of on-the-job training. Where it is permitted by contract, the objective of job rotation is to broaden an employee's experience and to provide backup staffing for vacation, illness, or increased production periods. Having workers familiar with more than one job, particularly those that include critical tasks, can ensure continued productive operations when the usual worker is not able to be on the job.

d. Sound Motion Pictures and Videotapes

Motion pictures and videotapes are good visual learning aids. They can show relationships among various functions, the operation of specific components, or the results of specific actions. A movie or videotape cannot teach an actual job skill, but it can provide an overview of the importance of a task to company objectives. Although the initial cost of motion pictures or videotapes may appear high, their usefulness for passing information on to large audiences at any time and as frequently as needed makes them inexpensive investments over the long run.

e. Demonstrations and Other Visual Aids

Demonstrations and visual aids, such as overhead transparencies or slides, offer a less expensive way to deliver some of the same information presented by a movie. These visual aids are more flexible than movies or videotapes since the order and content of the presentation can be adapted to the varying interests and background of the audience. This flexibility may be a disadvantage, however, in ensuring consistency in training across trainers. Slide-tape talks reduce this flexibility somewhat and represent a middle ground of convenience between movies and other visual aids.

3. Training Aids in the Workplace

Once a person has been trained initially on a job, it is often necessary to monitor and reinforce that training, particularly on tasks that are complex or occur infrequently. This is less problematic for jobs that change frequently because of major changes in the product line than it is for jobs that are similar across product lines and include long-service workers. Aids to reinforce earlier training may include:

- Annual, short, highly visual presentations (such as movies) provided at a department meeting or in the annual report (for example) that are tied to company goals and objectives.

- Provision of display posters or cards with relevant training information on them that are placed at the workplace or in a break area, wherever is more appropriate.

- Provision of a manual describing each task, based on the task analysis, that is available either through first-line supervision or directly to each worker.

- Use of quality audits and observation of workers to reinforce good performance and bring poor performance to the worker's attention. This should be constructive, not punitive, in nature.

Workplace aids to assist those doing inspection tasks where multiple, and often rare, defects must be identified are discussed in more detail in Chapter IV, Volume 1 (see also Harris and Chaney, 1969). These include photographic illustrations of defects and comparators where a good sample is placed alongside the one being inspected. Feedback on performance is probably the most effective training reinforcement approach for visual inspection tasks as well as for assembly and other types of jobs (Drury and Addison, 1973).

REFERENCES

Astrand, P.-O., and K. Rodahl. 1977. *Textbook of Work Physiology*. 2nd ed. New York: McGraw-Hill, 681 pages.

Blum, M. L., and J. C. Naylor. 1968. *Industrial Psychology: Its Theoretical and Social Foundations—A Revision of Industrial Psychology and Its Social Foundations*. New York: Harper & Row, 633 pages.

Crossman, E. R. F. W. 1959. "A Theory of the Acquisition of Speed-Skill." *Ergonomics*, 2: pp. 153-166.

Davis, L. E., and R. R. Canter. 1956. "Job Design Research." *Journal of Industrial Engineering, 7 (6)*: pp. 275-282.

Department of Employment. 1971. *Glossary of Training Terms*. 2nd ed. London: Her Majesty's Stationery Office. Cited in Stammers and Patrick, 1975, p. 46.

Drury, C. G., and J. L. Addison. 1973. "An Industrial Study of the Effects of Feedback and Fault Density in Inspection Performance." *Ergonomics, 16:* pp. 159-169.

Eckstrand, G. A. 1964. *Current Status of the Technology of Training*. AMRL-TDR-64-86. Wright-Patterson AFB, Ohio: Aerospace Medical Laboratories.

Harris, D. F., and F. B. Chaney. 1969. *Human Factors in Quality Assurance*. New York: Wiley, 234 pages.

Hymovich, L., and R. Lindholm. 1966. "Hand, Wrist and Forearm Injuries—The Result of Repetitive Motions." *Journal of Occupational Medicine, 8:* pp. 573-577.

Kilbridge, M. D. 1974. "Non-Productive Work as a Factor in the Economic Division of Labor." In Chapter 6 of *Humanism and Technology in Assembly Line Systems*, by T. O. Prenting and N. T. Thomopoulos. Rochelle Park, N.J.: Hayden, pp. 79-89.

Kottke, F. J., D. Halpern, J. K. M. Easton, A. T. Ozel, and C. A. Burrill. 1978. "The Training of Coordination." *Archives of Physical Medicine and Rehabilitation, 59:* pp. 567-572.

Müller, E. A. 1970. "Influence of Training and Inactivity on Muscle Strength." *Archives of Physical Medicine and Rehabilitation, 51 (8):* pp. 449-462.

Müller, E. A., and W. Rohmert. 1963. "Die Geschwindigkeit der Muskelkraft-Zunahme bei Isometrischen Training." *Internationale Zeitschrift Angwandte Physiologie, 19:* pp. 403-419.

Rohmert, W., and K. Schlaich. 1966. "Learning of Complex Manual Tasks." *International Journal of Production Research, 5 (2):*pp. 137-145.

Rutenfranz, J., and A. Iskander. 1969. "Untersuchungen über den Einfluss Verscheidener Arbeits- and Übungsbedingungen in Früher Lernstadien auf das Erlernen einer Einfachen Sensumotorischen Leistung" ("Influence of Different Working and Training Conditions Within the First Learning Periods on the Acquisition of a Simple Sensory-Motor Performance"). *Internationale Zeitschrift Eingewandte Physiologie, 27:* pp. 356-369.

Stammers, R., and J. Patrick. 1975. *The Psychology of Training*. London: Methuen, 144 pages.

PART V

Hours of Work

CONTRIBUTING AUTHORS

David M. Kiser, Ph.D., Physiology

Thomas J. Murphy, Ph.D., Psychology

Suzanne H. Rodgers, Ph.D., Physiology

A determination of appropriate job design must consider not only the intensity of effort required but also its duration. The longer a person is required to sustain an activity, the lower the percent of his or her capacity can be used. This has been discussed in Parts II and IV for both dynamic and static muscle work and for perceptual tasks such as visual inspection. The hours of work must be considered both in terms of their number and their distribution during a 24-hour day. The guidelines in the rest of this book are generally for eight-hour days or fractions thereof. In this section, the effects of extended hours or overtime and of different shift work schedules are discussed, and guidelines for the selection of appropriate work schedules are given.

CHAPTER 15

Overtime

CHAPTER 15. OVERTIME

A. Job Demands in Relation to Choosing Overtime Schedules

    1. Psychosocial Factors
    2. Physical and Environmental Job Demands
    3. Mental and Perceptual Job Demands

B. Other Factors Affecting Productivity in Overtime Schedules

    1. Measuring Productivity
    2. Type of Pay Plan—Incentive or Time-Based
    3. Worker- or Machine-Paced Jobs

C. Overtime by Increasing the Days Worked per Week

D. Guidelines for Overtime Scheduling

References for Chapter 15

There are two types of overtime used frequently in industry: increased hours per day and increased days per weeks. Overtime practices often result from seasonal or emergency demands that require increased production or maintenance service for a discrete time period. Since continued demands for these services or for increased production are not anticipated, extending the hours of people already trained on the job is preferable to taking on and training new employees or building additional manufacturing capacity. Occasional overtime on a smaller scale may also be brought about by the loss of an employee on a temporary basis due to vacation or illness. Unless another person is available to fill in, the other people in that department or job may have to work extended hours in order to fulfill production goals or provide the needed service.

The suitability and acceptance of overtime schedules are the results of many factors, such as:

- Duration of the overtime period.
- Frequency of occurrence of the overtime work.
- Financial needs of the worker.
- Feasibility of other options (besides overtime) to accomplish the task.
- The season of the year (less acceptable in summer for some people).
- Type of pay plan (incentive or time-based).
- Type of pacing (machine- or worker-paced).
- Physical demands of the job.
- Perceived need for the job to be done with some immediacy, for example, because of competitive contracts or because of wartime or emergency needs.
- Perceptual and mental demands of the job.

Although increased pay per week may be very satisfying to some people, loss of personal time can become dissatisfying as the overtime period continues (Best, 1933; Blum and Naylor, 1968). While a day a month, a week every six months, or a month every year may be acceptable overtime schedules for many workers, an extra day a week or a week each month may not be (Brachman, Eastman Kodak Company, 1972). These are based on information from people in maintenance and production jobs in the chemical industry and are rough guidelines. The factors mentioned above will modify the acceptability of overtime in specific cases. If these less acceptable overtime schedules can be expected in a job, this information should be included in the job description and advance notice given, whenever possible. Continuous overtime schedules should be avoided unless they are part of a shortened work week, as they can result in reduced productivity per hour or increased unit cost, and they can dissatisfy many workers (National Electrical Contractors Association, 1969; Poper, 1970;

U.S. Department of Labor, 1947). Absenteeism and accidents have been shown to increase in manufacturing plants when longer work weeks are scheduled (U.S. Department of Labor, 1947).

This chapter discusses the effects of increasing the hours of work per day in relation to the psychosocial, physical, environmental, perceptual, and mental demands of the jobs. The productivity effects are related to the compensation plans (incentive or time-based) and whether the work is paced by machine or by the worker. An alternative to daily overtime, adding an extra day to the work week, is also discussed.

A. JOB DEMANDS IN RELATION TO CHOOSING OVERTIME SCHEDULES

1. PSYCHOSOCIAL FACTORS

The acceptability of various work hours and schedules will be determined primarily by how they affect a person's life outside work. Schedules that reduce the amount of personal time for interacting with family and friends are generally less acceptable than those that take these needs into account. Table 15-1 shows how extended hours of work in a manufacturing plant can affect a person's personal time.

When a person works ten hours a day or more, personal time is minimal, and it is difficult to find recreational time for the family. This situation is accepted by the employee and family as long as it happens infrequently and for cause, and as long as the additional pay is important. As the frequency increases, or if other alternatives for satisfying production or service needs are available, the acceptance of long work hours will decline for many people, despite the increased pay.

2. PHYSICAL AND ENVIRONMENTAL JOB DEMANDS

The amount of energy required by a job or the presence of chemical and physical factors in the environment will determine the suitability of a specific overtime schedule for most people. Exposure or work levels that can be tolerated for eight hours may not be acceptable for ten to twelve hours, necessitating additional recovery time. In such cases, lengthening the workday can result in less increase in productivity than would be anticipated from the extended hours, since the workers will regulate their work rate to prevent fatigue. Figure 15-1 illustrates acceptable physical workloads as a function of the number of hours worked per day. A job that requires 33 percent of capacity averaged over eight hours could only be sustained on overtime by further reducing the effort level.

Appendix A categorizes some common activities and jobs by their energy requirements. Jobs that include tasks in the heavy, very heavy, and extremely heavy categories do not lend themselves to extended hours of work without job

Table 15-1: Effect of Extended Hours of Work on Leisure Time. The amounts of time spent on work (A), personal activities (B), and sleeping (C) for 8-, 10-, and 12-hour workdays are shown in columns 2 through 4. It is assumed that clothes change, travel, personal needs, and sleeping times are constant. Therefore, increased work time directly affects the amount of theoretically possible genuinely free time a worker has. A 12-hour workday cuts this free time to less than an hour. Subtotals for working and personal time are shown, as well as the grand total of 24 hours per day. *(Adapted from Rutenfranz, Colquhoun, and Knauth, 1976, Hours of Work and Shiftwork, presented at the 6th Congress of the International Ergonomics Association. Copyright 1976 by the Human Factors Society, Inc. and reproduced by permission.)*

Activity		Hours		
(A) Working time		8.00	10.00	12.00
Clothes change		0.25	0.25	0.25
Travel time		1.00	1.00	1.00
	Subtotal	9.25	11.25	13.25
(B) Time for personal needs		2.00	2.00	2.00
Theoretically possible, genuinely free time		4.75	2.75	0.75
	Subtotal	6.75	4.75	2.75
(C) Sleeping time		8.00	8.00	8.00
	Grand Total	24.00	24.00	24.00

redesign to reduce the effort. Those in the light and moderate categories may be suitable for overtime from a physical standpoint, but they should be further evaluated for their perceptual and mental demands and for the presence of environmental stressors.

Most guidelines for exposure to chemicals, dusts, and physical factors such as heat and noise are based on an eight-hour workday and a five-day week. Overtime and seven-day shift schedules may lower the acceptable exposure levels. The amount of lowering will be a function of a number of factors: the degree of exposure, the length and pattern of the work and rest cycles, and the amount of time needed to metabolize the substances in the body (for some chemical exposures). Until better data are available to quantify the acceptable exposure levels according to hours of work, one should proceed cautiously in areas where exposures are near to the permissible limits for an eight-hour shift (Brief and Scala, 1975; Hickey and Reist, 1979; Mason and Dershin, 1976).

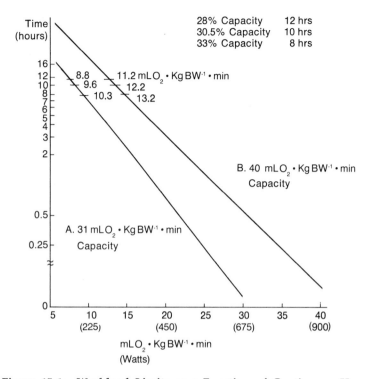

Figure 15-1: Workload Limits as a Function of Continuous Hours Worked   The average energy requirements of jobs or tasks, in milliliters of oxygen per kilogram body weight per minute and in watts, are shown on the horizontal axis. They are plotted against the continuous work time, in hours on a logarithmic scale, on the vertical axis. The values are based on one person with an aerobic work capacity of 31 mL of oxygen per kilogram of body weight per minute (A) and one person with a capacity of 40 (B). The first represents an average fitness level for female industrial workers and the second is an above average fitness level for male workers. Continuous work for periods exceeding eight hours requires the average energy requirement to be decreased proportionately from 33 percent to 30.5 percent to 28 percent of capacity, as is indicated for 10- and 12-hours shifts on each curve. This reduction in energy requirements is usually accomplished by working less rapidly or by taking additional recovery breaks during the shift. If a person is unable to adjust the energy requirements because, for example, the work is machine-paced, fatigue is more likely to result during extended working hours. (*Developed from information in Bink, 1962; Bonjer, 1962; Rodgers, Eastman Kodak Company, 1975*).

3. MENTAL AND PERCEPTUAL JOB DEMANDS

The capacities for mental work, such as problem solving, and perceptual work, such as inspection, are not easy to measure. Motivation and monotony confuse the measurement of task demands and may be primary determinants of productivity over extended hours of work. Visual tasks are those most frequently studied in the industrial setting, so this discussion will focus on them.

The phenomenon of visual fatigue is incompletely understood, but it can be aggravated by unsuitable lighting, small-sized objects, objects that contrast poorly with their background, poor definition of objects, illegibility, and movement of the object being viewed. Individual capacity for viewing an object, which is also affected by age, will be determined by eye and eye muscle characteristics, including color vision, and the use of corrective lenses (Ferguson, Major, and Keldoulis, 1974). If any of the environmental or task factors mentioned above is present, more of the worker's capacity will be needed to do the job. Consequently, there will be more potential for extended hours of work to result in lower production rates per hour. The harder the task, the more motivation necessary to sustain it. If this is already a problem in an eight-hour shift, more serious problems can be expected with extended hours. For example, during an eight-hour study of the performance of six textile workers on an afternoon shift, the time to make a reading increased 30 percent, the number of pieces of information processed dropped 25 percent, and the number of errors per reading increased sevenfold over the course of the shift. The data represent panel readings averaged over six weekdays for six female machinists. Although overtime was not part of this study, it can be anticipated that performance would stabilize at the lower level or fall off more if the shift were extended in this job.

Since work content will determine an individual's awareness of fatigue on a mental or perceptual task, and since there are very large differences among individuals, there is no easy way to determine which jobs should not be put on overtime schedules. Identification of worker-paced activities that show lower rates of productivity in the later, compared with the earlier, hours of the shift provide a starting point for evaluating ways to reduce visual or perceptual fatigue.

B. OTHER FACTORS AFFECTING PRODUCTIVITY IN OVERTIME SCHEDULES

1. MEASURING PRODUCTIVITY

In discussing productivity, we need to distinguish between the amount of work done or product produced per hour and the amount done per shift. A survey of 123 studies where hours were increased showed that productivity per hour dropped even though total units produced increased. The average output increase was 10 percent for a 25 percent increase in hours worked (Poper, 1970). Since one usually pays "time and one-half" for overtime, this drop in productivity per hour is especially costly. Individual motivation can overcome the factors that reduce output with extended hours, so there may be situations where productivity per hour is not negatively affected.

2. TYPE OF PAY PLAN—INCENTIVE OR TIME-BASED

Productivity per hour may be expected to decline somewhat as hours of work per shift increase when pay is based strictly on time at work. With an incentive system that continues to reward the worker for high performance throughout the extended hours, the fall in productivity may not be so great (Poper, 1970). Physical or perceptual fatigue may become the limiter in this instance.

3. WORKER- OR MACHINE-PACED JOBS

With increased hours of work per day, productivity per hour may fall if the system is paced by the worker (Poper, 1970). If the operation is machine-paced, the worker is forced to work at a fixed pace unless he or she can find ways to escape from the machine. Machine-paced operations should be carefully evaluated before being included in overtime schedules. If the pace is just acceptable for an eight-hour shift, it should be reduced if extra hours are worked. Based on productivity losses during overtime (Poper, 1970), approximately a 2.5 percent reduction in pace per additional hour (over eight) worked should reduce the potential for overload for most people. If the existing work pace on an eight-hour operation is fully acceptable for most people, it may not be necessary to reduce the pace as much when extended hours are worked.

C. OVERTIME BY INCREASING THE DAYS WORKED PER WEEK

The information from which one can predict the impact of working an extra day a week comes from studies on the effects of reducing the work week from 50–54 hours to 42–46 hours (Poper, 1970). Six-day weeks can result in reduced productivity per hour during the week, although the total output per week will be higher than that achieved on a five-day week. Cumulative fatigue, both physical and perceptual, can occur in some demanding jobs.

One can anticipate that the strongest dissatisfactions with extended days of work, especially if continued for a long period, will be psychosocial. The loss of personal time, especially family time on the weekend, may be acceptable infrequently or on an emergency basis. It is not desirable for extended periods, especially not for shift workers who have already lost social or family interaction through their work on the afternoon or night shifts.

D. GUIDELINES FOR OVERTIME SCHEDULING

Scheduled overtime is one way of dealing with a transient increased rate of production or the provision of service on an emergency basis in order to keep production levels constant, such as maintenance work. The following observations and guidelines should be considered when setting up an overtime schedule:

- If the workload (physical, perceptual, and/or mental) is at a level where people have difficulty sustaining performance over eight hours, extending the workday to 10 to 12 hours should be avoided.

- Scheduling an additional two to four hours per shift seldom results in 25 to 50 percent more productivity on a job. Frequently, productivity falls 5 to 10 percent below the expected increase (Brachman, Eastman Kodak Company, 1972; Poper, 1970).

- If the production line is machine-paced, extended hours of work will necessitate reevaluating the acceptable machine rate for most people. This will most often be a lower rate than that used over an eight-hour shift.

- Continuous overtime is not recommended. If the overtime period exceeds three months, productivity may fall even further in relation to hours worked than the 5 to 10 percent cited in the second guideline presented here.

- If frequent overtime is expected as a part of the job, it should be clear in the job description and at hiring.

- In production or service areas where volunteers are solicited for overtime work, it is usually preferable to rotate the extra work among several people. Specific situations may vary, but extended periods of eight or ten additional hours per week for one person can be expected to result in reduced productivity on the job by 5 to 10 percent (Brachman, Eastman Kodak Company, 1972).

- If continuous or long periods of overtime are required, alternative shift systems should be explored. These are discussed in Chapter 18.

- The preferability of extending each day by one or two hours versus extending the work week by one full day will depend on the type of work being done, as discussed earlier in this chapter. From a psychosocial viewpoint, most people prefer extending the usual workdays rather than giving up a traditional rest day with friends and family.

REFERENCES

Best, E. L. 1933. "A Study of a Change from 8 to 6 Hours of Work." *Bulletin No. 105,* Women's Bureau, U.S. Department of Labor. Washington, D.C.: U.S. Government Printing Office, 14 pages.

Bink, B. 1962. "The Physical Working Capacity in Relation to Working Time and Age." *Ergonomics, 5:* pp. 25-28.

Blum, M. L., and J. C. Naylor. 1986. *Industrial Psychology: Its Theoretical and Social Foundations.* Rev. ed. New York: Harper & Row, pp. 543-546.

Bonjer, F. H. 1962. "Actual Energy Expenditure in Relation to the Physical Working Capacity." *Ergonomics, 5 (1):* pp. 29-31.

Brief, R. S., and R. A. Scala. 1975. "Occupational Exposure Limits for Novel Work Schedules." *Journal of the American Industrial Hygiene Association, 36:* pp. 467-469.

Ferguson, D. A., G. Major, and T. Keldoulis. 1974. "Vision at Work: Visual Defect and the Visual Demands of Tasks." *Applied Ergonomics, 5 (2):* pp. 84-93.

Hickey, J. L. S., and P. C. Reist. 1979. "Adjusting Occupational Exposure Limits for Moonlighting, Overtime, and Environmental Exposures." *Journal of the American Industrial Hygiene Association, 40 (8):* pp. 727-733.

Mason, J. W., and H. Dershin. 1976. "Limits to Occupational Exposure in Chemical Environments Under Novel Work Schedules." *Journal of Occupational Medicine, 18 (9):* pp. 603-606.

National Electrical Contractors Association. 1969. *Overtime and Productivity in Electrical Construction.* A report on a continuing study by the National Electrical Contractors Association, 17 pages.

Poper, F. J. 1970. *A Critical Evaluation of the Empirical Evidence Underlying the Relationship Between Hours of Work and Labor Productivity.* Dissertation in the Department of Economics, New York University. No. 71-2329. Ann Arbor: University Microfilms, 260 pages.

Rutenfranz, J., W. P. Colquhoun, and P. Knauth. 1976. "Hours of Work and Shiftwork." In *Proceedings of the 6th Congress of the International Ergonomics Association, July 1976,* College Park, Maryland. Santa Monica, Calif.: Human Factors Society, pp. xlv-lii.

U.S. Department of Labor. 1947. "Hours of Work and Output." In *Bulletin No. 917.* Washington, D.C.: U.S. Government Printing Office. Cited in Blum and Naylor, 1968, p. 545.

CHAPTER 16

Shift Work—Use and Effects

CHAPTER 16. SHIFT WORK—USE AND EFFECTS

A. The Need for Shift Work

B. Patterns of Use of Shift Work

    1. Type of Industry
    2. Availability of Skilled Labor
    3. Nature of the Work
    4. The Economics of Shift Work and Product Profit Margin
    5. Flexibility for Supervision

C. Factors to Consider in Designing Shift Work Schedules

    1. Job Performance
       a. Measuring Productivity
       b. Error Rates
       c. Accident Rates
    2. Psychosocial Effects
       a. Worker Behavior
       b. Leisure Activity
       c. Family Life

References for Chapter 16

Some general needs for shift work in industry are discussed in this chapter, and performance and psychosocial effects of shift scheduling are reviewed.

A. THE NEED FOR SHIFT WORK

Shift work, defined as working other than daytime hours, is not a recent phenomenon. Night watches are referred to in earliest written history, and many a battle was initiated in the early hours to gain an advantage over a drowsy enemy. The widespread use of three-shift rotations in industry, however, has developed since the 1920s and represents about one-third of the shift work systems in use today (Sergean, 1971).

The earlier reasons for extending hours beyond the daytime shift were based on a need for continuous service or continuous processes. Examples of service needs are:

- Police.
- Fire.
- Hospital Staff.
- Security or night-watch people.
- Road gate or toll collectors.
- Military personnel.

These services are often standby, with routine duties being done until an emergency need for action arises. In some service jobs, it used to be possible for a person to "catch 40 winks" between calls, but this is seldom the case anymore. Night workers are expected to perform to the same standard as day workers in those tasks that are done on both shifts.

Continuous processes, such as operating electrical utilities, have also been a longstanding reason for night shift work. Bakeries often initiate their shifts in the early morning hours, as do many agricultural occupations. Ropemakers of the eighteenth century had to work nights because the technology for making rope prevented its being done when the sun was shining (George, 1966). Chemical manufacturing processes, including the making of antibiotics and other drugs, often require more than eight but less than 16 hours to complete. Consequently, a later shift has to be staffed in order to continue the manufacturing process once the chemical reaction is complete. Similar process-time determinants for shift work occur in the metal, paper, textile, and transportation industries, among others (Sergean, 1971).

In other instances, the need for shift or night work is determined by the time of delivery of the industry's service. For example, food service industries serving 6:00 a.m. breakfast require their employees to be at work in the early morning hours. Entertainers and communications workers have to fit their schedules to the times when people expect to have live entertainment or radio or television programs available. Laundries and photographic processing laboratories may offer overnight service, thereby necessitating night shift staffing. Computer programming and data analysis services often use night shift work because computer time is less expensive and more available in the late-night or early morning hours.

In the past 30 years, the growth of shift work has been linked to two additional factors that are more related to the economics of manufacturing. Rapid changes in technology and increased competition from overseas manufacturing companies with lower labor costs have stimulated American manufacturers to look for ways to lower unit costs and get maximum utilization of expensive manufacturing equipment (Owen, 1976). In less competitive times, production demands could be met with overtime and increasing capacity with additional equipment. By using equipment around the clock, however, a company recovers the equipment cost much earlier in the product's life cycle. Since many product life cycles have been dropping from more than ten years in the late 1960s to less than two years in 1983, it is important for industries to get paid back on their capital investments as quickly as possible. In addition to recovering equipment cost by using it over three shifts, this intensive use results in an earlier need to replace it. This permits the manufacturer to utilize improvements in technology that are rapidly making the older equipment obsolete (Sergean, 1971).

Two more reasons for an increase in shift work over the past 30 years are the social expectations of more access to services around the clock and the seasonal demands for products. These make occasional rotating shift work schedules preferable to simple overtime arrangements. As the shift work population has increased, so has the need for extended hours in supporting services, such as food stores and restaurants, transportation, and entertainment. Although this is a small part of the growth in shift work, it is a social expectation that will probably increase, not decrease, as more people work shifts in manufacturing (Murphy, Eastman Kodak Company, 1973). The reasons for using shift work instead of overtime to satisfy seasonal production or service demands are related to productivity levels; these are discussed earlier in this section under overtime.

B. PATTERNS OF USE OF SHIFT WORK

Obtaining figures on the numbers of workers engaged in some type of shift work is complicated by different data collection guidelines among countries and industry groups. In the United States, shift work has been defined in a Bureau of Labor Statistics survey (1975) as any work shift that starts other than between 7:00 to 9:00 a.m. On that basis, about 27 percent of the American work force participates in some kind of shift work (Colligan, 1981). Other countries, surveyed in different years and using different guidelines to choose the industry sample, give figures of 23 percent Japan in 1978 (Kogi, 1981), 18 percent in the United Kingdom in 1964, 20 percent in the Netherlands in 1969, and 22 percent in France in 1974 (Colquhoun, Ghata, and Rutenfranz, 1975). The percentage of shift workers varies among different industries, as can be seen in Table 16-1.

A 1964 survey of manual shift workers in the United Kingdom was made in manufacturing, mining, and quarrying (except coal), public utilities and national and local government services, laundries, dry cleaning, motor repairs and garages, and boot and shoe repair industries (Ministry of Labour, 1965a; Ministry of Labour, 1965b). The percentages of shift workers on each of the most common shift schedules were found to be:

- 41 percent on three-shift systems (22 percent on continuous 7-day and 19 percent on discontinuous 5-day)
- 17 percent on double-day shift systems.
- 23 percent on alternating day and night systems.
- 12 percent on permanent nights.
- 7 percent on part-time evening employment.
- Less than 0.5 percent on other systems.

In more recent years the number of shift workers has increased, and there has been more acceptance of rapidly rotating shift systems and extended hours per shift with shorter work weeks. Rapid rotation is found frequently in continuous three-shift systems, and extended hours of work are usually associated with alternating or permanent night and day shifts (Anon., 1970a). A more detailed discussion of shift schedules can be found in Chapter 17.

The frequency of selection of three-shift, two-shift-involving-night-work, and two-shift-daytime schedules is related to the type of industry, the availability of skilled labor, the nature of the work, the economics of each shift system relative to the product's profit margin, and the flexibility of the system for allowing changes as production levels vary. These are discussed briefly in the following subsections.

1. TYPE OF INDUSTRY

The type of industry determines the choice of shift work schedules for the reasons discussed in the first section of this chapter: the need to deliver a product or service at a specific time, the time it takes to complete a manufacturing cycle, the continuous nature of a manufacturing process, or the public's expectation of continuous service, as in hospital staffing. Tradition may also play an important part in the choice of shift work schedules, particularly in metal or mining industries where a long history of night work exists (Sergean, 1971). In some industries, such as textiles and food service, a need for part-time expansion ofthe work force at specific times of day dictates the choice of shift schedule. Table 16-2 summarizes the percentage of shift workers in three-shift, two-shift-involving-night-work, and two-shift-daytime schedules in Great Britain in the mid 1960s.

Table 16-1: Percent of Total Workers in Shift Work by Industry Type—United States, 1975   The total number of workers (column 2) and percentage (%) of shift workers (column 3) are shown for 30 types of industries (column 1). The data are based on a United States survey in 1975. Shift work is defined as starting work outside of the hours of 7 to 9 a.m. By this definition, industries with more than one-third of the work force on shift work include hospitals, transportation, food and kindred products, private households, primary metal industries, postal services, textile mill products, and rubber and plastic products. Not included in this tabulation are some services with large numbers of shift workers, such as fire and police departments and security services. See the text and Table 16-2 for a discussion of the types of shift schedules used in some of these industries. (*Bureau of Labor Statistics; reprinted by permission of the publisher from Table 1, p. 199, "Methodological and Practical Issues Related to Shiftwork Research" by M. J. Colligan in* Biological Rhythms, Sleep and Shiftwork *by L. C. Johnson, D. I. Tepas, W. P. Colquhoun, and M. J. Colligan, eds. Copyright 1981, Spectrum Publications, Inc. Jamaica, N.Y.)*

Industry Type	Total Workers (in thousands)	Percent Shift Workers
Hospital	1,117	36.9
Education	1,115	17.0
Other transportation services	763	39.6
Food and kindred products	593	42.7
Health	572	29.9
Private household	507	40.7
Transportation equipment	498	29.9
Primary metal industries	402	37.5
Nonelectrical machinery	363	18.9
Printing and publishing	327	28.5
Electrical equipment and supplies	278	14.8
Postal	277	45.8
Fabricated metal products	261	23.6
Other professional services	246	17.3
Welfare	221	21.8
Textile mill products	216	34.4
Chemical and allied products	199	19.7
Railroad and railway express service	177	32.6
Paper and allied products	176	32.4
Rubber and plastic products	174	35.0

(*continued*)

Table 16-1   (*Continued*)

Industry Type	Total Workers (in thousands)	Percent Shift Workers
Stone, clay, and glass products	154	28.5
Lumber and wood products	130	25.4
Instruments and related products	56	12.9
Apparel and other textile products	54	5.2
Miscellaneous durable goods industries	49	12.0
Petroleum and coal products	42	17.7
Furniture and fixtures	33	7.7
Ordnance	29	15.1
Tobacco	20	32.8
Leather and leather products	17	7.3

This summary of shift schedules by industry type is similar to a more recent Japanese survey (Kogi, 1981) where 84 percent of the shift workers in utilities were on three-shift continuous 7-day schedules, and 85 percent of the
mining shift workers were on three-shift discontinuous 5-day schedules. In
manufacturing, 38 percent of the shift workers were working three-shift continuous rotations, 32 percent were on day/night schedules, 14 percent on three-
shift discontinuous work, and 10 percent were working double-days (day and
afternoon shifts). The last shift schedule is often used in light manufacturing
where a large number of women are employed. Before the 1970s, there were
laws in many countries prohibiting the employment of women or children on
night shift work, so men still predominate in many of the three-shift continuous
rotation and day/night schedules being worked (Sergean, 1971).

2. AVAILABILITY OF SKILLED LABOR

The availability of skilled labor may determine the choice of shift schedule because of the psychosocial factors that make some schedules undesirable to certain people. These factors are discussed in the next section and are included in
the listings of advantages and disadvantages for each shift schedule described
in Chapter 17. The working population most likely to affect the choice of shift
schedule is the maintenance and repair staff. Skilled welders and electrical,
plumbing, and equipment maintenance personnel, for example, may be difficult
to find if a shift schedule requires continuous evening or night work or extended
hours for many weeks. The pay differential may attract them initially, but it may
not be sufficient to continue to justify the loss of weekends or evenings with
their families (Sergean, 1971).

Table 16-2: Percentage of Shift Workers on Three Schedules by Industry Type—United Kingdom, 1964. The percentage of shift workers (column 3) within nine different types of industries (column 2) on each of three shift schedules—three shifts, two shifts with night shift, and two shifts in daytime (column 1)—are shown. The 3-shift schedules are shown as a combination of all types of rotations or fixed shifts (see the text section on shift schedules). The 2-shift schedules are shown both as combined percentages and broken down into alternating (A) and permanent (P) day and night shifts, double-day shifts (D), and part-time evening and day shifts (E). The data are from a 1964 survey in the United Kingdom. This pattern of shift work can be expected to reflect current shift use patterns although the specific schedules in use may vary as new ones are developed. *(Adapted from information in Sergean, 1971).*

Shift Schedule	Industry Type	Percent of Shift Workers on This Schedule*
A. 3 shifts, all types	Utilities	83
	Chemicals	81
	Metal manufacturing	71
B. 2 shifts, with night shift	Shipbuilding and maritime engineering	94 (40A, 54P)
	Vehicle manufacturing	77 (62A, 15P)
	Engineering and electrical goods	62 (36A, 26P)
C. 2 shifts, daytime	Textiles	48 (35D, 13E)
	Clothing/footwear	47 (21D, 26E)
	Food/drink/tobacco	43 (24D, 19E)

*A = Alternating night and day shifts.
 P = Permanent night or day shift.
 D = Double-day shifts (daytime and afternoon/evening).
 E = Part-time evening shift and day shift.

3. NATURE OF THE WORK

The nature of the work will determine which shift schedules are appropriate; this has been discussed under overtime and will be mentioned again under job performance capabilities. Day/night and permanent night schedules often have extended hours, up to 12 hours per shift. Extended hours can also be found in three-shift schedules in situations where workers are absent, on vacation, or temporarily assigned to different work. The other workers fill in the openings on a temporary basis, often working an additional four to eight hours once or twice a week. As mentioned earlier, if the job is physically demanding for an eight-hour shift, extended hours of work on it are not recommended. The same is true of much mentally or perceptually demanding work. Seven-day continuous night shift schedules may result in an accumulated fatigue that can affect performance on a demanding task (Knauth, Rohmert, and Rutenfranz, 1979).

4. THE ECONOMICS OF SHIFT WORK AND PRODUCT PROFIT MARGIN

The economics of the shift system relative to the profit margin of the product or service must be considered, since shift schedules include premium pay that raises labor costs and, thus, may raise the unit cost of the product. However, because the increased utilization of capital equipment tends to reduce unit cost considerably, the impact of the increased labor costs may not be noticeable. Table 16-3 gives the results of a study of shift payment differentials in the United Kingdom in 1968.

Table 16-3: Shift Payment Differentials as a Percentage of the Minimum Hourly Rate—United Kingdom, 1968. Three commonly used shift systems are shown in column 1 with the number (n) of plants or departments surveyed using each schedule in column 2. The average percentage increase in shift pay for each schedule is compared to the minimum hourly rate and is shown in column 3, with the standard deviation of this average value (a measure of the amount of variance in the sample) in column 4. The range of percentage increases is shown in column 5; this gives a further indication of the variance of the sample. Permanent night shift schedules are about twice as expensive as double-days or 5-day 3-shift schedules. This can be attributed to the premium pay for weekend work. *(Sergean et al., 1969).*

		Shift Pay Differential		
Shift Schedule	n	Mean, or Average Percent Increase*	± 1 Standard Deviation	Range, Percent
Permanent nights	118	24.7	8.6	7.1 − 50.0
3-shift, discontinuous (5-day)	56	14.9	6.5	2.0 − 29.7
Double-days	71	12.2	6.3	2.0 − 33.3

5. FLEXIBILITY FOR SUPERVISION

The flexibility offered supervision in responding to seasonal or temporary rises and declines in the production rate will determine which shift schedule is preferable for a given department or industry. A 12-hour shift schedule, for example, essentially abolishes the option of overtime work by existing staff. If the nature of the work requires minimal skills so that temporary staff can be used as needed to meet high production demands, a 12-hour schedule will not be as problematic. If more skilled work is involved, this loss of the overtime option will limit the supervisor in his or her response to transient high production schedules.

There are several hundred variations on the shift schedules described in this section, many of which have been developed through dialogue between shift workers and management in attempts to balance the negative and positive aspects of each schedule. General guidelines for the design of shift schedules are given later. More detailed discussions of how to design and analyze shift and overtime schedules and assign employees to them can be found in the *Work Schedule Design Handbook* (U.S. Department of Housing and Urban Development, 1978) and in the work of several other investigators (Knauth, Rohmert, and Rutenfranz, 1979; Tejmar, 1976; Sergean, 1971).

C. FACTORS TO CONSIDER IN DESIGNING SHIFT WORK SCHEDULES

The most important considerations in shift schedule design are: the job performance capabilities of people at different hours, the physiological effects of night work, and the psychosocial effects of evening and night work on the workers and their families and friends. The physiological effects of night work and sleep loss have been discussed in Part II. Performance capabilities at different hours and the psychosocial implications of shift work are discussed in the subsections that follow.

1. JOB PERFORMANCE

Two types of influence on a person's ability to perform on the job can be associated with shift work schedules: performance changes as a function of the time of day or night that relate to the body's circadian rhythms (see Part II); and performance affected by an accumulated fatigue associated with inadequate quality and quantity of sleep while assigned to night shift work. These influences cannot be separated very effectively in studies of shift work. The usual measures of job performance capability in shift work studies have been productivity, error rates, and accident rates.

a. Measuring Productivity

Because many shift work jobs do not lend themselves to the common productivity measurements of units per time, time per unit, or total units per shift, it is not always possible to get productivity data for comparisons among shifts. In a study of the response time of teleprinter switchboard operators at different times of day (Browne, 1949), a significant lengthening of the response time was noted in the early morning hours (Figure 16-1). This could not be explained by a change in the call load, although more calls usually meant less delay at all times of day. The average difference in response time between day and night shifts became less in the last four days of the week on the late night shift, but

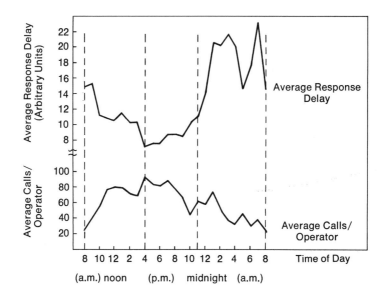

Figure 16-1: Response Time and Workload for Teleprinter Switchboard Operators on a Continuous Shift System. The work load or average calls per operator per hour (lower curve) and performance—measured as the average delay for calls during a given hour (upper curve)—are shown for teleprinter switchboard operators over a 24-hour period (horizontal scale). Shift changes are marked by dotted vertical lines at 4 p.m., 11 p.m., and 8 a.m. Longer response delays are interpreted to indicate lower operator attention levels or fatigue. Increases in work load create time pressure and are usually accompanied by shorter delays, which are expressed in arbitrary units to permit direct comparisons across hours of the day. The large increase in response delay from midnight to 8 a.m. cannot be explained by the lower call volume alone and is attributed to a reduced capacity of the operators to perform in the early morning hours. This delay does not necessarily translate into a decreased productivity rate on the late night shift for teleprinter switchboard operators because the volume of calls is then quite low. However, other operator-paced operations could be expected to show reduced productivity because of this early morning capacity effect. (*Adapted from Browne, 1949*).

it was still 25 to 30 percent longer on the night shift. Performance on each shift was best in the middle of the shift week. The author concluded that time of day accounted for the effect.

In a 1948 survey of United Kingdom industries (referred to earlier in Table 16-3), respondents were asked to assess productivity differences between day and night shift work. For those who had made comparative measurements, 58 percent observed no difference in productivity between the shifts. Slightly less productivity was observed in 26 percent, much less in 3 percent, and more in 13 percent of the respondents' studies (Anon., 1970b). Reductions in productivity, therefore, are not necessarily to be expected in going to a night shift schedule. The teleprinter shiftboard operators, for example, showed reduced performance, but the final productivity measure would not have indicated any serious performance failure. The delay was not sufficient to result in missed calls or to affect the total production rate for the shift. In operations where timing is critical, as in externally paced operations (see Part IV), the time of day would be more likely to produce a productivity effect.

b. Error Rates

The classic study of error rates by time of day was done in a Swedish gasworks (Bjerner, Holm, and Swensson, 1955). The number of errors made in over 175,000 recordings from 1921 to 1931, primarily by three men, was plotted against the time of day. The highest number of errors occurred at 3:00 a.m. with a second, lower peak occurring at 3:00 p.m. All three operators showed similar peaks in errors in the 1:00 a.m. to 4:00 a.m. period. Comparisons of the gasworks data with readings made in another gasworks and in a paper mill show that the same bimodal distribution of errors occurs in those plants. Figure 16-2 illustrates these error patterns for four studies, two from the first gasworks and one each from the second gasworks and a paper mill. All studies were done under continuous three-shift schedules.

In the Swedish gasworks study, the operators committed errors in readings. Errors of omission are also often higher in the night shift hours, as has been demonstrated in a study of locomotive drivers (Hildebrandt, Rohmert, and Rutenfranz, 1974). Failure to respond to a warning light and an auditory signal, each 2.5 seconds in duration, indicated a lack of attention by the driver. These signals were presented to the driver about every 20 minutes, and the time available to respond was five seconds. If there was no response to the signals, the "hooter" sounded for 30 seconds, and the driver had to act to prevent automatic braking of the locomotive. The number of times the "hooter" sounded gave an indirect evaluation of the alertness level of the driver (more "hooter" implying less alertness). The shape of the alertness curve mirrors the error studies above, with the alertness being lower just after lunch and in the early morning hours.

c. Accident Rates

The influence of shift work scheduling on accident rates in industry has been studied by many investigators, and the results have covered the range of pos-

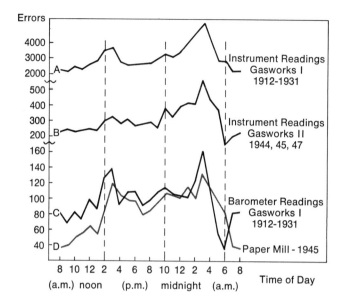

Figure 16-2: Number of Errors per Hour of the Day at Two Swedish Gasworks and a Paper Mill   The average number of errors made per hour of the day in reading instruments are shown for several years of work in two Swedish gasworks (A, B, and C) and one paper mill (D). In Gasworks I (curves A and C), the readings were made over a 20-year period, mostly by three people. In over 175,000 readings of the instruments, about 75,000 errors were found. In 7300 barometer readings in the period (C), 2350 errors were found. Later studies at another gasworks (B) showed 7417 errors in 26,300 readings over three years. A one-year study at a paper mill (D) found 2078 errors in 8800 readings. In all four studies there are clear peaks in the number of errors made at and around 3 a.m. with smaller, but distinct, increases in errors at 3 p.m. and 10 p.m. Shift changes are shown by vertical dotted lines at 2 p.m., 10 p.m., and 6 a.m. The consistency of these increased errors between plants suggests that it is a time-of-day effect and not specific to the gasworks industry. *(Adapted from Bjerner, Holm, and Swensson, 1955).*

sible effects from fewer night shift accidents (Vernon, 1920) to no difference or a slight increase on night shift work (Wyatt and Marriott, 1953) to a significant increase (Brandt, 1969). Confounding factors in these analyses of accidents are differences in reporting guidelines and changing characteristics of the work force, especially in terms of age. In the latter situation, a younger work force on night shift work may exhibit higher accident rates that confound any hour-of-the-day effect (Murphy, Eastman Kodak Company, 1980).

Because of the previously discussed effects of night work on mental work capacity (Bjerner, Holm, and Swensson, 1955) and alertness (Hildebrandt, Rohmert, and Rutenfranz, 1974), as well as the perception of some shift workers that work is "harder" in the early morning hours (Sergean, 1971), one can probably expect a slight increase in accidents in night shift work when it is compared with similar work on day shifts. The detectability of this effect should increase with more demanding work.

2. PSYCHOSOCIAL EFFECTS

Most studies of shift work have found that the primary determinants of its acceptability relate to psychological and social factors rather than to physiological problems (Gordon, McGill, and Maltese, 1981; Wedderburn, 1981; Murphy, Eastman Kodak Company, 1972, 1973, 1976, 1980; Rutenfranz, Colquhoun, and Knauth, 1976; Sergean, 1971; Mott et al., 1965). The seriousness of these psychosocial disruptions depends on the shift schedule worked as well as on the age, sex, marital status, number of children, and personality of the shift worker. The effects of shift work, especially three-shift systems, on worker behavior, leisure activity, and family life are discussed in the following subsections.

a. Worker Behavior

One of the few studies where the personalities of day and night shift workers were assessed showed that shift workers are not different from day workers in their degree of introversion or extroversion or in other personality characteristics (De la Mare and Walker, 1968). Any longstanding shift worker population will be more selected than a day shift population because people who do not tolerate night or evening shift work will voluntarily drop out as soon as they find more acceptable work schedules. This selection may be related to physiological difficulties in adjusting the body's circadian rhythms to night schedules, or it may be explained by psychological problems, most often the disruption of family life. Consequently, any comparison of the behavior of shift and day workers has to recognize that the shift workers are "survivors," and day workers will include those who could not tolerate the night work or have been restricted from it for medical reasons (Mott et al., 1965; Sergean, 1971).

These difficulties can be avoided by comparing the attitudes of the same group of shift workers towards two shift systems; a 5/2 (five days on/two days off) C-B-A rotation schedule for a 7-day department with four work crews (three-shift continuous); and a 2-2-3 rapid-rotation C-A-B four-crew schedule. The psychological, or behavioral, characteristics of interest are irritability and depression. A group of 117 shift workers was asked to check a symptom list that included these terms; follow-up interviews with 55 of them clarified their responses (Murphy, Eastman Kodak Company, 1976). During the C, or midnight to 8:00 a.m., shift, about twice as many of the shift workers had psychological complaints on the 5/2 as had them on the rapidly rotating schedule. Only about 22 percent of them considered the symptoms significant enough to report, however. These complaints related to irritability, attributed to accumulated fatigue on the five continuous night shifts, and to depression associated with isolation from their family and friends, concern about leaving a wife alone at night, and the loss of weekends. Depression could also be a factor during the B, or 4:00 p.m. to midnight shift, because of social isolation and the loss of weekends; irritability was not as prevalent during this shift rotation, however. In an earlier study of a 5-day, weekly rotating, three-shift discontinuous system (weekends off), 65 percent of the 29 workers interviewed expressed feelings of irritability near the end of the week on the C shift that affected their relationships with their families (Murphy, Eastman Kodak Company, 1972).

These studies suggest that about one-fifth of the shift worker population may experience the psychological symptoms of irritability and depression on 7-day continuous weekly-rotating schedules (5/2). Only about one-seventh of them have the same symptoms on a 2-2-3 rapidly rotating schedule (C-A-B). The other 80 to 90 percent of the shift workers apparently do not experience these symptoms strongly enough to consider them major problems with a given shift schedule (Wedderburn, 1967).

b. Leisure Activity

The influence of shift work on a person's ability to participate in community, recreational, and educational activities during his or her leisure time will depend both on the shift schedule and the local community. If the community has a large number of shift workers, provisions may be made to provide services outside the usual daytime working hours. Some of these include: scheduling some educational programs during both the day and evening so shift workers on rotating shifts can pursue further education; opening banks for an hour or two in the hours between 8:00 and 10:00 a.m. so night shift workers are accommodated; and keeping some restaurants and entertainment facilities open after midnight for the benefit of evening shift workers (Murphy, Eastman Kodak Company, 1972).

Rotating shifts make participation in any regularly scheduled recreational program, such as a bowling or golf league, very difficult. This may make a person feel less positive about a continuous three-shift/four-crew rotating shift schedule than about a fixed, or even a rapidly rotating, shift system (Wedderburn, 1967). Social interactions with friends on day work are usually curtailed when the worker is on the afternoon or night shift and may not be reactivated spontaneously when the day shift schedule is worked. The longer the shift rotation, the more the social isolation of the worker (Sergean, 1971).

Leisure time on the weekends is much more important to many shift workers than leisure time available during the week. Shift schedules that result in many calendar weekends becoming part of the regular work week are unpopular with most workers. Even with the premium pay for Saturday or Sunday work, some workers have difficulty in adjusting to the treatment of the weekend as just another workday. Saturday overtime, on the other hand, is socially more acceptable to them because they see it as partly voluntary and temporary (Wedderburn, 1967). Overtime on top of rotating shift work is very unpopular with some workers (Sergean, 1971).

c. Family Life

The social effects of shift work on the family are closely related to a worker's feeling of satisfaction or dissatisfaction on a given schedule (Murphy, Eastman Kodak Company, 1972, 1973, 1976, 1980). Young workers with school-age children are more reluctant to give up weekends and evenings than are older workers. Although rapidly rotating schedules, with no more than three days in a row on a shift, increase the opportunities for interaction with the family, the schedules are confusing for some spouses and friends to remember. Consequently, the social isolation of the shift worker may continue on the more rapid-rotation systems until others become accustomed to the schedules. Weekend work was mentioned by about 40 percent of the shift workers in one study (Murphy, Eastman Kodak Company, 1980) as a major source of dissatisfaction on an A-B-C rapid-rotation (2-2-3) shift schedule. Interviews established that the weekend was identified as a time for family recreation, church, and work around the house. About 60 percent of the wives of shift workers disliked the weekend work in rapidly rotating schedules (Banks, 1956). Women workers found shift work schedules difficult if they were raising a family, especially as a single parent, because of the need to make special arrangements for the children when they were at work and the children were at home (Wedderburn, 1981). This partially accounts for the small number of women in three-shift continuous schedules and the clustering of more of them in the double-day schedules (Sergean, 1971).

Shift work, especially 7-day continuous three-shift schedules and extended hours of work for several days at a time, tends to interfere with family life and produces some dissatisfaction because of that; it should not be entered into without consideration of these factors. The schedule that provides the most opportunity for family interaction, especially on the weekends, is likely to be most acceptable for many of the shift workers. It should be noted that up to 50 percent of shift workers do not find these interferences problematic (Wedderburn, 1981). The suitability of a given schedule will vary with other environmental and social factors, with individual differences in the working population, and with the type of work. Examples of several types of shift schedules are given in the next chapter, and the health and psychosocial determinants of their acceptability are identified, where possible.

REFERENCES

Anon. 1970a. "Shiftworking: The General Picture." Chapter 7 in *Hours of Work, Overtime, and Shiftworking*. National Board for Prices and Incomes Report No. 161, Cmnd 4554, December 1970. London: Her Majesty's Stationery Office, pp. 57-69. Included in Colquhoun and Rutenfranz, 1980, pp. 3-15.

Anon. 1970b. "The Economics of Shiftworking and Problems Facing Management." Chapter 8 in *Hours of Work, Overtime, and Shiftworking*. National Board for Prices and Incomes Report No. 161, Cmnd 4554, December 1970. London: Her Majesty's Stationery Office, pp. 70-80. Included in Colquhoun and Rutenfranz, 1980, pp. 427-437.

Banks, O. 1956. "Continuous Shift Work: The Attitude of Wives." *Occupational Psychology, 30:* pp. 69-75.

Bjerner, B., A. Holm, and A. Swensson. 1955. "Diurnal Variation in Mental Performance: A Study of Three-Shift Workers." *British Journal of Industrial Medicine, 12:* pp. 103-110. Included in Colquhoun and Rutenfranz, 1980, pp. 255-262.

Bureau of Labor Statistics. 1975. *Current Population Survey,* May 1975. Washington, D.C.: U.S. Bureau of Labor Statistics. Cited in Colligan, 1981.

Brandt, A. 1969. "Influence of Shift Work on the Health and Pathology of Workers." Original paper in German ("Uber den Einfluss der Schichtarbeit den Gesundheitzustand und das Krankheitsgeschehen der Werktatigen"). From the *Proceedings of an International Symposium,* January 31–February 1, 1969, Oslo, Norway, pp. 124-153. Translated by Information Company of America, 2204 Walnut Street, Philadelphia, Penn. 19103, 25 pages.

Browne, R. C. 1949. "The Day and Night Performance of Teleprinter Switch-board Operators." *Occupational Psychology, 23:* pp. 121-126. Included in Colquhoun and Rutenfranz, 1980, pp. 249-255.

Colligan, M. J. 1981. "Methodological and Practical Issues Related to Shiftwork Research." In *Biological Rhythms, Sleep and Shift Work,* edited by L. C. Johnson, D. I. Tepas, W. P. Colquhoun, and M. J. Colligan. New York: S.P. Medical and Scientific Books, pp. 197-203. Also published in *Journal of Occupational Medicine, 22 (3):* pp. 163-166.

Colquhoun, W. P., and J. Rutenfranz. 1980. *Studies of Shiftwork.* London: Taylor & Francis, 468 pages.

Colquhoun, W. P., J. Ghata, and J. Rutenfranz. 1975. "Shift Work. Biomedical and Psycho-Social Aspects." *Report of Consultation on Shift Work.* November 10–14, 1975. Geneva: W.H.O. (unpublished). Cited in Rutenfranz, Colquhoun, and Knauth, 1976.

De la Mare, G., and J. Walker. 1968. "Factors Influencing the Choice of Shift Rotation." *Occupational Psychology, 42:* pp. 1-21.

Folkard, S. 1981. "Shiftwork and Performance." In *Biological Rhythms, Sleep, and Shift Work,* edited by L. C. Johnson, D. I. Tepas, W. P. Colquhoun, and M. J. Colligan. New York: S.P. Medical and Scientific Books, pp. 283-305.

George, M. D. 1966. *London Life in the Eighteenth Century.* Hammondsworth, England: Peregrine. Cited in Sergean, 1971.

Gordon, G. C., W. L. McGill, and J. W. Maltese. 1981. "Home and Community Life of a Sample of Shift Workers." In *Biological Rhythms, Sleep, and Shift Work,* edited by L. C. Johnson, D. I. Tepas, W. P. Colquhoun, and M. J. Colligan, 1980. New York: S.P. Medical and Scientific Books, pp. 357-369.

Hildebrandt, G., W. Rohmert, and J. Rutenfranz. 1974. "Twelve and Twenty-Four Hour Rhythms in Error Frequency of Locomotive Drivers and the Influence of Tiredness." *International Journal of Chronobiology, 2:* pp. 97-110. Cited in Folkard, 1981.

Knauth, P., W. Rohmert, and J. Rutenfranz. 1979. "Systematic Selection of Shift Plans for Continuous Production with the Aid of Work-Physiological Criteria." *Applied Ergonomics, 10 (1):* pp. 9-15.

Kogi, K. 1981. "Research Motives and Methods in Field Approaches to Shift Work." In *Biological Rhythms, Sleep, and Shift Work,* edited by L. C. Johnson, D. I. Tepas,

W. P. Colquhoun, and M. J. Colligan. New York: S.P. Medical and Scientific Books, pp. 205-223.

Ministry of Labour. 1965a. "Shift Working." *Ministry of Labour Gazette, 63 (4).* Cited in Sergean, 1971.

Ministry of Labour. 1965b. "Shiftworking: Regional Analysis." *Ministry of Labour Gazette, 63 (6).* Cited in Sergean, 1971.

Mott, P. E., F. Mann, Q. McLoughlin, and D. Warwick. 1965. *Shift Work: Its Social, Psychological, and Physical Consequences.* Ann Arbor: University of Michigan Press, 351 pages.

Owen, J. D. 1976. "The Economics of Shift Work and Absenteeism." In *Shift Work and Health: A Symposium.* Sponsored by the National Institute of Occupational Safety and Health, U.S. Department of Health, Education and Welfare, June 12-13, 1975, Cincinnati, Ohio. Washington, D.C.: U.S. Government Printing Office, pp. 213-217.

Rutenfranz, J., W. P Colquhoun, and P. Knauth. 1976. "Hours of Work and Shiftwork." In *Proceedings of the 6th Congress of the International Ergonomics Association, July, 1976,* College Park, Maryland. Santa Monica, Calif.: Human Factors Society, pp. xlv-lii.

Sergean, R. 1971. *Managing Shiftwork.* London: Gower Press, Industrial Society, 242 pages.

Sergean, R., D. Howell, P. J. Taylor, and S. J. Pocock. 1969. "Compensation for Inconvenience: An Analysis of Shift Payments in Collective Agreements in the U.K." *Occupational Psychology, 43:* pp. 183-192. Included in Colquhoun and Rutenfranz, 1980, pp. 439-448.

Tejmar, J. 1976. "Shift Work Round the Clock in Supervision and Control. Schedule of Rewarded and Unrewarded Time." *Applied Ergonomics, 7:* pp. 66-74. Included in Colquhoun and Rutenfranz, 1980, pp. 406-414.

U.S. Department of Housing and Urban Development. 1978. *Work Schedule Design Handbook: Methods for Assigning Employees' Work Shifts and Days Off.* Capacity Sharing Program, Office of Policy Development and Research, No 624-521/1063. Washington, D.C.: U.S. Department of Housing and Urban Development, U.S. Government Printing Office, 368 pages. Also available from Institute for Public Program Analysis, 1328 Baur Blvd., St Louis, Mo. 63132.

Vernon, H. M. 1920. "The Speed of Adaptation of Output to Altered Hours of Work." *Reports of the Industrial Fatigue Research Board, No. 6.* London: His Majesty's Stationery Office, 33 pages.

Wedderburn, A. A. I. 1967. "Social Factors in Satisfaction with Swiftly Rotating Shifts." *Occupational Psychology, 41:* pp. 85-107. Included in Colquhoun and Rutenfranz, 1980, pp. 275-297.

Wedderburn, A. A. I. 1981. "How Important Are the Social Effects of Shift Work?" In *Biological Rhythms, Sleep, and Shift Work,* edited by L. C. Johnson, D. I. Tepas, W. P. Colquhoun, and M. J. Colligan. New York: S.P. Medical and Scientific Books, pp. 257-269.

Wyatt, S., and R. Marriott. 1953. "Night Work and Shift Changes." *British Journal of Industrial Medicine, 10:* pp. 164-168.

CHAPTER 17

Shift Schedules

CHAPTER 17. SHIFT SCHEDULES

A. Five-Day Shift Schedules

    1. Two Eight-Hour Shifts
    2. Three Eight-Hour Shifts—Three-Shift Discontinuous Schedules
    3. Twelve-Hour Shifts

B. Seven-Day Shift Schedules

    1. Eight-Hour Shifts—Fixed
    2. Eight-Hour Shifts—"Weekly" Rotation
    3. Eight-Hour Shifts—Rapid Rotation
    4. Eight-Hour Shifts—Combination Fixed and Rotating
    5. Twelve-Hour Shifts
    6. Combination Eight- and 12-Hour Shifts

C. Special Five- and Seven-Day Shift Schedules

    1. Five-Day with Consistent Saturday Overtime
    2. Reduced Hours per Week—Weekday/Weekend Plan
    3. Reduced Hours per Week—40 Hours or Less, Five or Nine Crews

References for Chapter 17

Changing attitudes towards job satisfaction and workers' expectations for job conditions make the design of good shift systems imperative. Choice of the optimal shift system for a given job or production department has to take into account the production needs, the specific type of work (physical, perceptual, mental), the desired population of skilled workers on the job and the physiological and psychosocial consequences of the schedules available. In addition, legal limitations on the number and distribution of hours worked per week, through the Fair Labor Standards Act and union contracts, must be considered when new shift systems are designed. No one system is likely to be optimal for all jobs.

In this chapter, three five-day and six seven-day shift schedules are presented and the advantages and disadvantages of each are summarized. Some special five- and seven-day schedules are also discussed. (The information presented here is based on studies by Knauth, Rohmert, and Rutenfranz, 1979; Murphy, Eastman Kodak Company, 1969-1981; Sergean, 1971.)

A. FIVE-DAY SHIFT SCHEDULES

These schedules are most commonly seen in manufacturing operations where production demands have increased beyond the amount that can be accomplished during the day shift. Management either has to commit additional capital expenditures for equipment or utilize the existing equipment better by extending the working hours. If production needs would mean routine overtime for the existing workforce, the addition of a second shift is preferable from the standpoint of long-range productivity (Poper, 1970). The choice of the proper five-day schedule from ones in Tables 17-1 through 17-3 will depend on the type of work being done and the demands of the manufacturing process.

The following assumptions are made in the schedules shown here:

- The organization needs to operate 80 to 120 hours per week and requires an equal number of employees on all shifts.

- Each employee works 40 hours per week and has a minimum rest period of 12 hours between work periods.

The shift and rest periods in the tables are defined as follows in this analysis:

$$
\begin{aligned}
\text{A shift} &= \text{8:00 a.m. to 4:00 p.m. (day shift)} \\
\text{B shift} &= \text{4:00 p.m. to midnight (afternoon or evening shift)} \\
\text{C shift} &= \text{midnight to 8:00 a.m. (night or early morning shift)} \\
\text{R} &= \text{rest day} \\
\text{D} &= \text{day shift, often 6:00 a.m. to 6:00 p.m.} \\
\text{N} &= \text{night shift, often 6:00 p.m. to 6:00 a.m.}
\end{aligned}
$$

Four weeks of the schedules for two or three crews are shown for each shift variant.

1. TWO EIGHT-HOUR SHIFTS

There are three common variations on a schedule of two eight-hour shifts:

- Fixed shifts (A and B only).
- Weekly alternation of A and B shifts.
- Weekly alternation of A and C shifts.

These are diagrammed in Table 17-1.

In the fixed or permanent shift schedule, seniority usually determines who will work the day shift; day work is considered preferable for social reasons and is a reward or prerogative of the longer service worker. An advantage of the fixed shift schedule is its predictability. Disadvantages of the schedule are that daytime supervision may be unfamiliar with the B shift workers and feel less in control of their production rate and quality levels. Younger workers who find themselves on straight B shift may experience dissatisfaction because of the interference with their social life that afternoon-evening work can have. However, some may welcome the fixed B shift if they are attending school during the day.

A and B shifts that alternate on a weekly basis (double-days) are frequently

Table 17-1: Five-Day Shift Schedules—Two Eight-Hour Shifts. Three examples of five-day shift schedules using two eight-hour shifts are shown in column 1. Columns 3 through 6 show which shift (A, B, or C) each of the two crews (column 2) works each day (MTWTF SSu) of weeks 1 through 4. Rest days are designated by an R, which is seen on every Saturday and Sunday in these schedules. Although starting hours may vary, for the purposes of these schedules the starting times are 8 a.m. for A shift, 4 p.m. for B shift, and midnight for C shift. (*Murphy, Eastman Kodak Company, 1981*).

Schedule	Crew	Week 1 MTWTF SSu	Week 2 MTWTF SSu	Week 3 MTWTF SSu	Week 4 MTWTF SSu
Fixed shifts, A	1	AAAAA RR	AAAAA RR	AAAAA RR	AAAAA RR
and B	2	BBBBB RR	BBBBB RR	BBBBB RR	BBBBB RR
Weekly alterna-	1	AAAAA RR	BBBBB RR	AAAAA RR	BBBBB RR
tion, A and B (dou-ble-days)	2	BBBBB RR	AAAAA RR	BBBBB RR	AAAAA RR
Weekly, A and C	1	AAAAA RR	CCCCC RR	AAAAA RR	CCCCC RR
	2	CCCCC RR	AAAAA RR	CCCCC RR	AAAAA RR

used instead of fixed shifts because the advantages and disadvantages of each shift are spread more equitably across the work force. Daytime supervision usually appreciates this weekly rotation as it gives them more familiarity with all of the workers. The disadvantage of this schedule is that the older workers may express some dissatisfaction with the periods of B shift work.

Weekly alternation of A and C shift is sometimes needed for processes that take more than eight hours to run or for operations that require heavy effort under thermal stress, which would be at its worst in the B shift. An example of the former would be a chemical reaction that requires 12 to 16 hours to complete, thereby making B shift staffing requirements low but increasing the need for C shift staff. An example of the latter would be cleaning and maintenance operations in hot environments, such as in the summer, which are best done during A and C shifts, since the accumulation of heat reaches its peak in the evening hours between 6:00 and 10:00 p.m. (Rodgers, Eastman Kodak Company, 1973). The advantage of A and C shift weekly rotation are the same as for A and B shift rotations; they spread the advantages and disadvantages of each shift more equitably across the work force, and the supervision becomes familiar with all of the workers. The disadvantage of the A-C rotation is that some people find it to be a very fatiguing schedule.

2. THREE EIGHT-HOUR SHIFTS—THREE-SHIFT DISCONTINUOUS SCHEDULES

Six examples are shown for three eight-hour shifts on a five-day shift schedule, also known as three-shift discontinuous schedules. These include fixed, partially fixed, weekly rotating, and rapidly rotating shift schedules. The patterns are illustrated in Table 17-2.

Fixed shifts on a schedule of three eight-hour shifts, are predictable and, therefore, liked by those who wish to work certain shifts. As in the earlier fixed shift schedule for two eight-hour shifts, the A shift jobs tend to be awarded to the longer service employees. C shift is the second easiest shift to fill in a fixed shift system, and B shift is most difficult to staff (Murphy, Eastman Kodak Company, 1972). Daytime supervision does not like the fixed shift schedule as much as a rotating one because B and C shift workers are not known as well. Younger workers tend to be assigned to the B and C shift and may express dissatisfaction with the social inconveniences of fixed shift work during the evening and early morning hours.

One approach to resolving the unpopularity of the B shift in a fixed shift schedule is to rotate A and B shifts weekly, but operate a fixed C shift, which can usually be staffed primarily with volunteers who can adapt to early morning work both physically and socially. This schedule spreads the advantages and disadvantages of A and B shifts more equitably among the workers and shows high acceptance for a shift pattern of three eight-hour days. Daily supervision remains unfamiliar with C shift workers, however.

A more common approach to staffing three eight-hour shifts in a five-day operation is to rotate the work force weekly. The two main ways to do this are

Table 17-2: Five-Day Shift Schedules—Three Eight-Hour Shifts. Six examples of five-day shift schedules using three eight-hour days are shown in column 1. The daily shift for each of the three crews (column 2) over a four-week period (columns 3 through 6) is indicated by A, B, or C under each day of the week (MTWTF SSu). Rest days (R) occur on the weekends except where C shift is worked one Saturday morning out of four. *(Murphy, Eastman Kodak Company, 1981).*

Schedule	Crew	Week 1 MTWTF SSu	Week 2 MTWTF SSu	Week 3 MTWTF SSu	Week 4 MTWTF SSu
Fixed shifts	1	AAAAA RR	AAAAA RR	AAAAA RR	AAAAA RR
	2	BBBBB RR	BBBBB RR	BBBBB RR	BBBBB RR
	3	CCCCC RR	CCCCC RR	CCCCC RR	CCCCC RR
Fixed C, weekly	1	AAAAA RR	BBBBB RR	AAAAA RR	BBBBB RR
alternating A	2	BBBBB RR	AAAAA RR	BBBBB RR	AAAAA RR
and B	3	CCCCC RR	CCCCC RR	CCCCC RR	CCCCC RR
Weekly rotating,	1	AAAAA RR	BBBBB RR	CCCCC RR	AAAAA RR
forward	2	BBBBB RR	CCCCC RR	AAAAA RR	BBBBB RR
	3	CCCCC RR	AAAAA RR	BBBBB RR	CCCCC RR
Weekly rotating,	1	CCCCC RR	BBBBB RR	AAAAA RR	CCCCC RR
backward	2	BBBBB RR	AAAAA RR	CCCCC RR	BBBBB RR
	3	AAAAA RR	CCCCC RR	BBBBB RR	AAAAA RR
Rapidly rotating,	1	AAAAA RR	CCBBB RR	BBRCC CR	AAAAA RR
2-3 for B and C,	2	CCBBB RR	BBRCC CR	AAAAA RR	CCBBB RR
1 week of A	3	BBRCC CR	AAAAA RR	CCBBB RR	BBRCC CR
Rapidly rotating,	1	AABBB RR	CCAAA RR	BBRCC CR	AABBB RR
2-3 for all three	2	CCAAA RR	BBRCC CR	AABBB RR	CCAAA RR
shifts	3	BBRCC CR	AABBB RR	CCAAA RR	BBRCC CR

forward (A-B-C) and backward (C-B-A) rotation. The differences between the two rotations are in the length of the weekend between the C and B shifts. The weekend is from 4:00 p.m. Friday to 4:00 p.m. Monday, or 3.3 days, in the backward rotation schedule. Otherwise, these weekly rotation three-shift schedules have similar characteristics. The positive and negative features of each shift are spread more equitably across the workforce, and daytime supervision can become familiar with all three crews. However, many workers, particularly the older ones, find it difficult to adapt socially and/or physically to a weekly rotation schedule. Some find it easier to adapt their sleeping patterns in the forward rather than in the backward rotation. Although some people have recommended extending the rotation to two or four weeks instead of one week (Czeisler, Moore-Ede, and Coleman, 1982; Winget, Hughes, and LaDou, 1978), interviews with shift workers suggest that the dissatisfaction of reduced social interactions with family and friends during the B and C shift rotation far outweigh the potential improvement in digestive system complaints and quality of sleep (Murphy, Eastman Kodak Company, 1976; Rutenfranz et al., 1977).

Another alternative to weekly rotation of shifts is rapid rotation, or the alternation among A, B, and C shifts within a week or two. The last two schedules in Table 17-2 illustrate two variations of rapid rotation. The first schedule involves a full week of A shift followed by two weeks alternating between B and C shift in a 2-3 schedule. The second schedule rotates in a 2-3 pattern among all three shifts. The advantages of both rapid-rotation schedules are that no one has to work five continuous days on B or C shift. This can be expected to result in less social dissatisfaction and less fatigue for many workers. Workers may prefer the rotation that includes A shift because that gives them some time each week when they can interact fully with their families.

The disadvantages of the rapid-rotation schedules are that they are constantly changing, so it is sometimes difficult for the worker or the family to know when rest time is scheduled. This can usually be alleviated by printing the schedules well in advance, but the schedule is still less predictable than straight or weekly rotating schedules. A serious administrative disadvantage is that the department using either of these five-day rapid-rotation schedules must either shut down on Wednesday C shift or use a short changeover, that is, extend the working hours for the other crews. The crew that gets Wednesday C shift off will work Saturday morning C shift, thereby shortening the usual weekend that week. These difficulties make rapid rotation less suitable for five-day departments than it is for seven-day departments, as will be seen later in this chapter.

3. TWELVE-HOUR SHIFTS

Two examples of schedules that use 12-hour shifts are shown in Table 17-3. The first is a 12-hour alternating schedule using three crews. It includes either three or four work periods per week, either night (N) or day (D), broken up by rest days (R). This schedule provides a seven-day (168-hour) rest period once every three weeks. Two of every three weeks, the crew works only three days instead of four; thus, more rest days occur than in other five-day schedules. Fatigue can be a problem for some workers on this schedule. This is particularly true if the jobs are physically moderate or heavy or make strong mental or perceptual demands on the workers. The 12-hour schedule will also be difficult for workers who have other responsibilities, such as running a household or raising children, because they cut across the times, often 6:00 a.m to 6:00 p.m., when those responsibilities must be met.

A second 12-hour shift schedule in a five-day week offers some advantages over the alternating 12-hour schedule by combining eight-hour and 12-hour shifts in a rapid-rotation schedule. As can be seen in the second part of Table 17-3, the two crews alternate between A, B, and C shifts and night and day 12-hour shifts. The C shift starts at 10:00 p.m. instead of midnight. The schedule allows a rapid rotation of shifts without a Wednesday shutdown or having to work on Saturday. This schedule has the advantage of only four workdays per week,

Table 17-3: Five-Day Shift Schedules—12 Hour Shifts. Column 1 gives two examples of 12-hour rotating shift systems for five-day departments. The shift schedule for each of the crews (column 2) is shown for each day (MTWTF SSu) over a four week period (columns 3 through 6). The C shift is marked with an asterisk (*) because it starts at 10 p.m. the previous day, instead of at midnight. In these schedules, A shift starts at 8 a.m., B shift at 4 p.m., N shift at 6 p.m., and D shift at 6 a.m. R designates a rest day. *(Murphy, Eastman Kodak Company, 1981).*

Schedule	Crew	Week 1 MTWTF SSu	Week 2 MTWTF SSu	Week 3 MTWTF SSu	Week 4 MTWTF SSu
12-hour alternat-	1	DDDRR RR	RRNNN RR	NNRDD RR	DDDRR RR
ing schedule	2	RRNNN RR	NNRDD RR	DDDRR RR	RRNNN RR
	3	NNRDD RR	DDDRR RR	RRNNN RR	NNRDD RR
Combination 8-	1	*CDBNR RR	NRADB RR	DNRCA RR	CDBNR RR
and 12-hour shifts,	2	NRADB RR	DNRCA RR	CDBNR RR	DNRCA RR
rapidly rotating					

and there are fewer 12-hour days than in the 12-hour alternating schedule. As in that schedule, there is a long weekend every three weeks. There are also fewer days in a row requiring daytime sleep. The major disadvantage of this schedule is that it is constantly changing, and some workers and their families may prefer a more predictable one.

B. SEVEN-DAY SHIFT SCHEDULES

A need for around-the-clock service and the existence of continuous-process manufacturing operations stimulated the development of shift schedules that provide continuous coverage seven days a week. These can be organized in ways similar to five-day shift schedules utilizing either fixed shifts, long-cycle (four-week) rotations, weekly rotations, or rapidly rotating shifts. They can use eight-hour, 12-hour, or combinations of eight- and 12-hour shifts. Three eight-hour shifts are also known as three-shift continuous schedules (Sergean, 1971). The differences among these schedules become more apparent when the usual weekend break is lost. The psychosocial effects of these schedules are the primary factors influencing worker satisfaction or dissatisfaction. Those that satisfy the psychosocial needs are usually best physiologically, too.

The assumptions stated above for five-day shift schedules are applicable here, too, except that the organization's operating hours are now 24 hours for seven days, or 168 hours per week. Each employee works an average of 42 hours a week. The shift definitions, A, B, C, N, and D remain the same, except where noted.

1. EIGHT-HOUR SHIFTS—FIXED

In order to staff all seven days and still give people two days off during a calendar week, you need 12 crews in a fixed shift schedule (Table 17-4). The fixed times are preferred by some people because they are more predictable than a rotating shift schedule, but the seven continuous days of work across weeks can result in an accumulating fatigue, especially for C shift workers. Younger workers may find seven days of B or C shift work socially isolating; they often have to work B and C shifts in a fixed shift schedule because workers with seniority take the A shift work. People working seven days of C shift usually revert to daytime schedules on their two rest days. So there is never a total physiological adjustment to night work, even when a person works it on a nonrotating shift schedule (Mott et al., 1965; Van Loon, 1963).

Other disadvantages of this shift schedule are that it may be difficult to find good maintenance people for the B and C shifts, and daytime supervision loses communication and familiarity with the people on those shifts. One of the least satisfying aspects of the fixed shift seven-day schedule is that the rest days only coincide with the calendar weekend once every four weeks. This interference with traditional weekend social interactions with family and friends is one

Table 17-4: Seven-Day Shift Schedules—Eight-Hour Shifts, Fixed, 12 Crews. A fixed, eight-hour shift schedule for a seven-day organization is shown. The shift to be worked each day of the week (MTWTF SSu) by each of the 12 crews (column 1) is shown for a four week period in columns 2 through 5. Rest days are designated by R. (Murphy, Eastman Kodak Company, 1981).

Crew	Week 1 MTWTF SSu	Week 2 MTWTF SSu	Week 3 MTWTF SSu	Week 4 MTWTF SSu
1	RAAAA AA	ARRAA AA	AAARR AA	AAAAA RR
2	ARRAA AA	AAARR AA	AAAAA RR	RAAAA AA
3	AAARR AA	AAAAA RR	RAAAA AA	ARRAA AA
4	AAAAA RR	RAAAA AA	ARRAA AA	AAARR AA
5	RBBBB BB	BRRBB BB	BBBRR BB	BBBBB RR
6	BRRBB BB	BBBRR BB	BBBBB RR	RBBBB BB
7	BBBRR BB	BBBBB RR	RBBBB BB	BRRBB BB
8	BBBBB RR	RBBBB BB	BRRBB BB	BBBRR BB
9	RCCCC CC	CRRCC CC	CCCRR CC	CCCCC RR
10	CRRCC CC	CCCRR CC	CCCCC RR	RCCCC CC
11	CCCRR CC	CCCCC RR	RCCCC CC	CRRCC CC
12	CCCCC RR	RCCCC CC	CRRCC CC	CCCRR CC

of the main reasons why alternative shift schedules have been developed for seven-day operations.

2. EIGHT-HOUR SHIFTS—"WEEKLY" ROTATION

The three schedules shown in Table 17-5 are for seven-day coverage using three eight-hour shifts. The first two schedules use four crews each and differ only in the direction of rotation through the shifts, forward or backward. The third schedule also uses four crews but follows a 20-week cycle and requires each person to work only five, instead of seven, days continuously before a rest day is provided. The seven-day continuous weekly rotating schedules share the advantages of distributing the positive and negative aspects of all three shifts more equitably across the workforce and making it easier for daytime supervision to know the workers. The forward rotation schedule appears to be preferable from the standpoint of sleep quality, it being easier to adapt to an advancing, rather than to a receding, bedtime hour (Dirken, 1966; Murphy, Eastman Kodak Company, 1972). The backward rotation schedule, on the other hand, provides an extra long weekend between the C and B shifts, as was discussed under five-day shift schedules.

The primary disadvantage of the weekly rotating schedules is that seven continuous days on the C shift may result in an accumulated sleep loss and gastrointestinal discomfort that produce irritability and depression in people who cannot adapt to the regimen. The seven continuous days of B shift result in social and family life interruptions that both the worker and his or her spouse may dislike. The third schedule in Table 17-5 attempts to address some of these problems by reducing the number of continuous days worked to five. However, although the physiological and some of the psychosocial problems are alleviated by this schedule, a new psychosocial problem is created. The rest days slowly rotate through the week, and there is a period of 15 weeks when they do not coincide at all with the traditional calendar weekend. Workers who need predictability in their lives may find this schedule difficult to remember until they have worked it for a few years.

3. EIGHT-HOUR SHIFTS—RAPID ROTATION

Four schedules for rapidly rotating shifts that cover a seven-day work week with three eight-hour shifts and four crews are shown in Table 17-6. Three of the schedules are variations of a 2-2-3 day schedule; workers rotate among A, B, and C shifts with no more than three days at a time on any shift. The fourth is a 6-workday schedule, designated 2-2-2; no more than two consecutive days of a shift are worked in this schedule. The same variations in shift order, that is C-A-B, A-B-C, and B-C-A, can be devised for the 2-2-2 schedule, although only C-A-B is illustrated in the table.

The advantages of rapidly rotating shifts over the fixed or weekly rotating shift schedules presented earlier are primarily psychosocial, although physiological problems are also addressed. The person working this schedule is not as

Table 17-5: Seven-Day Shift Schedules—Eight-Hour Shifts, Weekly Rotation. Three examples of eight-hour shift systems that rotate weekly and cover a seven-day organization are given in column 1. Simple forward and backward rotations are shown in the first two examples. The shift (A, B, or C) and rest days (R) are shown for each crew (column 2) for each day of the week (MTWTF SSu) over a four-week period (columns 3 through 6). The third schedule shows the 20-week cycle (column 2) for four crews by day of the week (column 3). The schedule is numbered for crew 1; crews 2, 3, and 4 start at the weeks marked 6, 11, and 16, as noted. At the end of 20 weeks each crew has rotated through each of the weekly schedules shown and has returned to the starting position indicated on this table. The starting times of the shifts are the same as indicated earlier, A shift at 8 a.m., B shift at 4 p.m., and C shift at midnight. *(Murphy, Eastman Kodak Company, 1981).*

Schedule	Crew	Week 1 MTWTF SSu	Week 2 MTWTF SSu	Week 3 MTWTF SSu	Week 4 MTWTF SSu
Four crews,	1	RAAAA AA	ARRBB BB	BBBRR CC	CCCCC RR
forward rotation	2	ARRBB BB	BBBRR CC	CCCCC RR	RAAAA AA
	3	BBBRR CC	CCCCC RR	RAAAA AA	ARRBB BB
	4	CCCCC RR	RAAAA AA	ARRBB BB	BBBRR CC
Four crews, back-	1	RBBBB BB	BRRAA AA	AAAARR CC	CCCCC RR
ward rotation	2	BRRAA AA	AAAARR CC	CCCCC RR	RBBBB BB
	3	AAAARR CC	CCCCC RR	RBBBB BB	BRRAA AA
	4	CCCCC RR	RBBBB BB	BRRAA AA	AAAARR CC

		Week	Day of the Week MTWTF SSu
Four crews,	Crew 1	1	CCCCC RB
5-day work week,		2	BBBBR RA
20 week cycle		3	AAAAR RC
		4	CCCCR BB
		5	BBBRR AA
	Crew 2	6	AAARR CC
		7	CCCRB BB
		8	BBRAA AA
		9	AARRC CC
		10	CCRBB BB
	Crew 3	11	BRRAA AA
		12	ARRCC CC
		13	CRBBB BB
		14	RRAAA AA
		15	RRCCC CC
	Crew 4	16	RBBBB BR
		17	RAAAA AR
		18	RCCCC CR
		19	BBBBB RR
		20	AAAAA RR Return to week 1

Table 17-6: Seven-Day Shift Schedules—Eight-Hour Shifts, Rapid Rotation. Three 2-2-3 rapidly rotating shift schedules and one 2-2-2 schedule are shown (column 1). The 2-2-3 rotations show the daily (MTWTF SSu) shift worked by each of the four crews (column 2) for a four-week period (columns 3 through 6). The differences among the schedules are determined by the shift (A, B, or C) that starts the work period after the two rest days (R). The 2-2-2 schedule is shown for a C-A-B rotation; it can also use A-B-C or B-C-A rotations, if desired. This schedule has a cycle time of eight weeks (column 2). The shift worked or rest day is given for each day of the week in column 4 for the full eight-week cycle. The numbers for the weeks are given for crew 1. Crews 2, 3, and 4 start the shift schedule at the weeks marked 3, 5, and 7, as noted. *(Adapted from Knauth, Rohmert, and Rutenfranz, 1979; Murphy, Eastman Kodak Company, 1981; Pocock, Sergean, and Taylor, 1972).*

Schedule	Crew	Week 1 MTWTF SSu	Week 2 MTWTF SSu	Week 3 MTWTF SSu	Week 4 MTWTF SSu
2-2-3 rotation, four crews					
C-A-B	1	RCCAA BB	BRRCC AA	ABBRR CC	CAABB RR
	2	BRRCC AA	ABBRR CC	CAABB RR	RCCAA BB
	3	ABBRR CC	CAABB RR	RCCAA BB	ABBRR CC
	4	CAABB RR	RCCAA BB	BRRCC AA	ABBRR CC
A-B-C	1	AABBR CC	CRAAB BB	RCCRA AA	BBRCC RR
	2	CRAAA BB	RCCRA AA	BBRCC RR	AABBR CC
	3	RCCRA AA	BBRCC RR	AABBR CC	CRAAB BB
	4	BBRCC RR	AABBR CC	CRAAB BB	RCCRA AA
B-C-A	1	BBRCC AA	ARBBR CC	CAARB BB	RCCAA RR
	2	ARBBR CC	CAARB BB	RCCAA RR	BBRCC AA
	3	CAARB BB	RCCAA RR	BBRCC AA	ARBBR CC
	4	RCCAA RR	BBRCC AA	ARBBR CC	CAARB BB

Schedule	Crew	Week	Day of the Week MTWTF SSu	
2-2-2 rotation, C-A-B,	1	1	CCAAB BR	
four crews, 8-		2	RCCAA BB	
week schedule	2	3	RRCCA AB	
		4	BRRCC AA	
	3	5	BBRRC CA	
		6	ABBRR CC	
	4	7	AABBR RC	
		8	CAABB RR	Return to week 1

subject to accumulated sleep loss as in the seven days of continuous work on C shift, nor is the social isolation from family and friends as likely to occur. Forward rotation appears to favor sleep adjustments, as has been discussed earlier. There are rest periods between each shift change that are more than the usual work break on a fixed shift, although they are not long enough to be counted as a day off. The direction or order of rotation between the shifts will determine how much of a break occurs between C and A shifts; the A-B-C schedule provides the most time off. As in other rotating shift schedules, daytime supervision can better interact with all of the workers; they particularly like rapidly rotating shift systems because they see each person every week or two (Murphy, Eastman Kodak Company, 1976).

The disadvantages of these rapidly rotating shift schedules are as follows:

- Rest days coincide with only one calendar weekend in four. In the 2-2-2 schedule, the coincidence of rest days on Saturdays or Sundays is even less.

- Some people do not easily adapt to the constantly changing shifts and feel that family and social life is difficult to plan. This feeling usually declines with increased experience with the schedule.

- Some people who rapidly adapt to shift changes feel that a slower rotation is better for them physically. They adapt quickly to sleep changes of a five- or seven-day C shift schedule, but do not like having to adapt to C shift work several times a month instead of once or twice.

- Some people feel that a 24-hour rest day after two to three C shifts is too little recovery for the remaining four to five days of the work week. They do not like the fact that they have to come in on C shift at the beginning of each work week in the C-A-B schedule. The A-B-C schedule is preferred because they start each work week on an A shift.

- The 48- and 72-hour rest breaks between the work weeks do not start and end at midnight, so many people do not consider them full days off.

Despite the disadvantages of the rapidly rotating schedules, they are usually preferred over weekly rotating schedules or seven-day fixed shift schedules because of increased opportunity for the worker to interact with family and friends (Murphy, Eastman Kodak Company, 1973, 1976). Special combinations of rapidly rotating shifts with other eight- and 12-hour rotations are given later in this chapter.

4. EIGHT-HOUR SHIFTS—COMBINATION FIXED AND ROTATING

Staffing three eight-hour shifts per day over a seven-day period while still having fixed shifts requires the addition of a fourth crew to fill in when the other crews are on rest days. This fourth crew works on a 2-2-3 C-A-B rapid rotation schedule, as can be seen in the first part of Table 17-7. A similar schedule can be generated for a slowly rotating shift, such as the three monthly-rotation crews and one rapid rotation crew of the second example in Table 17-7. A third variation in this table is one fixed C shift crew, two crews rapidly alternating between A and B shift, and one crew on a rapidly rotating C-A-B schedule.

The advantages and disadvantages of these combination systems include most of the benefits and problems discussed under fixed, rapidly rotating, and weekly rotating shift schedules. The fixed shifts may be difficult to staff, are difficult physiologically for people who experience accumulated sleep loss over the seven-day cycle on C shift, are problematic from a social standpoint because of extended periods on B and C shifts, and are disliked by daytime supervision because the B and C shift crews are not as accessible as they would be on a rotating schedule. Acceptance of the fixed schedule will depend on the number of people who volunteer for the B and C shifts, which may vary from time to time. However, the existence of a rapid-rotation schedule for the fourth crew means that this schedule usually accommodates a higher percentage of people than do the straight fixed shift or rapidly rotating schedules alone.

The schedule that uses three crews in a seven-day shift monthly rotation and one rapidly rotating 2-2-3 C-A-B crew has the advantage of being more predictable for three of the crews. However, the extended periods on B and C shift can result in strong feelings of isolation by the shift workers and dissatisfaction with their loss of contact with family and friends. When these schedules have been tried, the rapid-rotation assignment has often been preferred over the monthly rotation one (Murphy, 1979).

The combination of a fixed C shift, alternating A and B shifts, and a rapidly rotating 2-2-3 C-A-B shift has all of the positive and negative factors of each schedule and can accommodate many workers. The difficulty with this schedule is likely to be in finding the volunteers for a fixed C shift who have the needed skills. If this is resolved, the alternating A and B shift and the rapidly rotating C-A-B shift should not be difficult to staff. C shift workers will be isolated from the others. This could be a problem if information about a manufacturing process or data important for continuing care in a service job have to be passed from the C to the A or from the B to the C shifts.

5. TWELVE-HOUR SHIFTS

Two examples of 12-hour shift schedules with rotation between day and night shifts are presented in Table 17-8. Fixed shifts, one of the permanent night shift schedules (Sergean, 1971), may also be used and would have the same advantages and disadvantages of eight-hour fixed shift schedules except the fatigue would be greater for the susceptible person in a demanding job. The primary advantages of 12-hour shift schedules are that there are longer rest periods between workdays and about 50 percent of the rest days coincide with at least some part of the calendar weekend. Many of the physical and social advantages of rapidly rotating schedules are also present with the EOWEO, or "every-other-

Table 17-7: Seven-Day Shift Schedules—Eight-Hour Shifts, Combination Fixed and Rotating. Three combination fixed and rotating shift schedules are shown in column 1. The first and third schedules show the daily (MTWTF SSu) shift worked or rest days (A, B, C, or R) for each crew (column 2) for a four-week period (columns 3 through 6). The second schedule has a 12-week cycle (column 3). The daily shift or rest schedules for that period are shown for the three crews that rotate monthly (column 4) and for the one rapidly rotating crew (column 5). The schedule in column 3 is shown for crew 1; crews 2 and 3 start at the weeks labeled 5 and 9, as noted. *(Murphy, Eastman Kodak Company, 1981).*

Schedule	Crew	Week 1 MTWTF SSu	Week 2 MTWTF SSu	Week 3 MTWTF SSu	Week 4 MTWTF SSu
Three fixed crews,	1	RCCCC CC	CRRCC CC	CCCRR CC	CCCCC RR
1 rapidly rotat-	2	ARRAA AA	AAARR AA	AAAAA RR	RAAAA AA
ing crew, 2-2-3,	3	BBBRR BB	BBBBB RR	RBBBB BB	BRRBB BB
C-A-B	4	CAABB RR	RCCAA BB	BRRCC AA	ABBRR CC

Schedule	Crew	Week	Crews 1, 2, 3 Day of the Week MTWTF SSu	Crew 4 (Rapid Rotation) Day of the Week MTWTF SSU
Four crews, 3	1	1	RCCCC CC	RCCAA BB
monthly rotat-		2	CRRCC CC	BRRCC AA
ing, 1 rapidly		3	CCCRR CC	ABBRR CC
rotating		4	CCCCC RR	CAABB RR
	2	5	RAAAA AA	RCCAA BB
		6	ARRAA AA	BRRCC AA
		7	AAARR AA	ABBRR CC
		8	AAAAA RR	CAABB RR
	3	9	RBBBB BB	RCCAA BB
		10	BRRBB BB	BRRCC AA
		11	BBBRR BB	ABBRR CC
		12	BBBBB RR	CAABB RR
			Return to week 1	

Schedule	Crew	Week 1 MTWTF SSu	Week 2 MTWTF SSu	Week 3 MTWTF SSu	Week 4 MTWTF SSu
Four crews, 1	1	RCCCC CC	CRRCC CC	CCCRR CC	CCCCC RR
fixed C-shift, 2	2	BRRAA BB	BBBRR AA	ABBBB RR	RAABB BB
rapidly alternat-	3	ABBRR AA	AAABB RR	RAAAA BB	BRRAA AA
ing A and B, 1 rap-	4	CAABB RR	RCCAA BB	BRRCC AA	ABBRR CC
idly rotating C-A-B					

Table 17-8: Seven-Day Shift Schedules—12-Hour Shifts. Two examples of rotating 12-hour shift schedules to cover a seven-day organization are shown in column 1. The daily (MTWTF SSu) shift or rest schedule (N, D, or R) is shown for each of the four crews (column 2) over a four week period (columns 3 through 6). The weekend length and coincidence of the work week with the calendar weekend are factors that distinguish these shift schedules from other seven-day systems. The starting times for day (D) and night (N) shifts are 6 a.m. and 6 p.m., respectively. *(Murphy, Eastman Kodak Company, 1981).*

Schedule	Crew	Week 1 MTWTF SSu	Week 2 MTWTF SSu	Week 3 MTWTF SSu	Week 4 MTWTF SSu
EOWEO, every other	1	DRRNN RR	RDDRR NN	NRRDD RR	RNNRR DD
weekend off	2	RDDRR NN	NRRDD RR	RNNRR DD	DRRNN RR
	3	NRRDD RR	RNNRR DD	DRRNN RR	RDDRR NN
	4	RNNRR DD	DRRNN RR	RDDRR NN	NRRDD RR
The duPont sched-	1	DDDDR RR	RRRRN NN	NRRRD DD	RNNNR RR
ule, long weekend	2	RRRRN NN	NRRRD DD	RNNNR RR	DDDDR RR
	3	NRRRD DD	RNNNR RR	DDDDR RR	RRRRN NN
	4	RNNNR RR	DDDDR RR	RRRRN NN	NRRRD DD

weekend-off,'' schedule shown in Table 17-8. The duPont schedule, in the second part of the table, has the additional advantage of producing a rest break equivalent to seven-plus days every four weeks, a popular perquisite for those working the 12-hour schedule. Day shift is designated by D and night shift by N in these tables. Starting times are 6:00 a.m. and 6:00 p.m.

The disadvantages of 12-hour shift systems relate to the type of work being done and the length of time a person takes to adapt to the changing schedule. The long work shifts may interfere with a worker's family responsibilities—for example, with a mother's responsibilities at home after the children return from school. As was mentioned in the chapter on overtime, physical effort levels and environmental exposures must be considered when determining if a job is suitable for 12-hour shifts. In addition, when people are working 12-hour shifts, they are not as available for overtime work. This limits supervision's approaches for dealing with occasional elevated production demands. The shortened work week, in terms of days—not hours—per week in these 12-hour shift schedules, makes them quite popular for jobs where the workload is appropriate (Botzum and Lucas, 1981). The duPont schedule, with its long rest period once a month, has four continuous 12-hour workdays twice a month and six 12-hour workdays out of seven in another ''week.'' The seven-plus-day rest period is used by the worker partially to restore his or her energy after that intensive work period.

6. COMBINATION EIGHT- AND 12-HOUR SHIFTS

For jobs that need limited overtime on a regular basis and cannot justify going to a full 12-hour schedule, a combination of eight- and 12-hour shifts into one schedule may be appropriate. Two examples of these combination schedules are given in Table 17-9. In the first schedule there is one 12-hour day added to the schedule every other calendar week, a night shift (N) in one week and a day shift (D) at the end of the following week. The night shift starts at 6:00 p.m. on the previous Sunday evening. In this schedule, rest days fall on 50 percent of the calendar weekends, thereby reducing the dissatisfaction of shift workers who want to spend more time with their families. By putting only a few long days in the schedule, the work demands can be met without requiring constant overtime or a full 12-hour schedule. In these schedules the usual starting times are 6:00 a.m., 2:00 p.m., and 10:00 p.m. for the eight-hour shifts and 6:00 a.m. and 6:00 p.m. for the night and day shifts.

The second schedule shown in the table adds a second 12-hour day every

Table 17-9: Seven-Day Shift Schedules—Combination Eight- and 12-Hour Shifts. Column 1 gives two examples of shift schedules created by combining eight- and 12-hour shifts; these may be appropriate when a department is expanding but does not have enough work for a full 12-hour schedule. The daily (MTWTF SSu) shift or rest schedule (A, B, C, N, D, or R) is shown for four crews (column 2) across a four-week period (columns 3 through 6). Because of the interleaving of eight- and 12-hour shifts, the starting times for eight-hour shifts have been adjusted to 6 a.m., 2 p.m., and 10 p.m. for A, B, and C shifts, respectively. The night shift starts at 6 p.m. on the day before the one indicated in each of the weeks. *(Murphy, Eastman Kodak Company, 1981).*

Schedule	Crew	Week 1 MTWTF SSu	Week 2 MTWTF SSu	Week 3 MTWTF SSu	Week 4 MTWTF SSu
All 8-hour days	1	NRAAB BR	RCCRA AD	BBRCC RR	AABBR CC
except one 12-	2	RCCRA AD	BBRCC RR	AABBR CC	BRAAB BR
hour day every	3	BBRCC RR	AABBR CC	NRAAB BR	RCCRA AD
other week	4	AABBR CC	NRAAB BR	RCCRA AD	BBRCC RR
All 8-hour days	1	NRAAB RR	RCCRA DD	BBRCC RR	AABBR CN
except two 12-	2	RCCRA DD	BBRCC RR	AABBR CN	NRAAB RR
hour days every	3	BBRCC RR	AABBR CN	NRAAB RR	RCCRA DD
other week	4	AABBR CN	NRAAB RR	RCCRA DD	BBRCC RR

Starting times A = 6 a.m.
  B = 2 p.m.
  C = 10 p.m.
  D = 6 a.m.
  N = 6 p.m. (the day before the one marked)

other week, or a total of four such days per four weeks. This could be used if the required hours are greater than for the previous schedule and still do not merit a full 12-hour schedule. During any one calendar week, a person on this schedule can be on the job from 32 to 52 hours, averaging 42 work hours per week. This gives supervision more flexibility in requesting overtime work, if needed.

C. SPECIAL FIVE- AND SEVEN-DAY SHIFT SCHEDULES

Much interest has been shown in shift systems that modify or reduce the number of hours or days worked per week. If a department needs to work six days a week and overtime is not an appropriate approach, a schedule that alternates between five- and seven-day shift systems may be preferable to some of the modifications discussed earlier in this chapter.

1. FIVE-DAY WITH CONSISTENT SATURDAY OVERTIME

A five-day schedule with consistent Saturday overtime is shown in Table 17-10. Seven crews rotate A, B, and C shifts on a regular weekly schedule and then work a seven-day rapidly rotating shift schedule for four weeks. This pattern repeats every seven weeks. One advantage of this schedule is that it has more rest and weekend time available than does a regular five-day schedule with Saturday overtime. If a long-term overtime schedule is anticipated for a department, this approach should be considered. The primary disadvantage of this schedule is the strong objection by some five-day shift workers to working on Sundays. Some people would prefer to give up every Saturday than to give up their Sunday family activities (Murphy, Eastman Kodak Company, 1980).

2. REDUCED HOURS PER WEEK— WEEKDAY/WEEKEND PLAN

There are several shift schedule variations that reduce the number of hours worked per week by adding part-time crews or putting additional people on to permit the regular crews to take additional rest time. If a traditional five-day organization has to expand to continuous seven-day operations because of increased demand for service or the product, there are several schedules available. These have been discussed under special combinations in seven-day shift schedules. Another approach is to add two crews that work only on weekends, two 12-hour shifts. The three regular crews maintain their five-day schedule, and the two additional crews are paid for 24 hours plus the premiums for weekend work. This approach is best suited to situations where the weekend crews can be relatively unskilled and employed for short durations (Sergean, 1971). The longest time that most people will give up their Saturdays and Sundays is about two years (Murphy, Eastman Kodak Company, 1981). An organization with sea-

Table 17-10: Special Shift Schedules—Five-Day Shifts with Consistent Saturday Overtime.   A schedule that addresses the need for consistent Saturday overtime in a five-day organization is shown in column 1. The total cycle time is seven weeks, the first three weeks using a weekly rotation on five-day shifts and the last four weeks using a seven-day schedule with a 2-2-3 rapidly rotating schedule. The daily (MTWTF SSu) shift schedule for each of the seven crews (column 2) is shown in the columns labeled weeks 1 through 7. Starting times are 8 a.m., 4 p.m., and midnight for A, B, and C shifts, respectively. *(Murphy, Eastman Kodak Company, 1981).*

Schedule	Crew	Week 1 MTWTF SSu	Week 2 MTWTF SSu	Week 3 MTWTF SSu	Week 4 MTWTF SSu
Seven crews, alter-	1	AAAAA RR	BBBBB RR	CCCCC RR	AABBR CC
nating between 5-day	2	BBRCC RR	AAAAA RR	BBBBB RR	CCCCC RR
for 3 weeks and	3	RCCRA AA	BBRCC RR	AAAAA RR	BBBBB RR
7-day for 4 weeks	4	CRAAB BB	RCCRA AA	BBRCC RR	AAAAA RR
	5	AABBR CC	CRAAB BB	RCCRA AA	BBRCC RR
	6	CCCCC RR	AABBR CC	CRAAB BB	RCCRA AA
	7	BBBBB RR	CCCCC RR	AABBR CC	CRAAB BB

	Crew	Week 5 MTWTF SSu	Week 6 MTWTF SSu	Week 7 MTWTF SSu	
	1	CRAAB BB	RCCRA AA	BBRCC RR	
	2	AABBR CC	CRAAB BB	RCCRA AA	
	3	CCCCC RR	AABBR CC	CRAAB BB	
	4	BBBBB BR	CCCCC RR	AABBR CC	
	5	AAAAA RR	BBBBB RR	CCCCC RR	
	6	BBRCC RR	AAAAA RR	BBBBB RR	
	7	RCCRA AA	BBRCC RR	AAAAA RR	Return to Week 1

sonal production or service demands that are sustained for several months but well short of the full year might benefit from this type of augmentation of its workforce and shift coverage.

3. REDUCED HOURS PER WEEK—40 HOURS OR LESS, FIVE OR NINE CREWS

Reducing the work time from 42 to 40 hours a week for people on a continuous shift schedule may be considered advantageous in some jobs. About a 5 percent increase in staff is needed to reduce each person's workload by two hours per week. An equal number of people are scheduled for rest on each of the work-

Table 17-11: Special Shift Schedules—Reduced Hours per Week, Less than 40 Hours. Two reduced-hours schedules, one for nine crews and one for five, are shown in column 1. In the schedule using nine crews, cycle time is 18 weeks (columns 3 and 6). The daily (MTWTF SSu) shift or rest day schedule (A, B, C, or R) is shown in columns 4 and 7. The weeks are numbered for crew 1; the other crews start at weeks 3, 5, 7, 9, 11, 13, 15, and 17, as noted. They rotate through the schedule and return to this point at the end of 18 weeks. The five-crew schedule has a 20-week cycle; the daily schedules are indicated for each crew as in the first part of the table. The starting times of these shifts are 8 a.m., 4 p.m., and midnight for A, B, and C shifts, respectively. *(Murphy, Eastman Kodak Company, 1981).*

Schedule	Crew	Week	Day of the Week MTWTF SSu	Crew	Week	Day of the Week MTWTF SSu
Nine crews, 38.7	1	1	AAAAR RB	6	11	RRAAA AR
work hours per		2	BBBRR CC		12	RBBBB RR
week, on average	2	3	CCRRA AA	7	13	CCCCR RA
		4	ARRBB BB		14	AAARR BB
	3	5	RRCCC CR	8	15	BBRRC CC
		6	RAAAA RR		16	CRRAA AA
	4	7	BBBBR RC	9	17	RRBBB BR
		8	CCCRR AA		18	RCCCC RR
	5	9	AARRB BB			
		10	BRRCC CC			Return to Week 1

Schedule	Crew	Week	Day of the Week MTWTF SSu	Crew	Week	Day of the Week MTWTF SSu
Five crews, 33.6	1	1	CCCCR RB	4	13	RRBBB BR
work hours per		2	BBBRR RA		14	RRAAA AR
week, on aver-		3	AAARR RC		15	RRCCC CR
age		4	CCCRR BB		16	RBBBB RR
	2	5	BBRRR AA	5	17	RAAAA RR
		6	AARRR CC		18	RCCCC RR
		7	CCRRB BB		19	BBBBR RR
		8	BRRRA AA		20	AAAAR RR
	3	9	ARRRC CC			
		10	CRRBB BB			Return to Week 1
		11	RRRAA AA			
		12	RRRCC CC			

days in the four-week cycle, their work being covered by the additional staff. This produces slightly more rest days for most seven-day shift schedules. However, the additional rest days do not precede or follow another rest day or fall on a Sunday, thereby appearing less beneficial to the shift workers.

Two schedules that result in work weeks of less than 40 hours are shown in Table 17-11. Such a schedule might be used in an organization where continuous coverage is needed but contractual agreements for reduced working hours have been made. The first schedule uses nine crews; each person averages 38.7 hours per week on the job. There are never more than four continuous days of work before a rest day occurs, although three out of four calendar weeks include five eight-hour workdays. The social and physical benefits of rapidly rotating shifts are partially accomplished with this schedule with even more rest days. Its disadvantage, however, is that each shift is staffed by an ever-changing mix of people. Some operations might suffer from lack of teamwork because of this. Despite the shorter work week, most contracts negotiated for reduced hours include a compensatory increase in hourly pay; therefore, there are seldom labor savings in these schedules.

The second reduced-hours shift schedule uses five crews and averages 33.6 hours per week per person. The longest continuous work period is four days, and less than one calendar week out of four includes five working days. The long rest periods between work periods provide the physical and social advantages that are lacking in other seven-day shift schedules. As in the previous schedule, most contracts that include these shorter work weeks expect increases in hourly pay to balance out the difference in hours worked, so there are few financial incentives to stimulate the implementation of this schedule.

In the past two decades, there has been considerable discussion about shortening the work week without necessarily reducing the hours worked per week; one way is by extending the hours worked per day to 10 or 12 (Botzum and Lucas, 1981; Poor, 1971; Sergean, 1971). There can be physiological, social, and financial disincentives to using some of these schedules, but they have been successfully applied in some organizations. In considering the benefits or problems associated with shorter work weeks, however, you should note that moonlighting tends to increase as the hours or days of work are decreased below 40 and five, respectively (Greenbaum, 1963).

REFERENCES

Botzum, G. D., and R. L. Lucas. 1981. "9-3 Slide Shift Evaluation—A Practical Look at Rapid Rotation Theory." In *Proceedings of the 25th Annual Human Factors Society, October 1981, Rochester, N.Y.* Santa Monica, Calif.: Human Factors Society, pp. 207-209.

Colquhoun, W. P., and J. Rutenfranz. 1980. *Studies of Shiftwork.* London: Taylor & Francis, 468 pages.

Czeisler, C. A., M. C. Moore-Ede, and R. M. Coleman. 1982. "Rotating Shift Work Schedules That Disrupt Sleep Are Improved by Applying Circadian Principles." *Science,* 217: pp. 460-463.

Dirken, J. M. 1966. "Industrial Shift Work: Decrease in Well-Being and Specific Effects." *Ergonomics, 9 (2):* pp. 115-124.

Greenbaum, M. L. 1963. "The Shorter Work Week." In *Bulletin 50,* New York State School of Industrial and Labor Relations. Ithaca, N.Y.: Cornell University, 52 pages.

Knauth, P., W. Rohmert, and J. Rutenfranz. 1979. "Systematic Selection of Shift Plans for Continuous Production with the Aid of Work-Physiological Criteria." *Applied Ergonomics, 10 (1):* pp. 9-15.

Mott, P. E., F. Mann, Q. McLoughlin, and D. Warwick. 1965. *Shift Work: Its Social, Psychological, and Physical Consequences.* Ann Arbor: University of Michigan Press, 351 pages.

Murphy, T. J. 1979. Letter to the Editor. *Journal of Occupational Medicine, 21:* pp. 319-324.

Pocock, S. J., R. Sergean, and P. J. Taylor. 1972. "Absence of Three-Shift Workers: A Comparison of Traditional and Rapidly Rotating Systems." *Occupational Psychology, 46:* pp. 7-13. Included in Colquhoun and Rutenfranz, 1980, pp. 391-397.

Poor, R. 1971. *4 Days, 40 Hours.* Cambridge, Mass.: Bursk and Poor, 175 pages.

Poper, F. J. 1970. *A Critical Evaluation of the Empirical Evidence Underlying the Relationship Between Hours of Work and Labor Productivity.* Dissertation in the Department of Economics, New York University. No. 71-2329. Ann Arbor, Mich.: University Microfilms, 260 pages.

Rutenfranz, J., W. P. Colquhoun, P. Knauth, and J. N. Ghata. 1977. "Biomedical and Psychosocial Aspects of Shift Work: A Review." *Scandinavian Journal of Work, Environment, and Health, 3:* pp. 165-181.

Sergean, R. 1971. *Managing Shiftwork.* London: Gower Press, Industrial Society, 242 pages.

Van Loon, J. H. 1963. "Diurnal Body Temperature Curves in Shift Workers." *Ergonomics, 6:* pp. 267-273. Included in Colquhoun and Rutenfranz, 1980, pp. 35-41.

Winget, C. M., L. Hughes, and J. LaDou. 1978. "Physiological Effects of Rotational Work Shifting: A Review." *Journal of Occupational Medicine, 20:* pp. 204-210.

CHAPTER **18**

Alternative Work Schedules and Guidelines for Shift Workers

CHAPTER 18. ALTERNATIVE WORK SCHEDULES AND GUIDELINES FOR SHIFT WORKERS

A. Alternative Work Schedules

    1. Flextime
    2. Compressed Work Weeks
    3. Part-Time Employment and Job Sharing

B. Guidelines for the Design or Selection of Shift Systems

C. Guidelines for Shift Workers

    1. Sleeping
    2. Digestion

References for Chapter 18

There are many combinations and variations of the schedules described in the previous chapter that may be appropriate for organizations that need to provide coverage for more than 40 hours per week on a day shift schedule. Adjustments in the hours of work through "flextime" and work sharing are discussed in this chapter. The guidelines near the end of this chapter can be used to design or select a shift schedule for a specific situation that may not have been addressed here.

A. ALTERNATIVE WORK SCHEDULES

During the last decade, several alternatives to the traditional five-day, 40-hour work week have been developed. Several schedules for around-the-clock staffing were discussed in Chapter 17. Some of the more popular alternative schedules and patterns of work include flextime, compressed work weeks, part-time employment, and job sharing.

A recent survey of nonfarm wage and salary workers found 11.9 percent on flextime, 2.7 percent on compressed work weeks, and 13.9 percent on voluntary part-time schedules (Bureau of Labor Statistics, 1980). Fifteen to 20 percent of employers offer flextime (Rosow and Zager, 1983). A number of factors may be associated with the recent trend in the use of alternative work schedules:

- An influx of women into the work force, especially mothers with school-age children.
- An increase in multiple-worker and dual-career families.
- A greater number of single-parent families.
- A desire by some older workers to reduce hours but continue to work.
- Increased preference for leisure time and time to further one's education by some workers.
- The cost and time of commuting.
- A change in attitude towards work and an interest in improving the quality of working life.
- High productivity and capital utilization demands.

New approaches to work scheduling can also help in energy conservation, productivity growth, and in the avoidance of layoffs during periods of economic recession. A brief discussion of three alternative work schedules follows.

1. FLEXTIME

Flextime is a work-scheduling method in which starting and stopping times are established by employees within limits set by management. It is also called flexitime, plan-time, adaptable hours, variable hours, and gliding time. Flextime was introduced in Germany in 1967 and has since spread to most industrialized countries (Kuhne and Blair, 1978). Several variations of flextime exist, the variations depending on work and management requirements. Four types of flextime are described below (Rosow and Zager, 1983):

- Flexitour—employees can choose starting and stopping times, stay on the schedule for a predetermined time, and always work at least eight hours per day.
- Gliding time—employees may vary starting and stopping times daily, but they must work the company's set hours each day. These are usually the four or five hours in the middle of the work shift. Eight hours per day are required.
- Variable day—hours per day can be varied as long as the required number of hours are worked by the end of the week or month.
- Maxiflex—employees may vary hours daily and do not have to be present for a core, or set, time on all days.

Both employees and management attribute many benefits to the use of flextime. The major advantage for the employee appears to be the increased ability to minimize conflicts between personal or family needs and work requirements. Medical, legal, and other personal business can be attended to during the usual working hours without requiring special leave from work. Flextime may also optimize the physiological and psychological conditions for workers. "Morning" and "evening" people can arrange their work schedules to coincide with their most efficient working times. It has been suggested that this coordination of work time with the biological rhythms (see Part II) can improve individual productivity (Kuhne and Blair, 1978).

Flextime also appears to provide advantages to management. Reductions in time lost from tardiness or unofficial leave by workers may result in increases in productivity. Sick leave and absenteeism have been reported to decrease, probably by making time available during the day for personal business (Kuhne and Blair, 1978). Employees may also be able to adjust their work schedules to correspond to fluctuating workloads; thus, idle time should be reduced. Flextime has reportedly created a greater sense of responsibility and job satisfaction in employees, factors that contribute to better job performance.

Flextime can have other benefits not directly related to the job. Large-scale utilization of flextime can reduce commuter traffic congestion. Several cities have reported this where areawide flexible work scheduling is used (Kuhne and Blair, 1978; Rosow and Zager, 1983). Recreational and service facilities may also be better used in communities where flextime is instituted. Since flextime permits work times to be adapted to fit domestic and child care commitments, it is a beneficial schedule for many working mothers.

There are limitations to the use of flextime in industry especially in. manufacturing operations. Machine-paced and continuous-process operations may not be compatible with a flexible work schedule. Multiple shift operations may not be able to use flextime because of problems of scheduling and coordinating the staff. Flextime is not, however, restricted to white-collar workers. A primary requirement is that workers must be able to perform tasks independently. It has been used successfully in the electronics manufacturing industry (Nollen and Martin, 1978) where each worker, or a small group of workers, can work independently for at least two or three hours a day. The workers choose starting and stopping times within blocks of 2.5 hours each at the beginning and end of the shift.

2. COMPRESSED WORK WEEKS

Compressed work week schedules imply that a week's worth of labor is performed in fewer than five days. Several schedules have been suggested to reduce the number of days worked. These include:

- Four ten-hour days.
- Three 12-hour days.
- Four nine-hour days and one four-hour day (usually Friday).
- 5/4 - 9 plan: alternating five-day and four-day weeks, nine-hour days.
- Weekends, two 12-hour days paid at premium rates.

As with any work-scheduling approach, there are advantages and disadvantages to compressed work weeks. From a management perspective, there are several arguments for such schedules. Greater productivity and lower unit costs have been demonstrated (Hedges, 1971), for example. Higher output per labor hour and other advantages may be related to any of the following factors associated with short weeks:

- Reduced absenteeism, tardiness, and personnel turnover.
- Reduced start-up and shutdown time relative to operating time.
- Relatively less time spent in coffee and lunch breaks.
- Advantageous personnel recruitment possibilities.
- A better working environment due to reduced distractions when compressed weeks are staggered for five-, six-, or seven-day operations.
- More efficient office work because the time gained at the beginning and end of the day contains fewer interruptions, such as those from phone calls.

There are also definite benefits for employees working compressed week schedules. A major benefit is increased leisure time. Time off during the normal work week provides opportunities to do personal business and enhances recreational pursuits as well. Employees also gain time from reduced commuting time; better traffic conditions on work days because of altered working hours and fewer workdays per week result in significant reductions in travel time. Employees on these schedules may also save money because of reduced commuting costs and, in some cases, reduced child care costs.

Arguments against compressed work weeks center on possible adverse health and safety effects on workers. Compressed work schedules require workdays longer than eight hours. General fatigue due to longer working hours is one concern; jobs requiring heavy physical effort for several hours are not recommended for these schedules (see Parts II and IV for further discussion as well as the text on overtime and 12-hour shift schedules). Although improvements in technology have, in many cases, reduced physical effort requirements, increases in mental requirements may have resulted. Extended daily work hours may not be desirable in jobs with sustained and demanding mental or perceptual effort.

Some working mothers have found the longer hours of a compressed schedule to be undesirable. Child care and meal preparation can be significantly disrupted when an hour is added to the beginning and end of the day. Compressed work schedules may also not be appropriate for companies providing services for other companies on five-day schedules. Sales personnel are usually kept on five-day schedules for similar reasons.

3. PART-TIME EMPLOYMENT AND JOB SHARING

Part-time employment is not new, but the changing needs of the work force and changes in the economy have resulted in a steady increase in the use of this work schedule (Rosow and Zager, 1983). A group including women entering or reentering the work force, young professionals, single parents, and retirees comprises much of the part-time employment or job-sharing work force.

Recent retirees with significant amounts of experience and skill are often interested in working part-time to supplement their retirement income. One company has developed a program in which company retirees are hired part-time to fill jobs normally staffed by workers from temporary service agencies (Rosow and Zager, 1983). The retirees have shown greater productivity, perhaps due to knowledge of and loyalty to the company after years of service. Although the retirees are paid well, the program has resulted in a cost savings because the agency commissions have been eliminated.

The use of part-time professionals, such as lawyers, planners, and engineers, has also proved a successful job coverage technique. Job sharing, in which two individuals share a job, is becoming popular especially in the clerical, health care, social service, and teaching professions. The sum of the two individuals' work is usually found to be more than the output of a single employee in these programs although the same number of hours are worked.

There are several obstacles to the growth of part-time employment. The costs of such programs are of concern because of the fixed costs per employee, regardless of how much time is spent at work. Thus, part-time workers may cost more per hour than full-time workers. Possible solutions to this problem include prorated labor costs, cost sharing, and benefit waivers. Organized labor unions may oppose part-time employment and job sharing for fear of increased job competition and a possible impact on pay and benefit standards. Of particular concern is that part-time workers will be used to avoid overtime pay. Such problems can be avoided, and several unions have worked with management to develop flexible work scheduling practices that include part-time employment and job sharing (Roszow and Zager, 1983).

Other work scheduling programs, such as flextime and compressed work schedules, are affecting those desiring part-time employment or job sharing. Many workers with flexible schedules seek a second, part-time job, thus increasing the competition for part-time employment.

From the preceding discussions of flextime, compressed work weeks, part-time employment, and job sharing, you can see that the appropriate use of a program depends on many factors. The success of implementing one of these alternative work schedules will depend on how well all the above factors are considered and on the involvement of the affected personnel in the planning process.

B. GUIDELINES FOR THE DESIGN OR SELECTION OF SHIFT SYSTEMS

From the studies of worker attitudes and the responses to the shift work schedules discussed earlier, we have developed some basic guidelines for good schedule design or selection. (Based on Information from Murphy, Eastman Kodak Company, 1981; U.S. Department of Housing and Urban Development, 1978; Rutenfranz, Colquhoun, and Knauth, 1976; Sergean, 1971). These include:

- Arrange continuous shift systems to have rest days fall on some part of the weekend, when possible.
- Avoid multiple consecutive night shifts. One or two night shifts are preferable to more than four in a row in a rotating schedule.
- Provide a rest period of at least 24 hours after each night shift period. The more consecutive nights worked, the more rest time should be allowed before the next rotation occurs.
- Use short-cycle rotations on continuous shift work systems; these are generally preferable to long-cycle rotations, such as a monthly rotating system. The physiological advantages of longer rotations are usually overridden by the psychosocial disadvantages.
- Let the nature of the work determine the continuous hours per shift to be worked. Physically, mentally, or perceptually demanding work is often not well suited to 10- or 12-hour shift lengths. Exposures to chem-

ical or physical agents should also be considered when selecting a shift system.

- Provide combination shift systems to permit accommodation of more workers. The advantages and disadvantages of two or more schedules can be offered to crew members. Those who prefer fixed shifts can be accommodated, while those who prefer rotations may choose between the socially popular rapidly rotating schedule and the more physiologically appropriate longer period rotation system.

- If a shift schedule offers a choice to rotate forward and provide easier sleeping adjustments for the worker or to rotate backward and get a long (four-plus days) weekend every month, choose the latter.

- Institute a special shift system instead of continuing on overtime if production demands result in extended periods of overtime work (for example, more than three months) that are not strictly seasonal.

- Spread the positive and negative aspects of any shift schedule as equitably as possible across the crews.

- Avoid very complex schedules that make it difficult for the worker and his or her family and friends to know which are workdays and which are days off.

C. GUIDELINES FOR SHIFT WORKERS

The body adjusts to night work by changes in the amplitude and phase of the circadian rhythms (see Part II). Inexperienced shift workers may suffer distressing digestive and behavioral changes in the initial adjustment to night shift work. The guidelines in this section for dealing with the sleeping and digestive problems associated primarily with night or early morning shift work are designed to help smooth the adjustment for the less experienced or occasional shift worker (developed from material in Lennertz and Lennertz, Eastman Kodak Company, 1973; Mott et al., 1965).

1. SLEEPING

People who work permanent night shift, fixed C shift in a three shift rotation, or rotating three-shift continuous or discontinuous schedules have to learn how, when, and where to plan their nontraditional sleeping hours. During B shift (afternoon-evening) work periods, most workers go to bed a few hours after the end of the shift and sleep later in the morning. This does not present problems for most people. On C shift, however, all of the sleeping hours are at times of the day or evening when family, friends, and neighbors are awake, active, and inclined to be noisy. The following are guidelines to assist the shift worker in sleeping after night shift work:

- Plan your sleeping time and do not alter the scheduled hours except in very unusual circumstances. Make sure your family, friends, and neighbors are aware of the schedule.

- Consider breaking the sleeping time up into two periods, one in the morning after work, and one in the evening before returning to work. Many shift workers sleep five to six hours during the day and one to two hours in the middle of the evening. This gives them time to use daytime community resources and to be with their families for dinner and the early evening. Continuous sleep periods are fine if uninterrupted time can be ensured.

- Limit the size of the morning meal after work to help ensure uninterrupted sleep. Avoid stimulants, such as caffeine.

- Develop a good sleeping environment, including the following:
  - Heavy shades on the windows to darken the room.
  - A cool room without having to open windows that let in street noises. Use either a fan or an air conditioner.
  - Carpets, acoustical tiles, or the addition of masking noises from an air conditioner, sound equipment, or a tape recorder for soundproofing.
  - A comfortable bed.
  - An "on-off" switch for the telephone.

- Learn relaxation techniques (Benson, 1975) to assist you in falling asleep. Concentrate on relaxing your muscles and thinking about pleasant events.

- If noise from the surroundings is a problem, use earplugs or wear earphones and listen to quiet music or "white noise," a broad-band sound.

- Train your children to respect your sleeping hours during the day and to play in a more distant part of the living quarters, if possible.

- If you cannot fall asleep after an hour, read a book or listen to quiet music or the radio for a while. If sleep still does not come, reschedule your sleeping hours for later in the day—for example, from 2:00 p.m. to 10:00 p.m.

- Avoid getting locked in to too many commitments in the later afternoon during the periods on C shift. Leave time for napping if daytime sleep is inadequate or interrupted.

- If interruptions and noise disturb your sleeping time on a regular basis, discuss ways to reduce them with your spouse, children, neighbors, and friends. A sign on the outside door indicating that there is a "day-sleeper" on the premises may discourage unexpected intrusions from salespeople or passersby.

2. DIGESTION

Disturbances in the timing and amount of meals associated with evening and night work may cause digestive problems for many shift workers. These are often classified as heartburn, stomach cramps, constipation, loss of appetite, gas, and nausea. The following guidelines should help in alleviating some of these discomforting symptoms:

- When working B shift (afternoon-evening), have your main meal in the middle of the day instead of the middle of the work shift. Keep the other two meals lighter, and do not eat large amounts of food after work and before going to bed.

- On C shift, eat lightly during the whole work shift, especially for the first two to three nights of the rotation. Avoid gassy, greasy, and acidic foods in those hours. Do not drink large amounts of caffeinated beverages, especially on a relatively empty stomach.

- Have small- to moderate-size meals throughout the day and night shift during C shift rotations. Increase the number of meals, if needed, but avoid large ones. Drink four to six glasses of water a day, and include the usual balance of vegetables, fruits, lean meat, poultry, fish, dairy products, grains, and bread in your diet.

- Avoid alcoholic beverages, especially in the evenings before C shift work.

- Keep coffee or tea consumption at a minimal level, or use decaffeinated coffee or beverages.

- In the morning after C shift work, eat a moderate breakfast before going to bed. This should keep your hunger abated during your sleeping hours without keeping you awake with digestive discomfort.

- Avoid excessive use of antacids, tranquilizers, or sleeping pills. It is better to select the foods that do not aggravate digestive problems and to use relaxation techniques to assist with sleeping problems.

- Avoid replacing regular meals with candy bar and soft-drink snacks because of reduced appetite. Eat crackers or fruit in preference to candy during work breaks.

- If digestive complaints persist, seek medical help. In most instances, the problems can be reduced by simple attention to eating habits and food types. Regular exercise may also assist the worker who has elimination problems or a loss of appetite.

Since each shift schedule contains variations in the length of time a C shift is worked and the times it starts and ends, the new shift worker is advised to discuss ways of coping with sleeping and eating problems with veterans on the schedule. They usually will have found the best sleep and meal schedules for their bodies and will be happy to pass this wisdom on to the newer workers. For new shift systems, there may not be an advisory group of experienced workers to draw on. Here the worker is encouraged to start by eating small amounts of well balanced meals frequently throughout the day and then gradually establish the best pattern for one larger meal if it seems necessary. Sleeping schedules should be developed gradually, too; social commitments should be reduced until the best schedule is found so that napping time is available during the adjustment period.

With attention to these two areas of worker adjustment, sleeping and digestion, you can alleviate the physical problems that may make shift workers uncomfortable during the night shift rotation. Shift schedule design, discussed earlier, can address some of the psychological stresses of working outside of the usual daytime hours. The advantages of shift work, such as more time during the day to use community services and recreational facilities, increased pay, more responsibility, often with less direct supervision at work, and more variety on the job and outside of it, will then be featured.

REFERENCES

Benson, H. 1975. *The Relaxation Response.* New York: William Morrow, 222 pages.

Bureau of Labor Statistics. 1980. *Current Population Survey.* Washington, D.C.: Bureau of Labor Statistics.

Hedges, J. N. 1971. "A Look at the 4-Day Workweek." *Monthly Labor Review, October 1971:* pp. 33-37.

Kuhne, R. J., and C. O. Blair. 1978. "Changing the Workweek." *Business Horizons, April 1978:* pp. 39-44.

Mott, P. E., F. Mann, Q. McLoughlin, and D. Warwick. 1965. *Shift Work: Its Social, Psychological, and Physical Consequences.* Ann Arbor: University of Michigan Press, 351 pages.

Nollen, S. D., and V. H. Martin. 1978. *Alternative Work Schedules. Part 1: Flexitime.* American Management Association (AMA) Survey Report. New York: American Management Association, 53 pages.

Rosow, J. M., and R. Zager. 1983. "Punch Out the Time Clocks." *Harvard Business Review, 61 (March-April 1983):* pp. 12-30.

Rutenfranz, J., W. P. Colquhoun, and P. Knauth. 1976. " Hours of Work and Shiftwork." In *Proceedings of the 6th Congress of the International Ergonomics Association,* July 1976, College Park, Maryland. Santa Monica, Calif.: Human Factors Society, pp. xlv-lii.

Sergean, R. 1971. *Managing Shiftwork.* London: Gower Press, Industrial Society, 242 pages.

U.S. Department of Housing and Urban Development. 1978. *Work Schedule Design Handbook: Methods for Assigning Employees' Work Shifts and Days Off.* Capacity Sharing Program, Office of Policy Development and Research. Report No. 624-521/1063. Washington, D.C.: U.S. Department of Housing and Urban Development, U.S. Government Printing Office, 368 pages. Also available from Institute for Public Program Analysis, 1328 Baur Blvd., St. Louis, Mo. 63132.

PART VI

Manual Materials Handling

CONTRIBUTING AUTHORS

Waldo J. Nielsen, M.S., Industrial Statistics

Suzanne H. Rodgers, Ph.D., Physiology

Most industrial production workers are required to handle supplies or products on their jobs. The design of those handling tasks can affect the potential for injuries from overexertion, especially during lifting, lowering, pushing, and pulling activities. This part describes the characteristics of handling tasks and gives guidelines for their design in order to reduce the occurrence of manual handling injuries.

CHAPTER **19**

Manual Handling in Industry

CHAPTER 19. MANUAL HANDLING IN INDUSTRY

A. Manual Handling and Musculoskeletal Problems

B. Strategies for Reducing Manual Handling Injuries
1. The Systems Approach
2. Education—Teaching People to Lift
3. Selection
4. Redesigning Workplaces and Jobs

References for Chapter 19

Few industries are able to avoid the need for products or raw materials to be handled manually at some time during their production or use. Some observations on the relationship between manual handling and muscle and back problems and on ways industries have tried to reduce these problems are included in this chapter.

A. MANUAL HANDLING AND MUSCULOSKELETAL PROBLEMS

Approximately 25 percent of the accidents reported in industry each year are associated with the manual handling of bulk materials or intermediate-stage and finished products. A one-year analysis of injuries associated with manual handling shows the following (Rodgers, Eastman Kodak Company, 1980):

- 50 percent involve the back.
- 30 percent are shoulder, arm, or abdominal strains and sprains.
- 42 percent result in some lost time; 25 percent have more than five days lost from work.
- Mechanic, handler, production machine operator, and trainee are the jobs with the most lift, lower, push, and pull incidents.
- People with less than three years on the job have the most incidents.
- The leading tasks associated with these accidents are the handling of boxes, drums, carts, bulk materials, cans, metal, and pallets.

In studies where the relationship between back injuries and manual handling has been investigated, the following observations have been made:

- Lifting and lowering were associated with 49 to 60 percent of the low-back incidents (Brown, 1975; Nagira, Ohta, and Aoyama, 1979; Snook, Campanelli, and Hart, 1977; Stubbs and Nicholson, 1979).
- Awkward postures were associated with 12 to 19 percent of the incidents (Nagira, Ohta, and Aoyama, 1979; Snook, Campanelli, and Hart, 1977).
- "Job pressure" (usually time pressure) was associated with 6 percent of the reported incidents (Nagira, Ohta, and Aoyama, 1979).

Despite these findings, there is no clear relationship between the incidence of back pain in an industrial population and the level of effort in handling tasks. This is probably because of a high background rate of back pain in the population. The "incidence" of back pain is variously reported as from 3 to 60 percent, depending on the data source and period of measurement. The 3 percent rate was found in a study of industrial workers who experienced back pain severe enough to lose three days or more from work in one year (Friedlander and Newman, Eastman Kodak Company, 1981). The 60 percent rate comes from self-reporting of industrial workers in response to the question, "Have you ever had low-back pain?'' (Pederson, Petersen, and Staffeldt, 1975). A similar value was seen in a study of industrial men interviewed at retirement: 56 percent had reported an incident of back pain to the medical department during their working years, which were often more than 40 (Rowe, 1983). With this disparity in the collection of data, it is difficult to specify the role of physical effort in the development or aggravation of back pain. However, some additional observations are of interest:

- The 30- to 45-year-old population has the highest incidence of back pain (Brown, 1975; Rowe, 1983).

- A yearly recurrence rate of 40 percent has been measured in people who had three or more days absent for back pain (Friedlander and Newman, Eastman Kodak Company, 1981).

- People in heavier jobs, such as paper and chemical manufacturing handlers, reported more back pain (47 percent) than those in lighter jobs, such as clerks and light assembly workers (35 percent), in one study (Rowe, 1983). An additional 18 percent of lost-time incidents were recorded for people on heavier jobs in another study (Hult, 1953). These differences may be explained by the difficulty of continuing to perform a heavy job when a back problem is present; this results in more time away from the job and more reporting of symptoms (Rowe, 1983).

- In a one-year study of low-back pain incidents, 25 to 35 percent were associated with unusual or heavy physical effort. The remaining incidents occurred on tasks that were done regularly (Rowe, 1983). This finding is supported by other studies (Snook, Campanelli, and Hart, 1977). It suggests that not all back pain incidents are preventable by training people, selecting them according to their strengths, or designing improved handling workplaces.

B. STRATEGIES FOR REDUCING MANUAL HANDLING INJURIES

Although the amount of ''preventable'' low back pain may only be 33 percent of the cases observed (Rowe, 1983), several approaches have been used to try to reduce all types of musculoskeletal injuries associated with manual handling. These approaches include:

- A systems approach to materials handling in a new facility; that is, reducing the amount of manual product handling and using automatic transfer devices when possible.

- Teaching people safe lifting techniques.

- Selecting people for specific tasks according to their strengths or endurance.

- Redesigning the workplace, containers, and job to make the handling suitable for more people.

Another strategy, developing the workers' strength and endurance capacities through industrial fitness programs, has recently been used (Amoroso and Day, Eastman Kodak Company, 1978; Everett, 1979). However, the lack of extensive data from industry prevents its inclusion in the discussion at this time.

1. THE SYSTEMS APPROACH

Automated movement of bulk materials and intermediate- and final-stage product through a manufacturing system should be incorporated into the plant layout in order to minimize the need for manual handling. The systems approach to materials handling can best be accomplished during the initial planning of a facility since many workplaces are involved. A number of principles for design are given here. Their common goal is to create a manufacturing area that is part of the total system (from receiving the supplies to shipping the final product) and to use one handling and distribution system (Barnes, 1968; Schneider, 1975). The alternative is to treat each workplace separately and to provide a variety of materials handling aids or procedures to alleviate problems as they are identified.

The advantages of a systems approach to moving materials through the production process include (Woodson, 1981):

- Reduction of in-process inventory.
- Elimination of multiple rehandling of product.
- Less delay in moving materials between work stations.
- Improved control of production.
- Reduced damage due to handling.
- More efficient handling in distribution tasks for improved safety.
- Less opportunity for human error.

A systems design can be accomplished with application of one or more of the following principles of materials handling (Beck, Eastman Kodak Company, 1978):

- Unit Size, or Load, Principle: Increase the quantity, size, or weight of unit loads; handle them with powered equipment. A disadvantage of this approach is that it provides less flexibility in the use of storage space. Using drums instead of bags for handling dry chemicals is an illustration of this principle.
- Mechanization Principle: Mechanize handling operations. This would include the use of motorized conveyors or overhead hoists between workplaces.

- Standardization Principle: Standardize handling methods as well as types and sizes of handling equipment. Use equipment that is compatible across the system so the product does not have to be manually transferred between workplaces.

- Adaptability Principle: Use methods and equipment that can best perform a variety of tasks and applications.

- Dead Weight Principle: Reduce the ratio of dead weight of mobile handling equipment to load carried. Heavy metallic trays or cans, for example, represent dead weight in can and tray handling of the product; they should be used only if no lighter material is suitable for the task.

- Gravity Principle: Use gravity to move material wherever practical. Gravity feed conveyors are an application of this principle, as is lowering instead of lifting in manual handling.

- Automation Principle: Provide automation to include production handling and storage functions, such as automated stacker-retriever systems.

Manual materials handling should be routine only in operations where handling is occasional or where high flexibility is needed because of multiple product changes. Operations requiring continuous handling of one or a few products or intermediaries are best done by handling equipment such as industrial robots, automatic palletizers, or assembly machines. The use of computer-controlled equipment for automatic storage and retrieval operations solves the need for inventory control and handling. FIFO (first-in-first-out) procedures of inventory control are then possible with automatic stacker-retriever systems (Dehlinger, Nash, and Edwards, 1972; Schneider, 1975).

2. EDUCATION—TEACHING PEOPLE TO LIFT

The strategy of teaching people how to lift has been used for many years. Although some effects can be seen on a short-term basis, the rate of overexertion incidents has not decreased enough to suggest that training is sufficient in itself to prevent handling injuries. Lifting training is valuable to transmit information about handling techniques formally to new employees and to remind more experienced handlers about basic postures that avoid twisting of the trunk and help them control the load. New "kinetic" methods of handling are being taught that put more emphasis on rhythmic, freestyle lifting (Brown, 1975; Davies, 1972; Himbury, 1967; Jones, 1979).

Although the training of workers to lift safely makes them responsible for caring for their backs on the job, too much emphasis on "proper" lifting techniques can lead to a tendency to blame the people who develop back pain on the job during a handling activity. There is often no evidence that the back problem could have been prevented by using the "proper" lifting technique, so the blame may be inappropriate. Other factors related to job and workplace design—for example, the duration of handling and the heights of the lifts—may contribute to the accidents and force the handler to take an awkward posture or prevent him or her from making lifts or exerting forces most efficiently.

3. SELECTION

For many years there have been forms of selection in workplaces where heavy tasks are performed. Although there are few validated selection procedures that show a high correlation with ability to perform handling tasks, self-selection and "natural selection" are common. In the former, the individual decides not to do a job that appears to be too hard. The latter is the failure of a person to succeed on a job, usually because of inadequate strength, excessive fatigue, aggravation of an existing health problem, or injury. It is the last three events that industry wishes to prevent by looking for ways to screen candidates prior to placing them on a heavy job.

Selecting people for jobs according to their lifting strengths has been recommended by several investigators (Ayoub et al., 1979; Chaffin, Herrin, and Keyserling, 1978; Kamon and Goldfuss, 1978; Keyserling et al., 1980). Using strength-testing techniques to find people for heavy jobs implies that jobs can be described by their worst component, usually a strength requirement identified by a biomechanical model. Such a model includes information about joint position, muscle strength, and limb lengths from which forces on the back, knee, shoulder, wrist, and elbow can be estimated. The model does not consider the effects of lifting or exerting force repetitively; it deals with single events only. People who do not have enough strength in the most difficult lift, push, or pull are considered at risk for handling injuries. Although this technique is attractive as a tool for placing people on jobs where overexertion incidents may occur, it has not been shown to be predictive enough to meet Equal Employment Opportunity criteria or selection testing requirements in industry (Equal Employment Opportunity Commission et al., 1978). Strength-testing approaches make the following assumptions:

- That the job can be described by its greatest biomechanical stress, such as the force on the lumbar spine, on a joint, or on a muscle.

- That the worst stress is associated with the injury incidence on the job.

- That manual handling techniques are fairly stereotyped, with little variability among people doing them. This assumption is necessary to develop the biomechanical model of the strength requirements of jobs.

In industrial handling tasks, these assumptions often are not appropriate. People change their handling techniques according to their size and strength capabilities. Jobs include a large variety of tasks whose duration, intensity, frequency, and pattern during the shift will determine how much fatigue develops (see Part IV). However, there are some jobs where selection testing may be appropriate. Some physically demanding tasks, such as firefighting and some construction activities, are difficult to redesign, and it may be possible to validate the measured strength and endurance requirements to the individual's performance on a standard test.

4. REDESIGNING WORKPLACES AND JOBS

If a job is difficult for a large number of people, it is usually better to redesign it than to try to educate or select people for it. The latter approaches require continuous programs to maintain training and test new people for the job. Redesign, on the other hand, usually eliminates the problem and is a one-time investment. One study of compensible back injuries suggests that up to 67 percent of handling injuries might have been prevented if the tasks were designed to accommodate 75 percent of the potential work force (Snook, Campanelli, and Hart, 1977).

Not all jobs or workplaces lend themselves to redesign. The handling task may be determined by the product characteristics, as in large-size sheet materials; uncontrollable factors, such as the demands of firefighting; or by prohibitive costs, a problem often associated with older facilities. A majority of handling tasks, however, lend themselves to fairly simple alterations that make them suitable for more people and reduce the potential for manual handling injuries. They have been discussed throughout these volumes; a brief listing of some of these redesign strategies is included below:

- Provide ways to adjust the materials to be handled so that less lifting and more sliding can be done. For example, provide a levelator or scissors table to adjust the height of the load.
- Provide good handholds on containers or objects to be handled.
- Rotate people to a lighter job after one to two hours in a constant handling task.
- Provide carts and handling aids to support the weight of objects that have to be carried more than a few feet.
- Provide tools to help in applying forces with the hand.
- Provide space for in-process inventory in production line operations so time pressure does not drive the handler.

More expensive strategies are also worth investigating if the task is expected to continue for several years and if jobs done by people can be more appropriately done by machines. Highly repetitive tasks that do not require decision making, for instance, are being performed effectively by robots. Some of these stereotyped activities include putting cases on a pallet, sealing cases, and loading packaging or assembly machines (Callahan, 1982; Helander, 1982).

REFERENCES

Ayoub, M. M., R. Dryden, J. McDaniel, R. Knipfer, and D. Dixon. 1979. "Predicting Lifting Capacity." *American Industrial Hygiene Association Journal, 40:* pp. 1075-1084.

Barnes, R. M. 1968. *Motion and Time Study.* New York: Wiley, 689 pages.

Brown, J. R. 1975. "Factors Contributing to the Development of Back Pain in Industrial Workers." *American Industrial Hygiene Association Journal, 36:* pp. 26-31.

Callahan, J. M. 1982. "The State of Industrial Robots." *Byte, 7 (10):* pp. 128-142.

Chaffin, D. B., G. Herrin, and W. M. Keyserling. 1978. "Preemployment Strength Testing—An Updated Position." *Journal of Occupational Medicine, 20 (6):* pp. 403-408.

Davies, B. T. 1972. "Moving Loads Manually." *Applied Ergonomics, 3 (4):* pp. 190-194.

Dehlinger, J. R., G. S. Nash, and H. K. Edwards. 1972. "Materials Handling Improvement Needs Systems Approach." *Industrial Engineering, 4 (August):* pp. 10-13.

Equal Employment Opportunity Commission, Civil Service Commission, Department of Justice, and Department of Labor. 1978. "Uniform Guidelines on Employee Selection Procedures." No. 6570-06, Part 1607. *Federal Register, 43 (166):* pp. 38290-38345.

Everett, M. D. 1979. "Strategies for Increasing Employees' Level of Exercise and Physical Fitness." *Journal of Occupational Medicine, 21 (7):* pp. 463-467.

Helander, M. G. 1982. *Ergonomics in Automation.* A training program for IBM. Internal publication, 330 pages.

Himbury, S. 1967. *Kinetic Methods of Manual Handling in Industry.* Occupational Safety and Health Series No. 10. Geneva: International Labour Office, 38 pages.

Hult, L. 1953. "Munksforsundersokningen." *Nordisk Medicin, 50:* p. 1076. Quoted in American Industrial Hygiene Association Ergonomics Committee (1970), "Ergonomics Guide to Manual Lifting," *American Industrial Hygiene Association Journal, 31:* pp. 511-516.

Jones, D. F. 1979. *Back Pain and How to Prevent It the Dynacopic Way.* Publication of Dynacopics, Inc., 17 Brushwood Court, Don Mills, Ontario, Canada, M3A 1V9, 13 pages.

Kamon, E., and A. J. Goldfuss. 1978. "In-Plant Evaluation of the Muscle Strength of Workers." *American Industrial Hygiene Association Journal, 39:* pp. 801-807.

Keyserling, W. M., G. Herrin, D. B. Chaffin, T. J. Armstrong, and M. L. Foss. 1980. "Establishing an Industrial Strength-Testing Program." *American Industrial Hygiene Association Journal, 41:* pp. 730-736.

Nagira, T., T. Ohta, and H. Aoyama. 1979. "Low-Back Pain Among Electric Power and Supply Workers and Their Attitude Towards Its Prevention and Treatment." *Journal of Human Ergology, 8:* pp. 125-133.

Pederson, O. F., R. Petersen, and E. S. Staffeldt. 1975. "Back Pain and Isometric Back Muscle Strength of Workers in a Danish Factory." *Scandinavian Journal of Rehabilitative Medicine, 7:* pp. 125-128.

Rowe, M. L. 1983. *Backache at Work.* Fairport, N.Y.: Perinton Press, 122 pages.

Schneider, D. A. 1975. "You Can't See the 'New' in Handling with an Old Outlook." *Automation: The Production Engineering Magazine, 22 (11):* pp. 56-59.

Snook, S. H., R. A. Campanelli, and J. W. Hart. 1977. "A Study of Three Preventive

Approaches to Low-Back Injury." *Journal of Occupational Medicine, 20 (7):* pp. 478-481.

Stubbs, D. A., and A. S. Nicholson. 1979. "Manual Handling and Back Injuries in the Construction Industry: An Investigation." *Journal of Occupational Accidents, 2:* pp. 179-190.

Woodson, W. E. 1981. *Human Factors Design Handbook.* New York: McGraw-Hill, 1,049 pp.

Factors Impinging on the Manual Handling of Materials

CHAPTER 20. FACTORS IMPINGING ON THE MANUAL HANDLING OF MATERIALS

A. Grasp

A very critical factor determining the difficulty of a handling task is the interface between the person and the part, container, truck, or cart being handled. Objects with good handholds are considerably easier to lift, lower, carry, push, or pull. Very frequently, the type of grasp required to handle a container or object will be the factor determining how heavy it can be or how long it can be held. In this chapter, a discussion of several types of grasp and the guidelines for design of handles and containers are given. Part III described the biomechanics of grasp, and information about grip strength and how it is affected by wrist position can be found in Appendix A.

A. GRASP

In manual materials handling tasks, especially lifting, lowering, and carrying, the interface (boundary) between the person's hands and an object is a critical factor in the acceptability of a given load. Objects that can be handled with a power grip (as defined later in this section) can be heavier and held onto longer than those that must be controlled with a pinch grip. The dimensions of handles will determine the force per unit area on the hand and will, thereby, strongly influence the comfort with which a given object can be handled. In a study where voluntary maximum pulls were exerted against trays with several different handle configurations, the forces developed on trays with poor handles (those with edges less than 6 mm—0.25 in.—thick or with inadequate finger clearance) were about 55 percent of the forces exerted on trays with good handles—that is, with cut-through handholds or gripping blocks (Nielsen, Eastman Kodak Company, 1978). See the section on tray design in the next chapter for examples of good and poor handholds. Psychophysical studies of acceptable loads for lifting have demonstrated a 4 to 10 percent fall in acceptable weight (not maximal) if handholds are not provided (Garg and Saxena, 1980). Therefore, attention to grasp requirements when selecting or designing containers or objects can often increase the amount of weight that can be handled safely and comfortably.

Grasps can be classified according to the muscles used, the biomechanics of the hand or how forces are developed (see Part III), or the hand's functional use (Jacobsen and Sperling, 1976; Long et al., 1970; Napier, 1956). Some common types of grasps and their characteristics are discussed here.

1. PINCH OR PRECISION GRASP

Figure 20-1 illustrates pinch grasp. It is characterized by opposition of the thumb and the distal joints of the fingers. Because the palm and proximal finger surfaces (those nearest the palm) are not involved, pinch grip is only about 25 percent of power grip strength. For example, a person with 445 newtons (100 lbf) of power grip strength would have only about 110 newtons (25 lbf) of pinch grip strength in the same handling conditions. Pinch grip strength declines greatly at spans less than 2.5 cm (1 in.) or more than 7.5 cm (3 in.) (Jones, Eastman Kodak Company, 1974).

(a) Precision Tip

(b) Precision Adjustment

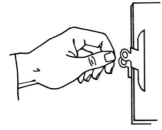

(c) Lateral

(d) Holding Tray on Rolled Rim

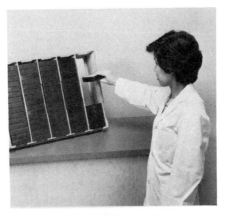

(e) Precision Palmar

(f) Procuring a Cartridge

Figure 20-1: Pinch or Precision Grasp   Three types of pinch or precision grasp are illustrated in drawings (*a, c, e*) and in workplace tasks (*b, d, f*). The precision tip grip illustrated in *a* and *b* is characterized by

Examples of the use of pinch grasp are:

- Stabilizing loose sheets or flat products that cannot be accommodated (that is, curled) in a power grip.

- Lifting bags placed in an upright position from the top seam if there are no side gussets in which to place the hands.

- Opposing one surface against another, as in holding a lid on a container or keeping two parts aligned.

2. POWER OR CYLINDRICAL GRASP

Figure 20-2 illustrates the power grasp, the maximum force that can be developed by the hand. Wrist orientation and grip span influence how much force the hand can develop for a given task. A cylinder 5 cm (2 in.) in diameter is considered optimal for power grip development (Petrofsky et al., 1980). The thumb is not needed when the power grasp is used to move an object vertically; it is needed to control the grasp if the motion is made across the body (transverse).

Examples of the use of power grasp are:

- Controlling the application of force, as in pushing and maneuvering a hand cart.

- Moving cylindrical materials transversely, such as pipes.

- Controlling lifts of objects with D-type handles, such as shovels.

- Gripping pliers and power tools—for example, an electric drill.

- Grasping hand brakes or circular control valves.

3. OBLIQUE GRASP

The oblique grasp is a variant of the power grasp; it is shown in Figure 20-3. It is characterized by gripping across a rectangular or curved surface or handle, and it has about 65 percent of power grip strength. Hand span strongly affects

---

having just the tips of the fingers and thumb in contact with the object or tool handle. In the lateral pinch shown in c and d, the thumb holds the object against the second joint of the index finger. This occurs in many tray-handling activities, especially when lifts are above 115 cm (45 in.). The precision palmar grasp shown in e and f is used in writing and some tool use, such as fine painting and soldering activities. It is also used in procuring products that are too wide to include in a power grasp. It is characterized by the object contacting most of the finger joints, but the fingers remain straight because they cannot curl. Pinch grasps like these can exert only about 25 percent as much force as power or cylindrical grasps because the hand is at a biomechanical disadvantage (see Part III). (Adapted from Jacobson and Sperling, 1976).

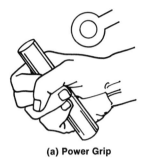

(a) Power Grip

(b) Carrying Pipe

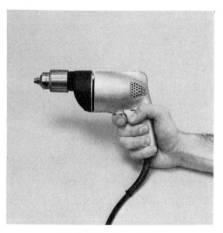

(c) Torquing a screwdriver
in a disassembly task

(d) Holding Power Drill

Figure 20-2: Power or Cylindrical Grasp   The cylindrical power grasp is illustrated by a drawing (*a*) and in three workplace applications (*b, c, d*). The power grip is characterized by direct opposition of the thumb to the fingers, all wrapped around a cylinder, sphere, or wedged-shaped object that fits comfortably in the hand. Power grip strength varies with the hand span or cylindrical diameter of the object being handled (see Appendix A). (*Adapted from Jacobson and Sperling, 1976*).

the force that can be applied; a span of 5 to 6 cm (2 to 2.5 inches) is recommended for maximum strength (Jones, Eastman Kodak Company, 1974). The thumb is very important for stabilization of the oblique grip.

Examples of the use of oblique grasp are:

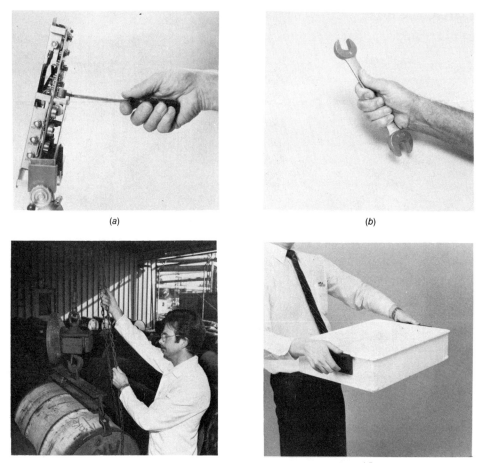

(a)

(b)

(c)

(d)

Figure 20-3: Oblique Grasp   The oblique grasp is like the power or cylindrical grasp except that the thumb is extended to stabilize the grasp rather than wrapped around the fingers. Four examples of workplace use of oblique grasp are shown, two during tool use (*a, b*) and two during manual handling tasks (*c, d*). The latter should be compared to Figure 20-1 *d* where a lateral pinch is used to support the tray by its rolled edge. Use of the block handle restores full power grip strength to the handler instead of leaving him or her with only 25 percent of that strength due to the pinch grip.

- Handling individual parts, such as cartridges, in inspection or assembly tasks.
- Gripping hand tools such as wrenches, screwdrivers, and paint scrapers.
- Holding a cover or lid on a container that has a designated handhold.

4. HOOK GRASP

Figure 20-4 illustrates the hook grasp. The thumb is passive in this grasp, the object weight being supported by the fingers. It is useful for handling at low heights where an object can be cradled in the hand. Hook grip strength is equal

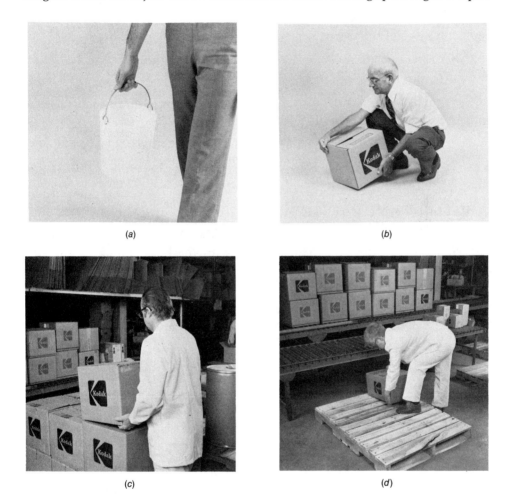

(a)   (b)

(c)   (d)

Figure 20-4: Hook Grasp   The hook grasp is characterized by a flat hand, curled fingers, and the thumb used primarily to stabilize the load rather than to grasp it. It is most effective when the arm is down at the body's side and the object handled can be carried with one hand (a). It is also used very commonly in lifting tasks (b, c, d), one hand lifting the object from the bottom with a hook grip, and the other hand pulling it across the opposite top corner with a flat palm supporting grasp (b). At higher lift locations, the hook grasp is more difficult to sustain because the wrist must be maximally extended to create the hook (c). Hook grasps require less muscular work than power grasps because the thumb is not active, so they are preferred for extended carrying tasks.

to power grip strength (Jones, Eastman Kodak Company, 1974) for objects about 5 cm (2 in.) wide. Very narrow objects or handles, such as a bucket's wire handle, can limit hook grasp strength as weight increases by pressing deeply into the hand and fingers. The load may exceed the maximum recommended skin pressure of 150 kPa (22 lbf/in.²), thereby limiting the amount of weight that can be handled comfortably for more than a few seconds during lifting.

Examples of the use of hook grasp are:

- Carrying objects to one side, as in carrying a suitcase.
- Lifting objects from floor level to 90 cm (35 in.).
- Pulling objects horizontally—for example, pulling a tray towards the edge of a shelf for rehandling or inspection of parts.
- Removing parts in maintenance activities, such as removing a pump motor from under a piece of equipment.
- Pulling down on a hook or strap.

5. PALM-UP AND PALM-DOWN GRASPS

In addition to the types of grasp already illustrated, there are grasps that approach an object palm down, and those that approach it palm up. Figure 20-5 illustrates these grasps.

Palm-down grasps are primarily used in precision activities, such as small-

(a)

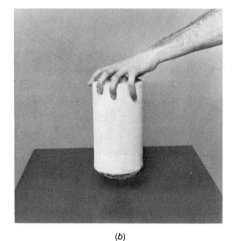

(b)

Figure 20-5: Palm-Up and Palm-Down Grasps   The orientation of the hand in handling small parts will affect how much weight can be supported. The palm-up grasp (a) permits the biceps muscles of the upper arm to support the weight so heavier objects can be handled. However, the object has to be supplied in a way that allows the worker to get his

parts assembly. Since the stronger arm muscles are not optimally positioned to exert force when the palm is turned down, objects weighing more than 0.5 kg (1 lbm) should not be handled repeatedly with this grasp (Rehnlund, 1973). Palm-up grasps are used in most low lifting, also shown by the corner grasp in Figure 20-4 *d*. This grasp is difficult to perform when lift height is above elbow height. At about 90 cm (35 in.) above the floor, the hands have to be reoriented from a low to a higher level on the object being lifted. As they move through this change point, the potential for losing control of the load is increased. The heavier the load being lifted, the more difficult the grasp change becomes. It is often this point that limits how heavy the object can be and how high it can be lifted in order to be acceptable for most people. Once the hand transition has been accomplished, the load can be boosted from below, where grasp strength will not be a limiting factor.

One way to aid the handler in traversing this grasp change point is to provide a shelf or platform at about 90 cm (35 in.) above the floor; the object can be placed on this while the handler regrasps it for lifts to higher levels. Because such a platform is not always available, however, guidelines for lifting in this chapter reflect the loss of strength and control as the hands are adjusted during a continuous lift from floor level to shoulder height or above.

B. DESIGN OF HANDLES FOR MANUAL HANDLING TASKS

Many materials and objects without designated handholds are transported in industry. In some instances, they have a component that acts as a handle for the person who has to move them. For example, pumps, power supplies, and racks have openings, projections, or support structures that can be used for a hook or power grasp during maintenance and cleaning tasks. Providing handles on these parts or molding the surfaces to permit a secure grip may prevent damage to them and reduce the potential for electric shock or hand or finger injury.

One has to consider handle orientation, clearance requirements for the hand, and the balance point of the load in order to determine the number of handles needed and their location. Also, the handle should be designed so that the force per unit area on the fingers and hand does not exceed 150 kPa (22 lbf/ in.^{2}) (Rehnlund, 1973). The guidelines given here are based on functional dimensions of the hand (see Appendix A). That information was developed from industrial (Champney, Eastman Kodak Company, 1977; Jones, Eastman Kodak

---

or her fingers underneath it, which is not the case in procuring many items off a counter or a horizontal conveyor belt. The latter activity requires a palm-down grasp (*b*), which reduces the effectiveness of the biceps muscle and, therefore, results in lower weight limits for one-handed lifts.

Company, 1974) and military (Garrett, 1971) studies. If people using the handles are wearing work gloves, additional clearance and width of 2.5 cm (1 in.) should be provided. Tray handles are discussed in Chapter 21.

Handles should be designed to accommodate the working hand as it supports a load and moves through several heights and distances in front of the body. The wrist joint provides great flexibility for moving a load, but strength is lost as the wrist moves away from its resting position (see Appendix A). Handle design that forces the wrist away from its strongest position will limit the amount of weight that can be handled in a task. Figure 20-6 illustrates a simple D-type handle. This handle is similar to ones seen on some types of hand trucks,

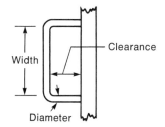

	Clearance		Width	
---	cm	inches	cm	inches
One Hand	6.4	2.5	12	4.8
Two Hands	6.4	2.5	24	9.5

Maximum diameter for full
  encirclement by hand = 3.8 cm (1.5 in.)
Exposed edges should be rounded to a minimum
  radius of 1 mm (0.04 in.)
Exposed corners should be rounded to 13 mm
  (0.5 in.)
Fingers should be able to curl to 120 degrees or more
  (as in a hook grip).

Minimum Diameter for Comfort:

Diameter		Range of Object Weight	
cm	inches	kg	lbm
0.6	0.25	Up to 7	Up to 15
1.3	0.50	7-9	15-20
1.9	0.75	Over 9	Over 20

Figure 20-6: Recommended Handle Dimensions   The minimum width and clearance values for D-type handles are given in centimeters and inches for one- and two-hand use. Other characteristics, such as the maximum diameter, are included in the middle of the figure. The lowest section of the figure includes information relating hand comfort at different weights to the minimum handle diameter. These dimen-

on cleaning or electroplating racks, on components such as pumps and power supplies, and on some tool boxes.

Some additional guidelines for handle selection and design are:

- Units weighing more than 4.5 kg (10 lbm) should have handles.

- Where feasible, bulky (more than 18 kg or 40 lbm) objects should have handles for two-person lifting.

- Handles should be located at or above the line passing through the center of gravity of the load.

- Handles should be textured (but not grooved) to reduce slippage of the fingers.

- Handles that have to lie flat on a surface should have a lock-in feature when they are folded out for use.

- T-bar handles, sometimes found on machinery components to aid in their removal for repair or maintenance, should have a center post no larger than 1.2 cm (0.5 in.) in diameter (Rigby, 1973).

REFERENCES

Garg, A., and U. Saxena. 1980. "Container Characteristics and Maximum Acceptable Weight of Lift." *Human Factors, 22 (4):* pp. 487-495.

Garrett, J. W. 1971. "The Adult Human Hand: Some Anthropometric and Biomechanical Considerations." *Human Factors, 13 (2):* pp. 117-131.

Jacobsen, C., and L. Sperling. 1976. "Classification of Hand Grip. A Preliminary Study." *Journal of Occupational Medicine, 18 (6):* pp. 395-398.

Long, C., II, P. W. Conrad, E. A. Hall, and S. L. Furler. 1970. "Intrinsic-Extrinsic Muscle Control of the Hand in Power Grip and Precision Handling." *Journal of Bone and Joint Surgery (American), 52A (5):* pp. 853-867.

Napier, S. 1956. "The Prehensile Movements of the Human Hand." *Journal of Bone and Joint Surgery, 38 (B):* pp. 902-913.

Petrofsky, J. S., C. Williams, G. Kamen, and A. R. Lind. 1980. "The Effect of Handgrip Span on Isometric Exercise Performance." *Ergonomics, 23 (12):* pp. 1129-1135.

Rehnlund, S. 1973. *Ergonomi.* Translated by C. Soderstrom. Stockholm: AB Volvo Bildungskonconcern, 87 pages.

Rigby, L. V. 1973. "Why Do People Drop Things?" *Quality Progress, 6 (9):* pp. 16-19

---

sions recognize the functional dimensions of the hand during lifting tasks and the need to provide a wide enough surface contact with the handle to avoid excessive tissue deformation. As object weight increases, handle diameter should increase so that the force per unit area on the hand is kept below 150 kPa (22 lb per in.^{2}). If gloves are worn during handling, an additional 2.5 cm (1 in.) of clearance and width is advisable. *(Human Factors Section, Eastman Kodak Company, 1976; Rigby, 1973).*

CHAPTER **21**

Design and Selection of Containers, Hand Carts, and Hand Trucks

CHAPTER 21. DESIGN AND SELECTION OF CONTAINERS, HAND CARTS, AND HAND TRUCKS

M any loads are transported in industry using containers, carts, or trucks. Their characteristics can determine how much weight can be put in them and how many people will be able to handle them safely. In this chapter, the design and selection of trays, cases or boxes, and hand carts and trucks are discussed.

A. TRAYS

Trays are containers characterized by a rectangular shape and are usually less than 15 cm (6 in.) deep. They are commonly used in industry to assist in the transport of multiple units of in-process or finished product. They may be used to supply parts to automated assembly processes, or they may be handled manually throughout the work area. Among the points to consider when designing or selecting trays for manufacturing or service areas are:

- Tray use—general or special purpose.
- Amount of product or number of items in each tray.
- Size, shape, and nature (such as solid or powder) of items or product in the tray.
- Manual or powered handling.
- Type of tray operations—for example, movement by conveyor, pallet or shelf storage, automated filling and emptying.
- Product protection needs.
- Workplace environment characteristics, including temperature and the presence of dust or oil.
- Nesting and stacking requirements.
- Use in-house only or outside as well.
- Cost restrictions.
- Technological limitations in the plastic molding process.
- Durability needs.

The final tray design should provide the best compromise of the functional and cost factors with the human interface needs.

There are four primary human interface concerns in tray design: weight, size, stability, and grasping characteristics. These interact strongly and influence the operator's perception of effort.

1. WEIGHT

Tray weight limits are strongly affected by the nature of the interface between the tray and the hands. With a poorly designed handhold, the pressure on the hands and fingers will determine the acceptable weight handled. A value of 150 kPa (22 lbf/in.^{2}) is the limit of acceptability for most people (Rehnlund, 1973).

The more ridge-like or sharper the edge by which the hand supports the tray, the less load the hand can support. Spreading the area of contact over more of the hand results in loads feeling more comfortable; heavier weights, therefore, become more acceptable to handlers. Guidelines for acceptable weights in tray handling are included in Chapter 23 under ''Lifting.''

2. SIZE

Tray dimensions are often determined by the amount of material the tray is designed to hold, which depends on the desired flow of parts or volume of bulk material needed in a given time period. Biomechanical and body size characteristics of workers should also be considered when selecting or designing trays. These considerations are discussed below in relation to tray dimensions:

- Tray width determines how far the load is supported in front of the body, which defines the load on the spinal column and lumbar discs, the arms, shoulders, and wrists (see Part III). The wider the tray, the more torque on these joint centers, and the more strain when handling it. A width of 36 cm (14 in.) or less is recommended, with an upper limit of 50 cm (20 in.). Tray width should not exceed length.

- Tray length determines how far apart the arms are when handling the tray, assuming the tray is handled across its length. The farther apart the arms are spread in a tray lifting task, the more the weight of the tray falls on the smaller, weaker muscles of the shoulder instead of being carried by the strong biceps muscles. This limits the amount of weight that can be handled safely by many workers. A length of 48 cm (19 in.) or less is recommended, with an upper limit of 80 cm (24 in.).

- Tray depth is usually kept below 15 cm (6 in.) because of the nature of the material being conveyed. ''Totes'' are often considerably deeper; they are most commonly used to collect waste or to supply parts to a hopper on automatic assembly equipment. The suitability of a given tray depth is highly associated with the location of its handholds and the way in which it is to be handled. A deep tray with handholds near the top may interfere with walking if it has to be carried. If the load is not stable, controlling motion of a deep tote or tray may put excessive demand on the small muscles of the hands, wrists and forearms. In general, the handholds should be placed above the line passing through the center of gravity of the tray when it is loaded to its most typical level. The handles should not be too close to the top of the tray so that the handler has to use a pinch grip instead of an oblique grip, however. Recommended tray depth for the whole range of handling and carrying tasks is 12 cm (5 in.) or less (Figure 21-1).

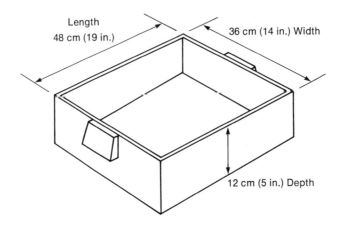

Figure 21-1: Recommended Tray Dimensions   The length, width, and depth of a general purpose tray for handling parts or products in a manufacturing area are indicated. Values for these dimensions are shown in centimeters and inches. These values provide considerable tray volume without exceeding dimensions that will load the smaller shoulder muscles statically (length), put excessive pressure on the lumbar spine discs (width), or interfere with carrying (depth).

3. STABILITY

An unbalanced load greatly increases the opportunities for spilling parts or product from the tray or for dropping the tray. A shifting load can cause the tray handler to take an awkward posture to regain control and may result in straining a muscle group. Tray stability is determined by the nature and distribution of the load in the tray. Some guidelines to improve stability are:

- Use the proper container for the material—for example, do not put liquids in a tray.
- Keep the load from shifting by using dividers or baffles, if necessary.
- Distribute the load uniformly in the tray.
- Size the tray appropriately for the nature of the materials it is to carry.
- Provide handles or handholds.
- Keep the center of the load below the handles.

4. GRASPING CHARACTERISTICS

Handholds on trays can provide a stable interface with the operator by reducing the need to grip the tray tightly. In addition, the opportunity for pinching the hand and fingers between trays is reduced. Handles that let the hand be used as a hook, instead of requiring gripping or pinching, are preferred for low lifting tasks. Examples of tray handholds are show in Figure 21-2.

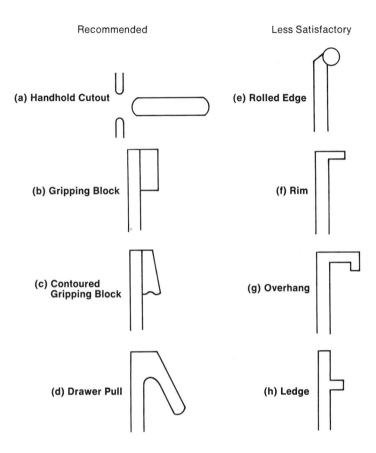

Figure 21-2: Examples of Tray Handholds   Four recommended (*a* through *d*) and four less satisfactory (*e* through *h*) handholds are illustrated. The recommended handholds permit the handler to use either a hook (*a, d*) or an oblique (*b, c*) grasp. The less satisfactory handholds require a lateral pinch grasp (*e, f, h*) or put pressure on a localized part of the finger (*g*) causing discomfort, particularly as tray weight is increased. The contoured gripping block handle (*c*) is superior to the straight gripping block (*b*) because it provides more grip security and stability when the weight is handled in high lifts. Handhold cutouts (*a*) and drawer-pull (*d*) handles are less satisfactory for high lifts (above shoulder height) than for low lifts (below chest height). (*Nielsen, Eastman Kodak Company, 1978*).

Four of these handholds were used in a psychometric study of tray handle design. People were asked to rate the overall level of comfort they felt when handling a standard industrial tote tray 30 times in low, intermediate, and high lift patterns. In addition, they gave ratings for grip comfort, stability, load balance, finger pressure, and wrist comfort. The results of this study are shown in Figure 21-3.

	Gripping Block	Gripping Block (Contoured)	Ledge	Overhang
1—Excellent 5—Terrible	⊓	⊓	⊢	⌐
A. Grip comfort	3.1	3.2	3.6	3.8
B. Balance	2.7	2.5	3.5	2.6
C. Secure grip	2.8	2.3	3.9	2.4
D. Finger pressure	2.7	3.2	3.6	4.4
E. Wrist comfort	2.9	2.9	3.7	3.1
In Order of Overall Preference	2	1	4	3

Figure 21-3: Comfort Ratings for Four Handle Configurations   Four of the handles illustrated in Figure 21-2 were rated for comfort and function in a psychometric study. Six separate ratings were made on a scale from 1 (excellent) to 5 (terrible). Five of the parameters rated are listed in column 1. The overall preference was not a straight average of the parameters looked at independently but reflected some degree of weighting of the discomfort felt with each handle type. The most favorable ratings were given to the two gripping block handles, and the most unfavorable rating was given to the ledge. *(Nielsen, Eastman Kodak Company, 1978).*

The contoured block grip was preferred (lowest numerical rating) in this study across the range of lifts. Its preferential rating extended across the parameters studied although pressure on the fingers was greater than it was with the straight block. Grip security and balance were improved by contouring the block. The ledge and overhang handholds were rated worse overall than their specific ratings suggest. The discomfort due to finger pressure on the edges of the handholds, especially in the high lift, made the overhang handhold unacceptable. The ledge combined finger pressure with instability. Wrist discomfort became very significant in the higher lifts as the subject tried to control the tray with a pinch grip while in maximal ulnar deviation at the wrist (see Appendix A).

Of the handles shown in Figure 21-2, the drawer pull is very comfortable for low lifts but becomes very uncomfortable at heights above elbow level. Tray weight has to be shifted from a stable hook grip to an unstable position where the fingers rest on the handle's edge when the wrists are turned for the higher lift. Cutout handles are not always practical if the product completely fills the tray or is likely to abrade the knuckles during handling. Recommended dimensions of drawer-pull, gripping block, and cutout handles are given in Figure 21-4. The gripping block has been found to be the most satisfactory approach for a wide range of tray-handling activities.

The following general guidelines for tray handle design are recommended:

- Provide handholds at both ends of the tray (across its length).

- Provide cutout handholds, if feasible, or drawer-pull handles if the tray is to be used only for lifts below 1 meter (40 in.).

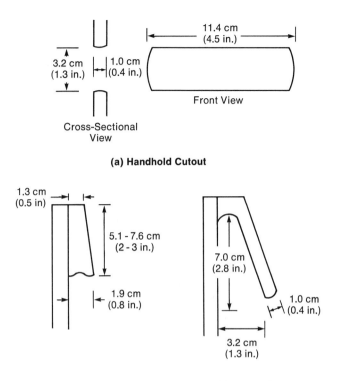

(a) Handhold Cutout

(b) Gripping Block   (c) Drawer Pull

Figure 21-4: Recommended Dimensions for Selected Tray Handle Types The length, clearances, and thicknesses of three of the recommended tray handholds from Figure 21-2 are shown. The handhold cutout (*a*) clearances (height and length) should be increased by one inch if work gloves are worn while handling the trays. The gripping block (*b*) thickness at the bottom should permit the curvature to be gradual and sloped, not be a V-shaped cutout that would cause fingertip discomfort. The length or height of the gripping block should not exceed 7.6 cm (3 in.) so grip strength will not be reduced very much for people with small hands. The drawer-pull handles should be curved at the top so the fingers do not get jammed into a crevice during the lift. If gloves are worn, the slope should be more gradual and the clearance increased about 1 cm (0.4 in.) from the tray to the bottom of the handhold. (*Garrett, 1971; Human Factors Section, Eastman Kodak Company, 1976; Nielsen, Eastman Kodak Company, 1978*).

- Provide a gripping block handhold if the tray will be lifted above 1 m (40 in.).

- Make the upper surface of the tray at the handle site at least 1.3 cm (0.5 in.) wide to permit thumb opposition.

- Provide for a nonslip surface for finger contact, and contour it for a secure grip. There should be no sharp edges, corners, seams, or ribbing at points of hand or finger contact. The surface in contact with the fingers preferably should be 1 cm (0.4 in.) thick and be rounded or beveled.

- Locate the handhold at or slightly above the line passing through the center of mass of the loaded tray.

- If the location of the handhold is not obvious, identify it by graphics or color coding.

5. LARGE TRAYS

A special tray is used in oven-drying, baking, and storage operations. In drying operations, for instance, chemicals are spread on trays to speed oven drying. The tray is usually characterized by being less than 5 cm (2 in.) in depth and greater than 51 cm (20 in.) in width and length (see Figure 21-5). These trays can be difficult to support and are often rested on the chest or abdomen during handling, which results in hyperextension of the back.

Table 21-1 presents the results of a study where 11 men and women were asked to rate effort levels and their comfort after lifting or lowering four large trays to three heights above the floor. The usual limiters to comfort were pain in the wrists, loss of control of the tray's load, or interference by the tray with taking a comfortable lifting posture. The four trays weighed 11.4, 12.7, 14.1, and 15.4 kg (25, 28, 31, and 34 lbm). These weights reflected their probable weights in a handling task if each tray was loaded to a depth of 2.5 cm (1 in.).

The combination of weight and dimension changes can be evaluated at several heights. For low lifts, tray dimension is more of a problem. It is difficult to keep the load close to the body and to bend the knees when handling a wide tray at 30 cm (12 in.) above the floor. For lifts close to the shoulder height, however, this dimension effect is less severe. The tray is usually propped on the chest to reduce its weight in the arms. The widest tray is still difficult to handle, but less difficult than in the low or higher lifts. For the highest lift studied, weight appears to be a problem, even with the smaller tray. Increased tray weight and dimensions increase the difficulty rating only for the largest tray, although all are difficult. Table 21-1 shows that trays over 76 cm (30 in.) long, 61 cm (24 in.) wide, and weighing more than 14 kg (31 lbm) will be difficult to handle, especially below the knees and above the shoulders. People with short arms will find large tray-handling very difficult at most heights. To keep these tray-handling tasks within the capabilities of most people, designers should keep tray size under 51 × 76 cm (20 × 30 in.), and trays should weigh 13.6 kg (30 lbm) or less when loaded. Handling heights should also be kept between 50 and 127 cm (20 and 50 in.) above the floor, where possible.

B. SHIPPING CASES AND BOXES

Much of the manual handling of a product in industry is done during its transportation and distribution. Corrugated shipping cases are commonly used in packaging the product or its components because they are compact, sturdy, dis-

Figure 21-5: Large-Tray Handling   A tray that is 91 cm (36 in.) long, 61 cm (24 in.) wide, and 5 cm (2 in.) deep is held at waist level in a handling task. The long dimension is held in front of the body as the tray would be oriented for placement in a drying rack. The tray is held against the abdomen, and the handler arches his back slightly to counteract the tendency of the tray to pull him forward. If the tray has to be lifted more than 140 cm (55 in.) above the floor, it will be supported farther from the body and there will be more arching of the back to keep it stable. At heights below 50 cm (20 in.), the tray will either have to be held in front of the bent knees or supported on them, making handling awkward.

posable, and efficient in forming trailer or railroad car loads. The size of the case in which materials are packaged is determined by several factors:

- Product size.
- Marketing considerations for the product.
- Cost of packaging materials.
- Production volumes.
- Best dimensions to ensure proper protection of the product in its transportation and distribution.
- Pallet size and most efficient pallet patterns for transporting the cases.

Table 21-1: Perceived Effort in Large-Tray Handling. The ease of handling four trays (columns 2 through 5) whose length and/or width exceed the recommended dimensions in Figure 21-1 is shown. The height of the shelf, in centimeters and inches, to which the tray is lifted or lowered is indicated in column 1. The three values represent below knee height, chest height, and above shoulder height for most people. The people who rated the difficulty of handling the trays were in the range of 160 cm (63 in.) to 193 cm (76 in.) tall and included men and women. The difficulty of handling the trays to the higher shelf was associated with stress at the wrists, which were in maximal ulnar deviation. The largest tray was difficult or very difficult to handle at all heights because of the need to support it away from the front of the body. Tray weight increased with tray size, and this also contributed to handling discomfort with the largest trays. *(Rodgers, J. Alexander, and Whispell, Eastman Kodak Company, 1980).*

Lift or Lower, Height from Floor in cm (in.)	Tray Size (width × length)			
	48 × 69 cm (18 × 27 in.)	51 × 76 cm (20 × 30 in.)	61 × 76 cm (24 × 30 in.)	61 × 91 cm (24 × 36 in.)
30 (12)	OK	OK	OK → D	D → VD
122 (48)	OK	OK	OK	D
>140 (>55)	D	D	D	VD

OK = Acceptable
D = Difficult
VD = Very difficult

The person who will be handling the cases is often not considered by the designer. Some distribution companies (such as United Parcel Service and the United States Post Office) have imposed maximum dimensions or circumferences on packages. This may affect case size decisions by industry. Computer models are being developed to help define optimal case size for specific products. It is important to include some constraints on case dimensions and weight, based on people's capacities, in order to help ensure safe and efficient manual handling of the cases throughout their distribution.

Recommended and maximum dimensions for the design of shipping cases and boxes are given here. The implications of exceeding these dimensions and the value of handholds in the manual handling of cases are also discussed.

1. CASE DIMENSIONS

For any given weight, the more compact the case, the easier it is to handle, especially if it is easy to grasp. As case length is increased, the arms are spread farther apart, and more stress is put on the shoulder muscles, rather than the stronger upper arm biceps. If width is increased, the center of the case's load moves away from the handler's body, increasing the stress on the trunk and arm muscles (see Part III). If case height (or depth) interferes with the legs during walking, the handler may be forced to carry the case higher, with bent arms. This requires a higher percentage of maximum arm strength and results in much faster fatigue development than when the case can be carried at the side or with the arms fully extended down.

a. Guidelines for New Case Design

The dimensions in Figure 21-6 are recommended as the upper limits when new cases are being designed. Cases that are designed within these guidelines will be less likely to exceed the lifting strength capabilities of a substantial part of the potential workforce when the cases are densely packed. The guidelines for lifting given in Chapter 23 assume that these case dimensions are used.

b. Implications of Exceeding the Recommended Dimensions

Large-size products, existing manufacturing procedures, and a trend towards more cost-effective bulk packaging all make implementation of a case dimension standard difficult. Studies of the effects of tray dimension on acceptable lifting loads can be used to measure the impact of larger dimensions on the amount of weight people can handle in cases. The reductions are based on an acceptable

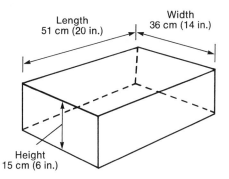

Figure 21-6: Maximum Dimensions Recommended for New Shipping Case Design. Recommended maximum dimensions in centimeters and inches for the design of new shipping cases are shown. Case design is largely determined by the size of the product, but these dimension guidelines should be considered when multiple-unit packaging is under consideration. More width is associated with more stress on the lumbar spine discs; more depth results in some interference with carrying. Depth may be increased substantially if there is little need to carry the case or if the other dimensions are reduced to form a cube. *(Developed from information in Ayoub et al., 1979; Ciriello and Snook, 1978; Garg and Saxena, 1980; Rigby, 1973).*

load of 18 kg (40 lbm) in a lift below 75 cm (30 in.) Table 21-2 illustrates some of these findings.

Although these reductions in acceptable load may appear quite small, they are based on average values. The impact of the dimension changes on people who are smaller or less strong will be considerably greater, limiting the people who can handle the cases manually. The weight of the case will also determine which dimensions are optimal; the heavier the case, the more impact its dimensions will have on the handler.

c. Effect of Lifting Frequency on Acceptable Weight and Case Dimensions

Lifting frequency also accentuates the problems of large-size cases. The more frequently cases are handled, the more opportunity there is for local muscle fatigue to build up. If the load is falling on the weaker muscles of the shoulder or hand and wrist, fatigue will occur sooner because a larger percentage of their capacity is needed to support the weight (see Part IV for further discussion of

Table 21-2: Impact of Increased Case Dimensions on the Acceptable Load for Manual Materials Handling. The impact of increases in shipping case dimensions (column 1) above the values given in Figure 21-6 (column 2) on the amount of weight that can be handled comfortably is shown in columns 3 through 5. The reduction in load is a factor to apply to the acceptable weight determined from the guidelines found in Figure 23-3 for occasional low lifts. Three increments of 10 cm (4 in.) each are used to calculate the acceptable weight reductions for changes in length, width, or height. Case width has the largest effect on load, although substantial increases in case depth can also have significant effects on the acceptability of some types of load. Maximum case dimensions, those that should not be exceeded in jobs where the cases must be handled manually, are given in the text. *(Developed from information in Ayoub et al., 1979; Ciriello and Snook, 1978; Garg and Saxena, 1980).*

Dimension	Maximum Dimension Recommended	Percent Reduction in Acceptable Load If the Dimension Is Increased by:		
		10 cm (4 in.)	20 cm (8 in.)	40 cm (16 in.)
Length	51 cm (20 in.)	2	5	7
Width	36 cm (14 in.)	8	15	17
Height	16 cm (6 in.)	2	5	14

muscle fatigue). In a psychophysical study (Ciriello and Snook, 1978), increasing the rate of lifting from one per minute to about six per minute resulted in a 15 percent reduction in acceptable weight for a 57 × 75 × 14 cm (23 × 29 × 6 in.) tote box lifted from the floor to knuckle height. When the box width was reduced from 75 to 49 cm (29 to 19 in.), the weight could be increased again almost to the one-per-minute frequency weight limit. This dimension effect was only seen at the lower lift height. In lifts between knuckle and shoulder height, the smaller width did not compensate for the increased frequency, and less weight was considered acceptable.

d. Maximum Dimensions

Defining an upper limit for case dimensions is somewhat impractical because some products, such as plates, posters, and strip metals, by their very nature exceed these dimensions. A more practical approach is to identify the case dimensions at which an aid should be provided to help the handler move the case. Based on the studies mentioned earlier and on observation of distribution activities where certain cases were identified as difficult to handle, the following upper limits are recommended for lifting cases without aids:

- Length, or longest dimension, usually between the hands—76 cm (30 in.).
- Width, or distance in front of the body—51 cm (20 in.).
- Height, or depth—46 cm (18 in.).

Products, such as fluorescent lamps, may be very long but small in other dimensions. If the cased product can be carried easily under the arm, the maximum length values can be exceeded without requiring handling aids. The values given in Figure 21-6 should be used to design new packages. The maximum limits indicate where a large part of the potential work force will have difficulty handling cases. It is assumed that the recommended weight limits for lifting and lowering given in Chapter 23 will also be followed in case design.

2. PROVISION OF HANDHOLDS

Most shipping cases are handled from the corners or bottom because they do not have handholds (see Figure 20-4). Cut-in handholds are provided on some cases where the contents permit access by the hands, as in cases of bottles, fruit, or similarly shaped objects without sharp edges. A flat product that packs very efficiently is not often packed in cases with handholds because the time and cost of fabricating an extra thickness in the case to form the handhold is prohibitive.

Although it is not anticipated that handholds on shipping cases will become standard, they should be considered for certain handling situations. Some of these are:

- For cases that weigh more than the recommended upper limit. Hand-holds improve grasping of the case, and they increase the acceptable load by up to 10 percent (Garg and Saxena, 1980; Nielson, Eastman Kodak Company, 1978).

- For cases that are handled constantly onto or off of pallets. The metabolic cost of lifting cases without handholds from floor or pallet levels is about 11 percent more than when they are lifted from handholds. The difference in work can be explained by the extra 12 to 15 cm (5 to 6 in.) the body must be lowered and raised if the case has to be lifted from the corners or bottom (Nielsen, Eastman Kodak Company, 1978).

- For a case that may have to be carried for short distances. A one-handed carry is often more efficient than a two-handed carry (see the discussion on carrying in Chapter 24).

- For a case that is wide or long enough to make it difficult to carry in front of the body or under the arm. For instance, a cutout handhold located on the flat surface of a package of posters or advertising materials with dimensions greater than 76 cm (30 in.) permits it to be carried under one arm.

Guidelines for cutout handhold design have been given earlier in this chapter under tray design.

C. HAND CARTS AND TRUCKS

These vehicles are defined as trucks, wagons, or carts that are pushed or pulled manually, that is, without mechanical assistance. These include platform trucks, drum trucks, hand lift trucks, pallet or skid trucks, mail carts, tea wagons and two-, three-, and four-wheel hand carts. Air pallets and other pneumatically-assisted transfer devices are not included here, but they are an excellent alternative to fully manual truck and cart design (see Chapter 22). In addition, many special purpose hand carts and trucks are fabricated in industry in order to store or transport products or parts between work areas. These may be closed cabinet trucks with racks or pegs, castered tanks or vessels that contain liquid or powdered chemicals, or other transfer devices used to move large-size products in intermediate stages—for example, paper stock rolls prior to being slit—from one production area to the next operation or to storage. Most of these trucks are fabricated by industry because they have to accommodate a specific product or process for which a general purpose cart may not be suitable. Figure 21-7 illustrates some common types of hand carts and trucks.

General guidelines are given here that apply to the selection, design, and redesign of hand carts and trucks. In addition, information is provided about the handling of trucks and carts and the identification of situations where a powered truck may be most suitable.

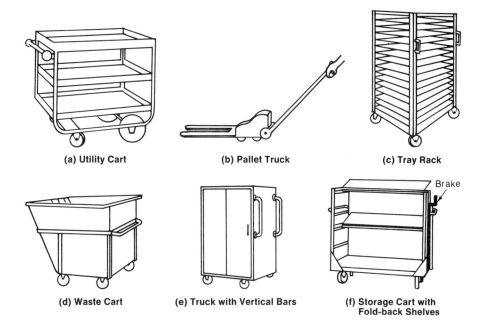

(a) Utility Cart   **(b) Pallet Truck**   **(c) Tray Rack**

(d) Waste Cart   **(e) Truck with Vertical Bars**   **(f) Storage Cart with Fold-back Shelves**

Figure 21-7: Examples of Hand Carts and Trucks   Six examples of industrial hand carts and trucks are shown. The utility cart (*a*) provides a way to move products or parts between departments or work stations so that carrying is not required. It has two fixed wheels (forward) and two swivel casters (rear, handle end). The lowest shelf is difficult to handle materials to and from and should only be used for very light or easy-to-grasp objects. The pallet truck (*b*) has a single T-shaped handle that moves from a vertical position at rest to less than 45 degrees above the floor for pulling or maneuvering the load. Its forks are inserted into a pallet and raised a few inches to clear the load from the floor. The tray rack (*c*) is about 127 cm (50 in.) tall and has two fixed wheels (front) and two swivel (rear, handle end) casters. The handles are D-type (see Figure 20-6) and are located at 102 to 114 cm (40 to 45 in.) above the floor for ease in maneuvering the rack. The waste cart (*d*) is a bin enclosed in a metal frame with a horizontal handle that projects about 20 cm (8 in.) out from the cart. This permits the handler to shift horizontal hand location during different cart motion patterns. The handle is located about 88 cm (35 in.) above the floor. The storage truck (*e*) has two vertical bars for handles so the handler can push or pull close to the load's center of gravity in different loading conditions. These handles are fixed horizontally and are recommended only for narrow, less than 50 cm (less than 20 in.), trucks. The open storage cart (*f*) is similar to the utility cart in *a* but has two additional features: The upper shelves fold back to permit easier loading and unloading of the bottom shelf, and there is a handbrake to arrest motion of the cart on an incline or while loading and unloading. The front caster is turned sideways to illustrate the braking mechanism. The front casters are normally fixed in the forward position, the brake being on the rear swivel-type casters. Although these carts are more expensive than utility carts, they provide more functional storage and are inherently safer in repetitive handling tasks.

1. SELECTION OF MANUAL VERSUS POWERED TRUCKS AND CARTS

A review of accidents associated with manual cart and truck handling shows three major concerns: fingers and hands caught in, on, or between the cart and a wall or piece of equipment; feet, heels, and the lower leg being bumped by or caught under the cart; and arm, shoulder, and back strains associated with slips, trips, and pushing and pulling of trucks (Rodgers, Eastman Kodak Company, 1978). With powered trucks, strains and sprains are considerably reduced, but hand and foot injuries are still seen. Attention to the design of trucks and carts should help to reduce the potential for these types of injuries.

In deciding which type of truck or cart is needed for a handling task, one usually considers the expected load, frequency of use, the duration of continuous use (closely related to the distance traveled), and characteristics of the work area, as such as aisle width, floor type, and other environmental factors that determine the suitability of using powered vehicles. In addition, the floor surface, material, and load-bearing characteristics must be considered. Table 21-3 summarizes the ranges of some of these characteristics for selected hand and powered trucks and carts. An alternative to vehicles for transporting products is a powered conveyor system. This should be considered carefully, especially in the initial design of a facility (see Chapter 19).

Table 21-3 can be summarized as follows:

- Two-, three-, and four-wheel hand carts generally should not be loaded with more than 227 kg (500 lbm) of materials. Hand pallet trucks can handle heavier loads. The load rating of a powered truck and of the floor in the area of interest must be considered when determining the weight limits for powered vehicles.

- Truck and cart tasks occurring less than 200 times a day are suitable for manual operations. At higher frequencies powered trucks are recommended, if feasible.

- If materials are frequently transported more than 33 m (100 ft), use of a powered truck should be considered.

- Powered lift trucks need aisles at least 2 m (6 ft) wide for maneuvering (Drury, 1974). Electric trucks generally need at least 1.3 m (4 ft) of aisle width.

2. DESIGN AND SELECTION FACTORS*

Although the load on the truck will, to a large extent, determine the forces needed to handle it in the workplace, there are several design factors that also influence the force requirements. These should be considered when purchasing

---

*Anon., 1983; Nielsen, Eastman Kodak Company, 1976.

Table 21-3: Recommended Limits in the Selection of Hand and Powered Trucks and Carts. Nine hand and powered trucks and carts (column 1) are described in terms of their maximum load, in kilograms and pounds (column 2); maximum distances they should be used to transport materials, in meters and feet (column 3); the maximum recommended frequency of use, or trips per eight-hour shift (column 4); the recommended minimum aisle width where these trucks are used, in meters and feet (column 5); and the type of transfer to and from the truck or cart (column 6). The transfers can be manual (Ma) or mechanical (Me), and may be done with the entire load (UL) or in parts (P). Most unit loads are pallets; all mechanically transferred materials are moved by powered equipment. These limits are recommended as a guide in the selection of manually controlled or powered trucks and carts. Aisle width values are based on the maneuverability of the cart and the tiering capacity—that is, the ability to raise a load to several meters (feet) above the truck. Some hand trucks and carts are illustrated in Figure 21-7. See Chapter II, Volume 1, for more information about aisle and corridor design. *(Developed in part from information in Anon., 1959; Anon., 1948).*

	Recommended Limits							
Type of Truck or Cart	Maximum Load		Maximum Transport Distance		Maximum Frequency Units (8 hrs)	Minimum Aisle Width		Type of Transfer to and from Truck*
	kg	lbm	m	ft		m	ft	
2-wheeled hand cart	114	250	16	50	200	1.0	3	Ma, P
3-wheeled hand cart	227	500	16	50	200	1.0	3	Ma, P
4-wheeled hand cart	227	500	33	100	200	1.3	4	Ma, P
Hand pallet truck	682	1500	33	100	200	1.3	4	Me, UL
Electric pallet truck	2273	5000	82	250	400	1.3	4	Me, UL
Electric handjack lift truck	2273	5000	33	100	400	1.3	4	Me, UL
Power low lift truck	2273	5000	328	1000	400	2.0	6	Me, P, UL
Electric handstacking truck	682	1500	82	250	400	1.3	4**	Me, UL
Power fork truck	2273	5000	164	500	400	2.0	6**	Me, UL

*Ma = Manual, Me = Mechanical, P = Parts, UL = Unit load.
**These trucks have tiering capability. In order to use it, ceilings must be more than 4 m (12 ft) high.

or fabricating a truck. Other design considerations are also discussed in the subsections that follow.

a. Wheels and Casters

- Diameter—Increasing the diameter generally decreases the necessary force.
- Composition—A harder caster or tire decreases the necessary force.
- Tread—A wider tread increases the wheel's rolling resistance with the floor; this increases the force required to move the cart and makes cornering more difficult. A crowned tread is easier to push and maneuver.
- Maintenance—Use good bearings and maintain them regularly. Tread wear should be monitored, as well as corrosion and other changes that might bind the wheels and increase the forces required to move the trucks.
- Swivel casters for all wheels should be used only when a cart or truck has to be maneuvered in very tight spaces.

b. Handle Type and Location

- If swivel casters are used, the operator should push and pull from the swivel end. To ensure this, place handles on the swivel end only (Figure 21-7a).
- If a truck or cart is primarily pulled rather than pushed (for example, a pallet truck), an adjustable T-bar handle is preferred. The handle should extend far enough out to prevent the operator from being struck on the heels when walking in front of the truck (Figure 21-7b). If a fixed handle is used, it should be located above 91 cm (36 in.). A minimum of 20 cm (8 in.) of horizontal extension should be allowed.

Since some carts are pushed more often than they are pulled, the following information relates to the design or selection of carts with handles used for pushing:

- In general, the handles should be placed so that they straddle the load's center of gravity, but at a height that permits comfortable posture as well as good biomechanical advantage (see Figure 21-7c).
- Handle type will be determined by cart dimensions and handling needs. It may be oriented vertically or horizontally and for one- or two-handed operation. Handles mounted flush on the cart surface should follow the guidelines given earlier in this chapter for handle design. Adequate clearance for the gloved hand is needed whether the handle is part of the truck or cart structure (such a vertical support), or attached to it (Figure 21-7c).

- The truck's width and length and the distance between the handles will determine its maneuverability in tight places. Bar handles, oriented horizontally (see Figure 21-7d), permit the handler to vary hand location to fit the task and accommodate his or her own size and strength needs. For nonadjustable handles that are mounted flush on the truck surface (Figure 21-7c), horizontal separations in excess of 46 cm (18 in.) are not recommended unless cart width or depth exceeds 91 cm (36 in.).

- Handle height should be in the range of 91 to 112 cm (36 to 44 in.). The higher the center of gravity of the loaded truck or cart, the higher the handle should be. Handles lower than 91 cm (36 in.) are not recommended because they force taller people to stoop when handling the truck.

- Vertical bars for handles resolve handle height problems by permitting the operator to choose the most comfortable position for the hands according to his or her biomechanical needs and the nature of the handling being done (Figure 21-7e). They are most satisfactory for narrow trucks, usually less than 51 cm (20 in.) wide.

c. Truck and Cart Dimensions

- Trucks that are longer than 1.3 m (4 ft) or wider than 1 m (3 ft) cannot be turned easily in many product-area aisles.

- The preferred heights for truck and cart shelves or pegs, from and to which products will be handled, should be between 51 and 114 cm (20 and 45 in.) above the floor (Figure 21-7f). If possible, handling heights less than 36 cm (14 in.) and greater than 274 cm (50 in.) should be avoided because of the strain they put on the knees and shoulders, respectively. The weight of the parts to be handled will determine the optimal shelf and peg heights (see the discussion on lifting in Chapter 23).

- Barriers to parts handling that result in awkward postures, such as very deep shelves requiring awkward extended reaches and small clearances between pegs or shelves that make access difficult, should be avoided. In some applications, a fold-back shelf permits easier loading and unloading of a cart (Figure 21-7f).

- Carts should not be too high to see over. A maximum height of 140 cm (55 in.) will accommodate most people. If this is not feasible, the cart should be designed for pulling.

- Carts and trucks that are loaded to more than 500 kg (1,100 lbm) and are operated in heavily populated work areas or up and down ramps should be provided with audio and/or visual alerting signals such as an intermittent tone or flashing light, and with a braking system to help prevent collisions with people in the aisles (Figure 21-7f).

3. HANDLING FACTORS*

The factors of most concern in the manual handling of trucks and carts relate to the forces required to start them, stop them, sustain their motion, and maneuver them. There is good agreement about the reasonable levels of horizontal force for starting, stopping, and maintaining trucks in motion, but few data are available on the safe level of forces for maneuvering them. The latter is strongly linked to the space available for turning and the precision with which the truck has to be located in a place. In most instances, the forces people can exert for maneuvering are considerably less than for pushing because the operator cannot get body weight behind the center of gravity of the load. Shoulder and arm muscle strength become the limiting factors in these situations. More discussion of pushing and pulling can be found in Chapter 22. The following information relates to the handling of trucks and carts:

- A truck should be moved at a walking pace that will ensure that it is under control at all times and that it can be stopped almost immediately. A walking pace of 3.2 to 4 km/hour (2 to 2.5 mph), covering a distance of about 60 m (200 ft) in one minute, is reasonable in this respect; for heavy loads a lower speed is desirable.

- The recommended upper limits of horizontal force to start, maintain, and stop the motion of a hand cart or truck are:
  - Start—225 N (50 lbf)
  - Maintain—less than 3 m (10 ft = 180 N (40 lbf); 1 min continuously = 112 N (25 lbf)
  - Emergency stop—within 1 m (3 ft) = 360 (80 lbf)

When these upper limits are exceeded, powered equipment should be considered.

- The characteristics of the surface over which a cart or truck is pushed or pulled will determine the forces and physical effort demands of the handling task. The values given here are not difficult to achieve if the truck is handled on well-maintained factory floors. Floors that are heavily etched or cracked or with depressions, drains, or caked materials on and in them make truck handling difficult. The coefficient of friction between the cart's wheels and the floor can increase threefold during turning and maneuvering (such as between a concrete and stone pavement), thereby requiring more physical effort from the operator to turn it (Anon., 1983). Large-wheel carts can overcome some of these handling problems on uneven surfaces, but they may not be practical in areas where aisle space

*Developed from information in Ciriello and Snook, 1978; Haisman, Winsman, and Goldman, 1972; Nielsen and Faulkner, Eastman Kodak Company, 1967; Strindberg and Petersson, 1972.

is limited. Powered trucks are generally preferable if the floor or other surface irregularities cannot be remedied.

- A heavy truck that has molded rubber wheels has a starting resistance of about 196 N per 1,000 kg total weight (40 lbf/ton) (Anon., 1983). The longer the truck stands between uses and the heavier the load, the more potential for the wheels to "flatten" and increase starting resistance. This supports the recommendation to use powered trucks for loads greater than 682 kg (1,500 lbm).

- Ramps are frequently found in areas where people are handling carts and trucks manually. They may be used for access to buildings, to connect buildings, or to provide entry to elevators or special work areas (see the section on ramps in Chapter II, Volume 1). The ramp often leads to or from a door that must be negotiated by the person controlling the cart. Thus, a one-handed pull is used to move the cart, and there may be awkward postures associated with holding the door open and pulling the cart through the opening. These handling conditions contribute to incidents where the person's hands, feet, and lower legs are caught between the cart and the door or a wall at the end of the ramp. It is recommended that carts with a total weight of more than 227 kg (500 lbm) be powered if ramps are regularly used. Ramps with more than a 2 percent grade are difficult since the cart's tendency to roll downhill adds to the force needed to move it. When handling a cart on a ramp, the operator should always be upgrade from the load, pulling it up the ramp or restraining it from above as it goes down the ramp. A braking system is recommended for small hand carts if ramps are regularly negotiated. Foot or hand brakes (Figure 21-7f) can be used to lock a cart in position or to restrain its motion. The latter should not require high forces or have to be held continuously for more than 30 seconds. Foot brakes may be useful for restraining heavier loads. They should not protrude much beyond the cart body in order to prevent them from striking the handler during cart movement.

REFERENCES

Anon. 1948. "How to Analyze and Solve Materials Handling Problems. Technique No. 1: Comparative Analysis. Exhibit G: Industrial Truck Selection." *Factory Management and Maintenance, 106 (1):* pp. 88-98.

Anon. 1959. "How Much Is Too Much to Push or to Pull?" *Engineering and Purchasing Planbook.* Chicago: Caster and Floor Truck Manufacturers Association, pp. 75-77.

Anon. 1983. *Material Handling Engineering Handbook and Directory.* Cleveland: Industrial Publishing Company, pp. A153-A156.

Ayoub, M. M., R. Dryden, J. McDaniel, R. Knipfer, and D. Dixon. 1979. "Predicting Lifting Capacity." *American Industrial Hygiene Association Journal, 40:* pp. 1075-1084.

Ciriello, V., and S. H. Snook. 1978. "The Effects of Size, Distance, Height, and Frequency on Manual Handling Performance." *Proceedings of the Human Factors Society, 22nd Annual Meeting, 1978, Detroit, Michigan.* Santa Monica, Calif.: Human Factors Society, pp. 318-322.

Drury, C. G. 1974. "Depth Perception and Materials Handling." *Ergonomics, 17 (5):* pp. 677-690.

Garg, A., and U. Saxena. 1980. "Container Characteristics and Maximum Acceptable Weight of Lift." *Human Factors 22 (4):* 487-495.

Garrett, J. W. 1971. "The Adult Human Hand: Some Anthropometric and Biomechanical Considerations." *Human Factors, 13 (2):* pp. 117-131.

Haisman, M. F., F. R. Winsmann, and R. F. Goldman. 1972. "Energy Cost of Pushing Loaded Handcarts." *Journal of Applied Physiology, 33 (2):* pp. 181-183.

Rehnlund, S. 1973. *Ergonomi.* Translated by C. Soderstrom. Stockholm: AB Volvo Bildungskonconcern, 87 pages.

Rigby, L. V. 1973. "Why Do People Drop Things?" *Quality Progress, 6 (9):* pp. 16-19.

Strindberg, L., and N. F. Petersson. 1972. "Measurement of Force Perception in Pushing Trolleys." *Ergonomics, 15 (4):* pp. 435-438.

CHAPTER 22

Exerting Forces

CHAPTER 22. EXERTING FORCES

The amount of force that can be exerted in a manual handling task depends on the posture the operator can take. Posture determines both muscle lengths and joint and body angles, which determine the total force that can be developed during pushing or pulling. By leaning into the push or away from the pull, the operator can apply more force to the object that must be moved. Other muscles stabilize the body to prevent motion that is neither desirable nor effective in moving an object. The frictional resistance between the handler's shoes and the floor is a major factor in the degree of body stabilization possible in handling tasks. For example, most people can move a heavy hand truck down a corridor because they can use the large muscle groups of the legs, trunk, and arms to push the load. If they have to maneuver the same truck in a tight space where they have to remain in an upright posture, much more arm and shoulder muscle work will be required to develop the forces. This may limit the number of people who can safely do the task, especially if shoulder strength requirements are high. The guidelines given for workspace dimensions in Chapter II, Volume 1, recognize this need for adequate space in handling areas.

In addition to hand cart and truck handling discussed in the preceding chapter, the following industrial tasks are characterized by the need to exert forces:

- Operating controls and tools (see Chapter III, Volume 1).
- Sliding containers, products, or materials on a flat surface, such as a table, the floor, a conveyor, or a shelf.
- Opening and closing doors and access ports.
- Forming or bending cardboard in casing and packing operations.
- Clearing jams in assembly or packaging.

It is difficult to describe these and other tasks in general terms because postural requirements may differ in each workplace during force exertions. For instance, it is more difficult to slide trays along a conveyor line that is 76 cm (30 in.) in front of the body (horizontally) than along one that is 51 cm (20 in.) away. The extended reach in the first situation makes it difficult for the stronger arm muscles to participate in the push or pull, putting the requirement for force development on the weaker shoulder muscles. The shorter distance allows the operator to bend the elbow and use the stronger arm muscles, as well as to turn sideways to get some body weight into the push or pull.

In addition to variable workplace situations, it is difficult to provide general guidelines for force exertion because the muscle strength data on industrial populations are limited and lack close agreement (Asmussen and Heebol-Nielsen, 1961; Davis and Stubbs, 1980; Kamon and Goldfuss, 1978; Keyserling et al., 1980; Laubach, 1976; Snook and Ciriello, 1974). Attempts to standardize strength measurements should improve the reliability and validity of these measurements in the future (Caldwell et al., 1974; Chaffin, 1975; Kroemer and Howard,

1970). Also, many of the existing strength data are on college and military men and thus do not reflect the older or female workforce in industry. Appendix A includes a compilation of muscle strength data; these values have been used, with some adjustments, to develop the guidelines for maximum force exertion seen in this section.

When using strength measures to assess the potential for overexertion on a handling task, you should identify the weakest muscle groups used in the task. These can be expected to fatigue first because they are stressed to a higher percentage of their maximum strength (see Part IV). This "weakest link" approach indicates that grasp and shoulder muscles are the primary limiters of many handling tasks. Researchers who develop computerized biomechanical models of force exertion and lifting tasks tend to focus more on the lower back (lumbar spine) and the compressive force on the lumbar intervertebral discs (Chaffin, 1974; Chaffin et al., 1977; Martin and Chaffin, 1972; Poulsen and Jorgensen, 1971). This is undoubtedly an important consideration in assessing the appropriateness of a handling task, but there is no agreement as to what constitutes a safe level of compressive force, and the hands and shoulders may limit the lift capability before spine stress is of concern. This is particularly likely when the task is a highly repetitive one and local muscle fatigue determines the acceptable load.

The guidelines in this chapter are derived from strength studies of industrial workers or military personnel in tasks that bear some resemblance to handling jobs. The values given represent upper limits for design so that the large majority of the potential work force can do the task without excessive fatigue. It should not be inferred that forces greater than the ones shown will put people at high risk of injury. Because people can usually alter posture or methods of applying force in the large variety of handling tasks seen in industry, these guidelines are more appropriately used for the design of new jobs than for the evaluation of the risk for injury in existing ones.

The information is organized according to the direction of the force that must be exerted: horizontal is toward and away from the body, vertical is up and down, and transverse is across the body in the horizontal plane. Grip forces, which are of interest especially in pulling tasks, are also treated briefly.

A. HORIZONTAL FORCES AWAY FROM AND TOWARD THE BODY

The exertion of push and pull forces in front of the body has been measured in studies of truck and cart handling (Ayoub and McDaniel, 1974; Nielsen and Faulkner, Eastman Kodak Company, 1967; Snook and Ciriello, 1974) and in a study of maximum horizontal pushes on different floor surfaces (Kroemer, 1970). From these studies, the following factors have been shown to influence the amount of force that can be developed in a horizontal push and pull:

- Body weight.
- Height of force application.
- Distance of force application from the body, or the amount of trunk flexion or extension.
- Frictional coefficient of the floor.
- Frictional coefficient of the shoes.
- Duration of force application or the distance moved.
- Availability of a structure against which the feet or back can push or prevent slippage.

A more complete discussion of hand cart- and truck-handling forces was given in the previous chapter. Table 22-1 summarizes the recommended upper force limits for the design of tasks where pushing or pulling horizontally in front of the body is required. Lower forces are more desirable. Higher forces are possible but are not recommended because they limit the available work force. Higher force requirements may increase fatigue and contribute to overexertion accidents.

If the time of force application exceeds three to five seconds, forces must be decreased (see Part IV). The forces in Table 22-1 are not the same as the weight of the object being moved. The force needed to initiate and sustain a cart or truck in motion can be measured with a push-pull, strain, or spring force gauge at the point where the hands would be during handling (see Part III for more information on the force measurement).

The values in Table 22-1 are for horizontal forces applied between waist and shoulder levels when standing or kneeling, and at chest and shoulder levels for seated operations. If the force has to be applied above or below these levels, the limits are decreased. For instance, if a horizontal force has to be exerted over a seated operator's head, as in moving items along an overhead conveyor, the upper limit of force is 54 N (12 lbf) (Kroemer, 1974) because the arm muscles are no longer in a position to develop high forces. The lower the point where force has to be applied, the more critical it becomes to have adequate space to take a posture where the large muscles of the legs and trunk can be used in the push or pull.

It is possible to develop considerably higher forces, in the range of 742 N (165 lbf), if the feet are supported against an immovable structure, and the strongest muscles can be used in a push (Kroemer, 1970). This technique is used to initiate movement of a cart or truck that has a wheel caught in a depression on the floor. Once the cart breaks free and starts moving, the handler has to shift posture to avoid losing balance. A seated operator with extended arms and the knees extended at 150 degrees can develop pull forces in the range of 630 N (140 lbf) (Caldwell, 1964). It is preferable to design tasks within the guidelines in Table 22-1, however, since these provide a margin of safety for extended durations of force application.

Table 22-1: Recommended Upper Force Limits for Horizontal Pushing and Pulling Tasks. Four upper limit forces, in newtons and pounds of force, are shown in column 2 for three postures: standing (A), kneeling (B), and seated (C) (column 1). There are two sets of limits for standing, one where it is possible to get the full body into the motion (A1) and one where primarily arm and shoulder muscles are involved and the arms are extended (A2). Examples of activities in the workplace where these types of forces may be required are shown in column 3. These forces are recommended for new designs to permit a large majority of the workers to do the job. Some individuals may be able to exert more force safely, but higher force requirements limit the work force available for the task. (*Developed from information in Caldwell, 1964; Davis and Stubbs, 1980; Keyserling et al., 1980; Kroemer, 1970; Nielsen and Faulkner, Eastman Kodak Company, 1967; Snook and Ciriello, 1974*).

Condition	Forces That Should Not Be Exceeded, in newtons (lbf)	Examples of Activities
A. Standing		
1. Whole body involved	225 (50)	Truck and cart handling. Moving equipment on wheels or casters. Sliding rolls on shafts.
2. Primarily arm and shoulder muscles, arms fully extended	110 (24)	Leaning over an obstacle to move an object. Pushing an object at or above shoulder height.
B. Kneeling	188 (42)	Removing or replacing a component from equipment, as in maintenance work. Handling in confined work areas, such as tunnels or large conduits.
C. Seated	130 (29)	Operating a vertical lever, such as a floor shift on heavy equipment. Moving trays or a product on and off conveyors.

B. VERTICAL PUSHING AND PULLING

Some examples of the use of vertical forces in the workplace are to operate controls, to package materials, and to clear jams on assembly machines. The height of force application determines which muscles will be available and how much strength is needed. Pulls down from above head level or up from 25 cm (10 in.) above the floor provide the greatest forces since body weight can be used in the former instance, and the strong leg and trunk muscles are used in the latter. It is assumed that grip strength is not a limiting factor in these pulls and pushes. Table 22-2 summarizes recommended upper limits for the vertical forces required in standing tasks; these should be within the strength capacities of a large majority of the potential work force. Lower forces would be even more desirable, especially in sustained or very repetitive work. Although many people can

Table 22-2: Recommended Upper Limits for Vertical Pushing and Pulling Forces in Standing Tasks. The recommended maximum force requirements, in newtons and pounds of force, for the design of vertical pulling (A-C in column 1) and pushing (D and E, column 1) tasks are shown in column 2. Examples of activities where these actions are used in the workplace are given in column 3. These recommended upper limits are selected in order to accommodate the large majority of potential workers in such tasks. Individual workers can exceed the limits safely, but the design should not require selection of those stronger people. *(Developed from information in Hunsicker, 1957; Keyserling et al., 1980; Kroemer, 1974; Yates et al., 1980).*

Conditions	Upper Limit of Force for Design, in newtons (lbf)	Examples of Activities
A. Pull down, above head height	540 (120)	Activating a control, hook grip; such as a safety shower handle or manual control.
	200 (45)	Operating a chain hoist, power grip; <5 cm (2 in.) diameter grip surface.
B. Pull down, shoulder level	315 (70)	Activating a control, hook grip. Threading up operations, as in paper manufacturing and stringing cable.
C. Pull up, 25 cm (10 in.) above the floor	315 (70)	Lifting an object with one hand.
Elbow height	148 (33)	Raising a lid or access port
Shoulder height	75 (17)	cover, palm up.
D. Push down, elbow height	287 (64)	Wrapping, packing. Sealing cases.
E. Push up, shoulder height ("boosting")	202 (45)	Raising a corner or end of an object, like a pipe or beam. Lifting an object to a high shelf.

develop greater forces, designing tasks at higher levels will limit the available work force.

Values for maximum forces in the seated workplace are less than those seen in Table 22-2. For downward pulls they are about 85 percent of the standing values (Asmussen and Heebol-Nielsen, 1961; Keyserling et al., 1980). Downward forces of the type needed in wrapping and packing tasks are much more difficult to develop from a seated posture. Work height in relation to elbow height will determine the amount of force that can be developed, as will the orientation of the hands and forearms (palms up or down, elbows out or in). In operations where forces greater than 45 N (10 lbf) must be routinely developed, it is advisable to provide a standing or sit-stand workplace (Chapter II, Volume 1).

C. TRANSVERSE OR LATERAL FORCES APPLIED HORIZONTALLY

In some workplaces, auxiliary equipment, such as ventilation ducts, supply pipes, and conduits, may prevent the handler from getting behind an object to be moved. It may be necessary for the operator to move the object across the body using shoulder and arm muscles before he or she can get into a position where the larger muscle groups can be used (Figure 22-1).

Pushing across the front of the body involves the weaker shoulder muscles; so the upper limits of force drop to about 68 N (15 lbf) at full arm's extension (Kamon and Terrell, Eastman Kodak Company, 1979). For lever operation while seated (as in crane controls), maximum transverse forces are only 50 to 70 percent of those that can be developed in horizontal pushes and pulls straight ahead of the body at the same elbow angle (Hunsicker, 1957).

D. FORCES DEVELOPED BY THE HAND

Grip and finger strengths are important components of manual assembly and handling tasks. Different types of grip were discussed in Chapter 20. The forces that have to be exerted in handling tasks can require a higher percentage of grip strength if handles are not available. The following guidelines are based on studies of hand strength (the strength data are found in Appendix A):

- Forces required to control handling by finger pinch should not exceed 45 N (10 lbf) and should be below 30 N (7 lbf) in very repetitive work. The provision of handles or other handling aids will relieve the need to use a pinch grip.

- Requirements for power grip forces greater than 225 N (50 lbf) should not be a regular part of a handling job. Such requirements include machine jam clearances where the larger arm and shoulder muscles cannot be used to help develop the forces.

- If only finger strength is available to extricate a part or pull an object,

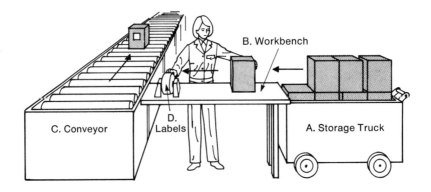

Figure 22-1: Transverse Force Application   A workplace is illustrated that requires the application of force with a transverse, or lateral, motion of the arm in order to move a case from the storage truck (A), across the workbench (B) where it is labeled (D), to the outgoing conveyor (C). Because the smaller muscles of the shoulder and arm are the primary contributors to this motion, less force can be generated than if the case is moved directly forward from the body. This reduced ability to exert force in the transverse plane can be overcome by providing a roller bearing or air table at the workbench. This reduces the sliding resistance, requiring much less work by the arm and shoulder muscles.

the forces should be kept below 40 N (9 lbf) (based on data from Josenhans, 1962). If the wrist can be brought into the force exertion, a force up to 144 N (32 lbf) is acceptable (Kroll, 1971). If the whole arm can be used, higher forces can be developed, depending on the height of application of force (see Table 22-2 for the design limits.)

REFERENCES

Asmussen, E., and K. Heebol-Nielsen. 1961. "Isometric Muscle Strength of Adult Men and Women." *Communication from the Danish National Association for Infantile Paralysis, 11:* pp. 1-44.

Ayoub, M. M., and J. W. McDaniel. 1974. "Effects of Operator Stance on Pushing and Pulling Tasks." *AIIE Transactions, 6:* pp. 185-195.

Caldwell, L. 1964. "Body Position and Strength and Endurance of Manual Pull." *Human Factors, 6 (5):* pp. 479-483.

Caldwell, L., D. B. Chaffin, F. N. Dukes-Dobos, K. H. E. Kroemer, L. L. Laubach, S. H. Snook, and D. E. Wasserman. 1974. "A Proposed Standard Procedure for Static Muscle Strength Testing." *American Industrial Hygiene Association Journal, 35:* pp. 201-206.

Chaffin, D. B. 1974. "Human Strength Capability and Low Back Pain." *Journal of Occupational Medicine, 16 (4):* pp. 248-254.

Chaffin, D. B. 1975. "Ergonomics Guide for the Assessment of Human Static Strength." *American Industrial Hygiene Association Journal, 36:* pp. 505-511.

Chaffin, D. B., G. Herrin, W. M. Keyserling, and A. Garg. 1977. "A Method for Evaluating the Biomechanical Stresses from Manual Materials Handling Jobs." *American Industrial Hygiene Association Journal, 38:* pp. 662-675.

Davis, P. R., and D. A. Stubbs. 1980. *Force Limits in Manual Work.* Materials Handling Research Unit, University of Surrey (England). Guilford, Surrey: IPC Science and Technology Press, 25 pages.

Hunsicker, P. A. 1957. *A Study of Muscle Forces and Fatigue.* WADC Report 57-586. Wright-Patterson AFB, Ohio: Wright Air Development Center, 47 pages.

Josenhans, W. C. T. 1962. "Physical Fitness, Muscle Force, and Endurance of Male Adults of Overweight." *Internationale Zeitschrift fur Angewandte Physiologie, (Einschliesslich Arbeitsphysiologie), 19:* pp. 173-182.

Kamon, E., and A. J. Goldfuss. 1978. "In-Plant Evaluation of the Muscle Strength of Workers." *American Industrial Hygiene Association Journal, 39:* pp. 801-807.

Keyserling, W. M., G. Herring, D. B. Chaffin, T. J. Armstrong, and M. L. Foss. 1980. "Establishing an Industrial Strength Testing Program." *American Industrial Hygiene Association Journal, 41:* pp. 730-736.

Kroemer, K. H. E. 1970. *Horizontal Static Forces Exerted by Men Standing in Common Working Postures on Surfaces with Various Tractions.* AMRL-TR-70-114. Wright-Patterson AFB, Ohio: Aerospace Medical Research Laboratory, 36 pages.

Kroemer, K. H. E. 1974. *Designing for Muscular Strength of Various Populations.* AMRL-TR-72-46. Wright-Patterson AFB, Ohio: Aerospace Medical Research Laboratory, 51 pages.

Kroemer, K. H. E., and J. M. Howard. 1970. "Towards Standardization of Muscle Strength Testing." *Medicine and Science in Sports, 2 (4):* pp. 224-230.

Kroll, W. 1971. "Isometric Strength Fatigue Patterns in Female Subjects." *Research Quarterly, 42 (3):* pp. 286-298.

Laubach, L. L. 1976. "Comparative Muscle Strength of Men and Women: A Critical Review of the Literature." *Aviation, Space, and Environmental Medicine, 47 (5):* pp. 534-542.

Martin J., and D. B. Chaffin. 1972. "Biomechanical Computerized Simulation of Human Strengths in Sagittal Plane Activities." *American Institutes of Industrial Engineers (AIIE) Transactions, 4 (1):* pp. 19-28.

Poulsen, E., and K. Jorgensen. 1971. "Back Muscle Strength, Lifting and Stooped Working Posture." *Applied Ergonomics, 23:* pp. 133-137.

Snook, S. H., and V. M. Ciriello. 1974. "Maximum Weights and Workloads Acceptable to Female Workers." *Journal of Occupational Medicine, 16:* pp. 527-534.

Yates, J. W., E. Kamon, S. H. Rodgers, and P. C. Champney. 1980. "Static Lifting Strength and Maximal Isometric Voluntary Contractions of Back, Arm, and Shoulder Muscles." *Ergonomics, 23 (1):* pp. 37-47.

CHAPTER 23

Lifting

CHAPTER 23. LIFTING

A. Factors Contributing to the Acceptable Weight for Lifting

B. The NIOSH Manual Lifting Guidelines and Industrial Tasks

    1. The Assumptions Behind the Guidelines
    2. The Guidelines for Occasional Lifts
    3. Frequent Lifting
    4. Other Limits to Lifting Capability

C. Occasional Lifts—Two Hands, Sagittal Plane, Compact Load

    1. Guidelines
    2. Load Location
        a. Distance in Front of the Body
        b. Height Above the Floor
    3. Actions Defined by Weight-Limit Curves

D. Guidelines for Repetitive Lifts

    1. Low Lifts
    2. Higher Lifts

E. Guidelines for One-Handed Lifting

References for Chapter 23

In some existing workplaces it is difficult to reduce manual handling by using automation to eliminate product rehandling. In newer workplaces, a very diverse product line requiring many product changes during a shift may make automation less satisfactory than using people. In these situations, manual handling may be necessary. The availability of handling aids, such as conveyors, levelators, and hoists, can minimize the actual amount of lifting and lowering needed to move raw material or the product. If these aids are provided, however, they must be accessible, easy to use, and take no more time than it takes to move the material by hand; otherwise, they will not be used.

The guidelines presented in this chapter are based on relatively stereotyped, two-handed, sagittal plane lifts and lowers. People have less capacity for asymmetric lifts, where the hands share the load unequally. Acceptable levels of weight for asymmetric lifts will be lower than the values recommended here.

The handling of materials in industry often includes having to lift them out of containers or off carts or pallets onto shelving, conveyors, workplaces, pallets, or vehicles. Lowering an object is somewhat less work than lifting it (Nielsen, Eastman Kodak Company, 1978; Passmore and Durnin, 1955). Therefore, guidelines for maximum acceptable weights are based on lifting capabilities. Although a case may only be lowered in one workplace if it is handled from a conveyor down to a pallet, it may have to be lifted at the next workplace. The guidelines in this chapter are based on lifting tasks with trays or cases. In addition, because the studies on which these guidelines are based used stereotyped lifting methods and a compact tray with good handholds, they assume a compact load, handled in the sagittal plane (with both hands in front of the body). Information on carrying, shoveling, and the handling of large-size sheets, bags, drums, and carboys can be found in Chapters 24 and 25. Attention is focused here on the factors affecting load suitability, guidelines for occasional two-handed lifts, and the effects of frequency on acceptable lifting weight. A short discussion of one-handed lifting is also included.

A. FACTORS CONTRIBUTING TO THE ACCEPTABLE WEIGHT FOR LIFTING

The most frequent question asked by the plant is, "How much can a person lift?" The question can be better phrased as "How much can people lift safely?" and "What is an acceptable load for handling tasks in industry?" The first of these rephrased questions has led to research on the biomechanical stresses of lifting, focusing especially on the lower back (Chaffin and Park, 1973; Tichauer, 1971). Investigations of handling injuries have also contributed to our understanding of how much is too much (Chaffin, Herrin, Keyserling, and Foulke, 1977; Davis, 1959; National Institute for Occupational Safety and Health [NIOSH], 1981), although the relationship between weight handled and back strains is not always clear. The question of how much is "acceptable" to industrial workers has been studied using psychophysical (scaled opinions and simulated tasks) methods (Garg and Saxena, 1980; Snook, 1978). The handler adjusts the weight in a tray or other container until it is acceptable for lifting over a specified time period. The individual integrates a number of factors relating to the object lifted, the work pattern, and his or her strength in deciding what is an acceptable weight.

Weight is only one of the determinants of the acceptable or maximum permissible loads a person can lift. A detailed analysis of the components of manual handling tasks has been developed to aid in directing research into injuries from handling (Herrin, 1978; NIOSH, 1981). Worker characteristics are one part of the system and include size, strength, fitness training, motivation, and risk-taking behavior. The last three characteristics can be modified through training programs; the first three are characteristics that should be considered in the design of handling tasks so that a large majority of the available workforce can do them safely.

Table 23-1 summarizes the most important system factors in the workplace. These should be considered when evaluating an existing handling task; guidelines for designing new systems that are within most people's capacities are found throughout this book.

The appropriateness of lifting a given load, then, will be determined by multiple factors. The complexity of this type of analysis may confuse the safety, engineering, or manufacturing department person who simply wants to know how heavy an object can be without risking injury to the handler. To answer these questions better, and without sacrificing too much accuracy, we present guidelines for object weight in lifting tasks as a function of the location of the lifts and the frequency of handling. The *Work Practices Guide for Manual Lifting* (NIOSH, 1981) summarizes information from psychophysical, biomechanical, and physiological studies of lifting and provides guidelines for the design of manual lifting tasks. A short discussion of the applicability of those guidelines to common industrial lifting tasks follows. Then we will propose a set of guidelines for occasional and repetitive lifts, which were developed from much of the same data as well as from data collected in industrial studies (Rodgers, Eastman Kodak Company, 1980). The occasional lift guidelines are similar to the NIOSH guidelines but the height and horizontal factors are included in the graph (see Figure 23-3). The repetitive lifting guidelines differ from the NIOSH frequency factor correction and more accurately reflect industrial experience with acceptable workloads.

B. THE NIOSH MANUAL LIFTING GUIDELINES AND INDUSTRIAL TASKS

1. THE ASSUMPTIONS BEHIND THE GUIDELINES

In addressing an issue as complex as how much people should be asked to lift in manual handling tasks, we must define the lifting conditions carefully. This has been done in the *Work Practices Guide for Manual Lifting* (NIOSH, 1981), which makes the following assumptions about the way the lift is made:

- Two-handed.
- Sagittal plane (in front of body)—no twisting.
- Good handholds.
- Balanced load.
- Less than 76 cm (30 in.) in width.
- Good footing.
- Good environment.
- Unrestricted posture.

In an analysis of how prevalent these conditions are in many industrial lifting situations, it was found that most objects are handled somewhat differently in the workplace (Rodgers, 1983). The assumptions most frequently violated relate to the availability of good handholds and to lifting being done in the sagittal plane. With the exception of trays, occasional boxes, bottles, and some tools, most objects lifted in industry do not have well-designed handholds. The most commonly handled items in small packaged-goods manufacturing are shipping cases or boxes, very few of which have a place to grasp with a power hand grip.

Sagittal-plane lifting is always recommended but is not often seen in industrial handling tasks. When heavy lifts are made infrequently, usually a few times a shift, most people will follow the guidelines for safe lifting with no twisting (Anon., 1979; Brown, 1975; Davies, 1972). If the object is being moved between shelves in a storage area or on transport carts, sagittal plane lifting will also often be used. Once the lifting frequency exceeds five or six lifts per minute, however, strict sagittal plane lifting is less common (Rodgers, 1983). This may relate to the energy expenditure differences between sagittal plane and "natural" lifting techniques. The former requires more energy at low lifting heights since the body is lifted with each object lifted off a pallet or low shelf. Some natural, or freestyle, lifting techniques require less energy and are preferred in short-duration but very repetitive handling activities (Davies, 1972; Garg, 1976; Jones, 1979; Nielsen, Eastman Kodak Company, 1978).

2. THE GUIDELINES FOR OCCASIONAL LIFTS

The *Work Practices Guide for Manual Lifting* (NIOSH, 1981) is based primarily on biomechanical analyses of the stress on the fourth and fifth lumbar vertebral discs of the spinal column and on lifting strength capabilities of the working population (Chaffin, Herrin, Keyserling, and Garg, 1977). This defines the upper limit MPL, or maximum permissible limit, of the three-zone guideline (Figure 23-1). Lifting tasks that exceed this limit are considered hazardous and should be avoided. The lower AL, or action limit, curve is defined by psychophysical studies of acceptable loads in lifting tasks (Snook, 1978). The AL is chosen to

Table 23-1: Factors Influencing the Suitability of Manual Handling Tasks. Six material or container characteristics and ten task and environment characteristics are shown that influence the ease of performing manual handling tasks. The material or container characteristics influence the way the handler controls the load and the stress placed on the lumbar spine and on specific muscle groups of the arm, shoulder, and hands. The task and environment characteristics influence worker posture, the potential for local muscle and whole body fatigue, and the ease of performing the task in general. Worker characteristics, not shown here, also influence the suitability of a given handling task for a given person. However, the design of handling tasks should try to accommodate the large majority of workers and not require selection by muscle strength, size, or general fitness levels. *(Based on Herrin, 1978; NIOSH, 1981).*

A. Material or Container Characteristics

- Length
- Width
- Height or depth
- Center of gravity
- Handles or grasp points
- Stability

B. Task and Environment Characteristics

- Height of lift or force application
- Forward reach
- Frequency (lifts/min)
- Duration
- Time pressure
- Availability of help
- Complexity
- Shift schedule
- Temperature
- Lighting

---

accommodate 75 percent of the women and 99 percent of the men in the workforce. Lifting tasks that fall below this curve should be suitable for most people providing that the guide's assumptions are met.

The zone between the design guidelines and the hazard level is marked as conditions where administrative controls, such as training, selection, or job accommodations to reduce the load, are needed. The closer a lifting condition

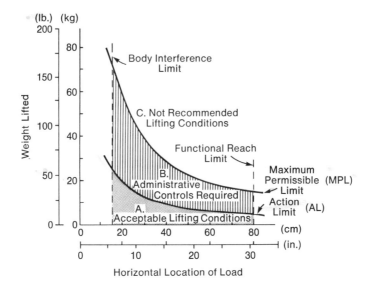

Figure 23-1: The NIOSH Guidelines for Occasional Lifts in the Sagittal Plane   The effect of horizontal load location in front of the spine, in centimeters and inches (horizontal axis), on the maximum weight, in kilograms and pounds (vertical axis), that should be handled in a low lifting task is shown. Vertical height of the lift is 76 cm (30 in.), and the frequency of handling is less than one lift every five minutes. Three zones are indicated. The "Acceptable Lifting Conditions" zone (A) is the area where most people should be able to perform the lift safely and marks the preferred design range for lifting tasks. The "Administrative Controls Required" zone (B) indicates weights that exceed the capabilities of many workers, so training and selection of workers are suggested. The "Not Recommended" zone (C) represents loads that exceed the lifting capacities of all but 25 percent of the male population. The zones are delineated by the "Maximum Permissible Limit" (MPL) and the "Action Limit" (AL) curves, which are described in the text. The guidelines are intended to be used for lifts in the sagittal plane of compact objects with handholds. Corrections must be made to the load limits if the object being handled does not meet these assumptions (see Table 21-3 and Figure 23-2). If lifts are made more frequently than once every five minutes, a frequency adjustment has to be made to the weight determined from this graph. Figure 23-5 gives data for adjusting medium height lifts and Figure 23-4 shows the weight and frequency guidelines for low lifting. (NIOSH, 1981).

is to the upper line, the more people will have difficulty with the lift.

The vertical location of a lift (V) and the vertical distance (D) traveled from the point where the object is procured to the point where it is disposed of determine what muscle groups are able to participate. There are factors to correct the value obtained from Figure 23-1 for heights above 75 cm (30 in.) and travel distances greater than 25 cm (10 in.). Figure 23-2 gives these in graphic form.

General formulas for calculating the action limit (AL) and maximum permissible limit (MPL) values for a given set of lifting conditions are given below. These formulas apply only to occasional (less than one every five minutes) lifts.

$$AL \text{ (kgm)} = 40 \ (15/H) \ (1 - 0.004 \ [V - 75]) \ (0.7 + 7.5/D)$$
$$AL \text{ (lbm)} = 90 \ (6/H) \ (1 - 0.01 \ [V - 30]) \ (0.7 + 3/D)$$
$$MPL = 3(AL)$$

H refers to the horizontal distance measured from the spine where the load is handled; it will be determined by the size of the object, the configuration of the workplace, or the nature of the task. Handling an object into a large, deep box or over a wide counter, for example, increases the H factor because the person cannot get close to the load. The vertical height (V) is determined by the initial hand location in the lift. It is adjusted for the 75 cm (30 in.) height assumption in the guidelines and the absolute difference is taken, as indicated by the brackets on either side of V − 75 or V − 30. D is the vertical travel distance, measured by the difference between the final and initial vertical locations of the hands during the lift. MPL is calculated by multiplying the AL value by 3, an approximation suggesting that the AL values represent about one-third of total lifting capacity for many people.

3. FREQUENT LIFTING

The interaction of workload limits with lifting strength limits requires that a correction be made to the occasional lift guidelines at lifting frequencies above one lift every five minutes, especially if above five or six lifts per minute (NIOSH, 1981; Rodgers, 1983). The frequency factor in the *Work Practices Guide for Manual Lifting* (NIOSH, 1981) is under revision because it does not reflect experience in industry with lifting rates greater than eight per minute. The material later in this chapter on repetitive lifting represents another approach to the frequency adjustment; it is based on some of the same physiological studies as are the NIOSH guidelines and is supplemented with data from work physiology studies in several workplaces (Petrofsky and Lind, 1978a; Petrofsky and Lind, 1978b; Nielsen, Eastman Kodak Company, 1962-1968; Rodgers, 1976).

4. OTHER LIMITS TO LIFTING CAPABILITY

The major determinants of the manual lifting guide's MPL values are biomechanical stresses on the lower back. AL values are primarily determined by psychophysical studies of acceptable lifts using a tray with good handholds, using sagittal-plane lifts, and following the other assumptions given earlier. In observations of many types of lifting tasks in the industrial setting, researchers have noted that grip strength, wrist stability, and shoulder flexor strength often limit a person's ability to lift before the pressure on the lower lumbar spinal discs reaches levels of concern. For this reason, applying the guidelines to lifting sit-

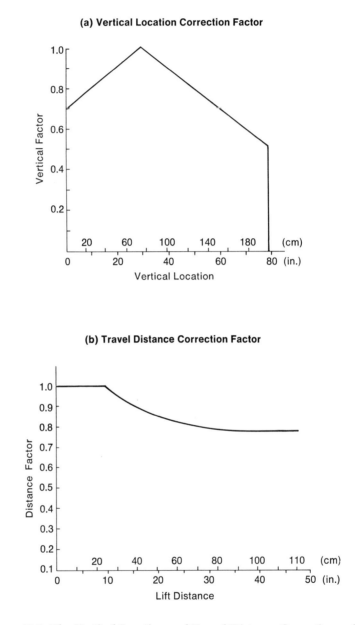

(a) Vertical Location Correction Factor

(b) Travel Distance Correction Factor

Figure 23-2: The Vertical Location and Travel Distance Corrections of the NIOSH Manual Lifting Guide. Two graphs illustrating factors for correcting the weight limits determined from Figure 23-1 for vertical heights above or below 76 cm (30 in.) (*a*) and for travel distances greater than 25 cm (10 in.) (*b*) are shown. The vertical and distance factors (vertical axes) are expressed as fractions of the weight obtained from Figure 23-1. Vertical location and lift distance (horizontal axes) are expressed in centimeters and inches. The shape of the vertical factor curve indi-

uations that do not meet the stated assumptions about good handholds, unrestricted postures, balanced loads, and sagittal plane lifts may give values that are not acceptable for most people. For example, many people seated at a workbench conveyor had difficulty lifting a 10 kg (22 lbm) tray onto a conveyor that was at 25 cm (10 in.) in front of the body and about 86 cm (34 in.) above the floor (Rodgers, 1983). The AL for the task would be 7 kg (15 lbm) and the MPL would be 21 kg (45 lbm) according to the formulas given earlier. Although this is a sagittal-plane lift using a tray with good handholds, the limiting factors of wrist and shoulder strength and the seated posture make it more difficult than the guide would predict.

Any attempt to give guidelines for lifting that can be used in a variety of working conditions will be liable to be misused and the assumptions neglected. The guidelines in this chapter, like the NIOSH guidelines, should be evaluated against the assumptions. Other factors that may be more limiting than lumbar disc pressure or whole-body lifting strength, such as grip or shoulder flexor strength, should also be evaluated before a weight limit for lifting is set.

C. OCCASIONAL LIFTS—TWO HANDS, SAGITTAL PLANE, COMPACT LOAD

1. GUIDELINES

There are two types of guidelines to consider when determining how much weight people should be asked to lift in the workplace. The first level is the weight that at least 75 percent of the potential work force should be able to handle. The potential work force in industry includes men and women, and old and young workers (see Chapter I, Volume 1). These guidelines represent the upper weight limit to be used when designing new lifting tasks or evaluating existing ones for redesign. The second level is the maximum weight limit or "not recommended" level; objects weighing more than this should not be manually handled by anyone. Fewer than 15 percent of the potential workforce can perform these lifts without difficulty (NIOSH, 1981).

---

cates that less weight can be handled both below and above 76 cm (30 in.). The sharp drop off in the factor at 200 cm (79 in.) reflects the upper limit of reach (fully stretched) for many people. At low lift heights, bent knees may interfere with bringing the load close to the body and the lifting posture may be unstable. With high lifts, the upper body musculature limits how much can be lifted. The distance factor changes more gradually and does not impact as much on the weight lifted. The change is largely associated with handling time—the larger the distance, the more time it takes to complete the lift. These factors are included in the formula for calculating the action limit (AL) and maximum permissible limit (MPL) values, and are described in the text. (NIOSH, 1981).

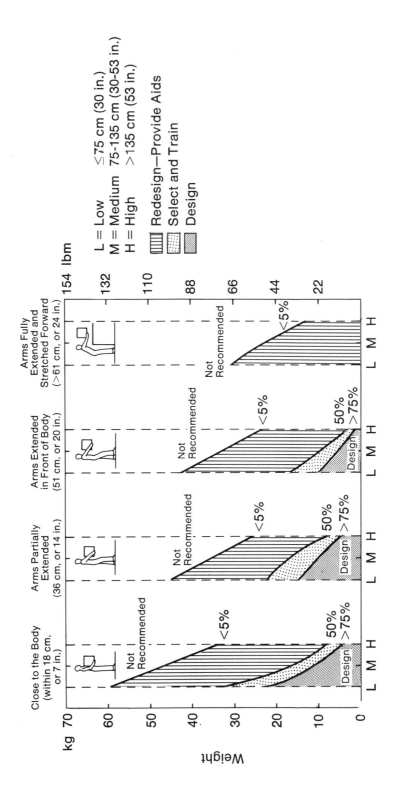

In the evaluation of existing handling tasks, many occasional lifts fall in the area between these two limits (Rodgers, 1983). Redesigning the task is usually more profitable than selecting people to do it if more than 50 percent of the potential workforce are not accommodated by the design. A third level is useful, therefore, to identify where 50 percent of the potential work force will find the lift difficult. Below this level, redesign is desirable, but it is not very difficult to find people who can lift the load safely. Thus, selection and training may be appropriate. Above this level, redesign is preferable in order to try to reduce the potential for overexertion injuries.

The guidelines presented in Figure 23-3 have been developed from four sources:

- Psychophysical studies of acceptable lift weights (trays) (Garg and Saxena 1980; Snook, 1978; Snook and Ciriello, 1974).

- Biomechanical analyses of the forces on the back and joints during lifts (Ayoub et al., 1979; Chaffin, Herrin, and Keyserling, 1978; NIOSH, 1981).

Figure 23-3: Guidelines for Occasional Lifts—Two Hands, Sagittal Plane, Compact Load   Four graphs show a modification of the NIOSH guidelines for occasional lifts (Figure 23-1) that incorporates the vertical height correction factor. It defines three vertical zones, low (L), medium (M), and high (H), shown for each of four horizontal lift locations across the horizontal axis. The horizontal lift locations are defined and illustrated at the top of the graphs. The vertical locations are defined at the right of the figure. The weights, in kilograms and pounds, are shown on the left and right vertical axes, respectively. Four zones are shown for each horizontal location of the load. As in the NIOSH guidelines, the lowest zone is the design recommendation where more than 75 percent of the work force is accommodated; the highest one is a "Not Recommended" level where less than 5 percent can lift safely. The NIOSH middle zone is divided into two zones here. The lower zone is where selection and training may be appropriate work practices because at least 50 percent of the potential work force will be capable of making the lift. The upper zone is where task redesign or the provision of handling aids is more likely to reduce the risk of overexertion injuries. Since more than half, and up to 95 percent, of the work force will have difficulty with them (between 5 and 50 percent can do them), most of these lifting tasks merit redesign. As horizontal distance increases, there is more need to redesign the task. These guidelines permit the ergonomist to identify quickly where an existing handling task falls in terms of work practice zones, and to help set a priority to act on those tasks that are difficult for more than 50 percent of the potential work force. The objects handled are assumed to be compact, have good handholds, and are lifted in the sagittal plane less than once every five minutes. See Figures 23-4 and 23-5 for the frequency modifications to these acceptable lift weights.

- Studies of maximum strengths developed in pulling up on a tray, and maximum strengths of specific muscle groups (Champney, Eastman Kodak Company, 1983; Kamon and Goldfuss, 1978; Keyserling et al., 1980; Pederson, Petersen, and Staffeldt, 1975; Yates et al. 1980).
- Observations of industrial handling tasks over a 20-year period.

Most of these studies were of two-handed, sagittal-plane lifts (in front of the body) that were sustained for less than a few seconds each. Containers with well-designed handholds were often used. Consequently, the guidelines in Figure 23-3 are appropriate for only some kinds of lifting tasks in industry. The acceptable load values are usually lower than those shown in Figure 23-3 if the lifts are performed differently, such as to the side instead of in front of the body, or if the load is less stable.

The relationship between object weight and lift height range (low, medium, high) is given for four distances in front of the body in these curves. Three weight-limit curves are shown for each horizontal distance. The weights that at least 75 percent of the work force should be able to handle without difficulty, or the design limit, is the lower curve; the weights that 50 percent of the work force should be able to handle, which can be used as a marker for evaluating existing lifting tasks, is the middle curve; and the weights that less than 5 percent of the work force can handle without difficulty, or the maximum load, is the upper curve.

Examples of the effect of load location on the amount of weight that can be lifted are given in the following subsection, and lift height ranges are defined. A discussion of recommended actions for lifts that exceed the design guidelines is also included.

2. LOAD LOCATION

a. Distance in Front of the Body

The effect of load location in front of the body on the acceptable weight for occasional lifts is significant. Many loads cannot be handled close to the body—that is, within 18 cm (7 in.) in front of the ankles. Factors that make it difficult to get the load close to the body include:

- Having to reach over equipment, containers, or workplaces to make a lift; for example, placing a product in a wide, deep container.
- Handling wide containers or products where the hands have to grasp the load with the arms extended in front of the body as with items that are more than 46 cm (18 in.) wide.
- Reaching across a counter or workplace to dispose of or procure an item from a conveyor or cart.
- Placing an item far back on a shelf, especially if it is a liquid in a container that cannot be slid easily across the shelf surface.

The farther in front of the body an object has to be lifted, the less weight is acceptable. Fifty and 75 percent population curves are not shown for the 76 cm (30 in.) extended reach because such lifts are not recommended. Only people with long forward reaches and well-developed shoulder muscles will find these lifts comfortable to perform.

To determine which location in front of the body should be used to assess the weight limits, observe whether the lifting requires extended arms or if the handler can keep the load close to the body. If extended arms are required, those limits (according to the amount of arm extension) should be used.

b. Height Above the Floor

The curves in Figure 23-3 are constructed by joining the maximum weights for each height of lift. These height ranges are measured as the distance between the floor and the hands supporting the load. They are:

- Low lifts—up to 75 cm (30 in.) above the floor. Very low lifts, under 25 cm (10 in.), are not desirable because it is difficult to maintain balance when squatting for such a lift, and stress on structural components of the back can be high.

- Medium lifts—from 75 to 135 cm (30 to 53 in.) above the floor. More strength is available in the lower than in the upper part of this range.

- High lifts—from 135 to 188 cm (53 to 74 in.) above the floor. In general, this range of lift heights is not recommended for new designs, except when items weighing less than 5 kg (11 lbm) are handled. Heavier items can be "boosted" from the bottom to reduce the forces on the wrists and shoulders in some workplaces. However, many lifting situations do not have a place to set the object down so the handler can change his or her grasp and boost it onto a high shelf or peg.

Most lifts are between conveyors, pallets, shelves, trucks, carts, and equipment where height ranges may be crossed. To determine the appropriate weight for new designs, or how much impact a given weight will have on the available workforce in an existing workplace, you should use the highest point of the lift. Thus, a lift from 64 to 125 cm (25 to 50 in.) —measured from hand height during the lift—should be evaluated by the medium-height range guidelines.

3. ACTIONS DEFINED BY WEIGHT-LIMIT CURVES

The 75, 50, and less than 5 percent population curves define four zones against which existing jobs and new designs can be measured.

- Design Zone—Lifts should be designed within this zone whenever possible.

- Selection and Training Zone—It is not too difficult to find people who can do these lifts. Redesign may be desirable, but is not essential. Training people how to lift and using validated selection tests for the job may be appropriate to help reduce the potential for overexertion injuries in less strong people.

- Redesign and Provide Aids Zone—More than half of the potential work force will find these lifts difficult. The load weight should be reduced, the lift location changed, and/or aids provided.

- Not Recommended Zone—These manual lifts should be eliminated and materials handling equipment used.

The values shown in these curves are based on the best data currently available. They have been developed by making assumptions that the potential workforce is composed of men, women, old, and young people, that design should be within "safe" limits for lifting by using acceptable load data and strength measures, and that the types of items lifted are compact, have handholds, are a stable and evenly distributed load, and are usually a tray or box. Many other types of handling tasks can be designed effectively if the weights are kept within the Design Zone limits shown in Figure 23-3. However, the less than 5 and 50 percent population weight limits will be overestimated in lifting tasks that do not meet these assumptions. As new data are developed, some of these curves may be altered; therefore, the absolute values should not be used to predict who can or cannot do a task based on a strength measure not related to the specific job.

Examples of how these curves can be used to determine the suitability of a given lift are included in "Problems" (Appendix C).

D. GUIDELINES FOR REPETITIVE LIFTS

The weight limits shown in Figure 23-3 are for occasional lifts. In many industrial jobs, items must be lifted frequently and repetitively. Some examples are:

- Putting products on, or unloading products from, pallets.
- Loading and unloading products and supplies from carts and trucks.
- Loading and unloading supplies and products from production machines.
- Performing weighing and inspection tasks—often, moving trays of supplies or products from a conveyor to a workplace.
- Handling products to and from storage shelves.

The amount of weight that can be handled per lift in repetitive lifting tasks is less than the occasional lift values. The rate of fatigue of the active muscles will depend on the percentage of their maximum strength required (which is related to the item's weight), the duration of holding, and the amount of recovery time between lifts. At higher lifting frequencies (lifts per minute), there is less recovery time between lifts, and fatigue will occur more rapidly. This is especially true if the muscles are loaded to more than 40 percent of their strength capacity (Rohmert, 1960; Scherrer and Monod, 1959). Guidelines for repetitive lifting at low and medium heights are given here. Frequent lifting above 135 cm (53 in.) is not recommended.

1. LOW LIFTS

Frequent lifting can contribute to whole-body fatigue as well as local muscle fatigue when items are handled at heights less than 76 cm (30 in.). The handler's body must be lowered and raised with each lift, resulting in increased energy expenditure (Garg, 1976; Nielsen, Eastman Kodak Company, 1978; NIOSH, 1981). This calisthenics-like work is unnecessary effort, and it limits productivity by limiting the acceptable continuous work duration. Handlers often control their workload by stooping over to lift instead of squatting. This risks excessive pressures on the lumbar spinal column discs, as has been discussed in Part III. Altering the height to require less bending will increase productivity by reducing the amount of recovery time needed and will leave an energy reserve for more sustained periods of lifting.

For low lifts, at lifting frequencies greater than one per minute lasting more than 20 minutes per hour, the weight level per lift is affected by the whole-body endurance, as well as by the local muscle requirements (NIOSH, 1981). At frequencies greater than five per minute and durations greater than 20 minutes per hour, it is primarily the energy requirements of the lifting that determine the acceptable weights. Muscle strength and endurance are not major limiting factors when the items weigh less than 14 kg (31 lbm) and less than five lifts per minute are made.

Figure 23-4 illustrates the upper weight limit for design of low, repetitive (frequent) lifting tasks. The curves are developed from studies of the energy requirements of lifting tasks (Nielsen, Eastman Kodak Company, 1978; Petrofsky and Lind, 1978a, 1978b), psychophysical studies of acceptable weights at several frequencies (Garg and Saxena, 1980; Snook, 1978), and information about the endurance, or aerobic, and strength capacities of industrial workers (Kamon and Goldfuss, 1978; Rodgers, Eastman Kodak Company, 1973).

For jobs where lifting must be done for a majority of the shift, the curve indicates that the most efficient combination of weight and frequency is about 14 kg (31 lbm) and five lifts per minute. If the lifts are from 25 to 76 cm (10 to 30 in.) above the floor and the total lifting task time per shift is about 300 minutes, it should be possible for one person to lift about 21,000 kg (23.1 tons) of product per shift (14 kg per lift × 5 lifts/minute × 300 minutes). Heavier items reduce the recommended frequency. For example, items weighing 18 kg (40 lbm) should not be lifted more than once per minute, averaged over the shift, if excessive fatigue is to be avoided. Therefore, only 5,400 kg (5.9 tons) could be handled per shift—potentially a 75 percent drop in productivity. Table 23-2 illustrates the weight per minute that can be lifted at the upper limit of design.

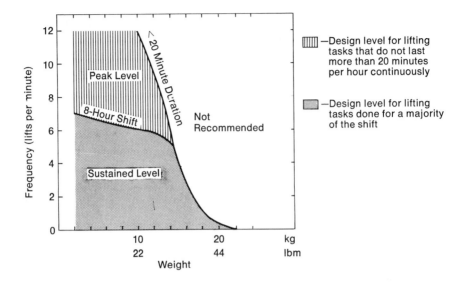

Figure 23-4: Recommended Upper Limit for Design of Repetitive Lifting Tasks Below 76 cm (30 in.)—Two Hands, Sagittal Plane, Compact Load   Two curves are shown that identify the design levels for combinations of weight (kilograms and pounds, horizontal axis) and lifting frequency (lifts per minute, vertical axis) over short (less than 20 minutes) and long (eight hours) durations. These guidelines should accommodate at least 75 percent of the potential work force. The "Peak Level" zone between these curves indicates where more than 20 minutes, but less than eight hours, of continuous lifting can be done. The area designated as "Sustained Level" can be performed for somewhat more than eight hours if extended shifts are worked, but other limits to workload acceptability will probably come in after 12 hours of continuous work (see Part VI). The "Not Recommended" zone is defined by the combinations of weight, frequency, and duration that result in local muscle fatigue or inadequate recovery time for the whole body work being done in low lifts. The curves are constructed by using a combination of psychophysical data that define acceptable lifts and metabolic data that define low lifting workloads. Local muscle effort is the main limiter of acceptable weight levels for short-duration work. For extended hours of work, the energy requirements of lifting and lowering the body with each lift must also be considered, especially for people with below-average work capacities. The curves suggest that the most efficient workload for an eight-hour shift would be met by keeping the weight of the object lifted at or below 14 kg (31 lbm) and the frequency of handling at five per minute. These guidelines are meant to be used for the design of lifting tasks. Heavier objects can be lifted safely, and faster frequencies can be used. However, the potential restriction of workers who do not have the appropriate work capacities and the need to provide adequate recovery periods can reduce productivity in these tasks (see Table 23-2).

Table 23-2: Weight Handled per Minute for Low Lifts as a Function of Item Weight and Lifting Frequency—Two-Handed Sagittal-Plane Lifts of Compact Objects. Values taken from Figure 23-4 illustrate how changing the object weight, in kilograms and pounds (column 1), and the frequency of handling it (column 2) can affect productivity. The most productive workload for lifts below 76 cm (30 in.) was identified as 14 kg (31 lbm) handled five times per minute; this is designated as 100 percent in column 4. Other combinations of weight and frequency that meet the design guidelines in Figure 23-4 are expressed as a percentage of this value after calculating the weight lifted per minute (column 3) for each. The reduction in productivity is related to the need for additional whole-body recovery time at higher lifting frequencies and to a need for local muscle recovery time at lower frequencies and higher object weights. Although more weight can be lifted at higher frequencies for short durations, the recovery time requirements eventually limit the productivity gains and actually reduce productivity over the shift. (*Developed from data shown in Figure 23-4*).

Object Weight in kg (lbm)	Lifting Frequency (lifts/min)	Weight Lifted in One Minute in kg (lbm)	Relative Productivity (Percent)
5 (11)	6.5	32.5 (72)	46
10 (22)	6.0	60.0 (132)	86
14 (31)	5.0	70.0 (165)	100
15 (33)	4.0	60.0 (132)	86
16 (35)	3.0	48.0 (105)	68
18 (40)	1.0	18.0 (40)	26
20 (44)	0.5	10.0 (22)	14

Values are also expressed as percentages of the most productive load, 14 kg (31 lbm) at five per minute.

Very light objects (less than 4 kg, or 10 lbm) are frequently handled two or three at a time when lifts are at about 76 cm (30 in.) above the ground (floor). The handler spares him- or herself the effort of raising and lowering the body with each item, usually trading off that effort for the increased stress on the arms and shoulders associated with trying to hold the boxes or objects together. This increases the productivity of handling, measured by the number of boxes handled per minute or per unit of energy expenditure, and may increase the amount of recovery time available as well.

If the guidelines for repetitive lifting in Figure 23-4 are followed, about 75 percent of the potential work force should be able to do the lifting task without excessive fatigue. Increased weight in a fixed frequency task such as unloading a production machine conveyor, or increased frequency per weight level, as in unloading pallets or shelves with objects of fixed weight, will result in faster fatigue for people with less endurance and lower strength capacities. It will also be more difficult to find people who can do the job over a full shift. Since there are different guidelines for shorter-duration lifting, one way to staff a task that is not within the guidelines is to use several people. For a repetitive lifting task that lasts a majority of the shift, people could be rotated in and out of the job from other, lighter effort tasks. They should not be expected to lift continuously for more than 20 minutes every hour, although some may find longer durations acceptable. For instance, one person could be expected to lift 10 kg (22 lbm) items about six times per minute for a full shift. If three workers in the area share the task, each doing it for no more than 20 minutes every hour, the lifting frequency and, potentially, the productivity could be doubled. See Part IV for more discussion of intensity and duration of work. Appendix C includes a work-load analysis problem that uses this information.

The guidelines in Figure 23-4 have been discussed in terms of lifting tasks where all objects weigh about the same. Very often the lifting task involves many different sizes and weights of product. If the lifting frequency is paced by the machine, the heaviest products should be evaluated first to determine how closely they approximate the upper limits in both Figures 23-3 and 23-4. The actual frequency of handling the heavier product may be very low if a large number of lighter cases are interspersed with it. If lifting is self-paced and not under time pressure from production demands or other social factors such as peer pressure or incentive pay, the guidelines for occasional lifts can be used.

2. HIGHER LIFTS

Frequency (lifts per minute) has less impact on the amount of weight that can be lifted between 76 and 135 cm (30 and 53 in.). The energy costs of higher-frequency lifting at these heights are not very great because the active muscle groups are not very large and less body weight is lifted (Astrand and Rodahl, 1977). If the weight handled is not reduced in proportion to the increased frequency, local fatigue of arm and shoulder muscles can occur. Figure 23-5 was developed from psychophysical studies (Snook, 1978), using several lifting frequencies. It indicates the factor by which the weight limits for medium-height range lifts in Figure 23-3 should be multiplied to correct them for lift frequency. For instance, a weight limit of 10 kg for occasional lifts in this range would be reduced to 8.2 kg (18 lbm) if it had to be lifted six times per minute.

In the medium-height range, the difference between lifting for less than 20 minutes per hour and lifting continuously throughout the shift is not great enough to warrant adding another adjustment curve. If most of the lifting is done in the upper portion of this range, from 115 to 135 cm (45 to 53 in.), how-ever, the shoulder muscles may fatigue from constant static loading (see the discussion of static work in Part IV). Wherever possible, frequent lifting should

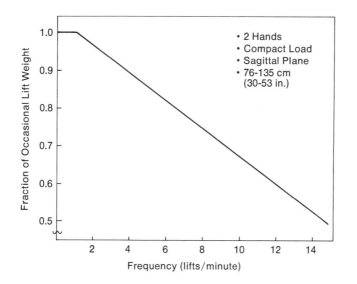

Figure 23-5: Frequency Adjustment for Medium-Height Range Lifts—Two Hands, Sagittal Plane, Compact Load. For lifts that are made above 76 cm (30 in.) but at or below 135 cm (53 in.), the frequency adjustment is based on local muscle fatigue instead of a combination of that and the total energy requirements of the task. The lifting frequency, in lifts per minute, is shown on the horizontal axis. The correction factor, as a fraction of the occasional (less than one lift per minute) lift weight taken from Figure 23-3, is shown on the vertical axis. The assumptions about the lift are shown in the upper right corner of the figure. Higher lifts are not included as they should not be done more frequently. However, an existing task that requires frequent high lifting can be evaluated roughly using these correction factors. *(Developed from data in Snook, 1978).*

not be required at heights greater than 115 cm (45 in.), as measured from the floor to the hands.

E. GUIDELINES FOR ONE-HANDED LIFTING

There is less research on one-handed lifting limits than on two-handed sagittal-plane lifting, so the information in this section is largely based on grip strength and fatigue rate studies. One-handed lifting is often seen at conveyor lines or assembly machines where parts have to be loaded or removed at a rate set by machine speed. The operator is standing or seated and may use one hand at a time or two hands alternately if the other hand is not needed to control the machine or perform another task. The criticality of the grasp in one-handed lifting tasks is apparent, since local fatigue of the hand and arm muscles will

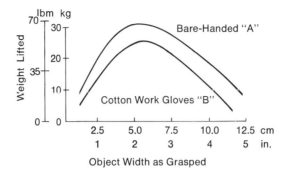

Object Width as Grasped

Figure 23-6: Guidelines for the Upper Limit of Weight Lifted with One Hand at Heights from 63 to 127 cm (25 to 50 in.) Above the Floor Two curves are shown that indicate the amount of weight (kilograms and pounds, vertical axis) that can be handled in an occasional one-handed lifting task (less than one per five minutes). The weight is related to the grasp span requirements and thereby to the dimension of the object where it is grasped (centimeters and inches, horizontal scale). Curve A is for a person working barehanded, and curve B is for a person wearing cotton work gloves. The maximum weights that can be lifted are found in those objects that permit the handler to use a power or hook grip. The grasp span should be between 4.5 and 5.7 cm (1.75 and 2.25 in.). Other factors, such as wrist angle, will affect total grip strength and may reduce the weight further. See Appendix C for more information on grip strength and hand span. *(Developed from information in Garg, 1983; Garg and Saxena, 1982; Kamon and Goldfuss, 1978; McConville and Hertzberg, 1966; Rohmert, 1960; Scherrer and Monod, 1959; SUNYAB-IE, 1982/1983).*

determine how much weight can be handled at each lifting frequency. Fatigue will be accentuated in lifting tasks that require pinch or where the items are large enough to spread the fingers and reduce power grip strength. The discussion on grasp in Chapter 21, the analysis of the biomechanics of grasp in Part III, and the data on grip strength in Appendix A should be reviewed for further information.

Figure 23-6 illustrates acceptable weights and frequencies for one-handed lifting tasks in standing operations. If the operation is done from a seated posture, object weight should not exceed 4 kg (10 lbm) (Rehnlund, 1973). These guidelines assume that the object can be grasped easily and that the lifting height is between 63 and 127 cm (25 and 50 in.) above the floor.

The metabolic demands of raising and lowering the body become as important in one-handed lifts below 63 cm (25 in.) as they are in two-handed lifts. If the objects have good handholds, the two-handed lift weights given in Figure 23-4 can be used as an approximation of the acceptable levels.

REFERENCES

Anon. 1979. *The ABC's of Moving Things Safely.* Greenfield, Mass.: Channing L. Bete Company, 15 pages.

Astrand, P.-O., and K. Rodahl. 1977. *Textbook of Work Physiology.* 2nd ed. New York: McGraw-Hill, pp. 291-329.

Ayoub, M. M., R. Dryden, J. McDaniel, R. Knipfer, and D. Dixon. 1979. "Predicting Lifting Capacity." *American Industrial Hygiene Association Journal, 40:* pp. 1075-1084.

Brown, J. R. 1975. "Factors Contributing to the Development of Back Pain in Industrial Workers." *American Industrial Hygiene Association Journal, 36:* pp. 26-31.

Chaffin, D. B., G. Herrin, and W. M. Keyserling. 1978. "Preemployment Strength Testing—An Updated Position." *Journal of Occupational Medicine, 20 (6):* pp. 403-408.

Chaffin, D. B., G. Herrin, M. Keyserling, and J. Foulke. 1977. *Pre-Employment Strength Testing.* DHEW/HIOSH Technical Report No. 77-163. Cincinnati: Department of Health, Education, and Welfare/National Institute for Occupational Safety and Health, 178 pages.

Chaffin, D. B., G. Herrin, W. M. Keyserling, and A. Garg. 1977. "A Method for Evaluating the Biomechanical Stresses from Manual Materials Handling Jobs." *American Industrial Hygiene Association Journal, 38:* pp. 662-675.

Chaffin, D. B., and K. S. Park. 1973. "A Longitudinal Study of Low Back Pain As Associated with Occupational Lifting Factors." *American Industrial Hygiene Association Journal, 34:* pp. 513-525.

Davies, B. T. 1972. "Moving Loads Manually." *Applied Ergonomics, 3 (4):* pp. 190-194.

Davis, P. R. 1959. "Posture of the Trunk during the Lifting of Weights." *British Medical Journal, 1:* pp. 87-89.

Garg, A. 1976. *A Metabolic Rate Prediction Model for Manual Material Handling Jobs.* Doctoral dissertation, University of Michigan, Department of Industrial and Operations Engineering. Ann Arbor: University Microfilms, 239 pages.

Garg, A. 1983. "Physiologic Responses to One-Handed Lifts in the Horizontal Plane by Female Workers." *American Industrial Hygiene Association Journal, 44 (3):* pp. 190-200.

Garg, A., and U. Saxena. 1980. "Container Characteristics and Maximum Acceptable Weight of Lift." *Human Factors, 22 (4):* pp. 487-495.

Garg, A., and U. Saxena. 1982. "Maximum Frequency Acceptable to Female Workers for One-Handed Lifts in the Horizontal Plane." *Ergonomics, 25 (9):* pp. 839-853.

Herrin, G. D. 1978. "A Taxonomy of Manual Materials Handling Hazards." In *Safety in Manual Materials Handling,* edited by C. G. Drury. DHEW/NIOSH Publication No. 78-185. Proceedings of an international symposium, July 18-20, 1976, Buffalo, N.Y. Cincinnati: National Institute for Occupational Safety and Health, Department of Health, Education, and Welfare, pp. 6-15.

Jones, D. F. 1979. *Back Pain and How To Prevent It the Dynacopic Way.* Publication of Dynacopics, Inc., 17 Brushwood Court, Don Mills, Ontario, Canada M3A 1V9, 13 pages.

Kamon, E., and A. J. Goldfuss. 1978. "In-Plant Evaluation of the Muscle Strength of Workers." *American Industrial Hygiene Association Journal, 39:* pp. 801-807.

Keyserling, W. M., G. Herrin, D. B. Chaffin, T. J. Armstrong, and M. L. Foss. 1980. "Establishing an Industrial Strength Testing Program." *American Industrial Hygiene Association Journal, 41:* pp. 730-736.

McConville, J. T., and H. T. E. Hertzberg. 1966. *A Study of One-Handed Lifting.* AMRL-TR-66-17. Wright-Patterson AFB, Ohio: Aerospace Medical Research Laboratory, 19 pages.

National Institute for Occupational Safety and Health (NIOSH). 1981. *Work Practices Guide for Manual Lifting.* DHHS/NIOSH Publication No. 81-122. Washington, D.C.: U.S. Government Printing Office, 183 pages. Available from American Industrial Hygiene Association, 475 Wolf Ledges Parkway, Akron, Ohio 43311.

Passmore, R., and J. V. G. A. Durnin. 1955. "Human Energy Expenditure." *Physiological Reviews, 35:* pp. 801-840.

Pederson, O. F., R. Petersen, and E. S. Staffeldt. 1975. "Back Pain and Isometric Back Muscle Strength of Workers in a Danish Factory." *Scandinavian Journal of Rehabilitative Medicine, 7:* pp. 125-128.

Petrofsky, J. S., and A. R. Lind. 1978a. "Comparison of Metabolic Circulatory and Ventilatory Responses of Men to Various Lifting Tasks and to Bicycle Ergometry." *Journal of Applied Physiology (REEP), 45 (1):* pp. 60-63.

Petrofsky, J. S., and A. R. Lind. 1978b. "Metabolic, Cardiovascular and Respiratory Factors in the Development of Fatigue in Lifting Tasks." *Journal of Applied Physiology (REEP), 45 (1):* pp. 64-68.

Rehnlund, S. 1973. *Ergonomi.* Translated by C. Soderstrom. Stockholm: AB Volvo Bildungskonconcern, 87 pages.

Rodgers, S. H. 1976. "Metabolic Indices in Materials Handling Tasks." In *Safety in Manual Materials Handling,* edited by C. G. Drury. DHEW/NIOSH Publication No. 78-185. Cincinnati, Ohio: Department of Health, Education and Welfare/National Institute for Occupational Safety and Health, pp. 52-56.

Rodgers, S. H. 1983. *Task 1: Applications of the NIOSH Work Practices Guide for Manual Lifting Guidelines to Fifteen Handling Tasks. Task 2: The Applicability of the NIOSH Work Practices Guide for Manual Lifting to Small Packaged Goods Manufacturing.* NIOSH (APEB) Contract No. 81-3281. Cincinnati: Department of Health and Human Services/National Institute for Occupational Safety and Health, 97 pages (Task 1) and 36 pages (Task 2). Not published.

Rohmert, W. 1960. "Ermittlung von Erholungspausen fur Statische Arbeit des Menschen." *Internationale Zeitschrift fur Angewandte Physiologie (Einschl. Arbeitsphysiologie), 18:* pp. 123-164.

Scherrer, J., and H. Monod. 1959. "Le travail musculaire local et la fatigue chez l'homme." *Journal de Physiologie (Paris), 52:* pp. 419-501.

Snook, S. H. 1978. "The Design of Manual Handling Tasks." *Ergonomics, 21:* pp. 963-985.

Snook, S. H., and V. M. Ciriello 1974. "Maximum Weights and Workloads Acceptable to Female Workers." *Journal of Occupational Medicine, 16:* pp. 527-534.

SUNYAB-IE. 1982/1983. Data from IE (Industrial Engineering) 436/536 laboratory experiments. Taught by S. H. Rodgers at the State University of New York at Buffalo. Paper in Preparation.

Tichauer, E. R. 1971. ''A Pilot Study of the Biomechanics of Lifting in Simulated Industrial Positions.'' *Journal of Safety Research*, 3: pp. 95-115.

Yates, J. W., E. Kamon, S. H. Rodgers, and P. C. Champney. 1980. ''Static Lifting Strength and Maximal Isometric Voluntary Contractions of Back, Arm, and Shoulder Muscles.'' *Ergonomics*, 23 (1): pp. 37-47.

Carrying and Shoveling

CHAPTER 24. CARRYING AND SHOVELING

The transport of materials by carrying and shoveling are special cases in manual handling. Carrying requires sustained effort and shoveling represents uneven loading of the arms and a somewhat unstable load. Guidelines for carrying and shoveling task designs are included in this chapter.

A. CARRYING

The forces developed while lifting an object usually last no more than a few seconds (Nielsen, Eastman Kodak Company, 1978). When an object is carried, the handler must sustain the force for as long as it takes to reach the destination. Thus, the weight that can be carried by hand is less than the amount that can be lifted. The longer the holding time, or the farther it has to be carried, the less weight the handler wishes to carry. Hand load carriage is an inefficient way of transporting materials (Datta and Ramanathan, 1971), but it is common in industries and service organizations. Other types of load carriage, such as backpack, shoulder harness, yoke, or on the head, will not be discussed in these volumes (Datta, Chatterjee, and Roy, 1973; Datta and Ramanathan, 1971; Lind and McNichol, 1968; Soule and Goldman, 1969).

Examples of job tasks that include carrying are:

- Package delivery.
- Sales.
- Warehouse expediting.
- Service repair.
- Inside-building message and mail delivery.

In addition, short carrying tasks exist in most jobs. The National Bureau of Standards (McGehan, 1977) studied one-handed carrying of a suitcase-like box in order to define an upper weight limit for portability. The subjects were a cross-section of the civilian population. The box had to be carried 122 meters (the equivalent of a city block), up a flight of stairs, and through a closed door. A maximum load of 10.7 kg (23.5 lbm) was identified as suitable for 90 percent of the population. Different maximum loads may be appropriate for each distance, type of carry (one- or two-handed), object size, and handhold type.

Figure 24-1 gives the recommended maximum weights to be carried in one or two hands as a function of travel distance. If objects are kept within these limits, a large majority of the work force should be able to carry them without excessive fatigue.

Separate curves are shown for compact and wide objects carried in one hand; wide objects load the shoulder muscles and reduce the amount of weight found acceptable. Curves are shown for both straight- and bent-arm carrying with two hands. Interference of an object with the legs during walking often makes it necessary for a person to carry the load with bent arms. Increased load on the forearm muscles limits the weight that can be carried this way. Most people accept about 2 to 3 kg (4 to 8 lbm) less weight when carrying a load with

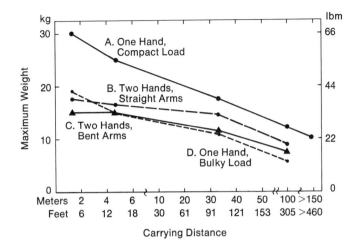

Figure 24-1: Guidelines for Carrying   The relationship between the amount of weight that can be carried (kilograms and pounds, left and right vertical axes, respectively) and the carrying distance (meters and feet, horizontal axis) is shown for four carrying conditions (A through D). In each instance, it is assumed that the object has a good handhold that allows the handler to use a hook or power grip. The one-hand carry can be considerably heavier if the object is compact (A) than if it is bulky (D). Bulky objects put more load on the weaker shoulder muscles. When an object is carried in two hands, more weight can be supported if the arms are straight down in front of the body (B) than if the elbows are bent (C), providing the load is not very deep and does not interfere with walking. These carrying conditions are illustrated in Figure 24-2. The reduced weight with extended distance of carrying can be accounted for by the amount of static loading of the grip, arm, and shoulder muscles. See Part IV for more information. *(Developed from information in Burse, 1978; Carlock, Weasner, and Strauss, 1963; Lind and McNichol, 1968; McConville and Hertzberg, 1966; McGehan, 1977; Rodgers, Eastman Kodak Company, 1969; Snook and Ciriello, 1974).*

bent arms than when they can hold it with their arms extended downward in front of the body (Snook and Ciriello, 1974). Figure 24-2 illustrates the four types of carry represented by the curves in Figure 24-1.

B. SHOVELING

The transport of bulk materials by shoveling is usually a short-term intensive activity in industry. Some examples of shoveling tasks are:

- Ditch digging and construction.
- Furnace stoking.
- Chemical manufacturing operations:

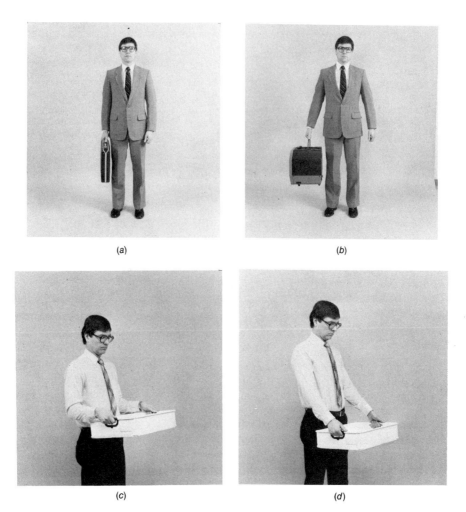

(a)   (b)

(c)   (d)

Figure 24-2: Examples of One-Handed and Two-Handed Carrying
Four illustrations of the carrying tasks referred to in Figure 24-1 are
shown. In *a*, the briefcase is compact and can be held close to the body
using a hook grasp. In *b*, the box is bulky and must be held at least 25
cm (10 in.) away from the side to prevent it from banging against the
legs. This puts a static load on the shoulder muscles and reduces the
amount of weight that can be carried comfortably. In *c*, the person is
supporting the tray with both hands with the elbows bent. Some of
the tray's weight is being supported against the chest, but the biceps
muscles of the arms are actively maintaining the elbow flexion. The
muscle work is considerably less if the tray is carried lower and rested
against the abdomen in a straight-arm carry (*d*). Because the tray is
quite shallow, this carry does not interfere with walking and can be
sustained for at least a minute.

- Loading chemical reactors.
- Preparing wet chemicals for drying.
- Removing chemicals from centrifuges and settling tanks.
- Snow shoveling.
- Landscaping operations.
- Cleaning operations such as ash removal.

1. FACTORS OF CONCERN IN SHOVELING

The development of local muscle strain and fatigue, including postural fatigue from bending over for extended periods, and the total effort required over time are the principal concerns in designing shoveling tasks. Local muscle fatigue will be determined by the percent of strength needed to do the shoveling or to maintain a stooped posture. The shovel used, the heights of the task and how high the shovel must be raised, and the weight of the material on it all affect local muscle fatigue. Material weight per shovel load will be determined by the amount of material to be transferred in the time available, thus determining the frequency of shoveling. Shoveling load, frequency, and height will determine the whole-body energy demands. These along with the demands of other activities in the job and the pattern of work and recovery tasks make up the total energy demands. Factors affecting local muscle and whole-body stress in shoveling are discussed here.

Shovels vary in blade design, handle type, and handle length. They should be selected according to task needs (Lehmann, 1962). Some guidelines for selecting shovels are:

- Use triangular blades and long handles for sand and dry earth.
- Use square blades and shorter handles for coal and stone.
- Since short-handled shovels result in stooping and fatigue of back muscles, provide a long-handled shovel whenever practical. The exception is for coal and stone shoveling where load stability is important.
- Determine the nature of the load and its stability on the shovel blade. A long-handled shovel may be preferred for leverage but may be unsatisfactory because of the muscle effort required to keep the load stable during transfer.
- Shovel weight is part of the total weight that determines the muscle strength requirements. Choose a shovel that is sufficiently strong for the task, but avoid unnecessarily heavy ones. Examples of typical shovel weights are:
  - Small coal shovel—2 kg (4.4 lbm).
  - Large coal shovel—3 kg (6.6 lbm).
  - Snow shovel—1.5 kg (3.3 lbm).

Lighter shovels are also available in aluminum and other alloys. Although there may be shovels that suit certain tasks, one versatile model should be chosen if more than one type of shovel is needed in a job or workplace. Specially designed shovels are appropriate only if the shoveling task is very stereotyped and the tool is not needed for other job duties.

2. GUIDELINES FOR SHOVELING WORK LOAD

The determinants of an acceptable shoveling work load include:

- Shovel weight.
- Height and distance of transfer.
- Nature of material being shovelled.

Table 24-1: Maximum Shoveling Workload for Three Conditions and Two Durations. Guidelines for maximum workload (in kilograms and pounds of material handled) are given for continuous shoveling of one minute (columns 1 and 2) and 15 minutes (columns 3 and 4) under three shoveling conditions (column 5). A range of workloads is given for the one-minute duration since this may vary substantially as frequency and load per blade change during the short task. High lifts reduce the amount of weight per lift and take more time, thereby reducing the maximum load handled in a given time period. Precise placement of the material shovelled also takes more time and may be accompanied by smaller loads per blade so less of the material is spilled. Durations greater than 15 minutes have not been included since shoveling tasks seldom last longer than this on a continuous basis, often being interrupted by another task after a few minutes. *(Developed from information in Kamon, 1973; Lehmann, 1962; Passmore and Durnin, 1955).*

Maximum Amount of Material Transported, Continuous Work				
Per Minute		Per 15 Mintues		
kg	lbm	kg	lbm	Conditions
80–90	175–200	750	1650	Lifts below 100 cm (40 in.), compact, noncritical load placement.
55–65	120–145	530	1165	Lifts routinely above 100 cm (40 in.).
22–33	50–75	245	535	Very precise load placement or shifting load (reduces possible lifting frequency and weight per lift).

- Precision with which the transfer must be made.

- Frequency of lifts per minute.

- Continuous duration of shoveling before a rest break can be taken.

- Amount (and weight) of material that can fit on the shovel blade.

Table 24-1 summarizes guidelines for maximum workloads that should reduce the potential for local arm and shoulder muscle, as well as whole-body, fatigue. A shovel weight of 2.3 kg (5.1 lbm) and a transfer distance of one meter (3.2 ft) were assumed in order to estimate workload suitability. The values per minute are given as ranges, indicating the level of precision of the available data.

The values for 15 minutes of continuous work represent the total amount of material that can be handled in that period of shoveling. Shoulder and arm muscle fatigue prevent people from being able to lift 15 times the amount that can be handled in one minute.

REFERENCES

Burse, R. L. 1978. "Manual Materials Handling: Effect of Task Characteristics of Frequency, Duration, and Pace." In *Safety in Manual Materials Handling*, edited by C. G. Drury. Cincinnati: U.S. Department of Health, Education, and Welfare/National Institute of Occupational Safety and Health, pp. 147-154.

Carlock, J., M. H. Weasner, and P. S. Strauss. 1963. "Portability: A New Look at an Old Problem." *Human Factors, 5:* pp. 577-581.

Datta, S. R., B. B. Chatterjee, and B. N. Roy. 1973. "The Relationship Between Energy Expenditure and Pulse Rates with Body Weight and the Load Carried During Load Carrying on the Level." *Ergonomics, 16 (4):* pp. 507-513.

Datta, S. R., and N. L. Ramanathan. 1971. "Ergonomic Comparison of Seven Modes of Carrying Loads on the Horizontal Plane." *Ergonomics, 14 (2):* pp. 269-278.

Kamon, E. 1973. Personal communication.

Lehmann, G. 1962. *Praktische Arbeitsphysiologie.* 2nd ed. Stuttgart: George Thieme Verlag, pp. 197-206.

Lind, A. R., and G. W. McNichol. 1968. "Cardiovascular Responses to Holding and Carrying Weights by Hand or Shoulder Harness." *Journal of Applied Physiology, 25 (3):* pp. 261-267.

McConville, J. T., and H. T. E. Hertzberg. 1966. *A Study of One-Handed Lifting.* AMRL-TR-66-17. Wright-Patterson AFB, Ohio: Aerospace Medical Research Laboratory, 19 pages.

McGehan, F. P. 1977. "When Is a Product Portable?" In *Dimensions/NBS.* Washington, D.C.: National Bureau of Standards, pp. 16-19.

Passmore, R., and J. V. G. A. Durnin. 1955. "Human Energy Expenditure." *Physiological Reviews, 35:* pp. 801-840.

Snook, S. H., and V. M. Ciriello. 1974. "Maximum Weights and Workloads Acceptable to Female Workers." *Journal of Occupational Medicine, 16:* pp. 527-534.

Soule, R. G., and R. F. Goldman. 1969. "Energy Costs of Loads Carried on the Head, Hands, or Feet." *Journal of Applied Physiology, 27 (5):* pp. 687-690.

Special Considerations in Manual Handling Tasks

CHAPTER 25. SPECIAL CONSIDERATIONS IN MANUAL HANDLING TASKS

A. Drum, Carboy, and Bag Handling

    1. Drum Handling
       a. Metal drums
       b. Nonmetal 208-Liter (55-Gallon) Drums
       c. Fiber Drums
    2. Carboy Handling
    3. Bag Handling
       a. Chemical and Food Bags
       b. Mail Bags

B. Wooden Pallet Handling

C. Large-Size Sheet Handling

D. Two-Person Handling

References for Chapter 25

There are handling tasks in industry that place unique demands on the person doing them. Among these are the handling of drums, carboys, bags, large sheets, and pallets. These are discussed in this chapter. A short analysis of two-person handling is also included.

A. DRUM, CARBOY, AND BAG HANDLING

Liquid and powdered chemicals are packaged in metal, plastic, or fiber drums, carboys, and heavy-walled bags in industry. Examples of items packaged in these containers are:

- Fertilizer and garden chemicals in bags.

- Bulk chemicals, such as water softeners and raw materials for chemical-making operations, in bags or fiber drums.

- Organic solvents in metal drums.

- Inorganic acids and bases in carboys.

- Grains and other foodstuffs in fiber drums or bags.

Figure 25-1 illustrates some of these containers. They come in a wide range of sizes and usually weigh from 11 to 295 kg (25 to 650 lbm).

How these containers are handled depends on their size, weight, number, and the time demands of the operations involved. Where large amounts of the materials are used, bulk containers that can be handled with automatic equipment are preferred. Many operations, however, cannot justify a fully automatic handling system, and handling aids such as tipsters, hand carts, and forklift truck attachments are used instead (Anon., 1979). In this section, some handling guidelines are given for drums. Research in this area has been limited. The recommended limits are based more on observations of where people have noted difficulty in doing a task than with objective measures of task strength requirements.

1. DRUM HANDLING

a. Metal Drums

In American industry, the standard metal drum for liquid chemicals has a capacity of 208 liters (55 gallons). It is usually made from steel and is approximately 0.6 m (2 ft) in diameter and 0.9 m (3 ft) high. As shown in Figure 25-1a, it generally has two hoops for rolling the drum on its side. The top and bottom are recessed, with the edge forming a rim or chime used in moving (chiming) it by hand and in transporting it by hand cart or powered truck. Empty metal drums weigh about 23 kg (50 lbm); filled drums weigh from about 45 to 295 kg (100 to 650 lbm), depending on the density of their contents (Himbury, 1967). One survey of drum weights showed that about 50 percent of the chemicals purchased in metal drums weighed less than 175 kg (385 lbm), and less than 10 percent of them weighed more than 227 kg (500 lbm). About 40 percent of the drums

(a)

(b)

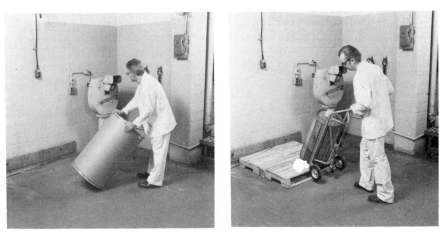

(c)   (d)

Figure 25-1: Examples of Chemical and Food Containers   Five examples of chemical and food containers are illustrated in several handling tasks. A standard 208-liter (55-gallon) metal drum (a) is shown as
it is being lifted on a hoist for weighing. The two chimes around its
circumference (the recessed rim at the top and the beaded rim at the
bottom) are useful places to secure handling equipment and for manually handling the drums (see Figures 25-2 and 25-3). The plastic drum
(b) is less easy to handle because it has a wide rim and few places to
attach a clamp or get a secure grip. The fiber drum (c) has metal rims
at its top and bottom, and at 91 cm (36 in.) is tall enough to chime
manually while the handler stands fairly erect. The carboy on a drum
cart (d) is being placed on a pallet for weighing. It is a heavy glass or
plastic bottle that is packed in an octagonal wooden case that is open
on one end. The case seldom has handholds, so the carboy is handled
from the bottom of the case and around the neck of the bottle.

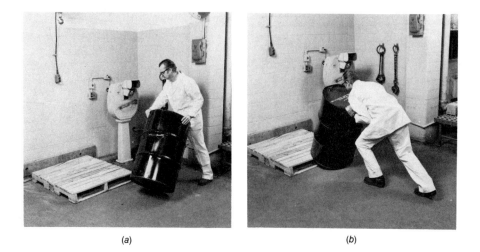

(a)   (b)

Figure 25-2: Manual Handling of Drums   Two examples of the manual handling of a 208-liter (55-gallon) metal drum are shown. In *a*, the handler has tipped the drum onto one side of its lower rim and is rotating, or chiming, it to another location. This is a skill that, once learned, requires little muscular effort. The skilled handler can often chime a metal drum into the appropriate location on a pallet by using the drum's momentum in a fast chiming action. If this is not done properly, however, the drum can get stuck on the edge of the pallet and require a whole-body push to move it into the desired position (*b*). Use of a drum cart (Figure 25-3) reduces the need to develop these large forces and can make drum handling more suitable for less strong workers.

weighed from 180 to 240 kg (396 to 530 lbm) (Alexander, Tennessee Eastman Company, 1978).

There are three primary ways in which metal drums are handled:

- Tipped on edge to get a hand cart underneath them.
- Moved from vertical to horizontal (or vice versa) to assist in emptying their contents or preparing to roll them on their side.
- Rolled on edge (chimed) on the level and on and off pallets, scales, and conveyors (see Figure 25-2).

Examples of hand carts for handling drums are shown in Figure 25-3. In order to get a drum on one of these carts, the handler can either tip the drum up and slide the cart's prongs ("toes") under it, or place the cart's latch over the top rim, which helps to pull the drum over as the cart is tilted back. The heavier the drum, the more force is needed to tip it, and the more space is needed to position the body so that maximal forces can be developed. The hand cart provides leverage and control for the handler. A four-wheel drum cart (Figure 25-3b) is designed for use in handling the drums up onto pallets, scales, or other raised surfaces.

Unless the drums are almost empty, turning them from a horizontal to a vertical orientation is not recommended as a manual task. A tipster, or a forklift truck with a drum-handling attachment, can be used to tip out the contents of the drum. Other techniques for emptying the drums that do not require tipping, such as using a siphon pump, may be preferred. Drum handling by rolling it on edge (chiming) is a skilled activity that requires little strength. Maintaining control of the drum is the primary concern, since the strength needed to retrieve it if it gets away is far in excess of most people's capabilities. Raising the drum 14 to 25 cm (6 to 10 in.) onto a pallet, conveyor, or weighing scale is a less controlled movement if done without an aid, and it is still difficult with a simple drum cart. Workplaces should be designed, wherever possible, to avoid situations where drums have to be chimed between levels. Hydraulic work tables built into the floor help to accomplish this.

The following guidelines are based on observations of metal drum handling tasks:

- Drums weighing more than 115 kg (253 lbm) should be handled with a hand cart or a similar aid.

- Drums weighing more than 227 kg (500 lbm) should be handled with powered equipment, such as a forklift truck with a drum handling attachment.

- Conveyors, pallets, and scales should be recessed in the floor or in a platform where 208-liter (55-gallon) drums are being filled or handled frequently. This reduces the requirement to tip heavy drums up onto the raised surface, and thereby reduces the risk of losing control of the load.

- Air pumps, and other techniques to siphon liquids out of metal drums while they remain upright, if they are safe to use with the chemicals involved, are preferable to having to turn drums on their sides for emptying. A hoist with a scale can be used to raise the drum for emptying; it permits the operator to weigh the amount of liquid delivered as well.

- People who have to handle drums should be trained with the type of drums and the situations they will encounter in the workplace. Frequent short training sessions are preferable to a single long session. Skill appears to develop as a function of the number of drums handled, with more learning picked up between sessions than within them.

(a)

(b)

(c)

Figure 25-3: Drum Carts   A two-wheel drum cart holding a fiber drum is illustrated in *(a)*. This cart has two handles, somewhat like a wheelbarrow. The drum is anchored by a latch that fits over the top rim and can be released once the drum has been transferred to its proper location. The four-wheel drum cart shown in *b* and *c* is more stable and offers handling options that are important for the heavier metal drums. The second set of wheels can be stood on (one foot) in order to tip the cart back for transport *(b)*. The latch that holds the drum on the truck is not visible, but it helps to stabilize the drum cart as the cart's prongs ("toes") are slid under the drum. A horizontal handle forward of the two main vertical cart handles also helps to stabilize the cart during the tilting and maneuvering required to put the drum on a scale for weighing *(c)*.

b. Nonmetal 208-Liter (55-Gallon) Drums

There are increasing numbers of plastic drums being fabricated to carry caustic or corrosive liquids that would deteriorate metal drums (Figure 25-1*b*). These are usually more barrel-like than metal drums, and they often do not have a distinct rim for handling them with a standard drum cart. The frictional resistance of the plastic is low, making them slippery and difficult to handle both manually and mechanically, especially if they are wet. These drums often do not chime as well as the metal ones, and they can be considerably heavier, often more than 227 kg (500 lbm), so manual handling of them is not recommended. The development of new handling equipment to fit these drums, along with improvements in their design, should make handling of them easier in the future.

c. Fiber Drums

Fiber drums (Figure 25-1*c*) are often used to hold dry chemicals and foodstuffs. They may have metallic rims (chimes) at each end or be fabricated entirely from heavy cardboard (Anon., 1979). The former, if approximately 91 cm (36 in.) high, can be handled on standard drum carts; the latter provide no gripping surface and are harder to control on or off a cart. Controlling the drum's motion with the hands produces increased forearm and shoulder muscle stress. This usually results in handlers accepting less weight in fiber than in metal drums.

The height of a fiber drum varies with the amount of chemical desired in the package. Drum diameter is partly determined by pallet size and trailer dimension considerations. Drums less than 76 cm (30 in.) high are difficult to chime; they are more frequently lifted onto carts for transport. Drums taller than 114 cm (45 in.) are also difficult to chime because the top surface is too far above elbow height for many people; these drums, when full, are better handled with aids.

The following guidelines are based on observations of fiber drum handling tasks:

- Drums that will have to be lifted frequently should weigh 18 kg (40 lbm) or less and should be compact. Most people can encircle a 30 cm (12 in.) diameter drum with their arms to provide stability in handling; larger-diameter drums, especially fiber drums that have no grasp points, put increased stress on the shoulders and forearms.

- Fiber drums weighing more than 115 kg (253 lbm) should be handled with aids. Those without rims and handholds should be handled with aids if their weight exceeds 104 kg (225 lbm).

- Where feasible, fiber drums should be securely strapped and transported and handled on pallets. If possible, they should be emptied by vacuum systems or unloaded from a tipster or a forklift truck with a similar attachment.

2. CARBOY HANDLING

Carboys (Figure 25-1d) are glass or plastic bottles encased in a wood or plastic frame, and they often contain inorganic acids and bases. The bottles have a capacity in the range of 7 to 95 liters (2 to 25 gallons), with the 19- to 38-liter (5- to 10-gallon) sizes being used most frequently. When handling a carboy, the operator has to support both the outside frame and the bottle neck; the former has to be positioned for stability and the latter has to be directed for pouring and controlling the flow. Because they often contain corrosive chemicals, precise control is needed.

Transporting carboys involves sliding and, occasionally, lifting them by the frame. They are usually short, less than 61 cm (24 in.) high, and cannot be handled easily without aids. The handling difficulty is further increased by the need to wear gloves to protect against chemical splashes and wooden frame splinters. Since carboys often weigh more than 18 kg (40 lbm), they should be strapped securely to and handled on pallets whenever possible. Siphons may be used to transfer their contents, where safety considerations are met, so that they do not have to be controlled manually. Tipping aids may also be used on occasion (Figure 25-4). Even carboys weighing less than 18 kg (40 lbm) may not be suitable for manual tipping. The time it takes to empty them can be long enough to fatigue the handler's arm and shoulder muscles seriously while he or she is controlling the pour.

3. BAG HANDLING

a. Chemical and Food Bags

Large amounts of powdered and pelletized chemicals and foodstuffs are handled in heavy paper or plastic bags. These bags are often designed to hold 15, 20, 23, 25, 38, 45, or 50 kg (33, 44, 50, 55, 80, 100, or 110 lbm) loads and to lie flat on a pallet or in a box. In the United States, the 23 kg (50 lbm) bag is most common. Bag length is usually in the 46 to 91 cm (18 to 36 in.) range, and widths are usually less than 51 cm (20 in.). The bag may be handled from the top at the gussets (see Figure 25-1e) where the handler can use a hook grasp for low transfers, or at the sides or corners, where more pinching or pressure application is needed (Figure 25-5). Burlap bags are also used for some food and agricultural products. They can sometimes be handled with a metal longshoreman's hook. This eliminates the need to grasp them around the neck or at a corner. Bags that have to be grasped by gathering the fabric or plastic and encircling it may be difficult for people with small grasp spans to handle, especially if the grip circumference exceeds 5 cm (2 in.). For further information, see the section on grip strength in Appendix A.

Bags offer some advantages over fiber drums for dry chemical handling:

Figure 25-4: Tipping Aid for Emptying Bottles or Carboys   A tipping aid is shown that permits a bottle or carboy to be turned towards the horizontal plane in order to empty its contents slowly without requiring its full weight to be supported by the operator. The frame is counterweighted so the operator can control the angle of the tip without having to grasp it tightly. If the transfer had to be made while holding the bottle or carboy, it would be interrupted several times to allow the weight bearing muscles to recover from the static muscle effort. Use of the aid, therefore, can increase productivity by making rest breaks less necessary.

- They are lighter and easier to handle close to the body.
- The lie flat on a pallet or in a trailer and provide efficient loading patterns.
- There are more options for the handler in lifting them: They can be handled from the top, the sides, or the corners.
- The corners can be opened for pouring; they provide a trough for controlling flow when they are emptied.

Figure 25-5: Bag Handling   A two-handed lift of a 23 kg (50 lbm) bag of powdered chemicals is illustrated. The contents of the bag will shift as it is lifted, shown by the indentation under the handler's left forearm. It is difficult to get a firm handhold on the shifting powder, so an open palm combined with a hook or pinch grip is most often used. The difficulty in gripping the bag makes high lifts especially difficult, and the tendency of the contents to shift adds to postural instability in very low lifts.

There are some disadvantages to bag handling, however, that relate to size, weight, and frequency of handling. The factors of most concern are:

- Handling large bags across their width is difficult because they are not rigid. They have to be supported according to their load distribution, not necessarily where the best handholds are.

- Lifting bags from floor level (when they are lying flat) requires a full squat; heavier bags may make it difficult to stand once the bag has been procured.

- Lifting bags over shoulder level is an unstable activity, especially if the load is shifting. There is potential for losing control of the load.

- Bags get damaged more readily than fiber drums.
- Plastic bags are slippery; they become especially difficult to handle if they are wet on the outside.
- The depth of a bag when lying flat is usually 7.5 to 15 cm (3 to 6 in.). This dimension is too wide for an effective power grasp (see Appendix A).
- Bags cannot be handled easily with special forklift truck attachments. Because of this, fiber drums may be more suitable for some operations.

The following guidelines for bag handling are based on studies of acceptable lifting levels and on observations of industrial bag handling:

- Where bags of any weight have to be handled more than 500 times during an eight-hour shift, the workplace should be designed to permit them to be slid. For example, provide roller-bearing tables to reduce the amount of lifting and lowering.
- Bags weighing less than 7 kg (15 lbm) can be handled comfortably by most people. When weight is increased to 11.4 kg (25 lbm), lifts above 127 cm (50 inches) become difficult for weaker people (Snook, 1978).
- Handling can be alternated among several people to decrease the amount each has to do. This will reduce the amount of whole-body fatigue in bag handling tasks. See Part IV for more information.
- Bags weighing 34 kg (75 lbm) or more should not be handled manually. If it is too difficult to handle them automatically, they should be handled below 102 cm (40 in.) and should not be lifted or lowered more than 50 times per eight-hour shift per person. They can be handled more frequently if they can be slid instead of lifted.
- Although bags are often handled from pallets or skids that are 14 cm (5.5 in.) above the floor, handling is much more difficult when an object has to be grasped less than 25 cm (10 in.) above the floor. A pallet load of bags can be placed on a platform or on a second, empty pallet to raise the lower levels up to at least 25 cm (10 in.).

Table 25-1 indicates the maximum number of bags a person should be asked to lift or lower during an eight-hour shift. Four weights and three height ranges are given. These values are meant as a guide only. If an existing situation exceeds these guidelines, it is not necessarily hazardous for the people doing it. However, it is quite probable that there are many people who cannot do the job as constituted, and that some "natural selection" has occurred.

All of the bag weights except 16 kg (35 lbm) shown in Table 25-1 exceed the design guidelines for occasional lifts presented in Chapter 23. Table 25-1 guidelines are for the evaluation of existing workplaces and jobs where it is

Table 25-1: Maximum Number of Bags to Be Lifted per Shift by One Person. The maximum number of bags that one person should be asked to lift per shift is defined in frequency categories ranging from less than 10 to 500 (columns 2 through 5) for four different bag weights (kilograms and pounds). Three lift height ranges are indicated in column 1. Lifts that cross more than one of these ranges should be rated against the lowest values. The number of bags that can be lifted per shift is reduced as bag weight and lift height increase. These maximum numbers are intended for use in evaluating existing jobs where bag weight cannot be altered easily. In designing a new job, either bag weight should be kept within the lifting guidelines given in Figure 23-3, or the workplace should be set up to permit sliding rather than lifting of the bags. *(Derived in part from data in Garg and Saxena, 1980; Snook, 1978).*

Vertical Height Above Floor** cm (in.)	Maximum Number of Bags to Be Lifted per Shift*			
	Weight of Bag, in kg (lbm)			
	16 (35)	25 (55)	34 (75)	45 (100)
25–102 (10–40)	500	250	50	< 10
103–127 (41–50)	250	100	< 10	N.R.
128–152 (51–60)	100	50	< 10	N.R.

*Fixed frequencies are given as an approximation of acceptable workload. They are: less than 10, 50, 100, 250, and 500 per eight-hour shift. To increase the number of units handled per shift, provide for horizontal transfer of the bags by sliding instead of lifting.

**The lift has some component where the hands are within this range. If it crosses two or three ranges, the lower frequency should be used. Lifts below 25 cm (10 in.) are not recommended because of their postural demands, such as squatting.

difficult to alter bag weight. When a new production station is designed and bag weight can be controlled, the following guidelines are recommended:

- Keep bag weight at or below 23 kg (50 lbm).
- Keep the lifts between 51 and 102 cm (20 and 40 in.).

This should make the handling task suitable for more than half of the potential work force. If lifts can be reduced by providing ways to slide the bags, even more people will be accommodated in the bag handling task.

b. Mail Bags

Mail bags deserve special consideration since they may be provided with a handle at the top and are more cylindrical than flat when filled. Lifting them from the floor to a 76 cm (30 in.) high sorting table a few times an hour is a common occurrence in many mail rooms. When asked to determine maximum acceptable loads for this lift, male college students chose an average of 27 kg (60 lbm) for a 56 cm wide by 89 cm long (22 by 35 in.) mail bag (Garg and Saxena, 1980). Based on tray-lifting acceptance studies with men and women, Snook (1978) has established that an 18 kg (40 lbm) load is the upper limit for handling mail bags to the sorting table. This level will accommodate the capacities of more than 75 percent of the potential work force. When heavier bags are used, provision should be made for workplace aids to minimize the bag lifting and lowering requirements.

B. WOODEN PALLET HANDLING

Wooden pallets are used throughout industry to move and store supplies between receiving, manufacturing, and distribution areas. Pallet size varies despite attempts at standardization (Cook, 1975). A 102 × 122 cm (40 × 48 in.) pallet is common in the United States, and 81 × 122 or 81 × 102 cm (32 × 48 or 32 × 40 in.) pallets are common in Europe. U.S. wooden pallets often weigh from 23 to 45 kg (50 to 100 lbm). Their weight and dimensions make them very awkward to lift and lower; splinters in the hands, contusions to the feet and lower legs, and shoulder and back strains and sprains have all been associated with manual pallet handling.

Most people handle pallets by sliding them on edge onto and off stacks (Figure 25-6). The higher the stack, the less controlled the drop or lift of the pallet, and the more potential for accidents. In Figure 25-6 a and b, the handler takes pallets off a stack by sliding them to the edge and controlling the drop. In Figure 25-6 c and d, the pallets are stacked by tilting them down from the upright position. The handler never has to support the full weight of the pallet when using these techniques.

Whenever possible, pallets should be stacked and unstacked using forklift trucks. When this is not feasible, it is usually the policy to require two people to handle them, especially when stacking them. This policy is appropriate but not always possible, since there is not always a person nearby when a pallet has to be moved. Pallet dispensers are available to provide consistent handling heights. These are helpful in areas where pallets are used frequently, such as the end of manufacturing and packaging lines. Smaller pallets, often plastic, can be used in some areas. They are often half the size and weight of the standard U.S. pallet, making them much easier to handle.

When wooden pallets are used infrequently (less than two per hour) and forklift trucks are not conveniently available for handling them, the following guidelines are recommended to reduce the potential for handling accidents (Stiner, Eastman Kodak Company, 1981):

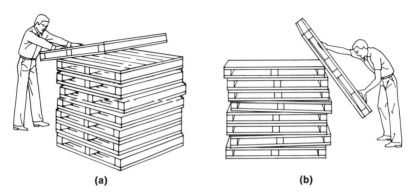

Unstacking Pallets

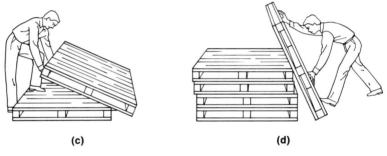

Stacking Pallets

Figure 25-6: Pallet Handling Techniques   Four illustrations of un-stacking (*a, b*) and stacking (*c, d*) wooden pallets are presented. The standard U.S. wooden pallet is 102 cm (40 in.) wide, 122 cm (48 in.) long, and 14 cm (5.5 in.) high. It often weighs 27 kg (60 lbm) or more and is very awkward to lift. The two open ends and the gaps between the slats of 2 to 5 cm (1 to 2 in.) provide the only handling sites. The techniques illustrated feature sliding rather than lifting (*a, c*), using hook grasps through the slat openings to control the pallet's motion (*b, d*). *(Stiner, Eastman Kodak Company, 1981).*

- Individuals should not stack pallets more than six high.
- Individuals should not unpile pallets from a stack more than nine high.
- If a second person is available, pallets can be stacked to nine high and unstacked from twelve high.
- Pallets should always be placed right-side up and with the wood grain aligned for easy sliding.

Alternatives to pallets are under development in order to reduce transportation costs and speed up loading of shipping containers, trucks, and railroad cars (Clark Equipment Company, 1980). Disposable and plastic pallets, slip sheets, triwall containers, and shrink-wrap packaging help to reduce the need for wooden pallets. Although the technology still needs improvement, these approaches are expected to replace the heavier pallets in time. By reducing heavy handling, they should help to reduce manual handling injuries. Product damage and difficulty in unloading products that are not on pallets are technological problems that must be solved before pallets are replaced in manufacturing and distribution operations. Metal skids (107 × 102 × 23 cm, or 42 × 40 × 9 in.) are used for the internal movement of a product by powered equipment. They are easier to slide the product onto and off of than pallets because they are not slatted.

C. LARGE-SIZE SHEET HANDLING

In construction and some manufacturing activities, large sheets of wood, metal, glass, cardboard, or plastic have to be handled. These sheets have no handholds and are too wide and too long to be carried under the arm or handled with a hook grasp. They are often less than 2.5 cm (1 in.) thick, making pinch grasp inefficient and inadequate to support their weight. Examples of materials presenting these handling difficulties are:

- Plate glass.
- Plywood, gypsum, dry wall, fiberboard, and similar construction materials.
- Wall panels.
- Metal and lithographic plates.
- Display signs, posters, and paintings.

If the weight of one of these items is less than 13 kg (29 lbm), a person should be able to carry it for short distances using a handling aid (Figure 25-7a). This aid provides support for the sheet while allowing the handler to support it with a power grip on a D- or J-type handle. Very long or wide sheets may not be easily controlled with this aid since the handle has to be placed at the balance point of the load.

Sheets weighing more than 13 kg (29 lbm) may fatigue the handler (Rigby, 1973). Other transport aids, such as carts or trucks, should be provided to handle heavier sheets when feasible. Two-person handling of sheet materials is also common. If sheet weight exceeds 20 kg (44 lbm), however, arm and shoulder fatigue may limit the effective carrying time and distance.

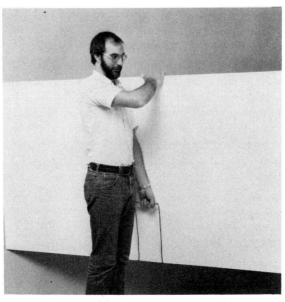

(a)

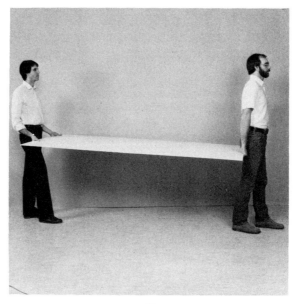

(b)

Figure 25-7: Large-Size Sheet Handling   The handling of large sheet materials, such as plywood, glass, or paneling, is illustrated for one- and two-person tasks (*a*, *b*). A handling aid is used in one-person carrying. It hooks under the sheet and provides a handle on which the handler can get a power or hook grasp. The other hand is used to steady

D. TWO-PERSON HANDLING

There are occasional handling tasks that cannot be done conveniently with materials handling aids, but where the objects to be handled are too bulky or too heavy to be handled by one person. Examples of these activities are:

- Handling long metal pipes and other supplies on construction sites.
- Handling large plates, sheets, or panels to and from transport trucks or work tables (see Figure 25-7b).
- Accurately positioning large parts in assembly tasks, as in vehicle assembly.
- Carrying a loaded stretcher downstairs.
- Moving large burlap bags of loose material weighing about 45 kg (100 lbm).
- Moving furniture.

It is preferable to improve the handling of large objects by eliminating the need for carrying or lifting them at all (see the section on the systems approach to handling in Chapter 19). Objects weighing more than 27 kg (60 lbm) (Snook, 1978) should be slid, rather than lifted, by the two-person team. For sheet handling, the upper weight limit is 20 kg (44 lbm) because of the difficulty of grasping them. The reasons for these weight limits are:

- A second person may not always be available. If the task needs to be done and help is not available, the individual will probably try to do it alone. Objects weighing more than 27 kg (60 lbm) would be outside of the strength capabilities of many people, and they may increase the risk of overexertion injury during handling.
- A two-person handling team may not be matched for size and strength. Thus, one person often will be carrying more than half of the load. When carrying a load downstairs or on the lower side of a sloped area, one person sometimes may be supporting almost all of the load.

---

the sheet at its top. This aid is useful for short-distance carrying in an area where air velocity is low. Special carts that have an A-frame against which the sheets can be rested are also available, but they cannot be used easily outside where the ground is uneven. The use of two people to handle the sheet material (b) may be preferable for outside work or when extended carries are needed. Lack of good handholds may limit the carrying time, however, especially if the task is done frequently over short time periods.

For extended carrying or highly repetitive lifting, objects should weigh 14 kg (30 lbm) or less. If the handling is done in a very dynamic or ballistic manner—that is, if the object is moving horizontally or upward when it reaches the handlers or if the period of strength exertion does not exceed one second per lift—heavier objects may be acceptable.

If feasible, the handles on objects lifted by two people simultaneously should be designed according to the guidelines given in Chapter 20. To be comfortably handled by two people the handles should be at least 25 cm (10 in.) apart, allowing the handlers to flex their elbows. If the same object is occasionally handled by one person, a separate handle should be located on it so that the load is balanced and will not interfere with walking during carrying (McGehan, 1977). If the objects do not have handles (for example, building supplies), the interface with the hands should be made as secure as possible. This may be done by providing protective gloves or aids like clamps or a piece of line to the handler that can reduce the need for gripping the object tightly and that will shift the weight to the arm and shoulder muscles instead. Gloves should be carefully selected because they reduce power grip strength (see Part IV).

In situations like construction work where two-person lifting is often done, it is important to match team members by size and strength. Placing a weak person on a team with a much stronger one is less appropriate than teaming two of the weaker people.

All team members should be trained in proper lifting techniques and taught to communicate throughout their handling tasks, starting and ending the lift on command. Extended carrying of heavy loads (more than one minute) by a two-person team is not recommended since the accumulated muscle fatigue may result in one person being disproportionately loaded.

REFERENCES

Anon. 1979. "Handling the 55-Gallon Drum." *Modern Materials Handling, 33:* pp. 66-73, 120-122.

Clark Equipment Company. 1980. "Slip-Sheet Handling Cuts Container Loading Costs - 95%!" *Modern Materials Handling, 34 (April 1980):* p. 111.

Cook, K. G. (USPS-PTR). 1975. *Human Factors and Related Design Aspects of Pallets.* Technical Note PTR-H-433-75-3. Research and Development Department, Postal Technology Research. Washington, D.C.: U.S. Postal Service, 51 pages.

Garg, A., and U. Saxena. 1980. "Container Characteristics and Maximum Acceptable Weight of Lift." *Human Factors, 22 (4):* pp. 487-495.

Himbury, S. 1967. *Kinetic Methods of Manual Handling in Industry.* ILO: Occupational Safety and Health Series No. 10. Geneva: International Labour Office, 38 pages.

McGehan, F. P. 1977. "When Is a Product Portable?" In *Dimensions/NBS.* Washington, D.C.: National Bureau of Standards, pp. 16-19.

Rigby, L. V. 1973. "Why Do People Drop Things?" *Quality Progress, 6 (9):* pp. 16-19.

Snook, S. H. 1978. "The Design of Manual Handling Tasks." *Ergonomics, 21:* pp. 963-985.

PART VII

Appendices

CONTRIBUTING AUTHORS

David M. Kiser, Ph.D., Physiology

Suzanne H. Rodgers, Ph.D., Physiology

This section contains information that can be used to evaluate biomechanical and metabolic workloads as well as methods for measuring human strength and aerobic work capacities and for studying physiological job demands. Case studies with suggested solutions are also included to illustrate how information in this book can be used to address workplace problems. Data on industrial populations have been used whenever possible, but they are not available for some of the strength and aerobic capacity measures of interest.

Application of these data to the evaluation of an existing workplace problem or the design of a new job or workplace is seldom straightforward. The guidelines found in Parts IV and VI for patterns of work and manual materials handling as well as those for workplace and equipment design in Volume 1 of this series have been developed from information about job demands and human capabilities.

Appendix A: Human Capacities and Job Demands

CHAPTER 26. APPENDIX A: HUMAN CAPACITIES AND JOB DEMANDS

A. Anthropometric Information for Biomechanical Analyses

    1. Body Size and Segment Length
    2. The Masses and Centers of Gravity of Body Segments
    3. The Ranges of Joint Motion
    4. Muscle Cross-Sectional Areas

B. Muscle Strength Data

    1. Forearm Flexion and Extension
    2. Forearm Rotation and Movements of the Hand at the Wrist
    3. Grip Strength
    4. Shoulder Flexion Strength
    5. Isometric Lifting Strengths and Push-Pull Forces
    6. Male-Female Strength Comparisons

C. Aerobic Work Capacity

    1. Whole-Body Maximum Aerobic Capacities
    2. Upper Body Maximum Aerobic Capacity
    3. Maximum Aerobic Capacities for Lifting Tasks

D. The Energy Requirements of Occupational Tasks

    1. Light Effort
    2. Moderate Effort
    3. Heavy Effort
    4. Very Heavy Effort
    5. Extremely Heavy Effort

References for Chapter 26

The first part of this chapter includes information about body size, segment weights, ranges of joint motion, and cross-sectional areas of muscles. This information can be used to estimate lever arms and counteractive forces in static and dynamic work postures and manual handling tasks.

The second part gives data on the strengths of several muscle groups that are actively involved in many industrial tasks, especially manual handling jobs. Torques are indicated for arm muscles involved in rotating the hand at the wrist. This information can be used to evaluate the percent of maximum voluntary strength that a particular task requires.

The third part summarizes data on the maximum aerobic work capacities of an industrial population. Values are given for whole-body work (treadmill) as well as for upper body work (a lifting task). Comparisons to data from non-industrial populations are given for bicycle ergometer and other lifting and arm cranking tasks.

The fourth part provides information on the energy expenditure demands of common industrial tasks. These tables show equivalent metabolic demands for a variety of tasks and can be used to estimate workload in situations where measurements cannot be made easily.

A. ANTHROPOMETRIC INFORMATION FOR BIOMECHANICAL ANALYSES

An introduction to the biomechanical analysis of work can be found in Part III. The techniques presented there are useful in evaluating many manual materials handling problems. Task variables that are difficult to measure directly must be estimated in order to calculate biomechanical stresses. Fortunately, these estimates can be obtained from basic anthropometric measurements, body-segment mass values, and the locations of body-segment centers of gravity. The range of motion of the major joints of the body and the cross-sectional area of several muscle groups are also included in this section.

1. BODY SIZE AND SEGMENT LENGTH

Some anthropometric data on body size are shown in Tables 26-1 and 26-2. (Volume 1 of this series contains a more thorough discussion of anthropometric data and should be consulted for additional information.) These data can be used to evaluate the muscles available for lifting at different heights above the floor. Below waist level, for example, most people use leg and back muscles to assist in lifting an object. Above shoulder height, most of the lifting is done with the arm and shoulder muscles.

Another way to use anthropometric information is in determining the length of lever arms in a biomechanical analysis of lifting tasks or work postures. Figures 26-1 and 26-2 show the lengths of various body segments as a proportion of height for males and females. It is important to recognize that these proportions are usually based on average-value percentages of the populations of interest. There is a large variation in individual proportions, so any calculations

Table 26-1: Anthropometric Size Data—Centimeters. Fifteen anthropometric measurements that are useful in the design of manual handling tasks and workplaces are shown in column 1. The average (50th percentile) values plus or minus one standard deviation (SD) are given for men (column 2) and women (column 3). The 5th, 50th, and 95th percentile values are given for a statistically combined population of men and women (50-50 mix) in columns 4 through 6. See also Table VIA-2, Appendix A, Volume 1. *(Adapted from NASA, 1978).*

Measurements	Males 50th ± 1 SD	Females 50th ± 1 SD	50-50 Mix		
			5th	50th	95th
Forward functional reach, acromial process to functional pinch	68.3 ± 4.3	62.5 ± 3.4	57.5	65.0	74.5
Eye height, standing	164.4 ± 6.1	151.4 ± 5.6	144.2	157.7	172.3
Shoulder height, standing	143.7 ± 6.2	132.9 ± 5.5	124.8	137.4	151.7
Elbow height, standing	110.5 ± 4.5	102.6 ± 4.8	96.4	106.7	116.3
Waist height, standing	106.3 ± 5.4	101.7 ± 5.0	94.9	103.9	113.5
Tibial height, standing	45.6 ± 2.8	42.0 ± 2.4	38.8	43.6	49.2
Functional overhead reach, standing	209.6 ± 8.5	199.2 ± 8.6	188.0	204.5	220.8
Upper leg length, seated	59.4 ± 2.8	57.4 ± 2.6	53.7	58.4	63.3
Elbow to fist length	38.5 ± 2.1	34.8 ± 2.3	31.9	36.7	41.1
Upper arm length	36.9 ± 1.9	34.1 ± 2.5	31.0	35.7	39.4
Popliteal height, seated	44.6 ± 2.5	41.0 ± 1.9	38.6	42.6	47.8
Hand length	19.0 ± 1.0	18.4 ± 1.0	17.0	18.7	20.4
Hand spread, digit one to digit two, second phalangeal joint	10.5 ± 1.7	8.1 ± 1.7	5.9	9.3	12.7
Grip breadth, inside diameter	4.9 ± 0.6	4.3 ± 0.3	3.8	4.5	5.7
Height	174.5 ± 6.6	162.1 ± 6.0	154.4	168.0	183.0

made can only be considered estimates. Designs based on these estimates should be simulated prior to implementation to test their appropriateness.

2. THE MASSES AND CENTERS OF GRAVITY OF BODY SEGMENTS

The masses of different body segments are also needed to estimate postural torques and lifting stresses on the body. Table 26-3 presents the slopes and constants for a series of regression equations that can be used to predict body-seg-

Table 26-2: Anthropometric Size Data—Inches. The information found in Table 26-1 is given in inches. See also Table VIA-3, Appendix A, Volume 1 of this series. *(Adapted from NASA, 1978).*

Measurements	Males 50th ± 1 SD	Females 50th ± 1 SD	50-50 Mix 5th	50-50 Mix 50th	50-50 Mix 95th
Forward functional reach, acromial process to functional pinch	26.9 ± 1.7	24.6 ± 1.3	22.6	25.6	29.3
Eye height, standing	64.7 ± 2.4	59.6 ± 2.2	56.8	62.1	67.8
Shoulder height, standing	56.6 ± 2.4	51.9 ± 2.7	48.4	54.4	59.7
Elbow height, standing	43.5 ± 1.8	40.4 ± 1.4	38.0	42.0	45.8
Waist height, standing	41.9 ± 2.1	40.0 ± 2.0	37.4	40.9	44.7
Tibial height, standing	17.9 ± 1.1	16.5 ± 0.9	15.3	17.2	19.4
Functional overhead reach, standing	82.5 ± 3.3	78.4 ± 3.4	74.0	80.5	86.9
Upper leg length, seated	23.4 ± 1.1	22.6 ± 1.0	21.1	23.0	24.9
Elbow to fist length	14.2 ± 0.9	12.7 ± 1.1	12.6	14.5	16.2
Upper arm length	14.5 ± 0.7	13.4 ± 0.4	12.9	13.8	15.5
Popliteal height, seated	17.2 ± 1.0	16.2 ± 0.7	15.1	16.6	18.4
Hand length	7.5 ± 0.4	7.2 ± 0.4	6.7	7.4	8.0
Hand spread, digit one to digit two, second phalangeal joint	4.1 ± 0.7	3.2 ± 0.7	2.3	3.6	5.0
Grip breadth, inside diameter	1.9 ± 0.2	1.7 ± 0.1	1.5	1.8	2.2
Height	68.7 ± 2.6	63.8 ± 2.4	60.8	66.2	72.0

ment mass from total body mass. Since muscle force and energy are required to move body segments, reasonably accurate biomechanical calculations require inclusion of segmental mass in the torque calculations. An estimate of the center of gravity is also required for torque calculations since the distance between the center of gravity of the segment and the joint where movement occurs defines the length of the moment arm for segmental torque calculations (see Part III).

The approximate locations of body-segment centers of gravity are shown in Figure 26-3 as a percentage of segment length from each joint. These values can be used to calculate the torque required to move the segment. They are especially important for calculating the load on the muscles during lifting tasks performed in different working postures. In general, the center of gravity of a segment is located slightly less than halfway (approximately 4/9) from the joint that is closer to the trunk.

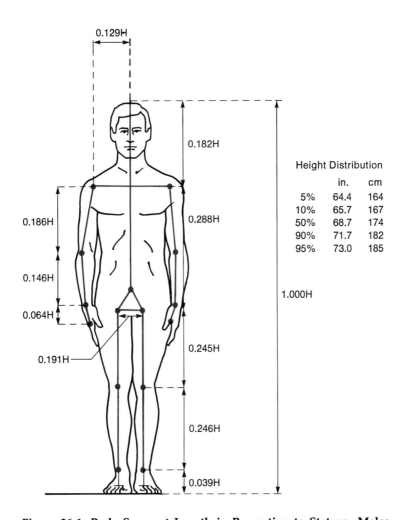

Figure 26-1: Body-Segment Length in Proportion to Stature—Males.
This illustrates the major body segments for males. The equations to
predict the length of arm and leg segments from height (H) for American males appear next to each segment. The height distribution for a
mixed population of males is shown in the upper right corner. An
estimate of body-segment lengths is required for calculations of torque.
Measured values should be used in workplace design instead of estimates whenever possible. *(Adapted from NASA, 1978).*

3. THE RANGES OF JOINT MOTION

Table 26-4 summarizes the motions of the major joints of the body and shows
the ranges of motion around these joints. The motions are illustrated in Figures
26-4 through 26-8. Muscle strength may be considerably less near the extreme

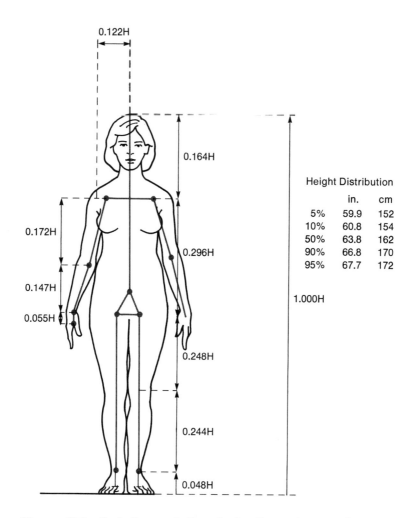

0.122H

0.164H

0.172H

0.296H

0.147H

0.055H

0.248H

0.244H

0.048H

Height Distribution

	in.	cm
5%	59.9	152
10%	60.8	154
50%	63.8	162
90%	66.8	170
95%	67.7	172

1.000H

Figure 26-2: Body-Segment Length in Proportion to Stature— Females This illustrates the major body segments for females. The equations to predict the length of arm and leg segments from height (H) for American females appear next to each segment.The height distribution for a mixed population of females is shown in the upper right corner. *(Adapted from NASA, 1978).*

of the range because of both physiological and biomechanical factors (see Parts II and III).

4. MUSCLE CROSS-SECTIONAL AREAS

The amount of strength that can be generated by a muscle is directly related to its cross-sectional area. Table 26-5 shows the approximate cross-sectional areas of six muscles. The cross-sectional area is estimated by measuring the girth of

Table 26-3: Regression Equations for Estimating the Mass of Body Segments. Column 1 shows the body segments of interest in biomechanical analyses. The slope (a), which is common to both kg and lbm calculations, is shown in column 2. Columns 3 and 4 give the constants for the equation in kg and lbm, respectively; columns 5 and 6 give the standard errors of estimate. Example: To estimate the mass of both upper extremities, in kg, the equation to use is 0.13 times the total body weight in kg minus 1.4 kg, with a standard error of ± 1.0 kg. (Adapted from Barter, 1957).

Body Segment	Slope (a)	Constant (b) kg	Constant (b) lbm	Standard Error of Estimate kg	Standard Error of Estimate lbm
Head, neck, and trunk	0.47	+5.4	+12.0	±2.9	±6.4
Total upper extremities	0.13	−1.4	−3.0	±1.0	±2.1
Both upper arms	0.08	−1.3	−2.9	±0.5	±1.0
Forearms plus hands	0.06	−0.6	−1.4	±0.5	±1.2
Both forearms	0.04	−0.2	−0.5	±0.5	±1.0
Both hands	0.01	+0.3	+0.7	±0.2	±0.4
Total lower extremities	0.31	+1.2	+2.7	±2.2	±4.9
Both upper legs	0.18	+1.5	+3.2	±1.6	±3.6
Both lower legs plus feet	0.13	−0.2	−0.5	±0.9	±2.0
Both lower legs	0.11	−0.9	−1.9	±0.7	±1.6
Both feet	0.02	+0.7	+1.5	±0.3	±0.6

Segment Mass = (a) × (Total Body Weight in kg or lbm) + (b)

the muscle at its widest part (the "belly" of the muscle) and calculating the area by assuming that the muscle is cylindrical at that point. The deltoid muscle is not longitudinal and does not fit these assumptions; its cross-sectional area has been estimated from a computerized axial tomography (CAT) scan.

B. MUSCLE STRENGTH DATA

The voluntary muscle system is made up of approximately 434 muscles of which about 75 pairs are involved in movement and postural control of the body (Rasch and Burke, 1978). This explains the complexity of analyzing muscular work and the difficulty of identifying the muscles that contribute to movement and of evaluating their capacities to develop force. As we will discuss in Appendix B under strength testing, the muscle strengths specific to a particular task should be measured. If this is not possible, one must rely on the measurements that have been published in the scientific literature. When using data from these studies, the same precautions should be applied as when designing a strength-

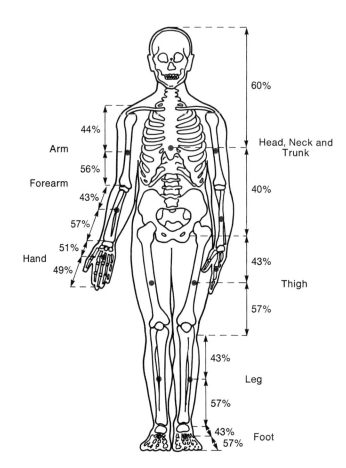

Figure 26-3: Estimated Body-Segment Centers of Gravity Expressed as a Percent of Segment Length The locations of body-segment centers of gravity are presented as percentages of segment length. These centers of gravity must be estimated for torque calculations since the mass of a segment results in a force acting through the segment's center of gravity. The distance from the proximal joint to the center of gravity defines the moment arm. For example, the center of gravity of the upper arm is 44 percent of the way towards the elbow from the shoulder. One can use this to define the moment arm for the upper arm and, thus, estimate the torque in handling tasks or other arm activities. (*Adapted from Dempster, 1955; Williams and Lissner, 1962*).

testing study. The following questions should be asked before using the data to solve an ergonomics problem:

- What subject population was tested?
- What was the subject's position during the measurements—that is, what

Table 26-4: Normal Ranges of Joint Motion. The average ranges of motion (columns 3 and 4, in degrees and radians) of eight joints (column 1) are shown for several types of motion (column 2). The motions are illustrated in Figures 26-4 through 26-8. (*Adapted from American Academy of Orthopedic Surgeons, 1965*).

Joint	Motion	Range of Motion*, degrees (radians)	
Elbow	Flexion to extension	150	(2.62)
	Hyperextension	10	(0.17)
Forearm	Pronation	80	(1.48)
	Supination	80	(1.48)
Wrist	Flexion	80	(1.40)
	Extension	70	(1.22)
	Radial deviation	25	(0.44)
	Ulnar deviation	40	(0.70)
Shoulder	Abduction	180	(3.14)
	Adduction	75	(1.31)
	Forward flexion	180	(3.14)
	Backward extension	60	(1.05)
	Horizontal flexion	130	(2.27)
	Horizontal extension	50	(0.87)
Cervical spine	Flexion	45	(0.78)
	Extension	45	(0.78)
	Lateral bending	45	(0.78)
	Rotation	60	(1.05)
Lumbar spine	Flexion	80	(1.40)
	Extension	20–30	(0.35–0.52)
	Lateral bending	35	(0.61)
	Rotation	45	(0.78)
Knee	Flexion	135	(2.36)
	Hyperextension	10	(0.17)
Ankle	Flexion (plantar flexion)	50	(0.87)
	Extension (dorsiflexion)	20	(0.35)

*The values represent averages; individual values may vary.

were the locations of the limbs? Was the person standing or seated?
- Were maximum measurements made?
- Was the measurement static or dynamic?
- If dynamic, what type? Isotonic (concentric or eccentric) or isokinetic?
- If dynamic, what was the velocity?
- How long was the effort sustained, and was the reading a peak or an average value?

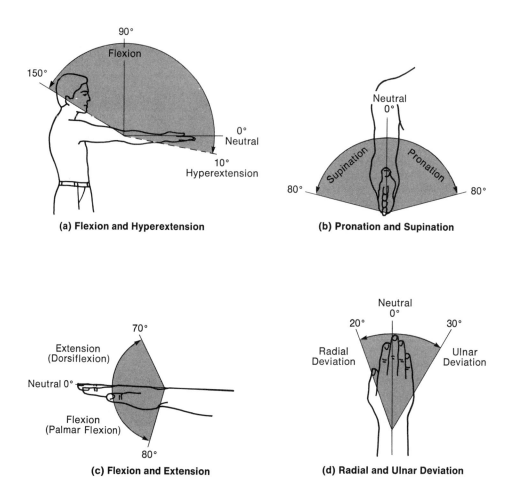

Figure 26-4: Ranges of Motion of the Forearm and Wrist Two illustrations of forearm motion (*a* and *b*) and two of wrist motion (*c* and *d*) are shown. Flexion and hyperextension of the elbow are shown in *a*, with 0 degrees marking the horizontal plane when the arm is stretched out in front of the body. Forearm pronation and supination are illustrated in *b*, where the neutral point (0 degrees) marks the vertical position of the hand. Palm down is pronation and palm up is supination. Flexion and extension (dorsiflexion) of the wrist are illustrated in *c*. The neutral point is the horizontal line across the top of the hand when it is fully extended. Flexion is towards the palm, and extension is towards the back of the hand. Radial and ulnar deviations of the wrist are illustrated in *d*, where the neutral point is through the center of the hand parallel with the middle finger of the outstretched hand. Radial deviation is towards the thumb, and ulnar deviation is towards the little finger. The wrist is kept in the neutral position with regard to flexion and extension during the testing of ulnar and radial deviations. (*Adapted from American Academy of Orthopaedic Surgeons, 1965*).

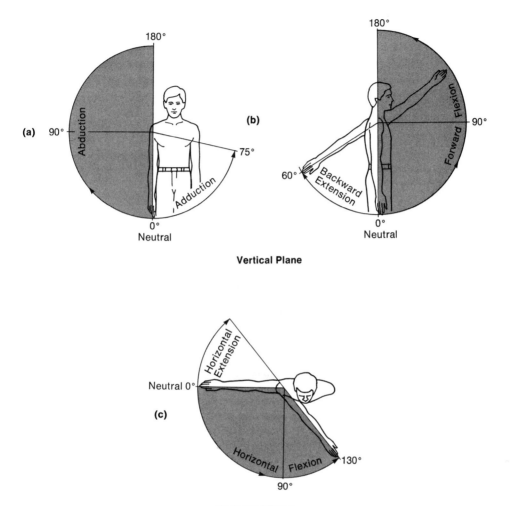

Figure 26-5: Ranges of Motion of the Arm and Shoulder   Two ranges of motion of the arm and shoulder in the vertical plane (*a* and *b*) and one range of motion in the horizontal plane (*c*) are shown. Abduction, or the movement of the arm away from the body and to the side, is illustrated in *a*. Adduction, raising the arm across the front of the body, is also shown in *a*. Forward flexion and backward extension of the arm and shoulder, where the movements are towards the front and rear (not the side) of the body, are illustrated in *b*. In both *a* and *b* the neutral point is the arm hanging down at the side of the body. Horizontal flexion and extension of the arm and shoulder, where the neutral point is with the arm extended to the side, are illustrated in *c*. Flexion is movement toward the front of the body and extension is movement toward the back with the trunk remaining in the forward-facing posture. (*Adapted from American Academy of Orthopaedic Surgeons, 1965*).

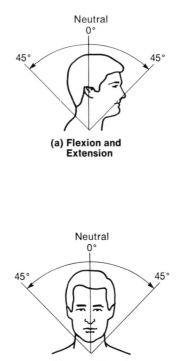

(a) **Flexion and Extension**

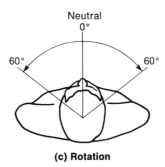

(b) **Lateral Bend**

(c) **Rotation**

Figure 26–6: Ranges of Motion of the Cervical Spine and Spine Rotation Three examples of motions of the cervical spine (head movements) are shown in *a*, *b*, and *c*. The neutral point is a vertical line representing erect posture. Flexion (bending the head forward) and extension (arching it backwards) are illustrated in *a*, and lateral bending of the cervical spine is shown in *b*. Rotation of the head at the cervical spine, as in shaking the head to signify "no," is illustrated in *c*. *Adapted from American Academy of Orthopaedic Surgeons, 1965).*

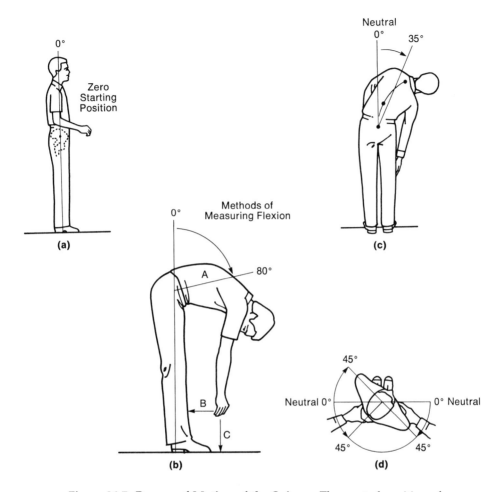

Figure 26-7: Ranges of Motion of the Spine   The neutral position of the spine is illustrated in *a,* with the vertical plane passing through the body's center of gravity. The ranges of motion for three levels of forward flexion of the trunk and its spinal segments are shown in *b.* A spinal flexion of 80 degrees is available if one measures the inclination of the trunk at the lumbar spine (A) from the neutral point (0 degrees). Additional information about spinal flexibility can be obtained if the horizontal distance from the finger tips to the legs (B) and the height of the finger tips above the floor (C) are also noted. Lateral bending of the thoracic and lumbar spine is illustrated in *c.* A range of 35 degrees from the vertical neutral plane passing through the body's center of gravity is available on each side of the spine. Rotation of the whole spine, similar to twisting the trunk while keeping the feet pointed directly ahead, is illustrated in *d.* The neutral point is represented by a vertical plane that passes through the body's center of gravity and divides the body into front and back halves. (*Adapted from American Academy of Orthopaedic Surgeons, 1965).*

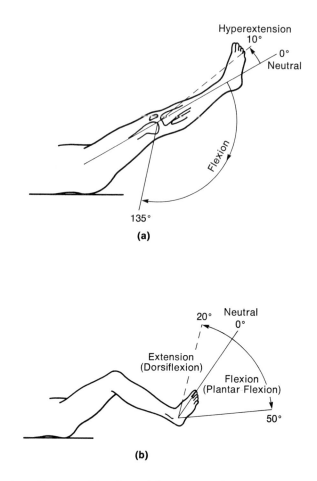

Figure 26-8: Ranges of Motion of the Knee and Ankle   The range of motion of the knee in flexion and hyperextension is illustrated in *a*. Using a neutral point of full leg extension, there are 135 degrees of flexion and 10 degrees of hyperextension ("locked knee") available. The ankle joint's extension and flexion range is illustrated in *b*; the neutral point is taken with the knee at about 90 degrees and the foot at 90 degrees from the lower leg. Flexion is towards the sole of the foot (plantar flexion), and extension (dorsiflexion) is towards the knee. (*Adapted from American Academy of Orthopaedic Surgeons, 1965*).

In many cases, a close evaluation will show that the published studies do not adequately describe the task conditions during data collection. However, by applying the physiological and biomechanical principles developed in Parts II and III, one can make adjustments to the literature values to make them more applicable to specific tasks in industry. This section includes summaries of selected muscle strength data both from the literature and from some industrial

Table 26-5: The Approximate Cross-Sectional Areas of Selected Muscle Groups. The estimated cross-sectional areas of three muscles of the arm and shoulder and three leg muscles (column 1) are shown in column 2. Average values plus or minus one standard deviation are given in square centimeters (cm²). The populations tested and the number in each group, where known, are given in column 3. The source of the information is cited in column 4. Note: The force per cross-sectional area has been reported to be between 20 N/cm² and 100 N/cm². Values of 30 to 40 N/cm² are most common. *(Brunnstrom, 1972; Haxton, 1944; Schantz, et al, 1983; Winter, 1979).*

Muscle	Cross-Sectional Area (cm²)	Population (n)	Reference
Upper Body			
Triceps brachii	31.1 ± 1.6	Male phys. ed. students (8)	Schantz et al., 1983
	19.0 ± 0.7	Female phys. ed. students (6)	Schantz et al., 1983
Biceps	23.5 ± 1.2	Male phys. ed. students (8)	Schantz et al., 1983
	14.4 ± 0.6	Female phys. ed. students (6)	Schantz et al., 1983
Deltoid (lateral)	25.3	Not given	Brunnstrom, 1972
Lower Body			
Quadriceps femoris	88.3	Male phys. ed. students (8)	Adapted from Schantz et al., 1983
	66.5	Female phys. ed. students (6)	Adapted from Schantz et al., 1983
Soleus	67	Not given	Winter, 1979
Gastrocnemius	35	Not given	Winter, 1979

studies not yet published. The intent is not to make a comprehensive review of the strength-testing literature but to provide information that will be useful in solving problems and improving the workplace. Where appropriate, attempts are made to suggest relationships that can be used to adjust the published strength values so they can better approximate industrial task requirements.

The muscle groups included are those that flex and extend the forearm, rotate the forearm and move the hand at the wrist, provide grip strength, and flex the arm at the shoulder, thereby raising the arms out in front of the body. Information about whole-body and upper body lifting strength, which involves large numbers of muscle groups, is also included. The chapter concludes with a summary of the strength and torque capacity differences between a 50th-percentile woman and a 50th-percentile man for several muscle groups.

1. FOREARM FLEXION AND EXTENSION

Many industrial tasks require arm movements with the elbow flexed or extended. Examples include most lifting and lowering tasks and extension and retraction of the arm within the seated work area (see Chapter II, Volume 1). The range of elbow flexion and extension is approximately 150 degrees (see Figure 26-4 a), depending on individual differences in anatomy and flexibility (Rasch and Burke, 1978).   Strengths and torques are shown in Tables 26-6 and 26-7

Table 26-6: Isometric Forearm Flexion and Extension Strengths—One Arm. Data from industrial, military, and college studies (column 1) of isometric forearm flexion and extension strength are shown in columns 2 and 3. The average values plus or minus one standard deviation are given for strength measured with the elbow bent at 90 degrees. Column 4 shows the estimated torque, calculated by assuming forearm lengths as shown in the footnote. Column 5 gives the source of data. Based on these studies, one can see that the average female has one-half to two-thirds the forearm strength of the average male, and forearm extensor strength is about one-half to two-thirds forearm flexor strength.

Population	Force, ± 1 SD		Torque,* Nm	Reference
	N	lbf		
Isometric Forearm Flexion Strength				
Industrial males (436)	276 ± 88	62 ± 20	70	Kamon and Goldfuss, 1978
Industrial females (136)	160 ± 51	36 ± 12	38	Kamon and Goldfuss, 1978
Industrial males (74)	336 ± 78	76 ± 18	85	Champney, Eastman Kodak Company, 1983
Industrial females, (18)	174 ± 56	39 ± 13	41	Champney, Eastman Kodak Company, 1983
Isometric Forearm Extension Strength				
Male soldiers (92)	159 ± 29	36 ± 6	40	Tornvall, 1963
College females (14)	106	24	25	Kroll, 1971

*Estimated from the force data by assuming average forearm lengths of: 0.238 m for women; 0.254 m for men. See Figures 26-1 and 26-2 for these estimates.

Table 26-7: Dynamic Forearm Flexion Strength and Torque—Two Arms. The maximum dynamic forearm flexion force and the estimated maximum torque are shown in columns 2 through 4. The mean plus or minus one standard deviation comes from measurements on industrial males (column 1). The female value is calculated using information from college studies that show the female/male ratio, on the average, to be 0.52 for this exertion. The dynamic forearm flexion was measured at a velocity of 0.73 meters/second using both arms, the values being taken at elbow angles between 90 and 110 degrees. Torque is estimated by assuming that the length of the forearm is 0.254 meters in males and 0.238 meters in females. These dynamic force values should not be compared to those for one-handed exertions in Table 26-6. *(Adapted from information in Kamon, Kiser, and Landa-Pytel, 1982).*

Population (n)	Force		Estimated Torque, 2 Arms, Nm
	N	**lbf**	
Industrial males (48)	324 ± 46	73 ± 10	82
Industrial females	168*	38	40

*The ratio female/male (0.52) was reported for college males and females during the same dynamic test (Landa-Pytel and Kamon, 1981).

for forearm flexion and extension. Graphs of the relationship between elbow angle and strength, as a percentage of the isometric strength at 90 degrees, are shown in Figures 26-9 and 26-10 for forearm flexion and extension, either as eccentric, concentric, or isometric muscle forces. The largest forces can be developed at elbow angles around 90 degrees for isometric and eccentric contractions, and at lesser angles for concentric contractions. (Concentric and eccentric contractions have been discussed further in Part II.)

2. FOREARM ROTATION AND MOVEMENTS OF THE HAND AT THE WRIST

Most assembly and packing tasks require repetitive movements of the hands, forearms, and wrist joints. Because repetitive motion tasks that require high force exertions may be associated with muscle, tendon, joint, and nerve irritation, it is important to know the maximum forces that can be developed by the muscles controlling these movements. Figure 26-4 *b* shows the motions of forearm pronation and supination that affect the force available from the biceps muscle. Isometric strength for handle and key pronation and supination is presented in Table 26-8.

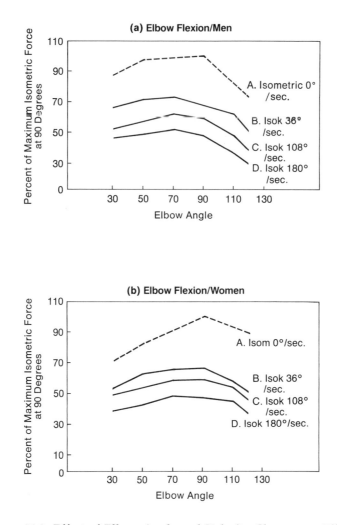

Figure 26-9: Effect of Elbow Angle and Velocity Changes on Elbow Flexor Torque in Men and Women   The percent change in flexor torque compared to isometric torque with the elbow at 90 degrees (vertical axis) is shown for elbow angles from 30 to 120 degrees (horizontal axis). Isometric strength is shown varying with elbow angle (A); three velocities of isokinetic forces are also given, ranging from 36 degrees per second (B) to 108 degrees per second (C) to 180 degrees per second (D). Similar graphs are given for men (*a*) and women (*b*). The amount of torque developed in isokinetic work is less than in isometric work, and the level decreases as the velocity of shortening increases. Isokinetic torque reaches its maximum at a 70-degree elbow angle. Elbow flexion is an important motion in most manual handling tasks involving lifting. (*Adapted from Knapik, Wright, and Mawdsley, 1983. Reprinted from Physical Therapy (63: 938–942, 1983) with the permission of the American Physical Therapy Association*).

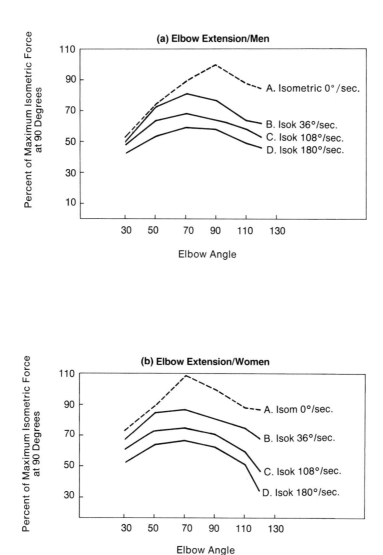

Figure 26-10: Effect of Elbow Angle and Velocity Changes on Elbow Extensor Torque in Men and Women   The percent change in extensor torque compared to isometric torque with the elbow at 90 degrees (vertical axis) is shown for elbow angles from 30 to 120 degrees (horizontal axis). This is similar to Figure 26-9 in that three velocities of isokinetic muscle work are represented (B, C, and D) and one curve (A) shows isometric work. Similar graphs are given for men (*a*) and women (*b*). As was seen in Figure 26-9, isokinetic torques are lower than isometric ones for elbow extension, and the peaks occur at 70 degrees of elbow angle rather than at 90 degrees. (*Adapted from Knapik, Wright, and Mawdsley, 1983. Reprinted from* Physical Therapy *(63: 938–942, 1983) with the permission of the American Physical Therapy Association*).

Table 26-8: Maximum Isometric Torques in Handle and Key Turning—Forearm Pronation and Supination. Data from a study of the torque-generating capabilities of male military recruits and female college students (column 2) using forearm pronation and supination are given in column 3; n is the number of subjects in each study. The torque is expressed in newton-meters (Nm); the average values plus or minus one standard deviation are given. Two types of torque conditions, handle and key, are shown in column 1. The major difference between them is that the key requires a pinch rather than power grip. This results in a torque capacity of about one-third as much for the key as for the handle. Forearm supination and pronation torque capacities are about the same. The female-male differences are greater for handle than for key pronation and supination, women's capacities being, on the average, about 40 percent less for the handle, 20 percent less for the key. *(Developed from information in Asmussen and Heebol-Nielsen, 1961).*

Type	Population (n)	Torque (Nm)
Isometric Forearm Pronation Strength		
Handle	Male, untrained military recruits (96)	14.1 ± 3.1
Handle	Female college students (81)	8.6 ± 1.6
Key	Male, untrained military recruits (96)	4.1 ± 0.6
Key	Female college students (81)	3.2 ± 0.5
Isometric Forearm Supination Strength		
Handle	Male, untrained military recruits (96)	15.0 ± 2.7
Handle	Female college students (81)	8.6 ± 1.5
Key	Male, untrained military recruits (96)	4.2 ± 0.7
Key	Female college students (81)	3.3 ± 0.5

Figure 26-4 *c* and *d* describe movements of the hand at the wrist. There is relatively little information available on the forces that can be developed during the various types of wrist movements; some values are presented in Table 26-9. The text on handgrip that follows shows how changes in wrist position can affect grip strength, an important factor in many handling and wrapping tasks.

3. GRIP STRENGTH

Because the handgrip is involved in almost all industrial tasks, whether controlling, pulling, lifting, lowering, wrapping, or holding items, data on the grip

Table 26-9: Maximum Isometric Torques Generated in Movements of the Hand Around the Wrist. The maximum isometric torques generated in wrist flexion and extension, in newton-meters, for male military recruits and female college students (column 1) are shown in column 2; n is the number of subjects in each study. The values shown are averages plus or minus one standard deviation. Wrist extension torque is about 25 percent greater than flexion torque, and the average female has about two-thirds of the torque capacity of the average male. These values are of importance in the design of repetitive handling or assembly tasks where force must be exerted during wrist rotation. *(Adapted from Asmussen and Heebol-Nielsen, 1961).*

Population (n)	Torque (Nm)
Isometric Wrist Flexion Strength	
Male, untrained military recruits (96)	8.0 ± 1.6
Female college students (81)	5.5 ± 0.9
Isometric Wrist Extension Strength	
Male, untrained military recruits (96)	10.1 ± 2.2
Female college students (81)	6.9 ± 1.2

strength capabilities of industrial workers are very important for ergonomic analysis. It is a relatively easy and inexpensive measurement to make, and, whenever possible, it should be measured on the worker population of interest. The absolute values will be heavily influenced by how it is measured. If the dynamometer is adjusted to each worker's optimum grip span, the maximum forces developed will be higher than those measured if the dynamometer is kept on a standard setting, such as 2 in. (5 cm) between the palm and fingers. Table 26-10 presents mean grip strength values taken from some representative sources in the literature and in industry.

The mechanics of the hand affect the amount of force that can be exerted on a dynamometer. Figure 26-11 illustrates the effect of changing hand span on the maximum force exerted on a dynamometer. If grip strength is an important factor in a task, it is recommended that the task hand span requirement be used when testing grip strength or when choosing values from Table 26-10.

Figure 26-12 shows that grip strength is also influenced by the position of the wrist. Task requirements that produce deviations from the neutral wrist position will reduce grip strength; therefore, this factor should be considered in an ergonomic analysis.

Table 26-10: Maximum Grip Strength Values. The isometric grip strengths, in newtons and pounds of force, for industrial and student populations (column 2) are shown in column 3. The values shown are averages plus or minus one standard deviation. The source of the data is given in column 4, and the method of selecting grip span is indicated in column 1. "Preferred span" means that the dynamometer span was adjusted by each individual to a distance that felt "comfortable." The fixed span setting of 5 cm (2 in.) is close to the optimal span for maximum force development for most people (see Figure 26-11)

Grip Span	Population (n)	Force, in newtons (lbf), ± 1 SD	Reference
Preferred span	Industrial males (463)	449 ± 105 (101 ± 24)	Kamon and Goldfuss, 1978
Preferred span	Industrial females (139)	268 ± 64 (60 ± 14)	Kamon and Goldfuss, 1978
Preferred span	Industrial males (74)	535 ± 97 (120 ± 22)	Champney, Eastman Kodak Company, 1979
Preferred span	Industrial females (18)	310 ± 59 (70 ± 13)	Champney, Eastman Kodak Company, 1979
Fixed span 5 cm (2 in.)	Students, male (18)	535 ± 110 (120 ± 25)	SUNYAB-IE, 1982/1983
Fixed span 5 cm (2 in.)	Students, female (8)	230 ± 52 (52 ± 12)	SUNYAB-IE, 1982/1983

The preceding discussion applies to power handgrips. Many items that are handled, such as plywood, boxes, bags, and other common items in the workplace, do not permit a cylindrical power grip to be used. These items require different degrees of pinch or oblique grip where hand mechanics prevent the development of the maximum forces seen in power grips (see Part III). The loss of strength observed in power grip studies can be explained in terms of the grip going from a true cylindrical one (Figure 20-2) to a wide pinch grip (Figure 20-1) as the object size becomes greater, and to a palmar precision grip (Figure 20-1) as the object size becomes smaller. Pinch grip strength is approximately 25 percent of power grip strength (Damon, Stoudt, and McFarland, 1966). The definitions of true and maximum grip span for cylindrical, oblique, and pinch grips are included in Appendix A, Volume 1, and in this volume's Glossary. Part VI includes more discussion of handgrip and how it is affected by workplace and job conditions.

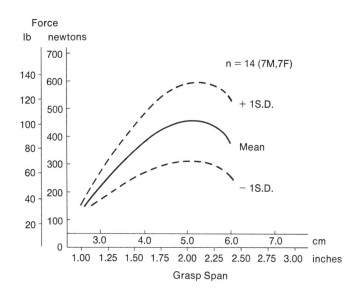

Figure 26-11: Effects of Grip Span on Maximum Isometric Grip Strength   Grip span, in centimeters and inches, (horizontal axes) was altered by changing the distance between the two handles of a Stoelting dynamometer. Force is shown in newtons and pounds (vertical scale). The solid curve gives the average (mean) values for grip strength at each grip span. The dotted curves show one standard deviation on either side of the mean. The greatest grip strengths are in the grip span range of 4.5 to 5.5 cm (1.75 to 2.25 in.). At narrower or wider spans, strength decreases because the hand is at a mechanical disadvantage for exerting force directly on the dynamometer handle. *(SUNYAB-IE, 1982/1983).*

4. SHOULDER FLEXION STRENGTH

The shoulder flexors are the muscles that make it possible for a worker to lift an object above waist level or from lower heights when the object is placed far in front of the body (Figure 26-5 *b*). They often limit the amount of weight a person can lift because they are weaker than the arm muscles or other muscles of lifting. Data on maximum isometric shoulder flexion strength and torque for two angles of shoulder flexion are given in Table 26-11. The force measurements in these studies were made from the wrist, keeping the elbow straight (see Figure 27-3c in the next chapter).

5. ISOMETRIC LIFTING STRENGTHS
   ## AND PUSH-PULL FORCES

Most manual handling tasks involve more than one muscle group. Although a single group such as the shoulder flexors or the grip muscles of the arm may

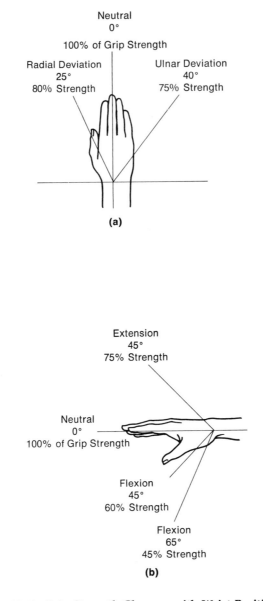

Figure 26-12: Grip Strength Changes with Wrist Position.   Wrist deviation from the neutral position (0 degrees) affects the maximum isometric grip strength that can be developed on a hand dynamometer. The degrees of deviation are indicated by lines converging on the neutral position line at the wrist. Radial and ulnar deviations are illustrated in *a*, and flexion and extension deviations in *b*. The significant losses in grip strength with changes in wrist position demonstrate the need to design handholds and workplaces so that the wrist can maintain its optimal neutral orientation. *(SUNYAB-IE, 1982/1983).*

Table 26-11: Maximum Isometric Shoulder Flexor Strength and Torque at Two Positions. The maximum isometric force and torque that can be developed by shoulder flexor muscles with the arm at 45 and 135 degrees from the side (column 1) are indicated in columns 2 through 4. Mean values, plus or minus one standard deviation, are given for the force measurements. The population tested is given in column 5, with the source in column 6. The torque values are calculated from the force measurements assuming moment arms for males and females as indicated in the footnote. In these studies, the average female's shoulder flexor strength is about 40 to 50 percent of the average male's. The shoulder flexors are stronger at arm positions below shoulder height, nearer to 45 degrees, but the percentage difference cannot be estimated because the populations tested are not comparable. *(Champney, Eastman Kodak Company, 1979; Yates et al., 1980).*

Position (angle)	Force		Torque,* Nm	Population	Reference
	N	lbf			
45 degrees	124 ± 40	28 ± 9	62	62 industrial men	Champney, Eastman Kodak Company, 1979
45 degrees	53 ± 22	12 ± 5	25	18 industrial women	Champney, Eastman Kodak Company, 1979
135 degrees	95 ± 21	21 ± 5	48	9 college men	Yates et al., 1980
135 degrees	44 ± 14	10 ± 3	21	9 college women	Yates et al., 1980

*Moment arms assumed (from Figures 26-1 and 28-2) for torque estimates were 0.5 m for men and 0.48 m for women.

limit a person's ability to lift a load, it is useful to measure the amount of force that can be developed in a simulated lifting, pushing, or pulling task when many muscle groups may be involved. Two studies of isometric force exertion are presented in the tables in this section. In one, an isometric upward pull is exerted on a tray that is anchored to a strain gauge at different heights and distances from the subject. This simulates a lifting situation where the hands remain in the same orientation throughout the lift. No regrasping of the tray is permitted as it is pulled up at both high and low locations. The method of testing this strength is found in Appendix B.

Although industrial lifting tasks are dynamic, rather than isometric, the illustration of how much muscle strength is lost as the tray is moved higher and away from the body makes it clear why high lifts with extended reaches are not recommended (see Tables 26-12 and 26-13). This information and data on the

Table 26-12: Maximum Isometric Pull Strength on a Tray.   Nine men and nine women students were studied with the isometric lifting test (described in the text) wherein a tray was pulled upward against a resistance at three distances in front of the ankles (columns 2 through 7) and at four heights above the floor (rows 1 through 4). The lift is a pull on a tray that is 50 cm (19.5 in.) long and 33 cm (13 in.) wide. The average newtons of force exerted (plus or minus one standard deviation) are given for men and women. The bolder values give the results of studies of an additional 38 men and 9 women. The highest pulls were different for men and women (188 cm versus 170 cm) because many of the women could not pull up at 188 cm above the floor. See Figure 27-4 for illustrations of some of these lifts. *(From information in Champney, Eastman Kodak Company, 1979; Yates et al., 1980).*

	Average Force (N) ± 1 SD					
	Distance in Front of the Ankles					
Height Above the Floor, in cm (in.)	18 cm (7 in.)		36 cm (14 in.)		51 cm (20 in.)	
	Men	Women	Men	Women	Men	Women
188 (74) (men) 170 (70) (women)	177 ± 59	66 ± 34	169 ± 44	54 ± 23	122 ± 26	41 ± 17
134 (53)	293 ± 77	104 ± 58	**253 ± 93**	**115 ± 39**	182 ± 69	**92 ± 41**
81 (32)	**607 ± 158**	**338 ± 146**	323 ± 96	**184 ± 80**	251 ± 30	118 ± 45
33 (13)	**744 ± 221**	**430 ± 190**	540 ± 144	248 ± 50	302 ± 91	131 ± 36

relationships between dynamic and isometric muscle strengths (see Figures 26-9 and 26-10) have been used to establish the limits for lifting and force exertions given in Part VI under manual handling activities.

The exertion of push and pull forces near waist level also involves many muscle groups. Figure 26-13 illustrates four common handling task force exertions. The force in newtons (average value plus or minus one standard deviation) is given for men and women in each of four pushes or pulls. These data in combination with other information were used to develop the guidelines for force exertions in Part VI.

6. MALE-FEMALE STRENGTH COMPARISONS

Data on men's and women's strength and torque capabilities have been presented in this section whenever available. Table 26-14 summarizes the average woman's strength as a percentage of the average man's strength for several of

Table 26-13: Loss in Tray Pull Strength with Increasing Lift Height and Distance from the Ankles. To illustrate the influence of lift location on lifting strength, we have expressed the data of the nine men and nine women students in Table 26-12 as a percent of each group's maximum isometric lift strength, measured at 33 cm (13 in.) above the floor and 18 cm (7 in.) in front of the ankles. Strength falls off rapidly as the tray is located higher and farther away from the body. Women experience a greater decrease in strength with height and distance of the lift than do men, which can probably be explained by anthropometric differences and the biomechanical advantages of the larger-sized person. *(From information In Champney, Eastman Kodak Company, 1979; Yates et al., 1980).*

	Percent of Maximum Lift Strength					
Height Above the Floor, in cm (in.)	Distance in Front of the Ankles					
	18 cm (7 in.)		36 cm (14 in.)		51 cm (20 in.)	
	Men	Women	Men	Women	Men	Women
188 (74)/170 (70)	25	15	25	15	20	10
134 (53)	35	25	40	25	25	20
81 (32)	85	75	45	45	35	30
33 (13)	100	100	80	70	45	35

these measurements. The isometric tray-pulling study shows the average woman with about 55 percent of the strength of the average man for lower lifts done closer to the body, but the value changes to 33 percent at higher, more distant locations. Isometric push and pull forces show the average woman to have about 70 to 75 percent of the strength of the average man.

The overlap in strength distributions is such that at least 33 percent of the men have strength capabilities that fall within the women's distribution. It is therefore not appropriate to use a person's sex as a selection criterion when strength is required on a job. The specific strength can be used as the selection criterion if it is not feasible to redesign the task.

C. AEROBIC WORK CAPACITY

The endurance capacities of healthy people are determined by their cardiovascular fitness. The ability of the heart to pump enough blood to the working muscles to satisfy the need for oxygen is a measure of that fitness, and measurements of heart rate and whole-body oxygen consumption can be used to

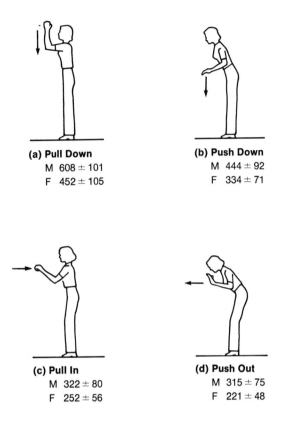

(a) Pull Down
M 608 ± 101
F   452 ± 105

(b) Push Down
M 444 ± 92
F   334 ± 71

(c) Pull In
M 322 ± 80
F   252 ± 56

(d) Push Out
M 315 ± 75
F   221 ± 48

Figure 26-13: Maximum Isometric Push and Pull Forces Measured with a Force Transducer—Two Hands Close Together   The maximum isometric forces that can be exerted in four pushing and pulling positions are given in newtons plus or minus one standard deviation for 309 men (M) and 35 women (F) under each illustration. A unidirectional force transducer was used for the measurements, with the hands held close together on a T-handle at the end of the transducer's cable. Vertical forces were measured for the pull-down and push-down activities (*a, b*), which are performed with the elbows at about 90 degrees. Horizontal pull-in and push-out forces (*c, d*) are measured with the arms partially extended and fully flexed, respectively. The feet are kept together, and the subject maintains an erect posture during these tests, so the measured values represent upper body rather than whole-body exertions. (*Adapted from Keyserling et al., 1980. Reprinted by permission of the American Industrial Hygiene Association Journal*).

predict the maximum aerobic work capacity for a given type of work. See Part IV for more discussion of aerobic work capacity and its relationship to job demands.

Several methods for measuring maximum aerobic work capacities are given

Table 26-14: Relative Muscle Strengths of Women and Men. The strength of an average woman, represented as a percentage of the strength of an average man (column 2) is presented for specific muscle actions (column 1). Relative muscle strength ranges from 52 percent for dynamic forearm flexion to 78 percent for isometric key pronation and supination; the mean value for all muscle actions is 65 percent. The tabulated values are calculated from information in Tables 26-6 through 26-11.

Muscle Action	Strength of an Average Woman as a Percentage of That of an Average Man
Isometric grip	60%
Isometric forearm flexion	58%
Isometric forearm extension	67%
Dynamic forearm flexion	52%
Isometric wrist flexion	69%
Isometric wrist extension	68%
Isometric handle supination	57%
Isometric key supination	78%
Isometric handle pronation	61%
Isometric key pronation	78%

in Appendix B. Some data on the capacities of industrial and other civilian populations for whole-body and upper body work are presented in this section.

1. WHOLE-BODY MAXIMUM AEROBIC CAPACITIES

The data presented in Table 26-15 and Figure 26-14 are from studies of industrial workers (Rodgers, Eastman Kodak Company, 1975). The test was submaximal, as described in Appendix B. Four to five levels of work on a treadmill were measured with a continuous electrocardiogram (ECG) recording and collection of the expired air for oxygen consumption measurements in the steady state of exercise. Two of the levels exceeded 40 percent of maximum predicted aerobic capacity as determined by the heart rate range. A predicted maximum heart rate was determined by subtracting the person's age from 220 (see Part III). The industrial population includes equal numbers of sedentary engineers and technicians and of production workers whose job demands ranged from light to very heavy.

The industrial population's maximum aerobic capacity levels (averages plus or minus one standard deviation) are compared to other studies of whole-body

Table 26-15: The Whole-Body Maximum Aerobic Work Capacities of Industrial Men and Women. Mean values, plus or minus one standard deviation, of whole-body maximum aerobic work capacities, expressed as milliliters of oxygen per kilogram of body weight per minute, are given in column 6 for industrial workers performing a submaximal treadmill test. The numbers of people studied (column 2), and their average ages (column 3) and weights (columns 4 and 5, in kilograms and pounds), plus or minus one standard deviation, are also given. In addition to studies on men and women, a statistical combination of data to simulate a 50-50 mix was also done (column 1). The data in the bottom half of the table are from the same subjects shown in the upper half, but they are organized by decades to show the influence of age on maximum aerobic work capacity. (Rodgers, Eastman Kodak Company, 1975).

Population	n	Age* (years)	Weight		Maximum Aerobic Capacity* (mL $O_2 \cdot$ kg $BW^{-1} \cdot min^{-1}$)
			kg	lbm	
Men	84	37 ± 12	78 ± 11	(172 ± 24)	38 ± 7
Women	37	33 ± 12	62 ± 9	(136 ± 20)	31 ± 6
50-50 mix	121	35 ± 12	70 ± 12	(154 ± 26)	34 ± 8
Men	27	24 ± 2	78 ± 11	(172 ± 24)	39 ± 8
	21	32 ± 2	77 ± 10	(169 ± 22)	39 ± 6
	20	46 ± 3	82 ± 12	(180 ± 26)	35 ± 7
	13	52 ± 2	78 ± 13	(172 ± 29)	37 ± 6
Women	20	24 ± 3	59 ± 6	(130 ± 13)	34 ± 5
	8	34 ± 3	61 ± 8	(134 ± 18)	30 ± 5
	4	44 ± 4	72 ± 16	(158 ± 35)	26 ± 4
	5	55 ± 3	66 ± 6	(145 ± 13)	25 ± 3

*Mean values ± 1 SD.

maximum aerobic capacity in Table 26-16. The maximum aerobic capacity levels of the bicycle ergometer test should be about 5 to 10 percent less than those of the treadmill tests (Astrand et al., 1973). A laddermill test (Kamon, 1973) gives values for $\dot{V}_{O_2 \; max}$ similar to those found in a treadmill test. The effect of age is clearly shown in this summary, the older people showing, on the average, about

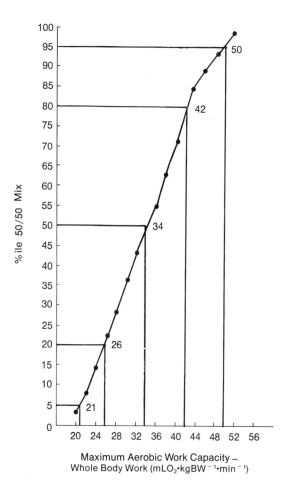

Figure 26-14: A Cumulative Frequency Distribution of the Maximum Aerobic Work Capacities Measured on a Treadmill Test   The maximum aerobic capacity for whole-body work is expressed in milliliters of oxygen used per kilogram of body weight per minute (horizontal axis). The percentiles of a 50-50 (male-female) mixed population, based on a weighted statistical combination of data from 84 men and 37 women, are listed on the vertical axis. Data from Table 26-15 are plotted as a cumulative frequency distribution so that the percentage of the population that has a given aerobic capacity or less can be read off the vertical scale. The lines forming right angles with the curve represent the 5th, 20th, 50th, 80th, and 95th percentiles for maximum aerobic capacity; the numbers at the intersection with the curve are the $\dot{V}_{O2\ max}$ values. The section on work-rest cycles and the problems demonstrate how this information can be applied in job design. (*Rodgers, Eastman Kodak Company, 1975*).

Table 26-16: Comparisons of Whole-Body Maximum Aerobic Work Capacities for Industrial and Civilian Populations. The mean maximum aerobic capacities for whole-body work are shown in column 5 in milliliters of oxygen per kilogram of body weight per minute. The number of subjects is shown in column 2, and their sex and average ages (plus or minus one standard deviation) in columns 3 and 4, respectively. The type of test is given in column 1. In submaximal tests, a predicted maximum heart rate is calculated, not measured, as it is in a maximal test, and the tests may therefore over- or underestimate the work capacity. The maximum aerobic capacities for industrial men are about 15 percent lower than the values reported for nonindustrial men; the industrial women are similar to a sedentary adult population studied by Profant et al. (1972), prior to an exercise program.

Test	n	Sex	Age (years)*	(mL $O_2 \cdot$ kg $BW^{-1} \cdot$ min^{-1})*	Reference
Treadmill,	84	M	37 ± 12	38 ± 7	Rodgers, Eastman
submaximal	37	F	33 ± 12	31 ± 6	Kodak Company, 1975
Treadmill,	28	M	20.7 ± 2.5	44.6 ± 8.0	Sloan, Koeslag, and
maximal					Bredell, 1973
Treadmill,	11	M	24.5	48.7	Robinson, 1938
maximal	10	M	35.1	43.1	
	9	M	44.3	39.5	
	7	M	51.0	38.4	
	8	M	63.1	34.5	
Treadmill,	35	M	37.1 ± 8.9	43.7 ± 10.2	Coleman and Burford,
submaximal					1971
Treadmill,	39	F	29 to 39	29.1 ± 3.5	Profant et al., 1972
maximal	47	F	40 to 49	26.9 ± 4.0	
	33	F	50 to 59	25.7 ± 3.0	
Bicycle ergom-	13	M	34.8	39.8 ± 7.3	Astrand, 1960
eter, submaximal	9	M	42.6	39.2 ± 5.5	
	66	M	53.3	33.1 ± 4.9	
	8	M	62.9	31.4 ± 5.3	
	8	F	25.0	39.9 ± 4.7	
	12	F	34.7	37.3 ± 5.2	
	8	F	43.9	32.5 ± 2.7	
	16	F	55.6	28.4 ± 2.7	
Bicycle ergome-	31	M	46.9	45.3 ± 4.0	Astrand et al., 1973
ter, submaximal	35	F	42.9	38.4 ± 4.0	
and maximal					

*Mean values ± 1 SD (where available).

60 to 80 percent of the maximum aerobic capacities of their colleagues in their 20s. A follow-up study of people in their 40s, whose maximum aerobic capacities had been measured in their 20s, showed about a 20 percent decline in capacity over the 20-year period (Astrand et al., 1973).

In general, the industrial population has lower maximum aerobic capacities than the other populations tested. This is probably explained by the differences in fitness of the populations since many of the civilian groups were tested at gymnasiums where they exercised once or twice a week.

For the workload guidelines given in Part IV, a maximum whole-body aerobic work capacity of 27 mL $O_2 \cdot$ kg $BW^{-1} \cdot min^{-1}$ was used, as 75 percent of the 50-50 male-female population mix would have that capacity or more. Jobs that require higher capacities can be done safely by some people, but there are fewer people suited for them.

2. UPPER BODY MAXIMUM AEROBIC CAPACITY

For jobs done while standing at a counter or next to a machine where the lower body is relatively stationary and all of the work is with the arms and shoulders, a measure of upper body maximum aerobic work capacity is needed. Table 26-17 summarizes data from one industrial lifting task and three studies of stan-

Table 26-17: Upper Body Maximum Aerobic Work Capacity as a Percent of Whole-Body Aerobic Work Capacity. Four reports of tests of upper body work capacity are shown in column 1; the tray lift was a three- or four-stage task using weights up to 12 kg (26 lbm) and is described in Appendix B. The upper body aerobic capacity is expressed as a percentage of the whole-body capacity (column 2). The number and sex of the people studied are in parentheses, and the source of the study is given in column 3. According to these studies, upper body aerobic capacity is about 64 to 78 percent of whole-body aerobic capacity.

Type of Upper Body Work	Average Upper Body As a Percentage of Whole-Body Capacity	Reference
Lifting tray at 24 lifts per minute, all lifts above 76 cm 30 in.	64% (10 F) 75% (11 M)	Rodgers, Eastman Kodak Company, 1973
Arm cranking	70% (13 M)	Astrand et al., 1965
Arm cranking	68% (3 M)	Reybrouck, Heigenhauser, and Faulkner, 1975
Arm cranking	78% (7 M)	Vokac et al., 1975

dardized arm cranking tasks. These indicate that upper body aerobic work capacity is about 70 percent of whole-body aerobic capacity.

3. MAXIMUM AEROBIC CAPACITIES FOR LIFTING TASKS

That the weight and frequency of a lift determine a person's capacity for lifting tasks is shown by studies of maximum aerobic work capacities during lifting at heights between 6 and 60 cm (2.5 to 23.5 in.) where box weight was varied (Lind and Petrofsky, 1978; Petrofsky and Lind, 1978). Table 26-18 summarizes the maximum aerobic capacities of four men in their 20s for several lifting tasks and indicates the percentage of whole-body maximum aerobic work that these represent.

D. THE ENERGY REQUIREMENTS OF OCCUPATIONAL TASKS

Some common occupational tasks are grouped according to their effort intensities, ranging from light to extremely heavy. Since duration will influence the acceptability of a task's intensity, the tasks have been grouped within each table according to their usual durations, using two hours as the maximum continuous time.

A listing of metabolic job demands enables the estimation of task workload when the equipment or expertise is not available to measure the demands directly. Workload estimates are needed in industry for the following purposes:

- Identification of energy requirements in conjunction with environmental surveys; for example:
  - In heat exposure studies, to assess the metabolic load, M, in the heat balance equation (see Chapter V, Volume 1).
  - In chemical exposure studies, to assess the metabolic and breathing levels of people working in the presence of chemicals. This may influence decisions about safe levels of exposure.
- Classification of jobs for industrial relations purposes; for example:
  - To identify heavy jobs or activities that merit redesign so more people can do them.
  - To evaluate differences in effort levels between jobs, which may be a factor in determining the appropriate wages.
  - To assess the potential difficulties of people returning to work after extended illnesses.
- Prediction of the staffing requirements for departments where these activities are done; identification of rest allowance requirements for heavier jobs.

Table 26-18: Lifting Task Maximum Aerobic Capacities as a Function of Box Weight. Mean values, plus or minus one standard deviation, of maximum aerobic capacity for low-height handling tasks with four different box weights (column 1) are shown in columns 3 and 4. Capacity is shown in liters per minute and in milliliters of oxygen per kilogram of body weight per minute. The lifting frequencies used by the four male subjects in this study to reach maximum levels of work are shown in column 2. Each maximum aerobic lifting capacity is expressed as a percentage of the whole-body capacity as measured on a bicycle ergometer (column 5). Lighter box weights and higher lifting frequencies may limit the available maximum aerobic capacities because of the difficulty of sustaining the rapid lifting rate and because of the negative effect the constant truck movement up and down at the waist can have on breathing efficiency. *(Based on studies by Lind and Petrofsky, 1978; Petrofsky and Lind, 1978).*

Box Weight kg (lbm)	Frequency (lifts/min)	Mean Maximum Aerobic Capacity, ± 1 SD		Lifting Capacity as Percentage of Whole Body Work Capacity
		(L/min)	(mL $O_2 \cdot$ kg $BW^{-1} \cdot min^{-1}$)	
0.9 (2)	64 – 72	1.99 ± 0.45	24.4 ± 5.5	54
6.8 (15)	52 – 62	2.38 ± 0.52	29.2 ± 6.4	64
22.7 (50)	24 – 30	2.79 ± 0.35	34.2 ± 4.3	75
36.4 (80)	20 – 24	3.01 ± 0.36	36.9 ± 4.4	80

There are, however, many factors that should be considered when using these tables to define the level of workload. They include:

- Recognition that people do not do tasks in the same way and that the data available in these tables may not be directly applicable to another industrial situation.

- An understanding that a job title in a given industry may not include the same activities or the same degree of effort as a job title in these tables.

- Recognition that within a single job or task there can be different levels of effort associated with varying product size, cycle times, or postural requirements.

- Acknowledgement that the way a person paces him- or herself will determine the metabolic demands of a task. This is especially important when evaluating the ability of an individual to return to work after an illness. The freedom the worker has to alter pace on the task should be considered, not just the energy requirements that have been established using a person in good health.

Another important consideration is that the metabolic demands are often not the limiting factor in determining job or task difficulty. A specific strength, reach, decision-making, or visual requirement may make a task too difficult for people even though it is within their aerobic capacities.

With these limitations in mind, Tables 26-19 through 26-23 show tasks and jobs for which oxygen consumption measurements have been made using healthy volunteers. The studies from which these tables are compiled include those by Aberg et al., 1968; Asmussen and Poulsen, 1963; Astrand and Rodahl, 1977; Belding, 1971; Brown and Crowden, 1963; Consolazio, Johnson, and Pecora, 1963; Davies et al., 1976; Davis, Faulkner, and Miller, 1969; Fordham et al., 1978; Garg and Saxena, 1979; Godin and Shephard, 1973; Goldman and Iampietro, 1962; Goldsmith et al., 1978; Hamilton and Chase, 1969; Hettinger, 1970; Kamon, 1973; Lehmann, 1962; McDonald, 1961; Moores, 1970; Nielsen, Eastman Kodak Company, 1962-1973; Oja, Louhevarra, and Korhonen, 1977; Passmore and Durnin, 1955; Raven et al., 1973; Rodgers, 1978; Snook and Ciriello, 1974; and Spitzer and Hettinger, 1958.

1. LIGHT EFFORT

There should be no aerobic capacity limitations for light work for at least 95 percent of the industrial population. These tasks require from 3 to 7 mL $O_2 \cdot kg\,BW^{-1} \cdot min^{-1}$. This is equivalent to 1 to 2.5 kcal per minute or 70 to 175 watts for whole-body work. For upper body work, the upper limits are 5 mL $O_2 \cdot kg\,BW^{-1} \cdot min^{-1}$ kcal/min, or 125 watts. Both of these guidelines assume continuous work periods of two hours. Table 26-19 gives examples of whole-body and upper body tasks that are classified as light effort.

2. MODERATE EFFORT

Moderately demanding tasks are not very difficult for 95 percent of the potential work force to sustain for two hours. However, longer continuous periods of work can be difficult for up to 45 percent of the work force. These tasks require from 7.1 to 10.7 mL $O_2 \cdot kg\,BW^{-1} \cdot min^{-1}$, which is equivalent to 2.6 to 3.75 kcal/min or 180 to 260 watts for whole-body work. For upper body work, moderate effort tasks have upper limits of 7.5 mL $O_2 \cdot kg\,BW^{-1} \cdot min^{-1}$, 2.6 kcal/min, or 185 watts. Table 26-20 shows moderate effort tasks.

3. HEAVY EFFORT

Heavy effort tasks include activities that 95 percent of the people can sustain for less than an hour continuously. Only about 55 percent of the potential work force will be able to sustain the work for two hours continuously. These tasks require from 10.8 to 17.1 mL $O_2 \cdot kg\,BW^{-1} \cdot min^{-1}$, which is equivalent to 3.8 to 6.0 kcal/min or 265 to 420 watts for whole-body work. For upper body work,

Table 26-19: Light Effort Tasks. Several occupational tasks or postures for whole-body and upper body effort are shown in column 1. Column 2 gives the usual continuous duration of each task in discrete categories of less than (<) 15 minutes, less than one hour, one to two hours, or greater than (>) two hours. These tasks are light, as defined in the text and in Table 26-24. *(See text for sources).*

Task	Usual Continuous Duration
Whole-Body Effort:	
Crouching	<15 minutes
Kneeling	<15 minutes
Lecturing, public speaking	1 - 2 hours
Nursing activities: patient care (except lifting), measurement	>2 hours
Sitting, feet and hands active	>2 hours
Sitting in a truck or car	up to 1 hour, >2 hours
Standing, light manual work	>2 hours
Stooping when standing	<15 minutes
Upper Body Effort:	
Canning paint	>2 hours
Coil winding	>2 hours
Drafting	>2 hours
Drilling machine trainees	>2 hours
Inspection of wooden separators	>2 hours
Hand sewing, fur industry	>2 hours
Light assembly work	>2 hours
Sitting, monitoring	>2 hours
Tailoring clothes, seated	>2 hours
Watch and clock repair	>2 hours

heavy effort has an upper limit of 12 mL $O_2 \cdot$ kg BW$^{-1} \cdot$ min^{-1}, 4.2 kcal/min, or 295 watts. Table 26-21 gives examples of heavy effort occupational tasks.

4. VERY HEAVY EFFORT

Tasks that require very heavy effort are difficult for more than 50 percent of the potential work force to sustain for a full hour. They are often done for 15 to 20 minutes continuously and followed by light work or rest. They require from 17.2

Table 26-20: Moderate Effort Tasks. This table is similar to Table 26-19 but the tasks in column 1 are more demanding. The lifting tasks give weight, lifting frequency, and lift location, sequentially. These tasks may vary widely between industries, so, whenever possible, it is preferable to measure oxygen demands rather than predict them. *(See text for sources).*

Task	Usual Continuous Duration
Whole-Body Effort:	
Boiler cleaning, power plant	1 – 2 hours
Building brick wall, 6 bricks per minute, 3.2 kg (7 lbm) per brick, 0 – 28 and 28 – 56 cm (0 – 11 and 11 – 22 in.)	>2 hours
Driving a truck (military)	1 – 2 hours, >2 hours
Expediting, paint production	>2 hours
Eyelet press operation	>2 hours
Forging	>2 hours
Joining, carpentry	>2 hours
Lifting 9 kg (20 lbm) boxes, 6/min, 50 – 102 cm (20 – 40 in.)	>2 hours
Lifting 10 kg (22 lbm) boxes, 10/min, 50 – 100 cm (20 – 39 in.)	>2 hours
Lifting 18 kg (40 lbm) boxes, 6/min, 50 – 102 cm (20 – 40 in.)	1–2 hours
Lifting 36 kg (80 lbm) boxes, 2/min, 50 – 102 cm (20 – 40 in.)	<15 minutes
Lifting 23 kg (50 lbm) boxes, 3/min, 0 – 50 cm (0 – 20 in.)	<15 minutes
Lifting 27 kg (60 lbm) boxes, 3/min, 0 – 50 cm (0 – 20 in.)	<15 minutes
Lowering 9 kg (20 lbm) boxes, 6/min, 102 to 50 cm (40 to 20 in.)	>2 hours
Machine tending	>2 hours
Machining parts	>2 hours
Plastic molding	>2 hours
Press machine operation	>2 hours
Push and pull, vertical, downwards, 20 cycles per minute, 75 cm (30 in.) high, 54 Nm (40 ft lb)	<15 minutes
Pushing cart weighing 590 kg (1300 lbm)	<15 minutes
Sheet metal work	>2 hours
Sorting scrap materials	>2 hours
Stock clerk, motor repair	>2 hours
Vehicle (truck) assembly	>2 hours
Walking, 4.2 km/hr (2.5 mph), 0% grade	<1 hour
Walking downhill, 5 km/hr (3 mph), 10 degree slope	<1 hour
Walking down stairs	<15 minutes
Upper Body Effort:	
Armature winding (electrical)	>2 hours
Assembly work, electrical	>2 hours
Bag and pack paper rolls	>2 hours
Bookbinding, seated	>2 hours
Bookkeeping	>2 hours

Table 26-20   (*Continued*)

Task	Usual Continuous Duration
Upper Body Effort:	
Building brick wall, 6 bricks/min, 3.2 kg (7 lbm) bricks, 56 to 104 cm (22 to 41 in.)	>2 hours
Camera assembly	>2 hours
Coal displacer crane operation	>2 hours
Finishing metal (shell) cases	>2 hours
Fixing rubber insulation to battery plates	>2 hours
Gasket inspection	>2 hours
Interleaving sheet film with paper	>2 hours
Lifting 10 kg (22 lbm) boxes, 10/min, 100 – 150 cm (39 – 59 in.)	<1 hour
Machine sewing, light machining	>2 hours
Machine sawing	>2 hours
Packaging, electronics	>2 hours
Punch press operation, electronics	>2 hours
Push and pull, vertical, downwards, 20 cycles/min, 175 cm (69 in.), 50 Nm (40 ft lb)	<15 minutes
Sitting, writing	>2 hours
Soldering lamps	>2 hours
Typewriter repair	>2 hours
Typing, key punching	>2 hours
Using screwdrivers, wrenches, 96 cm (38 in.) above the floor, 80 – 215 Nm (58 – 159 ft lb)	<15 minutes

to 28.6 mL O_2 • kg BW^{-1} • min^{-1}, which is equivalent to 6.1 to 10 kcal/min or 430 to 700 watts for whole-body work. The upper limits for upper body work are 20 mL O_2 • kg BW^{-1} • min^{-1}, 7 kcal/min, or 490 watts. A list of very heavy activities is found in Table 26-22.

5. EXTREMELY HEAVY EFFORT

Activities that require extremely heavy effort are difficult for approximately 70 percent of the potential work force to sustain for more than 15 to 20 minutes. Few people have the necessary aerobic fitness to sustain these tasks for a full hour without a break. These tasks are usually done intermittently, a few minutes of the extremely heavy effort interrupted by a much lighter task or a full rest break. Frequent repetition of the task without adequate recovery periods can result in fatigue over a shift, especially if the shift is an extended-hours schedule of 10 to 12 continuous hours. These tasks exceed the upper limits for very heavy effort tasks. That is, they require more than 28.6 mL O_2 • kg BW^{-1} • min^{-1}, 10 kcal/

Table 26-21: Heavy Effort Tasks. This table is similar to Tables 26-19 and 26-20, showing several occupational tasks in column 1 and the durations of continuous work in column 2. The definitions for heavy effort are given in the text and in Table 26-24. Because a given task (such as cabinetmaking) can have a variety of effort levels, these values should be used cautiously and in conjunction with estimates of effort levels such as those described in Part III. *(See text for sources).*

Task	Usual Continuous Duration
Whole-Body Effort:	
Ash removal in power plant	1 – 2 hours
Cabinetmaking	>2 hours
Carpentry	>2 hours
Casting lead balls in mold	>2 hours
Cleaning operations, including sweeping, mopping, scrubbing floor	1 – 2 hours
Digging trenches	<1 hour
Expeditor, machine shop	>2 hours
Emptying trash cans	>2 hours
Forge hammering (steel)	<1 hour
Galvanizing, unskilled	>2 hours
Gardening, landscaping	>2 hours
Gasket supply operation	>2 hours
Handling, production	>2 hours
Inspecting sheet steel	>2 hours
Joining floorboards	1 – 2 hours
Jolting (steel)	1 – 2 hours
Lead rolling in rolling mill (steel)	1 – 2 hours
Lifting 10 kg (22 lbm) boxes, 10/min, 0 – 50 cm (0 – 20 in.)	>2 hours
Lifting 13.5 kg (30 lbm) boxes, 12/min, 0 – 50 cm (0 – 20 in.)	<15 minutes
Lifting 16.4 kg (36 lbm) boxes, 9/min, 0 – 50 cm (0 – 20 in.)	<15 minutes
Lifting 20 kg (44 lbm) boxes, 10/min, 0 – 50 cm (0 – 20 in.,)	<15 minutes
Lifting 27 kg (60 lbm) boxes, 4/min, 0 – 50 cm (0 – 20 in.)	<15 minutes
Lifting 36 kg (80 lbm) boxes, 2/min, 0 – 50 cm (0 – 20 in.)	<15 minutes
Lifting 9 kg (20 lbm) boxes, 12/min, 50 – 102 cm (20 – 40 in.)	>2 hours
Lifting 18 kg (40 lbm) boxes, 12/min, 50 – 102 cm (20 – 40 in.)	1 - 2 hours
Lifting 27 kg (60 lbm) boxes, 8/min, 50 – 102 cm (20 – 40 in.)	<15 minutes
Lifting 36 kg (80 lbm) boxes, 4/min, 50 – 102 cm (20 – 40 in.)	<15 minutes
Loading and unloading carboys and 45 kg (100 lbm) bags	<15 minutes or <1 hour
Supplying production machines	>2 hours
Loading chemicals into mixer	<15 minutes
Loading trailers with cases	<1 hour
Loading plates into charging vat (steel)	<15 minutes
Lowering 9 kg (20 lbm) boxes, 6/min, 50 to 0 cm (20 to 0 in.)	>2 hours
Lowering 18 kg (40 lbm) boxes, 6/min, 102 to 50 cm (40 to 20 in.)	>2 hours
Mail delivery, nonmotorized	>2 hours
Making beds (hospital)	<15 minutes
Mixing cement	<15 minutes

Table 26-21 *(Continued)*

Task	Usual Continuous Duration
Whole-Body Effort:	
Operating plate shears (electronics)	>2 hours
Plastering walls	>2 hours
Polishing metal parts (locksmith)	>2 hours
Power truck driving	1 – 2 hours
Push and pull vertically down, 20 cycles/min, 75 cm (30 in.) high, 110 Nm (80 ft lb)	<15 minutes
Pushing and pulling boxes weighing 864 kg (1900 lbm)	<15 minutes
Pushing wheelbarrow at 4.5 km/hr (2.7 mph), 57 kg (125 lbm) load	<15 minutes
Repairing roads	>2 hours
Stacking lumber	<1 hour
Stockroom work	>2 hours
Stoking furnace	<15 minutes
Tool room work	>2 hours
Triphammer operation, partly seated	1 – 2 hours
Truck and automobile repair	>2 hours
Unloading wooden boxes from box car	1 – 2 hours
Walking 5.8 km/hr (3.5 mph), 0% grade	<15 minutes
Walking 4.2 km/hr (2.5 mph), 5% grade	<15 mintues
Walking 4 km/hr (2.4 mph), carrying 10 kg (22 lbm)	<15 minutes or <1 hour
Warehouse order picking	<1 hour
Waxing and buffing the floor	<1 hour
Welding	<1 hour
Upper Body Effort:	
Annealing	<1 hour
Assembly work, heavier items	>2 hours
Battery plate casting	>2 hours
Boot and shoe repair/fabrication	>2 hours
Carpentry, finishing, joining, sawing	>2 hours
Clerical work, filing	>2 hours
Coil assembly, electrical, heavier items	>2 hours
Cutting battery plates	>2 hours
Food preparation	>2 hours
Frame cutting	>2 hours
Gasket packing and supplying	>2 hours
Grinding, filing metal	>2 hours
Hand press operation	1 – 2 hours
House painting	>2 hours
Labeling, filling tubs, painting lids, paint industry	>2 hours
Lathe operation	>2 hours

Table 26-21 *(Continued)*

Task	Usual Continuous Duration
Upper Body Effort:	
Laundry operations	>2 hours
Lifting 4.5 kg (10 lbm) boxes, 9/min, 60 to 112 to 86 cm (24 to 44 to 34 in.)	<1 hour
Lifting 9 kg (20 lbm) boxes, 8/min, 102 to 152 cm (40 to 60 in.)	<1 hour
Lifting 20 kg (44 lbm) boxes, 10/min, 100-150 cm (39-59 in.)	<15 minutes
Lowering 18 kg (40 lbm) boxes, 6/min, 152 to 102 cm (60 to 40 in.)	<1 hour
Operate record-pressing machine (electronics), standing	>2 hours
Printing press operation	>2 hours
Punch press operation, standing	>2 hours
Push and pull, vertical, downwards on handle, 20 cycles/min, 110 Nm (80 ft lb), 175 cm (69 in.) height	<15 minutes
Spray painting, standing, machine work	>2 hours
Tapping and drilling	<15 minutes
Tool making	>2 hours
Turning handwheel, 50 cm (20 in.), diameter, 1.0 to 1.2 meters (39 to 47 in.) high, 88 Nm (64 ft lb) per revolution, 20 rpm	<15 mintues
Unloading rolls from slitter	<15 minutes
Woodworking machine operations	>2 hours
Wrapping and packing, large-size products	>2 hours

Table 26-22: Very Heavy Effort Tasks.   This is similar to the three previous tables, with very heavy occupational tasks listed in column 1 and their usual sustained durations in column 2. These tasks are seldom sustained more than an hour or two continuously. The energy demands for a particular task may be lower if frequent rest breaks are provided or taken (see Part IV). *(See text for sources).*

Task	Usual Continuous Duration
Whole-Body Effort:	
Break crust with jackhammer	<15 minutes
Burring, metal	>2 hours
Can handling, 23 kg (50 lbm)	>1 – 2 hours
Cane (sugar) cutting	>2 hours
Carboy washing	>2 hours
Chopping wood	<1 hour

Table 26-22   *(Continued)*

Task	Usual Continuous Duration
Whole-Body Effort:	
Climbing ladder, 50 degrees slope, 20 kg (44 lbm) load, 9 m/min (29 ft/min) vertical speed, 17 cm (7 in.) step height	<15 minutes
Climbing ladder, 70 degrees slope, no load, 11 m/min (36 ft/min) vertical speed, 17 cm (7 in.) step height	<15 mintues
Coal car unloading	1 – 2 hours
Cutting of sheet steel	>2 hours
Filling and stacking of 45 kg (100 lbm) bags	>2 hours
Furnace cleaning, heavy deposits	1 – 2 hours
Girdering and timbering, coal mine activities, including hewing	>2 hours
Handling and carrying heavy loads	<1 hour
Hopper loading, chemical or food bags	<15 minutes
Lifting 10 kg (22 lbm) boxes, 10/min, 0 – 150 cm (0 – 59 in.)	1 – 2 hours
Lifting 20 kg (44 lbm) boxes, 10/min, 0 – 100 cm (0 – 39 in.)	<1 hour
Lifting 30 kg (66 lbm) boxes, 10/min, 50 – 100 cm (20 – 39 in.)	<15 minutes
Lifting 36 kg (80 lbm) boxes, 6/min, 50 – 102 cm (20 – 40 in.)	<15 minutes
Lifting 9 kg (20 lbm) boxes, 8/min, 0 – 50 cm (0 – 20 in.)	1 – 2 hours
Lifting 18 kg (40 lbm) boxes, 8/min, 0 – 50 cm (0 – 20 in.)	<1 hour
Lifting 30 kg (66 lbm) boxes, 10/min, 0 – 50 cm (0 – 20 in.)	<15 minutes
Lifting 36 kg (80 lbm) boxes, 6/min, 0 – 50 cm (0 – 20 in.)	<15 minutes
Lowering 18 kg (40 lbm) boxes, 6/min, 50 to 0 cm (20 to 0 in.)	>2 hours
Pulling cart, 3.6 km/hr (2.2 mph), level, handle height = 1 m (39 in.), pull force = 12 kg (25 lbf)	<15 minutes
Punch press operation, heavy parts, standing	>2 hours
Push and pull, vertical, downward, 20 cycles/min, height = 75 cm (30 in.), 136 Nm (100 ft lb) torque	<15 minutes
Pushing tubs, carts, or wheelbarrows, 3.6 to 4.5 km/hr (2.2 to 2.7 mph)	<15 mintues
Sawing wood, 2 men, standing, 60 cycles/min	1 – 2 hours
Shoveling, up to 2 m (6.5 ft) height, 12/min, 8 kg (17 lbm) load	<15 minutes
Sledge hammer use, vertical, weight = 6.8 kg, 12 cycles/min	<15 mintues
Stacking concrete, 18.3 piles/hr	>2 hours
Stair climbing, 25 flights	<15 minutes
Stone masonry	>2 hours
Stretching of sheet steel	1 – 2 hours
Tree felling	1 – 2 hours
Unloading battery boxes from oven	<15 mintues
Walking at 4 km/hr (2.4 mph), 30 kg (66 lbm) load	1 – 2 hours
Upper Body Effort:	
Breaking apart diecut cardboard	>2 hours
Bookbinding, standing	>2 hours
Dusting, heavy cleaning, overhead	1 – 2 hours

Table 26-22   (*Continued*)

Task	Usual Continuous Duration
Upper Body Effort:	
Filing iron, 80 cycles/min	1 – 2 hours
Lifting 4.5 kg (10 lbm) boxes, 15 cycles/min, 60 to 112 to 86 cm (24 to 44 to 34 in.)	1 – 2 hours
Lifting 6.8 kg (15 lbm) boxes, 15 cycles/min, 60 to 112 to 86 cm (24 to 44 to 34 in.)	1 – 2 hours
Lifting 9 kg (20 lbm) boxes, 15 cycles/min, 60 to 112 to 86 cm (24 to 44 to 34 in.)	1 – 2 hours
Lifting 11.4 kg (25 lbm) boxes, 9 cycles/min, 60 to 112 to 86 cm (24 to 44 to 34 in.)	1 – 2 hours
Lifting 9 kg (20 lbm) boxes, 12 cycles/min, 102 to 152 cm (40 to 60 in.)	<1 hour
Lifting 18 kg (40 lbm) boxes, 12 cycles/min, 102 to 152 cm (40 to 60 in.)	<15 minutes
Lifting 27 kg (60 lbm) boxes, 8 cycles/min, 102 to 152 cm (40 to 60 in.)	<15 mintues
Loading and unloading commercial laundry washing machine	<15 minutes
Medium press operation	>2 hours
Operating square cutter	>2 hours
Packing and handling large-size product	<15 minutes
Spray painting in woodworking	1 – 2 hours
Turning a crank with 2 hands, radius = 40 cm (16 in.), pivot = 1 m (39 in.) above the ground, 26 rpm	<15 mintues
Turning handwheel, 50 cm (20 in.) diameter, height = 1 to 1.2 m (39 – 47 in.), work = 88 Nm (64 ft lb) per revolution, 30 rpm	<15 mintues
Using a hatchet to point a post, 51 – 58 cycles/min	<1 hour

Table 26-23: Extremely Heavy Effort Tasks.   This table shows a number of lifting and other handling tasks (column 1) that can be sustained only for very short periods (column 2). These tasks result in rapid fatigue if carried out continuously for more than a few minutes. (*See text for sources*).

Task	Usual Continuous Duration
Whole-Body Effort:	
Climbing ladder, 70 degrees slope, 20 kg (44 lbm) load, 11 m/min (36 ft/min), 17 cm (7 in.) step	<15 minutes
Climbing ladder, 90 degrees slope, 20 kg (44 lbm) load, 12 m/min (36 ft/min), 17 cm (7 in.) step	<15 minutes
Firefighting	1 – 2 hours

Table 26-23 *(Continued)*

Task	Usual Continuous Duration
Whole-Body Effort:	
Lifting 20 kg (44 lbm) boxes, 10/min, 0 - 150 cm (0 - 59 in.)	<1 hour
Lifting 30 kg (66 lbm) boxes, 10/min, 0 - 150 cm (0 - 59 in.)	<15 minutes
Lifting 6.8 kg (15 lbm) boxes, 6 to 60 cm (2 to 24 in.), 200 kpm/min	1 - 2 hours
Lifting 22.7 kg (50 lbm) boxes, 6 to 60 cm (2 to 24 in.), 400 kpm/min	1 - 2 hours
Lifting 36 kg (80 lbm) boxes, 6 to 60 cm (2 to 24 in.), 550 kpm/min	<15 minutes
Pulling or pushing a cart at 3.6 km/hr (2.2 mph), level, handle height = 1 m (39 cm), pull force = 16 kg (35 lbm)	<15 minutes
Sawing wood, kneeling, 2 men, 60 cycles/min	1 - 2 hours
Shoveling dolomite (foundry)	<15 minutes
Slag removal, iron and steel	1 - 2 hours
Trimming of felled trees with axe	1 - 2 hours
Walking up stairs, 3.8 flights/min	<15 minutes
Upper Body Effort:	
Lifting 9 kg (20 lbm) boxes, 15/min, 60 to 112 to 86 cm (24 to 44 to 34 in.)	<1 hour
Lifting 11.4 kg (25 lbm) boxes, 15/min, 60 to 112 to 86 cm (24 to 44 to 34 in.)	<1 hour
Lifting 20 kg (44 lbm) boxes, 10/min, 50-150 cm (20-59 in.)	<1 hour
Lifting 30 kg (66 lbm) boxes, 10/min, 50-150 cm (20-59 in.)	<15 minutes
Lifting 30 kg (66 lbm) boxes, 10/min, 100-150 cm (39-59 in.)	<15 minutes
Turning crank with both hands, radius = 40 cm (16 in.), pivot = 1 m (39 in.) above the ground, 30 rpm, load = 350 Nm (254 ft lb)	<15 minutes
Turning handwheel, 50 cm (20 in.) diameter, height = 1 to 1.2 m (39 - 47 in.), work = 88 Nm (64 ft lb), 50 rpm	<15 minutes

min, or 700 watts for whole-body work, and more than 20 mL $O_2 \cdot kg\, BW^{-1} \cdot min^{-1}$, 7 kcal/min, or 490 watts for upper body work. Table 26-23 shows examples of these tasks.

A summary of the whole-body and upper body aerobic work limits in three units of measurement is given in Table 26-24 for the five effort levels. These categories of effort correspond quite well to those reported by Brown and Crowden (1963).

Table 26-24: Effort Category Definitions in Terms of Aerobic Work Requirements. The five levels of effort represented by Tables 26-19 through 26-23 are shown in column 1. In each category, workload ranges are shown for whole-body work (WB) and for upper body work (UB). These values are given for oxygen consumption, in milliliters of oxygen per kilogram of body weight per minute, for kilocalories per minute, and for watts. The extremely heavy category of effort is defined as levels exceeding the very heavy effort upper limit (greater than 28.6 mL $O_2 \cdot$ kg $BW^{-1} \cdot min^{-1}$ for whole-body work).

Intensity	Aerobic Work Load Ranges					
	(mL $O_2 \cdot$ kg $BW^{-1} \cdot min^{-1}$)		kcal/min		watts	
	WB*	UB**	WB*	UB**	WB*	UB**
Light	3.0 - 7.0	≤ 5	1 - 2.5	≤ 1.8	70 - 175	≤ 125
Moderate	> 7.0 - 10.7	> 5 - 7.5	> 2.5 - 3.8	> 1.8 - 2.6	> 175 - 260	> 125 - 185
Heavy	> 10.7 - 17.1	> 7.5 - 12	> 3.8 - 6.0	> 2.6 - 4.2	> 260 - 420	> 185 - 295
Very Heavy	> 17.1 - 28.6	> 12 - 20	> 6.0 - 10.0	> 4.2 - 7.0	> 420 - 700	> 295 - 490
Extremely Heavy	> 28.6	> 20	> 10.0	> 7.0	> 700	> 490

*WB = Whole-body work.
**UB = Upper-body work.

REFERENCES

American Academy of Orthopaedic Surgeons. 1965. *Joint Motion. Method of Measuring and Recording.* Chicago, Ill.: American Academy of Orthopaedic Surgeons, 87 pages.

Aberg, U., K. Elgstrand, P. Magnus, and A. Lindholm. 1968. "Analysis of Components and Prediction of Energy Expenditure in Manual Tasks." *International Journal of Production Research, 6 (3):* pp.189-196.

Asmussen, E., and K. Heebol-Nielsen. 1961. "Isometric Muscle Strength of Adult Men and Women." *Communications from the Testing and Observation Institute of the Danish National Association for Infantile Paralysis, NR-11:* pp. 1-41.

Asmussen, E., and E. Poulsen. 1963. "Energy Expenditure in Light Industry. Its Relation to Age, Sex, and Aerobic Capacity." *Communications from the Testing and Observation Institute of the Danish National Association for Infantile Paralysis, 13:* pp. 3-13.

Astrand, I. 1960. "Aerobic Work Capacity in Men and Women with Special Reference to Age." *Acta Physiologica Scandinavica, 49 (Suppl. 169):* pp. 1-92.

Astrand, I., P.-O. Astrand, I. Hallback, and A. Kilbom. 1973. "Reduction in Maximal Oxygen Uptake with Age." *Journal of Applied Physiology, 35 (5):* pp. 649-654.

Astrand, P.-O., B. Ekblom, R. Messin, B. Saltin, and J. Stenberg. 1965. "Intra-Arterial Blood Pressure During Exercise with Different Muscle Groups." *Journal of Applied Physiology*, 20: pp. 253-256.

Astrand, P.-O., and K. Rodahl. 1977. *Textbook of Work Physiology. Physiological Bases of Exercise*. 2nd ed. New York: McGraw-Hill, 681 pages.

Barter, J. T. 1957. *Estimation of the Mass of Body Segments*. WADC-TR 57-260. Wright-Patterson AFB, Ohio: Wright Air Development Center, p. 6.

Belding, H. 1971. "Ergonomics Guide to the Assessment of Metabolic and Cardiac Costs of Physical Work." *American Industrial Hygiene Association Journal*, 32: pp. 560-565.

Brown, J. R., and G. P. Crowden. 1963. "Energy Expenditure Ranges and Muscular Work Grades." *British Journal of Industrial Medicine*, 20: pp. 277-283.

Brunnstrom, S. 1972. *Clinical Kinesiology*. Philadelphia: F. A. Davis Company, p. 319.

Coleman, A. E., and C. L. Burford. 1971. "Aerobic Capacity in Sedentary Adults Participating in an Adult Fitness Program." *American Corrective Therapy Journal*, 25 (2): pp. 48-51.

Consolazio, C. F., R. E. Johnson, and L. J. Pecora. 1963. *Physiological Measurements of Metabolic Functions in Man*. New York: McGraw-Hill, pp. 329-333.

Damon, A., H. W. Stoudt, and R. A. McFarland. 1966. *The Human Body in Equipment Design*. Cambridge, Mass.: Harvard University Press, 360 pages.

Davies, C. T. M., J. R. Brotherhood, K. J. Collins, C. Dore, F. Imms, J. Musgrove, J. S. Weiner, M. A. Amin, H. M. Ismail, M. El Karim, A. H. S. Omer, and M. Y. Sukkar. 1976. "Energy Expenditure and Physiological Performance of Sudanese Cane Cutters." *British Journal of Industrial Medicine*, 33: pp. 181-186.

Davis, H. L., T. W. Faulkner, and C. I. Miller. 1969. "Work Physiology." *Human Factors* 11 (2): pp. 157-166.

Dempster, W. 1955. *Space Requirements of the Seated Operator*. WADC Technical Report No. 55-159, July 1955. Washington, D.C.: U.S. Department of Commerce, Office of Technical Services.

Fordham, M., K. Appenteng, R. Goldsmith, and C. O'Brien. 1978. "The Cost of Work in Medical Nursing," *Ergonomics*, 21 (5): pp. 331-342.

Garg, A., and U. Saxena. 1979. "Effects of Lifting Frequency and Technique on Physical Fatigue with Special Reference to Psychophysical Methodology and Metabolic Rate." *American Industrial Hygiene Association Journal*, 40: pp. 894-903.

Godin, G., and R. J. Shephard. 1973. "Body Weight and the Energy Cost of Activity." *Archives of Environmental Health*, 27: pp. 289-293.

Goldman, R. F., and P. F. Iampietro. 1962. "Energy Cost of Load Carriage." *Journal of Applied Physiology*, 17 (4): pp. 675-676.

Goldsmith, R., C. O'Brien, G. L. E. Tan, W. S. Smith, and M. Dixon. 1978. "The Cost of Work on a Vehicle Assembly Line." *Ergonomics*, 21 (5): pp. 315-323.

Hamilton, B. J., and R. B. Chase. 1969. "A Work Physiology Study of the Relative Effects of Pace and Weight in a Carton Handling Task." *AIIE Transactions*, 1: pp. 106-111.

Haxton, H. A. 1944. "Absolute Muscle Force in the Ankle Flexors of Men." *Journal of Physiology (London)*, 103: pp. 267-273.

Hettinger, T. 1970. *Angwandte Ergonomie*. Frechen, DDR: Bartmann Verlag, p. 122.

Kamon, E. 1973. "Laddermill and Ergometry: A Comparative Summary." *Human Factors,* 15 (1): pp. 75-90.

Kamon, E., and A. Goldfuss. 1978. "In-Plant Evaluation of the Muscle Strength of Workers." *American Industrial Hygiene Association Journal, 39:* pp. 801-807.

Kamon, E., D. Kiser, and J. Landa-Pytel. 1982. "Dynamic and Static Lifting Capacity and Muscular Strength of Steelmill Workers." *American Industrial Hygiene Association Journal, 43:* pp. 853-857.

Keyserling, W. M., G. D. Herrin, D. B. Chaffin, T. J. Armstrong, and M. L. Foss. 1980. "Establishing an Industrial Strength Testing Program." *American Industrial Hygiene Association Journal, 41:* pp. 730-736.

Knapik, J. J., J. E. Wright, R. H. Mawdsley, and J. Braun, 1983. "Isometric, Isotonic, and Isokinetic Torque Variations in Four Muscle Groups Through a Range of Joint Motion." *Journal of the American Physical Therapy Association 63 (6):* pp. 938-942.

Kroll, W. 1971. "Isometric Strength Fatigue Patterns in Female Subjects." *Research Quarterly, 42 (3):* pp. 286-298.

Landa-Pytel, J., and E. Kamon. 1981. "Dynamic and Static Lifting Capacity and Muscular Strength of Steelmill Workers." *American Industrial Hygiene Association Journal, 43:* pp. 853-857.

Lehmann, G. 1962. *Praktische Arbeitsphysiologie.* 2nd ed. Stuttgart: Georg Thieme Verlag, 409 pages.

Lind, A. R., and J. S. Petrofsky. 1978. "Cardiovascular and Respiratory Limitations on Muscular Fatigue During Lifting Tasks." In *Safety in Manual Materials Handling,* edited by C. G. Drury. Report on the international symposium at the State University of New York at Buffalo, July 18-20, 1976. Washington, D.C.: Superintendent of Documents (USHEW/NIOSH), pp. 57-62.

McDonald, I. 1961. "Statistical Studies of Recorded Energy Expenditure of Man." *Nutrition Abstracts and Reviews, 31:* p. 739-762.

Moores, B. 1970. "A Comparison of Work Load Using Physiological and Time Study Assessments." *Ergonomics, 13 (6):* pp. 769-776.

NASA. 1978. *Anthropometric Source Book, Volumes I, II, and III.* (Reference Publication 1024.) Edited by the staff of the Anthropology Research Project, Webb Associates. Yellow Springs, Ohio: NASA Scientific and Technical Information Office, 1167 pages.

Oja, P., V. Louhevaara, and O. Korhonen. 1977. "Age and Sex as Determinants of the Relative Aerobic Strain of Nonmotorized Mail Delivery." *Scandinavian Journal of Work, Environment, and Health, 3:* pp. 225-233.

Passmore, R., and J. V. G. A. Durnin. 1955. "Human Energy Expenditure." *Physiological Reviews, 35:* pp. 801-840.

Petrofsky, J. S., and A. R. Lind. 1978. "Comparison of Metabolic and Ventilatory Responses of Men to Various Lifting Tasks and Bicycle Ergometry." *Journal of Applied Physiology: Respiration, Environment, and Exercise Physiology, 45 (1):* pp. 60-63.

Profant, G. R., R. G. Early, K. L. Nilson, F. Kusumi, V. Hofer, and R. A. Bruce. 1972. "Responses to Maximal Exercise in Healthy, Middle-Aged Women." *Journal of Applied Physiology, 33 (5):* pp. 595-599.

Rasch, P. J., and R. K. Burke. 1978. *Kinesiology and Applied Anatomy.* 6th ed. Philadelphia: Lea and Febiger, 496 pages.

Raven, P. B., M. O. Colwell, B. L. Drinkwater, and S. M. Horvath. 1973. "Indirect Calorimetric Estimation of Specific Tasks of Aluminum Smelter Workers." *Journal of Occupational Medicine, 15 (11):* pp. 894-898.

Reybrouck, T., G. F. Heigenhauser, and J. A. Faulkner. 1975. "Limitations to Maximum Oxygen Uptake in Arm, Leg, and Combined Arm-Leg Ergometry." *Journal of Applied Physiology, 38 (5):* pp. 774-779.

Robinson, S. 1938. "Environmental Studies of Physical Fitness in Relation to Age." *Arbeitsphysiologie, 10:* pp. 251-323.

Rodgers, S. H. 1978. "Metabolic Indices in Materials Handling Tasks." in *Safety in Manual Materials Handling,* edited by C. G. Drury. DHEW/NIOSH Publication No. 78-185. Cincinnati, Ohio: Department of Health, Education, and Welfare/National Institute for Occupational Safety and Health, pp. 52-56.

Schantz, P., E. Randall-Fox, W. Hutchinson, A. Tyden, and P.-O. Astrand. 1983. "Muscle Fiber Type Distribution, Muscle Cross-Sectional Area, and Maximal Voluntary Strength in Humans." *Acta Physiologica Scandinavica, 117:* pp. 219-226.

Sloan, A. W., J. H. Koeslag, and G. A. G. Bredell. 1973. "Body Composition, Work Capacity, and Work Efficiency of Active and Inactive Young Men." *European Journal of Applied Physiology, 32:* pp. 17-24.

Snook, S. H., and V. M. Ciriello. 1974. "Maximum Weights and Work Loads Acceptable to Female Workers." *Journal of Occupational Medicine, 16 (8):* pp. 527-534.

Spitzer, H., and T. Hettinger, 1958. *Tafeln fur Kalorienumsatz bei Korperlicher Arbeit.* Darmstadt, DDR: REFA Publication. Cited in Astrand and Rodahl, 1977.

SUNYAB-IE. 1982/1983. Data from student laboratory projects for Industrial Engineering 436/536 (Physiological Basis of Human Factors) at the State University of New York at Buffalo, S. H. Rodgers, instructor.

Tornvall, G. 1963. "Assessment of Physical Capabilities." *Acta Physiologica Scandinavica, 58 (Suppl. 201):* 102 pages.

Vokac, Z., H. Bell, E. Bautz-Holter, and K. Rodahl. 1975. "Oxygen Uptake/Heart Rate Relationship in Leg and Arm Exercise, Sitting and Standing." *Journal of Applied Physiology, 39 (1):* pp. 54-59.

Williams, M., and H. R. Lissner. 1962. *Biomechanics of Human Motion.* Philadelphia, Pa.: Saunders, 147 pages.

Winter, D. A. 1979. *Biomechanics of Human Movement.* New York: Wiley, p. 61.

Yates, J. W., E. Kamon, S. H. Rodgers, and P. C. Champney. 1980. "Static Lifting Strength and Maximal Isometric Voluntary Contractions of Back, Arm, and Shoulder Muscles." *Ergonomics, 23:* pp. 37-47.

Appendix B: Methods to Evaluate Maximum Capacities and to Measure Job Demands

CHAPTER 27. APPENDIX B: METHODS TO EVALUATE MAXIMUM CAPACITIES AND TO MEASURE JOB DEMANDS

A. Strength Testing in Industry

    1. Static Strength-Testing Techniques
        a. Standard Procedures for Static Muscle Strength Testing
            (1) Measurement of Static Strength
            (2) Subject Description
            (3) Instructions to Subjects
            (4) Test Procedures
            (5) Reporting the Test Results
        b. Tray-Lifting Strength
    2. Dynamic Strength Testing
        a. Psychophysical Methods
        b. Isotonic Dynamic Tests
        c. Isokinetic Strength Testing in Industry

B. Aerobic Work Capacity Tests

    1. Whole-Body Aerobic Capacity Tests
        a. General Design of Tests
        b. Examples of Capacity Tests
    2. Upper Body Aerobic Capacity Tests

C. Cardiovascular and Metabolic Measurements

    1. Heart Rate
    2. Blood Pressure
    3. Minute Ventilation
    4. Oxygen Consumption and Carbon Dioxide Production
        a. Oxygen Consumption
        b. Carbon Dioxide Production

D. Review of Trigonometry for Biomechanical Analyses

References for Chapter 27

$\mathbf{M}$ethods for testing the lifting strength capabilities of industrial workers and for measuring their maximum aerobic capacities for work with different muscle groups are covered in this chapter. Brief summaries of techniques for measuring heart rate, blood pressure, and oxygen consumption are also given. These methods and techniques may be used to measure the physiological demands of jobs and to relate these demands to the capacities of the work force to determine what percentage of that population will find the job appropriate.

Since physiological job demands can be quantified only if a person is measured while doing the work, the methods for measuring maximum aerobic capacity can also be used to identify how fit that worker is compared to the general population. Data on strength and aerobic work capacities are found in Appendix A. Part IV should be reviewed for information on acceptable workloads.

A. STRENGTH TESTING IN INDUSTRY

Information on the strength capacities of workers is very valuable when designing new tasks or evaluating existing ones. Strength data are available from many sources, some of which are presented in this chapter. However, it may be preferable to measure directly the muscle strengths of the population for which a specific task is being designed. The many biomechanical variables which affect muscle strength were discussed in Parts II and III. It is important to take these variables into consideration whenever strength testing is undertaken. Similarly, when strength data from other sources are used, the conditions under which the measurements were taken must be considered.

1. STATIC STRENGTH-TESTING TECHNIQUES

Static strength has been defined as the capacity to produce torque or force by maximal voluntary isometric exertion (Chaffin, 1975). Static strength measurements can be made for many muscles or groups of muscles in the body. A complete description of type and direction of effort must be included in any strength survey. There are several other less obvious factors that must also be controlled when measuring static strength. These are described in detail in papers on techniques for evaluating static strength (Caldwell et al., 1974; Chaffin, 1975). These papers should be reviewed before any static strength-testing programs are instituted; their major recommendations are summarized later in this chapter.

a. Standard Procedures for Static Muscle Strength Testing

Many variables affect the measurement of static strength. A thorough understanding of all factors involved, biological as well as procedural, is necessary if valid and reliable strength norms are to be established. This section summarizes the major considerations that should be given to designing an applied strength-testing program. This information should also be useful in evaluating published strength studies.

(1) Measurement of Static Strength

Due to psychomotor, physiological, and biomechanical factors, a time delay exists between the command to exert force and the development of peak force. Figure 27-1 shows this force-time relationship. The value recorded for static strength depends on when the force is measured. A single measure during the first second will be different from a measure at two seconds. An average over the first four seconds will result in a strength measurement less than the peak seen between the second and third seconds. Also, if the force is a true maximal effort, it cannot be held for more than six seconds; the values will decline if the effort is sustained longer. There is therefore a need to standardize static strength measurements. Static strength should be measured between the first and fourth seconds of the effort, as indicated in Figure 27-1, and averaged. Measurement during this period will give a good estimate of maximum static strength since the averaged value will not include those parts of the force curve where the force is changing rapidly.

(2) Subject Description

A complete description of the test population should always be available. Static strength varies greatly, depending on such factors as:

- Age.
- Sex.
- Current health status.

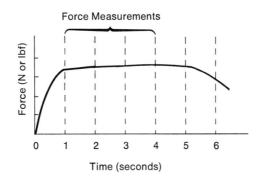

Figure 27-1: Static Force Development and Recommended Measurement Time   The profile of a force curve for an isometric muscle contraction is shown with force measured in newtons (N) or pounds of force (lbf) on the vertical axis and time, in seconds, on the horizontal axis. The area for measurement of the force that can be generated by a specific set of muscles is marked between the first and fourth seconds of the effort. Earlier measurement includes the transient development of force; measurements after four seconds risk some muscle fatigue effects.

- Anthropometry.
- Level of muscle development.
- Experience in exerting static efforts.

When describing an experiment or when evaluating and interpreting a published study, you must know these factors in order to have confidence when using the data in ergonomic designs and analyses.

(3) Instructions to Subjects

Any individual volunteering for a strength testing study has a right to know the intent of the test as well as any risk associated with participation. The confidentiality of individual performance must also be ensured unless permission is given otherwise. Subjects should not be coerced to participate, and, after testing begins, all subjects should be permitted to discontinue participation at any time.

During testing, the procedures must be designed to minimize variation due to motivation. The following are recommended as procedural guides:

- Test instructions should be consistent for all subjects.
- Instructions should be clear and concise.
- Instructions should not include motivational appeals.
- The subject should be told to increase the effort to maximum rapidly but without jerking.
- The maximum effort should be maintained for at least four, and preferably five or six, seconds.
- Qualitative performance feedback should be provided in positive terms, but comparative information should not be given.
- The test environment should be controlled so that variables such as spectators, noise, or other distracting or motivating factors do not influence the results.

(4) Test Procedures

Measurements of static strength are very sensitive to test procedures. As static strength testing is actually the measurement of static torque, many of the biomechanical factors discussed in Part III apply here. In order to ensure that a relationship exists between the muscle strength measured and the strength required for job performance, you need to know the following conditions:

- The specific muscles and body segments tested.
- The body posture during testing.
- The body support and reaction forces available.
- The type of measuring and recording devices used.
- The coupling of the measurement devices to the body.

(5) Reporting the Test Results

Static strength test results should be consistently reported. Generally, the following statistics are considered important and should be reported:

- Mean.
- Standard deviation.
- Skewness.
- Minimum and maximum values (range).
- Sample size.

b. Tray-Lifting Strength

The most important considerations in applied strength-testing programs are the validity and reliability of the tests used. Specific tests should be designed to evaluate the muscles or groups of muscles that contribute to task performance. However, at the same time, the tests should also incorporate the controls necessary to ensure test reliability. Presented here is one approach to static strength testing that has been used successfully in an industrial setting. The purpose of this strength testing program was to predict tray-lifting capacities from anthropometric and strength measurements of workers. A strength-testing device was used for measuring the maximal isometric strengths of specific muscle groups and for measuring maximal isometric tray-lifting capacity. Figure 27-2 shows the device used for all strength tests. A similar device has been used by other investigators for industrial strength testing studies (Kamon and Goldfuss, 1978; Kamon, Kiser, and Landa-Pytel, 1982; Yates et al., 1980).

The design and construction of the strength-testing device was based on two requirements. First, the unit had to be versatile and adjustable so that it could test many muscle groups and accommodate workers of various anthropometric dimensions. Second, the unit had to be portable so that it could be taken out of the laboratory setting and into the plant. The unit pictured in Figure 27-2 proved to be a good design. Forces were measured with a spring-loaded Stoelting handgrip dynamometer that was modified and attached to the base of the testing unit. A potentiometer attached to the dynamometer dial provided a force signal that was integrated during the last three seconds of a four-second effort and was displayed on a digital readout. A load cell or strain gauge could also have been used for the measurement. The dynamometer was attached to a band on the subject or to a tray by a steel cable so that the attachment was always perpendicular to the body segment or the bottom of the tray. Upon the command of an experimenter, the subject would perform a maximal isometric contraction lasting at least four seconds. These maximum contractions were performed for isolated muscle groups and for simulated tray lifts.

Illustrations of the test positions are found in Figure 27-3 for specific muscle groups and in Figure 27-4 for tray lifts. The muscle groups in Figure 27-3 were chosen because of their importance in tray lifting tasks. Figure 27-3a shows

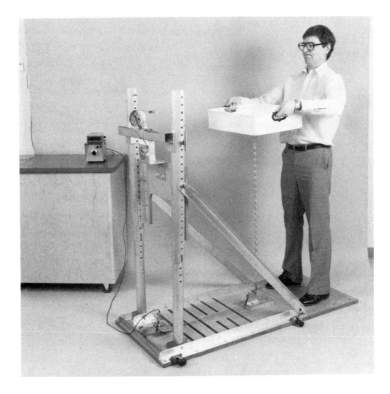

Figure 27-2: Equipment for Measuring Static Strength Capacities
The primary equipment is a Stoelting hand dynamometer that is attached by cable and pulley system to the bottom of a metal tray. Cable length can be adjusted so that strength at different vertical locations can be measured. The subject grasps the tray handholds and pulls directly upward; the experimenter checks that the cable is vertical and not slanted. Horizontal lift position is adjusted by having the subject place his or her feet at specific locations on the apparatus baseboard. A potentiometer output is displayed on a digital readout after being processed in an integrator. The integrator has a time delay so that the average force sustained from the beginning of the second second to the end of the fourth second of the pull is taken, as recommended in standard strength testing protocols. The apparatus can also be used to measure the strength of specific muscle groups. *(Champney, Eastman Kodak Company, 1979; Kamon and Goldfuss, 1978; Yates et al., 1980).*

the measurement of back extension strength. A strap was placed around a subject's shoulders as he or she flexed (90 degrees) at the waist. Upon command of the experimenter, the subject would make a maximal attempt to straighten (extend) the back, while the dynamometer measured the force of the back extension. Figure 27-4*b* shows the test position for elbow flexion strength. For this

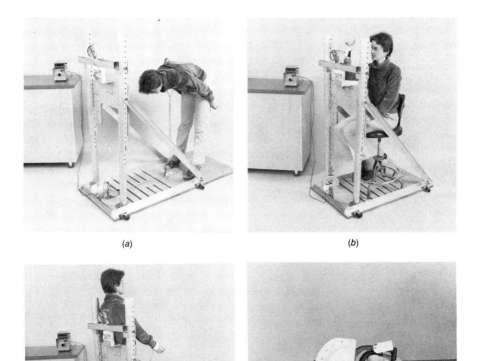

(a)

(b)

(c)

(d)

Figure 27-3: Measurement of Muscle Group Strengths for a Static Lifting Strength Study   Four muscle groups that are important in lifting tasks were tested using a static strength testing device (see Figure 27-2). Measurement of back extension strength is shown in *a*, forearm flexion strength in *b*, shoulder flexion strength at a 45-degree angle in *c*, and handgrip strength in *d*. (*Champney, Eastman Kodak Company, 1979; Yates et al., 1980*).

measurement, the subject was comfortably seated with the elbow bent at 90 degrees and resting on a padded support. The forearm was in a vertical, mid-supinated position. The cable from the dynamometer was attached perpendicularly to an adjustable strap around the subject's wrists.

Figure 27-3c demonstrates the position for shoulder flexor strength mea-

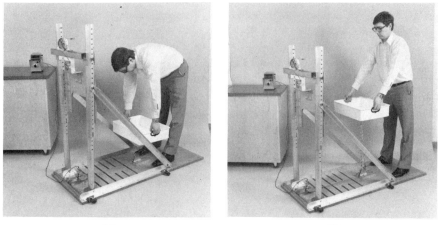

(a)   (b)

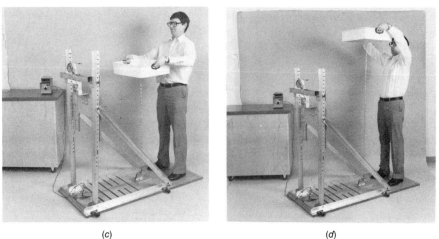

(c)   (d)

Figure 27-4: Vertical Lifting Locations for a Static Tray Lifting Force Study   The vertical testing positions used during a tray-lifting strength study are illustrated. Adjusting the chain linking the dynamometer to the tray allowed variable vertical positions to be established; a, b, c, and d show static tray lifts performed at vertical heights of 33, 81, 134, and 188 cm (13, 32, 53, and 74 in.), respectively. Tests at these positions were repeated for three different horizontal distances between the ankles and the center of the tray. Measurements in these 12 positions were made in the pilot study; five of these were chosen for the industrial study based on their predictability and importance in lifting tasks. See Table 26-12 for a summary of the results of this testing. (*Champney, Eastman Kodak Company, 1979; Yates et al., 1980*).

surements, which were made with the arm at 45 degrees above the vertical from the arm's resting position. This position was chosen because of the involvement of shoulder flexors early in lifting, especially when the load is not close to the body. The subject stood with his or her back towards the dynamometer. The position of the shoulder was determined by measuring the angle between the upper arm and the trunk with a goniometer. The axis of rotation was assumed to be the gleno-humeral joint in the shoulder. The cable from the dynamometer was attached to the wrist in the same manner as for elbow flexion, and the subject pulled up maximally, keeping the feet flat on the apparatus' baseboard. Figure 27-3d demonstrates the measurement of grip strength using a handgrip dynamometer. Each subject adjusted the dynamometer handle to a preferred span. Maximum grip strength was assessed with the subject standing and holding the dynamometer in the preferred hand with the arm at its resting position at the side of the body.

The second phase of the strength-testing program was to measure maximal isometric tray-lifting strength. For these measurements the same lifting apparatus was used. The cable from the dynamometer was attached to the center of a metal tray 50 cm (20 in.) long (between the hands), 33 cm (13 in.) wide, and 10 cm (4 in.) deep. The tray had cut-through handholds 11 cm (4.5 in.) long and 5 cm (2 in.) high, with a beveled edge that increased the area of force exertion on the hands during the upward pull. A pilot study had revealed that tray handle design was an important factor in measured maximal lifting strength, and the cutout handle proved to be an efficient design.

A matrix of tray lift positions was selected so that strength changes associated with changes in the horizontal or vertical location of the tray could be observed. Maximum tray-lifting strength was tested at all 12 positions of the matrix in the pilot study and in five of those positions in the subsequent study of industrial workers. Figure 27-4, a through d, demonstrate tray-lifting strength tests at each of the vertical positions of the matrix with the tray close to the body. The measurements were repeated for two more horizontal distances where the tray was farther away from the body. The results of these studies are shown in Table 26-12 in the previous chapter.

2. DYNAMIC STRENGTH TESTING

Many tasks in industry require muscle effort to move an object in contrast to static muscle efforts during which no movement occurs. The primary reasons static muscle strength tests are recommended is that they permit control of most of the variables that influence muscle strength. Dynamic strength tests, while more valid from the industrial perspective, are difficult to standardize. The primary problems are load acceleration and muscle length changes. However, dynamic strength tests do have a role in an industrial strength-testing program providing their relative positive and negative features are understood. In this section, three types of dynamic tests will be described: psychophysical methods, isotonic dynamic tests, and isokinetic dynamic tests.

a. Psychophysical Methods

Psychophysics is a field of psychology that deals with the relationship between sensations and physical stimuli. The dependent variable usually measured is the perceived exertion. This can be measured in response to physical work, such as lifting, pushing, and pulling. The effects of industrial environmental factors, such as light, temperature, noise, or machine pacing, can also be evaluated using psychophysical techniques. A more extensive discussion of these techniques has been given in Appendix B, Volume 1, of this series.

Psychophysical methods have been used by several investigators to study acceptable weights for manual lifting tasks (Ayoub, Dryden, and Knipfer, 1976; Snook, 1978; Snook and Ciriello, 1974). In these studies the subject (worker) can select the weight that is acceptable to lift under the existing experimental conditions. From an industrial standpoint, there are many job or task variables that can be studied separately or in combination; these might include the following:

- Frequency.
- Duration.
- Size.
- Distance.
- Height.
- Pacing.
- One or two hands.
- Type of handholds.
- Ambient temperature.
- Time of day.

The worker, when presented with the lifting conditions, adjusts the weight of the lift to match his or her personal feelings of fatigue or perception of what can be sustained for several hours. As the characteristics of the lift or work environment are varied, the acceptable weight may change. These studies permit a person to evaluate acceptable workload in dynamic tasks, but the results cannot be extrapolated too far beyond the conditions of the test. What is an acceptable load to lift in a well-designed tray with handholds is not necessarily acceptable if the load is unstable or in a container with rough edges.

b. Isotonic Dynamic Tests

Isotonic dynamic tests refer to tests of muscle force development in which the muscle length changes. This is in contrast to static tests in which muscle length is relatively constant. Isotonic tests of strength are relatively easy and inexpensive to conduct. Weighted objects such as trays, boxes, push-pull carts, or other objects can be used as test loads. The subjects attempt to move the test objects the desired direction and distance on a trial-and-error basis. Weight can be added or removed from the containers so that an estimate of the maximum load or effort can be established.

While these tests are easy to conduct, they are not easy to standardize. Because the human capacity for force generation is very sensitive to minor changes in the characteristics of the effort, isotonic tests tend to be confounded by the techniques used and the velocity and acceleration of the dynamic movement. Useful data bases on dynamic efforts have rarely been established due to the high degree of specificity associated with these types of tests.

c. Isokinetic Strength Testing in Industry

New approaches to dynamic strength testing have been reported that utilize isokinetic strength-testing devices. These devices control the speed of movement to preset values, regardless of the effort exerted. Thus, they represent constant velocity strength measurement systems, and, as such, eliminate most acceleration during the dynamic effort. Recently, the use of a commercially available isokinetic device for measuring dynamic lifting strength has been described (Landa-Pytel and Kamon, 1981). A picture of a modified isokinetic device is shown in Figure 27-5. The modifications to the device include a load cell mounted on the handle to measure instantaneous force, and a speed sensor to measure the velocity of the rope as it is pulled from the device.

Peak dynamic strength, as measured on the isokinetic device, has been shown to correlate well with the actual dynamic lifting capacity of a group of men and women (Landa-Pytel and Kamon, 1981). In this study, the group whose isokinetic strength was measured also selected a maximum acceptable weight for lifting six times per minute. For both men and women, the selected weight from the psychophysical test was 22 percent of the maximum dynamic lift on the isokinetic device. Thus, it appears that isokinetic tests can be used to predict maximum lifting capacity and, perhaps, to determine acceptable loads for frequent lifting. This can be done easily and quickly because only one measurement is required.

The isokinetic technique can also be used for applied lifting studies. An isokinetic device can be attached to a tray or other object to simulate a lifting task. The force and velocity signals are directed to a computer where they are digitized and stored. During data analysis, the velocity data are integrated to give the displacement data. Thus, during the simulated lifts, a record of instantaneous force at incremental vertical heights can be obtained (Figure 27-6).

Since the isokinetic device eliminates most acceleration, the force generated at a vertical location depends on the force-generating capacity of the body at that position. This measurement system effectively eliminates ballistic lifts in which very large forces are generated at low heights where the lifting capacity is high and the lifter relies on momentum to assist the lift at higher positions.

Table 27-1 and Figure 27-6 present examples of a dynamic lifting profile that can be obtained from the isokinetic strength-testing system. This system can also be used to detect positions of maximal force generating capacity. In an experimental test of isokinetic lifting strength, ten subjects were asked to per-

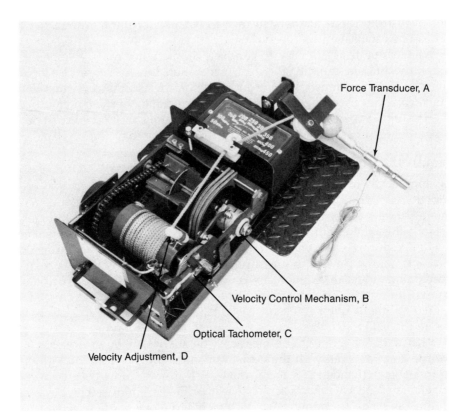

Figure 27-5: A Modified Isokinetic Device for Force and Velocity Measurements   This commercially available, isokinetic muscle-training device has been modified by adding a transducer (A) to measure force and velocity. Force is measured using a piezoelectric load cell that is attached to the end of a nylon cord. An optical tachometer (C) monitors the rate of drum rotation on which the nylon cord is wound; the velocity of the unwinding cord is calculated from the tachometer measurements. Velocity of movement can be adjusted by changing the position of the velocity adjustment lever (D) that is attached to a velocity control mechanism (B). *(Kiser and Corl, Eastman Kodak Company, 1983)*.

form maximal isokinetic tray lifts. The maximum forces were measured, and the vertical positions for maximum force generation were calculated. Table 27-2 presents the maximums for the ten subjects. Other values, based on percentages of maximum capacity, can also be calculated. These types of measurements can be very useful in establishing guidelines for vertical lifting.

Since the resistance that must be overcome during lifting is frictional as opposed to gravitational, the isokinetic strength-testing system is not restricted

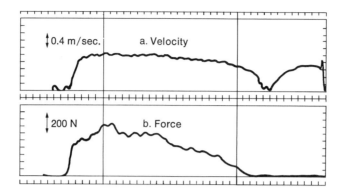

Figure 27-6: Force and Velocity Curves During a Simulated Isokinetic Tray Lift   Two strip chart recordings of continuous velocity (*a*) and force (*b*) measurements are presented for a simulated isokinetic tray lift from 23 to 150 cm (9 to 60 in.) above the floor and within seven inches of the front of the body. Since the lift is isokinetic, the velocity rises quickly at the beginning of the lift to a plateau at a preset velocity. This velocity is relatively constant throughout the lift. The force rises to a peak early in the lift when the strong leg and back muscles are active. As the lift becomes dependent on the arms and shoulders, the forces decrease in a linear fashion, indicating the lower-force-generating capacity of the upper body muscles. (*Kiser and Corl, Eastman Kodak Company, 1983*).

to vertical force measurements. The system can also be used to profile pushing, pulling, and oblique lifting capacities. The use of an isokinetic testing approach does not eliminate the need for standardization of testing procedures. Of particular importance is the velocity of testing because the force that can be developed by a group of muscles decreases with increasing velocity (see Part II). Therefore, simulated isokinetic lifting at high velocities will result in maximal strength values that are less than those found in tests performed at slower speeds.

B. AEROBIC WORK CAPACITY TESTS

A person's maximum aerobic work capacity is the maximum amount of work that can be done when there is enough oxygen to supply the muscles. It may also be called the physical work, or endurance, capacity. The muscles can contract without adequate oxygenation by using anaerobic metabolism, but very long recovery periods are needed. Therefore, the aerobic work capacity is of most interest to ergonomists when evaluating the suitability of a given workload. Knowledge of human physiological capacity under working conditions can help in evaluating work intensity and in developing recommendations for new

Table 27-1: A Dynamic Lifting Profile for One Subject Lifting at a Velocity of One Meter per Second. The values were calculated on a computer from force and velocity data gathered during a simulated lift. Columns 1 and 2 give the vertical heights selected for analysis, while column 3 presents each vertical height in column 1 as a percentage of the subject's shoulder (acromial) height. Columns 4 and 5 list the absolute forces in newtons and pounds of force that were generated at each vertical position. Column 6 shows the relative forces at each position, calculated by dividing each measured force by the maximum force generated in any position and multiplying by 100. In this example, the maximum occurred between two of the vertical positions selected for analysis, so 100 percent does not appear in column 6. *(Kiser and Corl, Eastman Kodak Company, 1983).*

Vertical Height		Relative Height (Percent Shoulder Height)	Absolute Force		Relative Force (Percent Maximum)
cm	(in.)		newtons	(lbf)	
23	(9)	16	184	(41)	60
30	(12)	21	172	(39)	57
40	(16)	28	282	(63)	93
50	(20)	35	286	(64)	94
60	(24)	43	282	(63)	92
70	(28)	50	258	(58)	84
80	(32)	57	243	(55)	80
90	(35)	64	194	(44)	64
100	(39)	71	110	(25)	36
110	(43)	78	32	(7)	10
120	(47)	85	42	(9)	14
130	(51)	92	30	(7)	10
140	(55)	99	38	(8)	13
150	(59)	106	28	(6)	9

Maximum force: 305 N (68 lbf)

Height at maximum force = 43 cm (17 in.)

Table 27-2: Maximum Simulated Lift Forces and Corresponding Vertical Heights. Ten subjects (column 1) performed maximal isokinetic tray lifts at a lift velocity of one meter per second. Maximal forces were recorded from a piezoelectric force transducer (columns 2 and 3); computer integration of the velocity transducer signal allowed calculation of the vertical displacements at maximum force (columns 4 and 5). With two exceptions, the peak lifting force occurred between 40 cm (16 in.) and 50 cm (20 in.) above the floor. *(Kiser and Corl, Eastman Kodak Company, 1983).*

Subject	Maximum Lift Force		Vertical Height at Maximum Force	
	newtons	(lbf)	cm	(in.)
1	305	(68)	46	(18)
2	440	(99)	48	(19)
3	493	(111)	48	(19)
4	365	(82)	46	(18)
5	272	(61)	48	(19)
6	793	(178)	32	(13)
7	317	(71)	50	(20)
8	432	(97)	40	(16)
9	467	(105)	57	(22)
10	233	(52)	50	(20)

Median maximum lift force = 432 newtons (97 lbf)
Median vertical height at maximum force = 48 cm (19 in.)

work practices and workplace designs. Some of the tests used to generate the data given in Appendix A on maximum aerobic work capacities are described in this section.

1. WHOLE-BODY AEROBIC CAPACITY TESTS

Because there are many types of work capacity, there is no one specific protocol for the administration of work capacity tests. Tests have been developed by many research groups, and most fulfill the basic requirements of a good exercise test. Some of the more commonly used exercise protocols are presented here. Some of the factors that influence the design of a work capacity test protocol include:

- The purpose of the test.
- The muscle groups to be tested.

- The time available per test.
- The equipment available.

Specific tests have been developed to determine the working capacity of people in industries where heavy work is common, such as coal mining. An indication of test specificity is seen in the following list:

- Walking.
- Running.
- Crawling.
- Stooped walking.
- Cycling.
- Arm cranking.
- Arm cranking and cycling.
- Rowing.
- Laddermill climbing.
- Shoveling.
- Bench stepping.
- Swimming.

However, when information on maximum aerobic capacity ($\dot{V}_{O_2}$ max) is desired, several standardized tests are used. These tests are usually administered using a motorized treadmill, a stationary cycle ergometer, or a bench-stepping test (see Figure 27-7). Upper body tests using arm crank ergometers are also frequently used.

In work capacity testing, some general types of test designs include:

- A single-stage test during which the subject works at a constant work rate for the entire test, *or* a multi-stage test (graded) during which the subject works at progressively higher work rates as the test progresses.

- An intermittent (discontinuous) test in which the subject is given rest periods between the work stages of a multi-stage test, *or* a continuous test in which the subject exercises continuously until the test is terminated.

- A submaximal test that calls for termination of the test at a predetermined point, usually at some percentage of predicted maximum heart rate (see Part III), *or* a maximal test that lasts until the subject can no longer continue or until signs or symptoms develop that could place the subject at risk.

a. General Design of Tests

A single-stage test was one of the earliest and most convenient work capacity tests used. Two examples are the Master's Two-Step Test and a similar test from the Harvard Fatigue Laboratory, both still used by some general practitioners to support their diagnoses of possible cardiovascular disease in patients (Brouha, 1943; Master, 1934). For maximal work capacity evaluations, multi-stage or graded exercise tests (GXT) are the appropriate protocols to consider. GXTs are generally used for the following purposes:

- Evaluating an individual's capacity for physical activity, whether work or exercise.

- Determining a level of physical stress at which symptoms of physiological impairment occur—for example, to identify the functional capacity of a person with cardiovascular disease.

- Evaluating the effectiveness of a physical conditioning or rehabilitation program.

From an ergonomics standpoint, the GXT is most useful for evaluating work capacity. Medical personnel should be present during the capacity test to monitor the electrocardiogram (ECG) and blood pressure and to ensure that the person's response to the exercise test is normal. Individuals interested in administering GXTs should refer to the American College of Sports Medicine (1980) publication, *Guidelines for Graded Exercise Testing and Exercise Prescription*.

Many different GXT protocols are being used for work capacity testing. The main differences between protocols include:

- The initial starting speed and grade (stage 1).
- The duration of each stage.
- The work intensity increment between stages.

A basic protocol for three types of GXTs is shown in Figure 27-7. The values are based on recommendations of the International Committee for the Standardization of Physical Fitness Tests (Ahta, 1974).

It may be desirable to modify these standard protocols or to develop a unique protocol to determine the work capacities of interest. When doing this, you should incorporate some basic GXT design principles into the test. For example:

- A warm-up stage should be built into the progressive increase in work intensity. This permits the muscles to adjust to the blood flow requirements and energy demands as they are increased at higher workloads. See Part II for further information on muscle blood flow regulation.

- The starting level and intensity increments between stages should be designed so that several stages can be completed before a test end point is reached. At least one, and preferably two, of those stages should be at workloads of more than 40 percent of capacity.

Some additional test design guidelines are:

- Design the test to last at least 12 minutes but usually not more than 20 minutes. Each stage should last at least two to three minutes for continuous tests and four to five minutes for intermittent or discontinuous tests.

- Start healthy individuals on the work capacity test at an energy expenditure approximately three times resting expenditure (3 METs). The initial intensity can be modified for individuals with very high or very low capacity levels.

- Change the intensity of the work at each stage by 2 METs (for healthy,

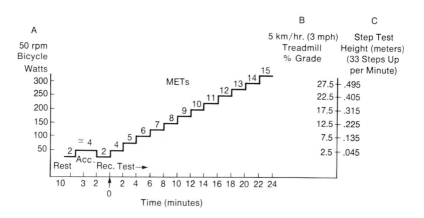

Figure 27-7: Whole-Body Aerobic Capacity Test Protocols   The basic protocol for three types of whole-body aerobic capacity tests is illustrated. The workload in watts on a bicycle ergometer needed to achieve the given MET values (on the stepped line) are shown on vertical scale A. Scale B gives the percent grade on a treadmill running at 5 km/hr (3 mph) that will produce the same MET workloads. For stepping up on a platform at 33 steps per minute, scale C gives the necessary step height in meters to provide the appropriate MET workloads. The horizontal scale shows the time in minutes with zero marking the start of the test. Two minutes are spent at each work level in this test protocol. There is also a ten-minute rest period before the test begins ("Rest"), a low-level trial exercise period for four minutes ("Acc."), and two minutes of recovery time ("Rec.") before the official test begins. The guidelines for design of capacity tests used in this basic protocol are discussed in more detail in the text. (*Adapted from Atha, 1974*).

sedentary individuals) and by not more than 3 METs. For individuals with poor conditioning, 1 MET increments may be more appropriate.

b. Examples of Capacity Tests

The recommended protocol for work capacity testing shown in Figure 27-7 incorporates the basic principles of GXT design. The approximate aerobic demands for four staged capacity tests are given in Tables 27-3 through 27-6. Individuals studied on each test will not necessarily have the exact metabolic responses shown in these tables since other factors, such as their degree of coordination, can influence the oxygen consumption measurement. These tests can be used, however, to identify an individual's cardiovascular response at specific workloads and to classify his or her fitness level.

The equipment needed to measure energy requirements during work capacity tests is not always available. However, reasonably good estimates of the energy cost can be determined by knowing the conditions of work. Table 27-7

Table 27-3: Aerobic Demands of Step Test Work Levels. The aerobic demands of six step heights (column 1) and four stepping rates (columns 2 through 9) are given in METs and liters of oxygen per minute ($\dot{V}_{O_2}$). The values are estimates for a 70 kg person and will differ for people of other body weights. A submaximal whole-body aerobic capacity test with several levels can be developed by using this table to identify a combination of heights and stepping rates that will gradually increase the load on the person's cardiovascular system. See Figure 27-7 for a suggested step test protocol. (*Adapted from American College of Sports Medicine, 1980*).

Step Height cm (in.)	Steps per Minute							
	12		18		24		30	
	METs	$(\dot{V}_{O_2})^*$	METs	$(\dot{V}_{O_2})^*$	METs	$(\dot{V}_{O_2})^*$	METs	$(\dot{V}_{O_2})^*$
0 (0)	1.2	(0.28)	1.8	(0.42)	2.4	(0.56)	3.0	(0.70)
8 (3.2)	1.9	(0.44)	2.8	(0.65)	3.7	(0.86)	4.6	(1.07)
16 (6.3)	2.5	(0.58)	3.8	(0.88)	5.0	(1.16)	6.3	(1.47)
24 (9.4)	3.2	(0.75)	4.8	(1.12)	6.3	(1.47)	7.9	(1.84)
32 (12.6)	3.8	(0.88)	5.7	(1.33)	7.7	(1.79)	9.6	(2.24)
40 (15.8)	4.5	(1.05)	6.7	(1.56)	9.0	(2.10)	11.2	(2.61)

*Based on a 70 kg person. One MET is one kilocalorie per kilogram body weight per hour. Oxygen consumption ($\dot{V}_{O_2}$) is expressed in liters per minute.

Table 27-4: Aerobic Demands of Treadmill Levels. The aerobic demands of walking on a treadmill ergometer at ten grades (column 1) and six speeds (columns 2 through 13) are given in METs and liters of oxygen per minute. The percent grade is the slope of the treadmill above the horizontal plane. The values shown are based on a 70 kg person. Oxygen consumption will vary with differences in body weight, and $\dot{V}_{O_2}$ and METs can vary with differences in a person's skill at walking on a treadmill. A graded exercise protocol that meets the requirements of a safe and reliable test can be constructed from the data in this table. See Figure 27-7 for a suggested treadmill protocol. (*Adapted from American College of Sports Medicine, 1980*).

Speed, in km/hr (mph)

Percent Grade	2.8 (1.7)		3.3 (2.0)		4.2 (2.5)		5.0 (3.0)		5.7 (3.5)		6.25 (3.75)	
	METs	$\dot{V}_{O_2}$*	METs	$\dot{V}_{O_2}$*	METs	$\dot{V}_{O_2}$*	METs	$\dot{V}_{O_2}$*	METs	$\dot{V}_{O_2}$*	METs	$\dot{V}_{O_2}$*
0	2.3	0.54	2.5	0.58	2.9	0.68	3.3	0.77	3.6	0.84	3.9	0.91
2.5	2.9	0.68	3.2	0.75	3.8	0.88	4.3	1.00	4.8	1.12	5.2	1.21
5.0	3.5	0.82	3.9	0.91	4.6	1.07	5.4	1.26	5.9	1.37	6.5	1.51
7.5	4.1	0.96	4.6	1.07	5.5	1.28	6.4	1.49	7.1	1.65	7.8	1.82
10.0	4.6	1.07	5.3	1.23	6.3	1.47	7.4	1.72	8.3	1.93	9.1	2.12
12.5	5.2	1.21	6.0	1.40	7.2	1.68	8.5	1.98	9.5	2.21	10.4	2.42
15.0	5.8	1.35	6.6	1.54	8.1	1.89	9.5	2.21	10.6	2.47	11.7	2.73
17.5	6.4	1.49	7.3	1.70	8.9	2.07	10.5	2.45	11.8	2.75	12.9	3.00
20.0	7.0	1.63	8.0	1.86	9.8	2.28	11.6	2.70	13.0	3.03	14.2	3.31
22.5	7.6	1.77	8.7	2.03	10.6	2.47	12.6	2.94	14.2	3.31	15.5	3.61

*Based on a 70 kg person. One MET is one kilocalorie per kilogram body weight per hour. Oxygen consumption ($\dot{V}_{O_2}$) is expressed in liters per minute.

Table 27-5: Aerobic Demands of a Staged Treadmill Task Used in Industrial Stress Testing. The METs and oxygen demands of walking on a treadmill at four speeds (columns 2 through 9) and at six grades (column 1) are given. These values are averages from industrial studies on 40 people (columns 2, 3, 6 to 9) and 140 people (columns 4 and 5); the values in parentheses in the body of the table are estimates. When these values are compared to those in Table 27-3, a higher oxygen usage is found at the lower workloads for the industrial population. This probably reflects the inexperience of the industrial population with treadmill walking; the data were taken from their first test only. The protocol for this test is similar to the one shown in Figure 27-7 except that more time was spent at the first few levels of work and a majority of the data were collected at walking rates of 4 km/hr (2.5 mph). *(Rodgers, Eastman Kodak Company, 1973).*

Percent Grade	Speed, in km/hr (mph)							
	3.3	(2)	4.2	(2.5)	5	(3)	5.8	(3.5)
	METs	$\dot{V}_{O_2}$*	METs	$\dot{V}_{O_2}$*	METs	$\dot{V}_{O_2}$*	METs	$\dot{V}_{O_2}$*
0	2.8	9.8	3.3	11.4	(3.9)	(13.5)**	(4.3)	(15.1)**
5.0	3.9	13.5	4.6	16.1	(5.4)	(19.1)	(6.3)	(21.8)
7.5	4.4	15.4	5.3	18.4	(6.3)	(21.9)	(7.2)	(25.2)
10.0	(4.9)	(17.2)**	5.9	20.8	(7.0)	(24.6)	8.3	28.8
12.5	(5.4)	(19.1)	6.6	23.1	(7.8)	(27.4)	(9.1)	(32.0)
15.0	(6.0)	(21.0)	7.3	25.5	8.7	30.3	10.1	35.4

* $\dot{V}_{O_2}$ is expressed in milliliters of oxygen per kilogram of body weight per minute.
**Values in parentheses are estimated from the relationship between oxygen consumption and grade changes at constant speed where one to three points on the curve were measured.

presents equations that can be used to calculate the expected energy cost during a work capacity test. Because various units of measurement are used to describe the work performed during testing, Table 27-8 shows the conversion factors among the most commonly used units.

2. UPPER BODY AEROBIC CAPACITY TESTS

The maximum aerobic capacity for a person who is standing at a workplace where most of the work is performed with the arms will be lower than that possible if the larger muscles of the legs and lower trunk are dynamically active. The data in Table 26-17 in Appendix A show that upper body aerobic capacity is about 70 percent of whole-body aerobic capacity. These data have been gen-

Table 27-6: Aerobic Demands of Bicycle Ergometer Work Levels. The energy requirements (in METs) and the oxygen requirements, ($\dot{V}_{O_2}$ liters per minute), of pedaling on a stationary bicycle at seven different workloads, expressed in watts, are shown across the top of columns 2 through 8. People of six different body weights (column 1) are included to show how the METs are affected at each workload. Local muscle fatigue can be a problem for people untrained to bicycle riding. A recommended bicycle ergometer test protocol is shown in Figure 27-7. *(Adapted from American College of Sports Medicine, 1980).*

Body Weight kg (lbm)	Energy Requirements, in METs						
	Exercise Rate, $\dot{V}_{O_2}$						
	50 (0.90)	75 (1.20)	100 (1.50)	125 (1.80)	150 (2.10)	175 (2.40)	200 watts (2.70) $\dot{V}_{O_2}$, L/min
50 (110)	5.1	6.9	8.6	10.3	12.0	13.7	15.4
60 (132)	4.3	5.7	7.1	8.6	10.0	11.4	12.9
70 (154)	3.7	4.9	6.1	7.3	8.6	9.8	11.0
80 (176)	3.2	4.3	5.4	6.4	7.5	8.6	9.6
90 (198)	2.9	3.8	4.8	5.7	6.7	7.6	8.6
100 (220)	2.6	3.4	4.3	5.1	6.0	6.9	7.7

erated from two types of upper body capacity tests: arm cranking on a modified bicycle ergometer known as an arm ergometer, and an upper body lifting test, wherein the weight of a tray is increased in four stages while it is lifted at 24 times per minute around a series of shelves. The protocols for each of these tests of upper body capacity are diagrammed in Figure 27-8.

The lifting task test of upper body capacity appears to be limited by the discomfort felt at the wrists when the tray is lifted to 127 cm (50 in.) above the floor. The heavier the tray, the more discomfort was felt, so some people were unable to handle the 11.5 kg (25 lbm) tray for the three to four minutes required for the test. The arm cranking test is more standardized, but there are not many industrial tasks that use the muscles of the trunk and arms in the way that they are used in a cranking task. Therefore, the test may predict a higher aerobic capacity than is actually available in materials handling, packing, or assembly tasks done primarily with the upper body.

As was discussed in Appendix A under aerobic work capacities, the aero-

Table 27-7: Equations for the Estimation of Energy Expenditures During Various Activities. The amount of oxygen, in milliliters per kilogram of body weight per minute (mL O_2/kg BW • min), required to walk, run, pedal, or step up at a given speed or grade can be calculated using these six sets of prediction formulas. The equations yield estimates only and should not be used to determine individual suitability for a given job. They are useful in predicting the appropriate staging of a whole-body aerobic capacity test protocol for an individual. *(American College of Sports Medicine, 1980).*

A. Horizontal Walking

$\dot{V}_{O_2}$ (mL/kg·min) = speed (m/min) × 0.1 mL O_2/kg·min + 3.5 mL O_2/kg·min

---

B. Grade Walking (50 to 100 m/min)

---

1. Horizontal Component: Calculate as above.
2. Vertical Component:
   $\dot{V}_{O_2}$ (mL/kg·min) = % grade* × speed (m/min) × 1.8 mL O_2/kg·min
3. Total for Grade Walking: Horizontal Component + Vertical Component

---

C. Horizontal Jogging and Running (greater than 134 m/min)

$\dot{V}_{O_2}$ (mL/kg·min) = speed (m/min) × 0.2 mL O_2/kg·min + 3.5 mL O_2/kg·min

---

D. Grade Running (Treadmill)

---

1. Horizontal Component: Calculate as above.
2. Vertical Component:
   $\dot{V}_{O_2}$ (mL/kg·min) = % grade* × speed (m/min) × 1.8 mL O_2/kg·min × 0.5
3. Total for Grade Running: Horizontal Component + Vertical Component

---

E. Bicycle Ergometer

$\dot{V}_{O_2}$ (mL/min) = work rate (kpm/min) × 2 mL.kpm + 300 mL/min

---

F. Bench Stepping

---

1. Stepping Up and Down:
   $\dot{V}_{O_2}$ (mL/min) = height (m/lift) × rate per minute × 1.33 × 1.8 mL(m/lift)kgm × body mass (kg)
2. Stepping Back and Forth on Level:
   $\dot{V}_{O_2}$ (mL/min) = (rate per minute/10)** × 3.5 mL O_2/kg·MET × body mass (kg)
3. Total for Bench Stepping:
   Stepping Up and Down + Stepping Back and Forth

---

1 mph = 26.8 m/min = 1.6 km/hr.

*% grade = fraction of vertical distance climbed per minute divided by the belt speed.

**Approximate energy cost in METs.

Table 27-8: Equivalents for Standard Units in Capacity Testing. Several constants that demonstrate the relationship between power and work and energy units of measurement are given. Although the SI system uses the watt as the unit of power, the MET is a standard unit in exercise physiology capacity testing. The kilocalorie has been used traditionally by nutritionists, ergonomists, and industrial engineers as a measure of energy; milliliters of oxygen per kg body weight per minute is used by physiologists. The kilopond-meter (kpm) is only used in bicycle ergometry where watts are more commonly used to describe workload. The 5 kcal equivalent for one liter of oxygen is approximate and depends heavily on the metabolic pattern of the subject and the type and duration of the muscular work being done. *(Astrand and Rodahl, 1977).*

		Power
1 watt	=	0.01433 kcal/min
1 watt	=	1 joule/sec
1 watt	=	6.12 kpm/min
1 kcal/min	=	69.767 watts
1 kcal/hr	=	1.163 watts
1 MET	=	3.5 mL O_2/kg·min
1 MET	=	1.0 kcal/kg·hr

		Work and Energy
1 kcal	=	4,186 joules
1 kcal	=	426.85 kpm
1 kg-m	=	2.34×10^{-3} kcal
1 L O_2	≅	5 kcal

bic capacity for a manual handling task will vary with the weight, location, and frequency of lifting. Therefore, the preferred approach for measuring lifting capacity is to simulate the workplace conditions and to vary load and frequency of lifting to produce a staged workload similar to the treadmill and ergometer tests discussed earlier in this chapter.

C. CARDIOVASCULAR AND METABOLIC MEASUREMENTS

Most industrial evaluations of job demands measure the worker's performance but not the physiological cost of performing. Measures of heart rate, blood pres-

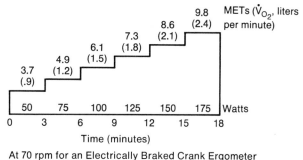

At 70 rpm for an Electrically Braked Crank Ergometer
At 50 rpm for a Friction Crank Ergometer

(a)

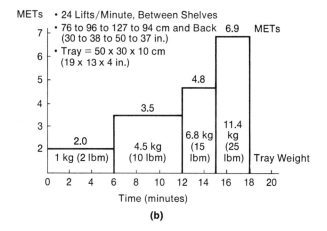

(b)

Figure 27-8: Upper Body Aerobic Capacity Test Protocols   Two examples of graded tests to evaluate upper body aerobic capacity are shown. In *a*, an arm crank ergometer is used to increase the load sequentially on the upper body musculature; each stage lasts three minutes. The work required is shown in watts above the time axis. The aerobic demands are shown above the line as liters of oxygen per minute ($\dot{V}_{O_2}$) and METs. In *b*, a protocol for a submaximal capacity lifting test used in an industrial setting is indicated at the top of the graph. Tray weight was varied from 1 to 11.4 kg (2 to 25 lbm) as shown in each time period. The work in METs is indicated above each block of time. The longer work periods at lower workloads reflect the need to "warm up" and settle into a steady work pattern. Oxygen consumption was measured during the last two minutes of each time block. (*Rodgers, Eastman Kodak Company, 1974; Sawka, Foley, and Timental, 1983*).

sure, minute ventilation, and metabolism (oxygen consumption and carbon dioxide production) can quantify the cost of performing at a given level on the job, especially in jobs where physical effort is present. Part III's chapter on survey methods should be reviewed for examples of when these measurements are best used to quantify job demands.

As technological improvements in physiological monitoring equipment are occurring rapidly, there is no attempt made here to recommend a specific instrument for these measurements. The basic techniques are described briefly and the appropriate cautions and controls are included. With this information, the interested reader should be able to evaluate current equipment and choose the most suitable system for the needs of the plant and the finances available. Part III's chapter on heart rate interpretation methodology should be consulted for more information about these measures.

1. HEART RATE

The most convenient measure of the physiological cost of work is the heart or pulse rate. This is the number of heartbeats per minute, which reflects the need for delivery of oxygen via the blood flow to the working muscles. There are several ways to measure the heart rate. For example:

- Radial pulse—the pulsations of the radial artery are counted by putting the index and middle fingers in the groove on the thumb's side of the wrist.

- Carotid pulse—the pulsations of the carotid artery are counted by putting the index and middle finger on the neck near the angle of the jaw but more towards the chin. Light pressure is required as the artery contains stretch receptors to control blood pressure and heavy pressure may change the heart rate or result in the subject feeling dizzy, especially after heavy exercise.

- Radiotelemetry of a signal corresponding to each R wave of the electrocardiogram (ECG) (see Figure 27-9) after the signal has been processed by a transmitter. This permits monitoring the heart rate without interfering with the worker's activities. The FM receiver (see Figure 27-10) can be wired into a recording device to keep a running heart rate record during the work period.

- Radiotelemetry of the ECG, which permits both heart rate and ECG analysis to be done during the person's work and rest periods.

- Continuous taping of the electrocardiogram for storage of heart rate information and subsequent analysis of the intervals between beats. Figure 27-9 demonstrates how heart rate can be read from an ECG trace. This is done automatically by some forms of tape scanning equipment (see Figure 27-10), providing a continuous heart rate trace for the job as done

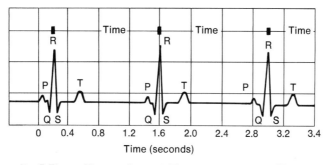

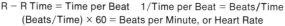

R − R Time = Time per Beat   1/Time per Beat = Beats/Time
(Beats/Time) × 60 = Beats per Minute, or Heart Rate

The R − R intervals should be measured over at least 3
beats to reduce the variability in heart rate associated
with respiration and other factors.

Figure 27-9: Heart Rate Calculations from the Electrocardiogram
The electrocardiogram (ECG) trace measures the heart's electrical activity (voltage) against time, which is determined by the speed of the
recording paper when the ECG is made. The electrocardiogram has
three distinct components. The P wave represents the depolarization
of the atria or upper chambers of the heart. The QRS complex represents the depolarization of the ventricles. The T wave represents the
repolarization of the ventricles. The R waves can be used to measure
the time between beats of the heart; the transformation of the intervals
between several spikes to the heart rate in beats per minute is shown
below the trace. This technique permits changes in heart rate to be measured over short intervals and is more accurate than taking a pulse rate
over the same time period.

by one person. Part III's chapter on heart rate interpretation methodology illustrates several of these traces.

When taking a manual pulse rate or a heart rate by telemetry, it is usual
to count the beats over a 15-second period and multiply by four to get the heart
rate per minute. For resting or light work heart rates, it may be more accurate
to count the beats over a 30-second period, as a miscount of one beat will have
less impact with the longer measurement time. If the pulse is counted manually
and the worker is interrupted while doing a task, the beats should be counted
over ten seconds and the result multiplied by six to get a more accurate measure
of the task demands. It is also important to take the pulse rates in the steady
state of work, not in the very first minute of a task. For very short activities,
this is not a satisfactory way of evaluating the heart rate demands of a task. See
Parts II and III for more discussion of heart rate.

The most critical heart rate range for studies of occupational task demands

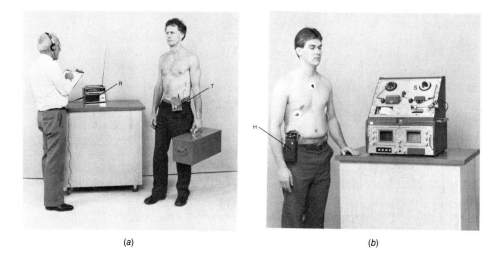

(a)   (b)

Figure 27-10: Heart Rate Monitoring Equipment   Two types of equipment used to monitor the heart rates of people in the workplace are shown. In *a,* a telemetry apparatus is being used to collect the beat-by-beat heart rate at a distance from the worker. The worker wears a set of electrodes that pick up the signals from the heart. The transmitter (T) converts the R wave (Figure 27-9) into a sound signal that can be received by an FM receiver several feet away from the worker. The observer's headphones are plugged into the receiver (R); heartbeats are counted over a 15-second period and the value is multiplied by four to get the heart rate. This is an inexpensive system for monitoring heart rate; more expensive telemetry systems permit direct recording of the ECG at the receiver. In *b,* a tape recording system (Holter monitor) is used to capture a continuous electrocardiogram. The recorder (H) is worn at the worker's belt, and electrodes pick up the ECG from three locations on his chest. After the study is done, a scanner (S) is used to read back the tape at high speed and to collect the heart rate data. Figures 9-1 through 9-10 illustrate the processed scanner output from a recorder. With both types of equipment, the observer can collect physiological data without interfering with the employee's work pattern.

is from 60 to 180 beats per minute. For job studies, it is desirable to be able to detect changes in heart rate of five and ten beats per minute with some work tasks. Some scanners provide heart rate output traces with little scale resolution in this range. It may be preferable to use the raw ECG output signal and to process it so it can be fed into a recording device with more resolution (Evan, Eastman Kodak Company, 1972). The heart rate traces in Part III have been developed in this manner and have a scale of 20 beats per minute per 2.5 cm (1 in.)

2. BLOOD PRESSURE

Blood pressure measurements are made with a blood pressure cuff and gauge (sphygmomanometer) and stethoscope (Figure 27-11). Two readings are taken: systolic blood pressure, which is the highest pressure recorded during the heart's cycle; and diastolic pressure, which is the pressure in the heart and large arteries when the heart is relaxed and filling with blood between beats. Systolic pressure is detected when the pressure cuff is slowly deflated from a pressure above 200

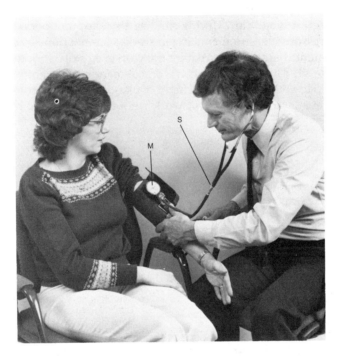

Figure 27-11: Blood Pressure Measurement The technique for measuring blood pressure with a sphygmomanometer (blood pressure cuff and meter) is illustrated. The observer wraps the cuff firmly but not tightly on the worker's upper arm just above the elbow, with the bulb on the medial (towards the body) side of the arm when the palm is up. The cuff is inflated and then slowly deflated as the observer listens through a stethoscope (S) at the inside of the elbow to the sounds of blood flowing through the brachial artery. The first sounds indicate systolic pressure, which is recorded from the meter (M). As the cuff is further deflated, the sounds become muffled and disappear; the diastolic pressure reading is taken at this point. Use of this equipment interferes with the work pattern because the worker must stop the task to have a measurement made. Blood pressure measurements are most useful in determining the cardiovascular stresses of heavy dynamic and static muscle effort.

mm Hg. When the brachial artery pulse at the elbow bend is first heard, the gauge reading is systolic blood pressure. As the cuff is further deflated, the pulse sound disappears. The point at which it becomes "muffled" and then can no longer be heard is when the pressure gauge reading reflects diastolic blood pressure. This is a more difficult reading to make than the systolic pressure, and a high ambient noise level can make it almost impossible to detect.

In addition to difficulties in detecting the diastolic pressure because of ambient noise levels, the heart rate level can influence its reading. Figure 27-12 illustrates how the diastolic blood pressure reading is higher at higher heart rates and lower at slower heart rates because the level is determined by the "run-off" time for blood flow during the heart's relaxation period.

Blood pressure measurements in job studies are not done as often as heart rate measurements because they require the worker to stop work while the pressure is measured. This may alter the work demands sufficiently to make the study invalid. However, the measurement of blood pressure during work is very important in evaluating the stress of work where postural static work is signif-

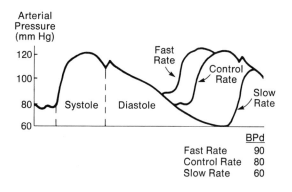

	BPd
Fast Rate	90
Control Rate	80
Slow Rate	60

Figure 27-12: Effect of Heart Rate on the Diastolic Blood Pressure Reading   The arterial blood pressure curve is shown during the contraction (systole) and relaxation (diastole) of the heart. The pressure was measured directly by placing a small tube, or catheter, in an arm artery. Systolic pressure (BPs) is the peak pressure during heart contraction; diastolic pressure (BPd) is the lowest pressure during relaxation. The effect of heart rate on the BPd reading is shown on the right side of the curve. The downward slope of the pressure curve, or diastolic "run off," is constant, depending on the amount of resistance in the peripheral blood vessels. If the heart is beating fast, the next systole will occur at a higher point on the diastolic pressure curve, giving a higher diastolic pressure than at rest. When the heart rate is very low, diastolic pressure is lower because more run off can occur before the next systole. These measurement problems make diastolic pressure more difficult to interpret during work, so systolic pressure is usually used to assess workload. (*Adapted from Rushmer, 1972*).

icant (Lind and McNichol, 1967). Blood pressure changes may be very high in static muscle work while heart rate changes remain minimal.

3. MINUTE VENTILATION

The volume of gas breathed out per minute ($\dot{V}_E$) gives a measure of the respiratory demands of work. The amount breathed in per minute ($\dot{V}_I$) is sometimes measured instead in work and exercise physiology and is a good indicator of $\dot{V}_E$. The expired volume is measured by breathing through a gas meter that mechanically determines the volume per breath (tidal volume) and accumulates the volume measurement over several breaths (respiratory rate) for a fixed time period, such as one minute. To collect the expired air, the worker must wear a partial face mask or a mouthpiece and nose clip, which can become a problem when communications and good visual control are needed on the job. Newer forms of equipment to measure expired gas flow rates and oxygen consumption are being developed, but the problems of collecting expired air without producing some discomfort for the worker have not yet been resolved satisfactorily.

The readings of minute ventilation are usually made in conjunction with oxygen consumption measurements of the job demands. It is very important that the samples chosen reflect the work pattern, since oxygen consumption must be measured over several minutes. The worker is not in a steady state if the ventilation rate is changing frequently during the sample. The ventilation would reflect an average of all of the activities performed during the time of the sample. Instantaneous readings of minute ventilation can be made using a special flow transducer. These peak values should not be used to calculate oxygen consumption values unless a fast-responding oxygen electrode reading can be made in the same time period.

The volume of gas breathed out during minute ventilation is less than the volume that gas occupied when it was in the lungs. This is a function of the gas laws, which show that volume increases with increased temperature or with decreased barometric pressure. To determine the volume that the gas occupied in the lungs, one must convert the values from ATPS (ambient temperature and pressure, saturated) to BTPS (body temperature and pressure, saturated). These volumes are corrected again to STPD (standard temperature and pressure, dry) in the oxygen consumption equation shown later in this section. The interested reader should consult a sourcebook on pulmonary physiology for more detailed instructions of how to measure minute ventilation and collect expired gas for subsequent oxygen analysis (Comroe et al., 1962; Consolazio, Johnson, and Pecora, 1963; West, 1974).

4. OXYGEN CONSUMPTION AND CARBON DIOXIDE PRODUCTION

a. Oxygen Consumption

The estimation of oxygen uptake in the lung is made by measuring the amount of air exchanged per minute (the minute ventilation) and determining the amount of oxygen in the inspired and expired air. The equation describing this relationship is:

$$\dot{V}_{O_2} = \frac{\dot{V}_E \, (F_{I_{O_2}} - F_{E_{O_2}})}{100}$$

where $\dot{V}_{O_2}$ is the oxygen uptake in the lung in liters per minute, STPD; $\dot{V}_E$ is the minute ventilation in liters per minute, BTPS; $F_{I_{O_2}}$ is the fraction of oxygen in the inspired air (20.9 percent if the room air is breathed); and $F_{E_{O_2}}$ is the fraction of the oxygen in the expired air sample.

The difference between the amount of oxygen breathed in and the amount breathed out is a measure of the amount consumed.

A sample of expired gas taken from a mixing chamber or a sample bag is analyzed for oxygen partial pressure using a polarographic electrode or another O_2 sensing instrument (Davies, 1962). Figure 27-13 illustrates a Koffrani-Michaelis respirometer that has been used successfully in field measurements of the oxygen consumption demands of jobs (Davis, Faulkner, and Miller, 1969; Lehmann, 1962). At higher workloads, the amount of oxygen removed from the

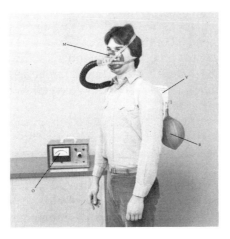

Figure 27-13: Oxygen Consumption Measurement   The Koffrani-Michaelis oxygen consumption respirometer is illustrated in use. The backpack consists of a gas meter (V) that measures the volume of air expired per breath and accumulates volume over time so that the ventilation ($\dot{V}_E$) can be calculated. The worker wears a partial face mask (M) with a two-way breathing valve and flexible conductive tubing that directs the expired air to a sampling port to which a gas bladder (B) is attached. The gas bladder sample is then passed through an oxygen meter (O) to determine the expired gas values. This equipment, although not the most recent, is inexpensive, reliable, and permits measurement of job demands with minimal interference in most jobs.

alveoli by the blood will increase, lowering the amount of O_2 found in the expired air. The O_2 extraction, or $(F_{I_{O_2}} - F_{E_{O_2}})$, from the lungs is therefore increased.

b. Carbon Dioxide Production

The carbon dioxide formed in muscular work must be removed from the body in order to allow work to continue. This CO_2 production $(\dot{V}_{CO_2})$ can be estimated by measuring the minute ventilation and the amount of CO_2 in the expired air (Rahn and Fenn, 1955), or:

$$\dot{V}_{CO_2} = \frac{\dot{V}_E (F_{E_{CO_2}})}{100}$$

where $\dot{V}_{CO_2}$ is expressed in liters per minute, STPD; $\dot{V}_E$ is expressed in liters per minute, BTPS; and $F_{E_{CO_2}}$ is the percentage of CO_2 in an expired air sample, measured with an infrared detector or other analytic instrument such as a mass spectrometer.

The strong influence of carbon dioxide (CO_2) on the regulation of breathing and its role in the acid/base balance of the body make it a less reliable measurement than oxygen consumption for quantifying physical effort. It is very reliable, however, if the work is steady and is sustained for at least five minutes so adequate amounts of an expired air sample can be collected.

Choosing when to take a sample for oxygen consumption and carbon dioxide production measurements is an important part of studying metabolic job demands. If a particular task is thought to be a problem for many people to perform, it should be measured directly to get a working average value of the job demands. However, it should also be looked at in the context of recovery or rest periods available during the time the task is being done (see Part IV). It is wise to have a time standard rating of the job made during the time of oxygen consumption measurements in order to determine whether the work rate is at, above, or below the expected rate.

In reporting the oxygen consumption values for a job or task, one should include the average demands over the shift, the working demands over the shift weighted for the amount of time spent in each activity, and the peak demands (Davis, Faulkner, and Miller, 1969).

D. REVIEW OF TRIGONOMETRY FOR BIOMECHANICAL ANALYSES

In the biomechanics section of Part III, techniques for solving force and torque problems were presented. Many of these techniques rely on trigonometric relationships that are based on the ratios of the sides of a right triangle and their enclosed angles. A brief review of the basics is included here to benefit readers who do not routinely use such calculations.

For problem solving in biomechanics we are concerned with the right triangle, a triangle containing an internal angle of 90 degrees. The sum of a tri-

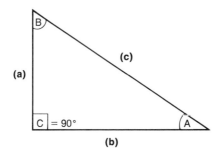

Figure 27-14: A Right Triangle for Trigonometric Analysis   A right triangle is shown with its three internal angles marked A, B, and C. The right angle of 90 degrees is marked as C. The three sides of the triangle are indicated by a, b, and c. The hypotenuse (c) is the side opposite the right angle. Sides a and b are defined as the "opposite" and "adjacent" sides to angle A, respectively. The sum of the angles in a right triangle is 180 degrees. See the text and Part III for more information about using these trigonometric relationships to analyze postures and lifting tasks.

angle's three internal angles always equals 180 degrees. The right triangle consists of six parts, three angles and three sides (Figure 27-14). The longest of the three sides is always opposite the right angle and is called the hypotenuse. In force vector or torque problems using trigonometric relationships, the lengths of the sides are proportional to the magnitudes of the forces.

There are six trig functions, defined as follows:

Sine A (sin A) = side opposite angle A/hypotenuse
Cosine A (cos A) = side adjacent angle A/hypotenuse
Tangent (tan A) = side opposite angle A/side adjacent angle A
Cosecant A (csc A) = hypotenuse/side opposite angle A
Secant A (sec A) = hypotenuse/side adjacent angle A
Cotangent A (cot A) = side adjacent angle A/side opposite angle A

Using the labels in Figure 27-14, we can express the most frequently used relationships as:

Sin A = a/c   Cos A = b/c   Tan A = a/b
Sin B = b/c   Cos B = a/c   Tan B = b/a

REFERENCES

American College of Sports Medicine. 1980. *Guidelines for Graded Exercise Testing and Exercise Prescription.* Philadelphia: Lea and Febiger, 151 pages.

Astrand, P. -O., and K. Rodahl. 1977. *Textbook of Work Physiology. Physiological Bases of Exercise.* 2nd ed. New York: McGraw-Hill, 681 pages.

Atha, J. 1974. "Physical Fitness Measurements." Part VII in *Fitness, Health, and Work Capacity: International Standards for Assessment,* edited by Leonard A. Larson, International Committee for the Standardization of Physical Fitness Tests. New York: Macmillan. pp. 449–492.

Ayoub, M. M., R. D. Dryden, and R. E. Knipfer. 1976. "Psychophysical-Based Models for the Prediction of Lifting Capacity of the Industrial Worker." Paper presented at the Automotive Engineering Congress and Exposition (SAE), in Detroit, Mich. Cited in *Work Practices Guide for Manual Lifting,* National Institute for Occupational Safety and Health, 1981. Washington, D.C.: U.S. Department of Health and Human Services/National Institute for Occupational Safety and Health, 183 pages.

Brouha, L. 1943. "The Step Test: A Simple Method of Measuring Physical Fitness for Muscular Work in Young Men." *Research Quarterly, 14:* pp. 31–36.

Caldwell, L. S., D. Chaffin, F. Dukes-Dobos, K. H. E. Kroemer, L. Laubach, S. Snook, and D. Wasserman. 1974. "A Proposed Standard Procedure for Static Muscle Strength Testing." *American Industrial Hygiene Association Journal, 35:* pp. 201–206.

Chaffin, D. B. 1975. "Ergonomics Guide for the Assessment of Human Static Strength." *American Industrial Hygiene Association Journal, 36 (7):* pp. 505–511.

Comroe, J. H., Jr., R. E. Forster II, A. R. Dubois, W. A. Briscoe, and E. Carlsen. 1962. *The Lung.* 2nd ed. Chicago: Year Book Medical Publishers, 390 pages.

Consolazio, C. R., R. E. Johnson, and L. J. Pecora. 1963. *Physiological Measurements of Metabolic Functions in Man.* New York: McGraw-Hill, pp. 329–333.

Davies, P. W. 1962. "The Oxygen Cathode." Chapter 3 in *Physical Techniques in Biological Research—Volume IV: Special Methods,* edited by W. L. Nastuk. New York: Academic Press, pp. 137–179.

Davis, H. L., T. W. Faulkner, and C. I. Miller. 1969. "Work Physiology." *Human Factors, 11 (2):* pp. 157–166.

Kamon, E., and A. Goldfuss. 1978. "In-Plant Evaluation of the Muscle Strength of Workers." *American Industrial Hygiene Association Journal, 39:* pp. 801–807.

Kamon, E., D. Kiser, and J. Landa-Pytel. 1982. "Dynamic and Static Lifting Capacity and Muscular Strength of Steelmill Workers." *American Industrial Hygiene Association Journal, 43:* pp. 853–857.

Landa-Pytel, J., and E. Kamon. 1981. "Dynamic Strength Test as a Predictor for Maximal and Acceptable Lifting." *Ergonomics, 24 (9):* pp. 663–672.

Lehmann, G. 1962. *Praktische Arbeitsphysiologie.* 2nd ed. Stuttgart: Georg Thieme Verlag, 409 pages.

Lind, A. R., and G. W. McNicol. 1967. "Circulatory Responses to Sustained Hand-Grip Contractions Performed during Other Exercise, Both Rhythmic and Static." *Journal of Physiology, 192 (3):* pp. 595–607.

Master, M. A. 1934. "Two Step Test of Myocardial Function." *American Heart Journal, 10:* p. 495.

Rahn, H., and W. O. Fenn. 1955. *A Graphical Analysis of the Respiratory and Gas Exchange: The O_2-CO_2 Diagram.* Washington, D.C.: American Physiological Society, 41 pages.

Rushmer, R. F. 1972. *Structure and Function of the Cardiovascular System.* Revised reprint. Philadelphia: W. B. Saunders Company, p. 154.

Sawka, M. N., M. E. Foley, and N. A. Timental. 1983. "Determination of Maximal Aerobic Power during Upper Body Exercise." *Journal of Applied Physiology (Respiratory, Environmental, and Exercise Physiology), 54:* pp. 113-117.

Snook, S. H. 1978. "The Design of Manual Handling Tasks." *Ergonomics, 21:* pp. 963–985.

Snook, S. H., and V. M. Ciriello. 1974. "Maximum Weights and Work Loads Acceptable to Female Workers." *Journal of Occupational Medicine, 16 (8):* pp. 527–534.

West, J. B. 1974. *Respiratory Physiology: The Essentials.* Baltimore: Williams and Wilkins, 185 pages.

Yates, J. W., E. Kamon, S. H. Rodgers, and P. C. Champney. 1980. "Static Lifting Strength and Maximal Isometric Voluntary Contractions of Back, Arm, and Shoulder Muscles." *Ergonomics, 23:* pp. 37–47.

CHAPTER 28

Appendix C: Problems

CHAPTER 28. APPENDIX C: PROBLEMS

This chapter includes 13 industrial work problems that can be addressed using information found in this book. Each problem is stated, the information used to evaluate it is described, and some conclusions are drawn about the difficulty of the task, the appropriateness of the design, or some other aspect of job design. The suggested approaches to solving these problems are not necessarily appropriate for similar jobs or tasks, since each situation has its own unique set of qualifiers. These problems illustrate how to use the data and information in this book to carry out ergonomic analyses of occupational tasks.

A. EVALUATION OF JOB DEMANDS

1. PROBLEM 1—PULLING SUPPLY ROLLS FROM STORAGE SHELVES TO A TRANSFER TRUCK

a. Background

Large rolls of paper weighing 45 kg (100 lbm) each are stored on specially constructed shelving units. As needed, the rolls are transferred from storage to manufacturing areas on a specially designed transfer truck. The support surface of the hand truck is adjustable so that it can be aligned with each storage shelf and the rolls can be slid from the storage shelf to the transfer truck for transport to the production area. The starting force to move a roll was measured as 256 N (57 lbf). Some of the workers have complained that the rolls on the top shelf, 152 cm (60 in.), are too difficult to pull onto the truck.

b. Ergonomic Principles and Information Used to Solve the Problem

Part VI includes guidelines for exerting force in different body positions, and Appendix A gives additional information on the forces that can be exerted by specific muscle groups. These data can be used to estimate the forces that could be generated in pulling the rolls from the storage shelves. The pulling force required to move the rolls, 256 N (57 lbf), can be compared to values presented in Table 22-1. Nearly 75 percent of the male population and more than 50 percent of the female population should have the capacity to generate the force necessary to pull the roll.

    The reports of difficulty had been received from more workers than this data analysis would suggest. A closer evaluation of the actual pulling task showed that an important fact had not been considered. Since the transfer cart is placed in front of the storage shelves, the worker has to pull the rolls while standing to one side of the truck. The direction of pull is therefore not directly in front of the roll but at an angle of approximately 45 degrees. In the section on biomechanical analysis of work in Part III, the effect of change in the angle of pull on a lever system is discussed. A similar relationship can be used to evaluate the task requirements in this case, and Figure 28-1 illustrates this analysis.

    The force required to pull the roll directly out from the shelf is 256 N (57 lbf). Since a direct pull cannot be used, a higher muscle force is needed to ac-

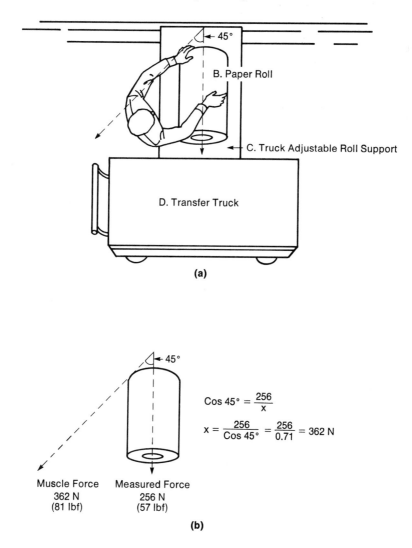

A. Storage Shelving

← 45°

B. Paper Roll

C. Truck Adjustable Roll Support

D. Transfer Truck

(a)

← 45°

$$\text{Cos } 45° = \frac{256}{x}$$

$$x = \frac{256}{\text{Cos } 45°} = \frac{256}{0.71} = 362 \text{ N}$$

Muscle Force   Measured Force
362 N   256 N
(81 lbf)   (57 lbf)

(b)

Figure 28-1: Paper Roll Handling Task   The top view of a paper roll transfer operation, from storage shelf (A) to transfer truck (D) on an adjustable roll support (C), is shown in *a*. The handler stands to the side of the roll (B), pulls it off the shelf, and slides it into the truck. The handler's muscles act at a 45-degree angle to the direction of motion. The direct pull force, measured standing in front of the roll, is 256 N (57 lbf), as shown in *b*. The actual muscle force needed to move the roll is calculated using trigonometric relationships. The force the muscles exert at 45 degrees to the roll's line of motion is 362 N (81 lbf).

complish the transfer. This force can be calculated using the trigonometric relationships given in Appendix B. The results are shown in Figure 28-1b. This force vector analysis indicates that a pulling force of 362 N (81 lbf) is required to move the roll onto the transfer cart from the side. Based on the strength values presented in Figure 26-13, this force is greater than the strength capacity of almost 75 percent of the males and more than 90 percent of the females. Thus, it becomes clear why people complained of the task's difficulty.

Two possible solutions to this problem are to change the method of transfer so that the transfer truck does not interfere with a direct pull on the roll or to reduce the force needed to pull the roll out from the shelf and onto the truck. In this example, the latter approach is likely to be more cost effective. To design the system to accommodate most workers, we need to determine the maximum allowable force for pulling out the roll. This force can be calculated using trigonometric relationships and information from Figure 26-13. The calculations show that a muscle force of 186 N (42 lbf) could be developed by nearly 90 percent of the females and almost all of the males performing the pulling task. This is the value for the force developed at 45 degrees from the roll's motion. The force vector diagram in Figure 28-1 is now used to calculate the direct pulling force requirements:

$$\cos 45° = \frac{X}{186}$$
$$X = (0.71)(186) = 132 \text{ N or } 30 \text{ lbf}$$

c. Conclusion

If we want to modify the roll transfer task to accommodate most workers, the pulling force should not exceed 132 N when measured from the front of the roll. The pulling force at 45 degrees, the handler's position, then does not exceed the recommended muscle force limit of 186 N (42 lbf). These desired pulling forces can be obtained by adding roller bearings to the storage shelves and to the transfer cart extension.

2. PROBLEM 2—AIR GUN DESIGN AND STATIC MUSCLE LOADING

a. Background

Several workers are required to clean molds between molding runs. The cleaning is done using a compressed air gun that blows high pressure air into the mold, removing residue from the previous run (Figure 28-2). Each worker cleans five molds per minute, and averages about 2,150 molds per shift.

An ergonomic evaluation of the task reveals several problems. The location of the mold and the design of the tool force the worker to use undesirable working postures. In order to blow air into the mold, the worker has to point the barrel of the air gun down into the mold. This can only be accomplished by abducting the shoulder (the arm away from the body) and flexing the wrist (Fig-a). Fatigue and discomfort can result from the excessive muscle tension required

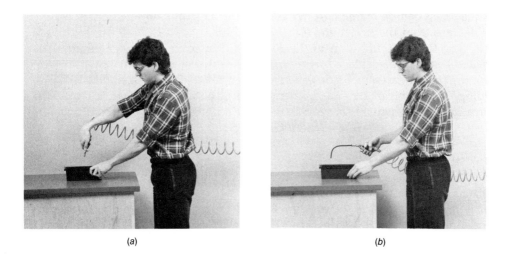

(a)   (b)

Figure 28-2: Air Gun Design and Static Muscle Loading   Two versions of air gun design are shown in use cleaning a mold between molding runs. In *a*, the air gun barrel and handle are straight. This forces the worker to abduct the shoulder, moving the upper arm away from the body, and to deviate the wrist to the radial side while flexing it. This posture is uncomfortable and reduces gripping strength, which results in the force applied to the trigger being a greater percentage of maximum strength. In *b*, the air gun barrel has been bent down at a 90-degree angle; the handle is also bent down at about 70 degrees. The trigger has been replaced with a levered bar that requires less force to operate. This design eliminates the shoulder abduction and much of the wrist deviation, thereby reducing static muscle fatigue.

to maintain these static postures. The air gun trigger design poses an additional problem. The very thin, wire-like design of the trigger results in relatively high forces being distributed over a small area on the index finger.

b. Ergonomic Principles and Information Used to Solve the Problem

Part III's chapter on the biomechanical analysis of work points out the biomechanical problems associated with shoulder abduction. The first step in reducing the stress of the task is to eliminate that postural requirement; however, when the arm is not abducted, the barrel of the air gun cannot be directed into the mold. This problem can be circumvented by bending the tip of the barrel. Although this is an improvement, the straight gun handle still produces an awkward deviation of the wrist, and this stress may lead to overuse symptoms. The section on repetitive work in Part IV and Appendix A's data on grip strength loss and stress on the wrist tendons with wrist deviations apply here. The source of the stress can be eliminated by bending the air gun's handle approximately 70 degrees from the horizontal (see Figure 28-2b).

The third modification to the tool is a change in its trigger design. The trigger wire can be replaced by a bar that has a larger and smoother surface. The trigger bar can also be designed as a second class lever (see Part III), providing a mechanical advantage when it is pressed to release the air. Thus, the force needed to actuate the gun is reduced, and this force is spread over the greater surface area of one or more fingers.

c. Conclusion

The discomfort and fatigue of repetitive use of the air gun for mold cleaning is related to the static postures of the shoulder and wrist. By angling the tip of the air gun barrel, bending the air gun's handle about 70 degrees down from the horizontal, and replacing the wire trigger with a flat bar, the ergonomist can improve the worker's gripping posture and substantially reduce his or her discomfort during repetitive use of the gun.

B. PATTERNS OF WORK

1. PROBLEM 3—SHIPPING DOCK WORKLOAD

a. Background

Mr. Jones is a 50-year-old shipping dock worker who is responsible for loading four trucks per day with cases of product that range from 10 to 30 kg (22 to 66 lbm) each. His electrocardiogram (ECG) has been monitored on the job and the following heart rate (HR) values were recorded:

$$\text{Resting HR} = 70 \text{ beats/min}$$
$$\text{Work} = 120 \text{ beats/min.}$$
$$\text{Continuous work duration} = 60 \text{ min.}$$
$$\text{Average HR for full shift} = 105 \text{ beats/min.}$$
$$\text{Highest HR} = 160 \text{ beats/min. for five min.}$$

In a subsequent treadmill stress test, he was found to have an average fitness level for the industrial population using a 50-50 mix of men and women.

This job is not easy for Mr. Jones and has been difficult to staff with women. You are asked to recommend a more appropriate workload.

b. Ergonomic Principles and Information Used to Solve the Problem

The heart rate data can be used to estimate the percent of aerobic capacity (in this case, percent HR range) that is needed to do the work. From the work-rest cycles section on dynamic work, we can use the relationship between heart rate and percent capacity, or:

$$\frac{HR_{work} - HR_{rest}}{HR_{max} - HR_{rest}} = \% \text{ maximum aerobic capacity or } \% \text{ HR range}$$

HR max = 220 − Mr. Jone's age or 170 beats per minute.

HR rest = 70 beats per minute.

HR work = either the average for the shift (105 beats per minute), the working average for 60 minutes (120 beats per minute), or the five-minute average (160 beats per minute). Each should be tested against the graph of percent maximum aerobic capacity versus duration in Figure 11-1.

From this analysis, one can determine that Mr. Jones is working at:

1. 35 percent of his predicted maximum aerobic capacity for the eight-hour shift.

2. 50 percent of his predicted maximum aerobic capacity for one hour of work.

3. 90 percent of his predicted maximum aerobic capacity for five minutes of work.

From the data in Figure 11-1, we can see that this represents a marginally excessive workload for the shift and a maximum acceptable workload for the shorter periods. The most intense work will produce significant amounts of lactic acid (see Table 11-2) and require extended recovery time.

Since Mr. Jones was found to have an average maximum aerobic capacity (50th percentile) for this type of whole-body work, it is clear that at least 50 percent of the potential work force will be working at or near their maximum acceptable workload levels (see Figure 26-14). To bring the job within the acceptable levels of work for more people, the total workload should be reduced by providing more light work or rest periods. Some ways in which this might be done are:

• Reduce the amount of manual lifting by using pallets, triwall containers, or other aids through which cases can be handled in groups by equipment instead of singly by people.

• Rotate the handling activity frequently (every 20 minutes) with a co-worker who drives the forklift truck or handles the pallet truck. This provides a recovery break and reduces the amount of lactic acid formed in the muscles.

• Interrupt the handing task every 15 to 20 minutes with short periods of paper work or communications with other workers. This also reduces the amount of lactic acid produced.

• Use the forklift truck or other equipment to permit the handler to lower the cases into place and reduce the amount of lifting required. This reduces the workload by about 10 to 15 percent. Sliding the cases off the pallet instead of lifting them can also reduce the load on the cardiovascular system and thereby reduce local muscle fatigue.

c. Conclusion

The workload is slightly above, at, and just below the maximum acceptable levels for aerobic work for eight-hour, one-hour, and five-minute continuous lifting. It merits redesign since at least half the work force will find the job too difficult. Techniques to reduce the workload by including more light activity or rest periods and reducing the length of the continuous work periods are suggested.

2. PROBLEM 4—DETERMINING STAFFING REQUIREMENTS FOR A PRODUCTION MACHINE OPERATION

a. Background

Several new packaging machines are being placed in a very large work area (Figure 28-3). Trays are loaded into one end and taken off the other end at the rate of two per minute. The in and out conveyors hold two trays each, and it is important to keep the machines supplied from and remove the finished product to pallets on the floor at each end. The trays weigh 10 kg (22 lbm) each initially and 12.5 kg (27.5 lbm) each at the end of packaging. They are 36 × 25 × 15 cm (14 × 10 × 6 in.) in length, width, and height, respectively, and have good handholds. They enter and exit the machine at a height of 75 cm (30 in.) above the floor. It takes three seconds to lift a tray to or from the conveyors.

There are six machines going into the work area and the labor requirements for running them are being determined. The major question is whether one worker can tend two or three machines. Figure 28-3 gives distances and travel times determined from a study of a similar production area.

b. Ergonomic Principles and Information Used to Solve the Problem

Two factors have to be evaluated in determining the number of machines one operator can tend. The workload can be measured in terms of lifts per minute and weight per lift. The demands are as follows:

- Two machines per operator: total of eight lifts per minute, four at each end of the machines with weights of 10 and 12.5 kg (22 and 27.5 lbm) each.

- Three machines per operator: total of twelve lifts per minute, six at each end of the machines with weights of 10 and 12.5 (22 and 27.5 lbm) each.

Using the guidelines in Figure 23-4, we can see that the workload is acceptable for both conditions, only if it is kept to less then 20 minutes every hour.

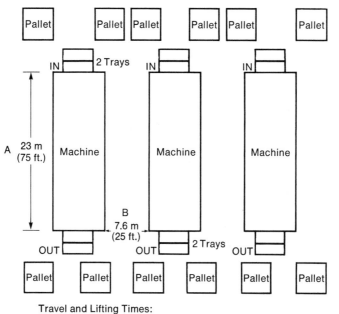

Travel and Lifting Times:
12 Seconds In to Out Ends of One Machine
4 Seconds In to In or Out to Out Ends of Different Machines
3 Seconds to Lift or Lower a Tray; Maximum of 2 Lifts
    at One Time per Work Station
2 Trays In and 2 Trays Out per Minute per Machine

Figure 28-3: Production Machine Layout The layout of three identical production machines is shown in a large production area. Machine length (A) and the distances between machines (B) are indicated as are the locations of the pallets for loading (top) and unloading (bottom) trays at the machines. There is room for two trays on the conveyors at each end of a machine. Information needed to calculate the time pressure of running several machines is given below the layout.

If the pallets are raised so the lifts are made at 76 cm (30 in.), the task would be more appropriate for about half of the workforce (from Figures 23–3 and 23–5).

The second consideration is the amount of time it takes to cover the machines, which depends on the distances between them. This can be analyzed as follows:

- Maximum lifts at one time per end = 2.
- Three seconds per lift = 6 seconds at each machine end.
- Travel time between ends of the same machine = 12 seconds.
- Travel time between same ends of different machines = 4 seconds.

- Optimum loading of three machines = 6 seconds × 3 = 18 seconds plus 20 seconds travel time to load three machines and go to the other end = 38 seconds.
- Optimum unloading time for three machines = 6 seconds × 3 = 18 seconds plus 20 seconds travel time between the ends of the machines and from one end to the other of a machine in preparation for the next loading = 38 seconds.
- Total cycle time = 76 seconds to load and unload.

At a packaging rate of two trays per minute, the machines will be looking for the third tray at the loading end before the full cycle of loading and unloading is complete.

From this analysis it is clear that, if three machines must be supplied and unloaded throughout the shift, there will be no time for the operator to do auxiliary tasks or to trouble-shoot the machine and he or she will have to run between ends of the machines to keep them full. This would be extremely difficult for most people to sustain for a full eight-hour shift (see the text on variability in work rate in Part IV.) There are a number of ways to make the job more acceptable even with three machines being tended by one operator. These include:

- Providing additional conveyor space at both ends of the machine. Giving the operator three to five minutes of "inventory" on either end (or room for six to ten trays) releases him or her from the machine's time pressure and makes auxiliary tasks more possible.
- Designing the production machines in a curve instead of a straight line and bringing the ends together so they can be loaded and unloaded without long travel distances being required.
- Providing an automatic palletizing device or robot at the ends of the machines (especially if they are grouped as described in the previous suggestion).
- Including a worker in the area who can respond to unusual situations or do the trouble-shooting as needed and who can back up the primary machine operator.
- Raising the pallets off the floor and permitting horizontal transfer of the trays at each end of the machine, thereby reducing the whole-body workload and the time to make a lift.

c. Conclusion

The job requirements for tending the packaging machine suggest that three machines per operator will be too difficult, both because of the external machine-pacing pressure and because the workload is above the upper limit for repetitive lifting. Providing more conveyor space to relieve the machine-pacing pressure and raising the pallets to reduce the whole-body work in tray handling would make the tending of three machines more acceptable to the total workforce.

3. PROBLEM 5—REPETITIVE MOTION AND KNIFE HANDLE DESIGN

a. Background

Meat cutters in a food processing plant have experienced some pain in their hands and wrists following several hours of cutting up partially frozen sides of beef. They are using a power grip on a knife with a handle 2.5 cm (1.0 in.) deep and 1.5 cm (0.6 in.) wide, as illustrated in Figure 28-4. To protect their hands from the cold meat, they wear cotton knit gloves on both hands.

The force to cut through the partially frozen meat has been estimated to be about 110 N (25 lbf) using a straight cut and transmitting optimal force application from the grip to the edge of the knife. The people who appear to be having the most difficulty with the job have less than 445 N (100 lbf) of grip strength. A request has been made that something be done to reduce the job demands so more people can perform it without developing sore arms and hands.

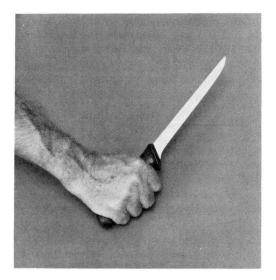

Figure 28-4: Meat Cutting Knife Handle and Grip Style. A power grip is illustrated as it would be used to control a butcher's knife during meat cutting. The handle is 2.5 cm (1.0 in.) deep and 1.5 cm (0.6 in.) wide. During the cutting motion, it is held in an oblique grip with radial deviation of the wrist. Grip style and handle size both influence the percentage of maximum grip strength available for this work.

b. Ergonomic Principles and Information Used to Solve the Problem

The factor of concern in this problem is the grip strength and how it is affected by the knife handle dimensions, the cotton gloves, and the wrist angles during the cutting activity. Figure 26-11 demonstrates how the grip span, represented by the knife handle depth in the power grip, affects grip strength. At 2.5 cm (1.0 in.), the grip strength falls to about 40 percent of the maximum value observed at a 5.0 cm (2.0 in.) span. Table 13-5 indicates that cotton gloves reduce grip strength about 26 percent. Thus, a person with an initial grip strength of 445 N (100 lbf) may only have about 135 N (30 lbf) of functional grip strength when wearing gloves and gripping a 2.5 cm (1.0 in.) deep knife. Small deviations of the wrist in the radial direction may be seen as the knife is pushed through the meat, and these may reduce grip strength even further (see Figure 26-12). The 110 N (25 lbf) required to cut the partially frozen meat then becomes very close to a maximum force. Repeating this several times per minute for several hours each shift could result in muscle soreness and fatigue for many workers since most of the women and about 50 percent of the men would have less than 445 N (100 lbf) of grip strength in optimum conditions (see Table 26-10).

To improve the situation, the depth of the knife handle can be increased so it is closer to the 5 cm (2 in.) optimum grip span shown in Figure 26-11. This will return the functional grip strength to values closer to 310 N (70 lbf). The grip strength required to cut the meat will be closer to 40 percent of maximum functional strength while wearing gloves. A rubber glove could also be considered as a way to gain some additional strength, providing it met the other need of keeping the hand warm.

c. Conclusion

The knife handle depth is too little to permit a good power grip, reducing the functional grip strength to about 40 percent of the maximum value at a 5 cm (2 in.) grip span. Increasing the depth, without widening the handle significantly or changing its relationship to the blade, should improve the grip strength available and make repetitive cutting exertions a much smaller percentage of maximum functional grip strength.

4. PROBLEM 6—REPETITIVE MOTIONS IN A PACKAGING JOB

a. Background

About one-third of the people working on a packaging task have complained of sore fingers, hands, and wrists. A movie made of the job requirements showed that the task consists of placing sheets of paper in a protective foil-lined bag, folding over the top several times, and taping it down securely. The workers perform the wrapping task throughout the shift, forming about 800 packages per shift. They wear cotton gloves to protect the paper from skin oils and perspiration.

Many small hand motions are used while exerting pressure to fold the protective paper in the required manner. At the completion of the wrap, the package is held with a pinch grip with one hand while tape is obtained from a manual tape dispenser with the other. The tape is pressed firmly over all exterior folds to secure the wrap, taking about three pieces of tape per package. The hand gripping the package is extended at the wrist about 25 degrees from the neutral position as is the hand using the tape dispenser.

b. Ergonomic Principles and Information Used to Solve the Problem

The text on repetitive motions in Part IV discusses the factors that are associated with muscle and joint soreness in repetitive tasks. These factors include the effect of cotton gloves on grip strength. Appendix A shows the loss of grip strength with changes in the positions of the fingers and wrist, including the effect of using pinch grip instead of power grip to hold a package. From this information, and by evaluating the type of grip (pinch) and angles of the wrist (extension) in this task, one can calculate that the functional grip strength is substantially reduced, probably to less than 25 percent of the normal grip strength. The losses in strength are associated with:

- The need to use pinch grip (P) instead of power grip; this drops the functional task strength to 25 percent of the maximum value.

- The use of cotton gloves (G); these reduce power grip strength to 75 percent of maximum but probably have much less effect on pinch grip strength, and the reduction may be closer to 95 percent of the maximum value.

- Wrist extension (E) of 25 degrees above the wrist's neutral point; this reduces grip strength to about 85 percent of the maximum value for pinch.

These changes in strength can be summarized as follows, using 100 percent as the maximum power grip strength with the wrist in the neutral position and barehanded (no gloves):

$$100\% \times \underset{P}{0.25} \times \underset{G}{0.95} \times \underset{E}{0.85} = \begin{array}{l}20\% \text{ of maximum} \\ \text{power grip} \\ \text{strength}\end{array} = \begin{array}{l}\text{task's functional} \\ \text{grip strength}\end{array}$$

The average grip strength for women at their preferred span minus one standard deviation yields a maximum power grip strength of about 200 N (45 lbf) as the 100 percent point. The task's functional grip strength, then, would be 40 N (9 lbf). The forces exerted during packaging are probably close to this level, making people with less grip strength more susceptible to local muscle fatigue and possible overexertion problems.

The forces required to dispense the tape and hold the package together

are also relatively high compared to the available grip strength. The highly repetitive nature of the task makes it less suitable for people with a history of
traumatic disorders of the wrist, arm, or hand.

c. Conclusion

Several solutions to the problem can be developed. These include:

- Reducing the number of folds necessary in the paper wrapping task.
- Replacing manual tape dispensers with automatic dispensers.
- Rotating workers frequently among tasks.
- Scoring the packaging materials where folds must be made so less force
  is needed to bend and secure them.
- Using more pliable packaging materials.

When some of the wrapping requirements were eliminated and automatic
tape dispensers were provided, many of the complaints about sore arms dis-
appeared and productivity increased.

C. HOURS OF WORK

1. PROBLEM 7—OVERTIME VERSUS INCREASED LABOR TO ACCOMPLISH A TASK

a. Background

A landscape firm has been asked to fill a rush order for the planting of 200
bushes and small trees at a corporate headquarters within three days. It takes
15 minutes to do each planting and the measured oxygen consumption for the
landscaper is 12 mL O_2/kg BW • min. The plants are brought to the work site in
lots of 50 and it takes about two hours of additional work per 50 plants to prepare them for planting and to unload the truck. During these activities the oxygen consumption averages 6 mL O_2/kg BW • min.

The landscaping firm has an option to put three people on the job for eight-
hour shifts until the work is done, or to put two people on the job and have
them work overtime to complete it in time. Assuming the workers have maximum aerobic capacities of 40 mL O_2/kg BW • min, which schedule would you
recommend?

b. Ergonomic Principles and Information Used to Solve the Problem

The guidelines for extended hours of work and total aerobic job demands are
used to evaluate the two options. Figure 15-1 indicates that the acceptable average oxygen consumption for eight hours of work is 13.2 mL O_2/kg BW • min for
a person with a 40 mL O_2/kg BW • min maximum aerobic work capacity (whole-
body work). With an additional two hours of work, this value drops to 12.2,
and with a total of 12 hours of work per shift, it falls to 11.2 mL O_2/kg BW • min.

Thus, it is necessary to determine how long it will take two people to plant 200 trees and bushes, and to see whether the workload exceeds these guidelines.

Since it takes 12.5 hours to plant a truckload of 50 bushes and trees plus an additional two hours of accessory work, each truckload will require a total of 14.5 hours or 7.25 hours of one person's time per eight-hour shift if two people are sharing the task. With 200 bushes and trees, this is 29 hours of work per person in a three day period, or about five hours of overtime in three days, not counting meals or other disturbances. If an hour of recovery pauses at 4 mL O_2/kg BW • min is included per truckload, the average oxygen consumption during that time will be as follows:

$$\text{Work breaks} = 1 \text{ hour at 4 mL } O_2/\text{kg BW} \cdot \text{min}$$
$$\text{Planting} = 6.25 \text{ hours at 12 mL } O_2/\text{kg BW} \cdot \text{min}$$
$$\text{Preparation and unloading} = 1 \text{ hour at 6 mL } O_2/\text{kg BW} \cdot \text{min}$$

This averages 10.3 mL O_2/kg BW • min over the 8.25 hours spent on each truckload of plants. During each of the three days, the two landscapers would have to do the equivalent of 1.33 truckloads, and the total time worked would be 11 hours per shift (including the one hour of breaks per truck). Assuming that the distribution of the lighter activities would be broken up equally across the three days for the fourth truckload, one can estimate that the workload is close to the upper acceptable level, but could be done by two people with 40 mL O_2/kg BW • min aerobic work capacities (see Figure 26-14). The upper limit for this type of work would lie between 11 and 12 mL O_2/kg BW • min, so 10.3 would be within the guidelines for overtime work.

Looking at the other option, one can see that it would take each of three people about four hours of planting time and somewhat less than two hours of other activities (including one hour of breaks) to plant a truckload of bushes and trees. Assuming six hours per truckload per person and 1.33 truckloads per day, they could do the work in three eight-hour shifts at an average oxygen consumption of 10.3 mL O_2/min/kg BW. This is well below the 13.2 mL O_2/min/kg BW given as the guideline for eight hours of whole-body work for a person with 40 mL O_2/min/kg BW maximum aerobic capacity.

c. Conclusion

Either work schedule would be acceptable for people with an above-average whole-body maximum aerobic work capacity. The use of three people on eight-hour shifts would bring the job demands within the capacities of more people and reduce the time pressure on individual workers.

2. PROBLEM 8—SCHEDULING OF A MAINTENANCE PROJECT

a. Background

An automatic storage and retrieval system is exhibiting some erratic failures. The maintenance crew has been asked to find a time when they can troubleshoot the system and try to get it working reliably again. The job involves complex computer checks as well as repairs on the mechanical storage and retrieval equipment. It is estimated that the repairs will take about 15 hours to complete. Two maintenance workers have the requisite skill to do the job, but they are tied up on a higher-priority project during the day shift for the next few weeks. They have offered to come in on another shift or on the weekend to do the job. Physical workload is not of concern in this work, but mental workload is. In order not to interfere very much with the normal production schedule in the area, the choice of time to make the repairs has been narrowed down to the midnight to 8 a.m. shift (early morning) or the daytime Saturday and Sunday shifts.

You have been asked to advise the department head which of these alternatives is preferable for the performance and well-being of the workers.

b. Ergonomic Principles and Information Used to Solve the Problem

The choice of whether to extend the hours per day or the days per week to accomplish this repair work is influenced by many factors. Each maintenance worker will work a full day shift as well as the early morning shift if that option is chosen, giving them a 16-hour work day and ensuring them of inadequate sleep and contact with their family and friends. If they work on the weekend, they give up their rest days and would have a continuous period of 12 work days before the next weekend. Some of the psychological and physiological fac-

Table 28-1: Factors Influencing Selection of an Overtime Schedule. Positive and negative factors that can influence the choice of two different overtime schedules for a 15-hour maintenance task are shown in columns 2 and 3. The decision whether to work double shifts for two days or to give up a weekend will be influenced by the nature of the work, psychosocial factors relating to the worker's needs for rest and relaxation with his or her family and friends, and physiological effects of the extended work schedules, such as fatigue.

	Early Morning Shift, 16 Hours per Day	Saturday and Sunday Work
Negative factors	Reduced alertness. More liable to err. Less backup help. Reduced access to parts. Less sleep, more fatigue. Isolation from family. Digestive discomfort, irritability.	Reduced recovery time (0 rest days). Less time with family and friends. Reduced access to parts and help.
Positive factors	Few work interruptions. Overtime pay. Weekend free.	Double pay on Sunday. Few physiological symptoms or adjustments.

tors discussed in Part V should be related to each of the options, as shown in Table 28-1.

The performance and physiological adjustments to shift work are discussed in Part II under biological rhythms. The increased potential for errors in mental tasks and the digestive and sleep disorders associated with early morning shift work make the morning double-shift schedule less satisfactory than the weekend one. Although the maintenance workers will lose their weekend to do the work, this is a rare situation and is not likely to recur.

c. Conclusion

The maintenance workers should be requested to repair the storage and retrieval system on Saturday and Sunday during the day shift. Their performance should be better then, permitting them to work more effectively and accurately. The physiological discomforts associated with early morning shift work and working double shifts will be minimal in weekend work. Although the workers have less weekend time with their families, they should be alert and rested enough to enjoy their hours off, which may not be the situation if they had been working two 16-hour days prior to the weekend and getting little sleep.

D. MANUAL MATERIALS HANDLING

1. PROBLEM 9—JOB DESIGN FOR A PACKAGING LINE

a. Background

Eight people, all in the same pay classification, work together on a packaging line to produce 1,680 cases of product per eight-hour shift. The job titles and handling loads (for an eight-hour shift) are:

- Four wrappers and packers—handle 9 kg (20 lbm) units, two per minute.
- One labeler and taper—handles 9 kg (20 lbm) units, eight per minute.
- Two case packers—handle 18 kg (40 lbm) units, two per minute.
- One case sealer and palletizer—handles 18 kg (40 lbm) units, four per minute, utilizing low lifts.

There are no barriers to alternating the employees among these job titles. You have been asked to recommend the best job design to minimize fatigue and keep productivity high.

b. Ergonomic Principles and Information Used to Solve the Problem

There are several ways to evaluate the workload in each of the jobs listed above. Figure 23-4 indicates that the wrapper-packers handling the 9 kg (20 lbm) units are well within the recommended workloads. The labeler-taper and the people handling the 18 kg (40 lbm) units are outside those limits for an eight-hour shift. Estimates of the workload from the tables in Appendix A on the energy ex-

Table 28-2: Analysis of Workload for Packaging Line Operators. A summary of the manual lifting demands (column 2) for four job titles (column 1) on a packaging line is given. The level of effort (column 3) is determined from Tables 26-19 to 26-21. The acceptable continuous duration of the task (including scheduled work breaks) is given in column 4. This analysis allows evaluation of various job designs to spread the effort more evenly among the packaging line workers.

Job Title	Workload	Effort Level	Acceptable Continuous Duration
Wrapper-packer	9 kg (20 lbm) 2 per minute	Light - Moderate	8 hours
Labeler-taper	9 kg (20 lbm) 8 per minute	Moderate	<8 hours
Case packer	18 kg (40 lbm) 2 per minute	Moderate	<8 hours
Case sealer	18 kg (40 lbm) 4 per minute	Heavy	1 hour

penditure of several occupational activities and estimation of the energy demands of jobs in Part III lead to the same conclusions, as Table 28-2 illustrates.

Since the case sealer and palletizer job is the most difficult, it is advisable to rotate the workers between it and the lightest job (wrapper-packer) so that no one has more than one hour of continuous heavy work. The case packers should rotate to the lighter job on a less frequent schedule, perhaps once every two to four hours.

In addition to the whole-body energy requirements of these tasks, the local muscle activity should be evaluated and possible overexertion workloads identified for small muscle groups. The wrapper-packer and labeler-taper jobs are done at about elbow height and have quite heavy continuous activity of the shoulder, arm, and hand muscles. The case packer's job is done below elbow height and involves some leaning with static loading of the back. The case sealer-palletizer job is a very dynamic task using most of the muscles of the body. The jobs needing the most local muscle relief are therefore case packer and labeler-taper; workers should be rotated out of these every hour or two.

c. Conclusion

Based on an analysis of the energy requirements and of local muscle fatigue, the ergonomist would recommend that people be rotated among these jobs so they do not have to do more than one hour of continuous work on the case sealer-palletizer job or more than two hours of continuous work in case wrapper-packer and labeler-taper jobs. Raising the height of the pallet so that palletizing requires less whole-body work will reduce the energy requirements of that job, but the potential muscle fatigue from the 18 kg (40 lbm) load would make a two-hour rotation desirable even with the workplace improvement. An automatic label and tape dispenser would reduce local muscle and joint stress in the labeler-taper job by eliminating the repetitive forceful pinch grip and wrist rotation (to tear off tape).

2. PROBLEM 10—LOADING PARTS INTO A HOPPER ON AN ASSEMBLY MACHINE

a. Background

A machine operator has to keep a hopper supplied with small parts needed in the assembly of a cartridge. The parts come in a cardboard box that is 60 × 51 × 51 cm (24 × 20 × 20 in.) in length, width, and height, respectively, and weighs 14 kg (31 lbm). The hopper, on top of the machine, is 229 cm (90 in.) above the floor and 25 cm (10 in.) in from the front surface. A platform is being built to permit access to the hopper, but safety concerns have led to its being only 71 cm (28 in.) high with two steps.

In about a year, the parts will be supplied to the hopper by an automatic conveyor system. In the meantime, you are asked to evaluate whether this lift, done at most six times per shift, is acceptable and, if not, what should be done to improve it.

b. Ergonomic Principles and Information Used to Solve the Problem

Even with the platform, the hopper height is 158 cm (62 in.) and the worker's hands will have to be about 165 cm (65 in.) high during the loading task. The 25 cm (10 in.) forward reach plus the dimensions of the box and the need to tip it up mean that the box will be held at least 51 cm (20 in.) in front of the body. Figure 23-3 can be used to identify how difficult this lift is. The third set of curves should be used because of the horizontal distance, and the acceptable weights are read from the right edge (H) of that set of curves, which represents lifts more than 135 cm (53 in.) above the standing surface.

The box weight of 14 kg (31 lbm) intersects the H line about halfway between the 50th and the 95th percentiles of the 50-50 male-female workforce. In other words, about 20 to 25 percent of the workforce would find the lift acceptable and the remaining 75 to 80 percent would not. That indicates that redesign or the provision of handling aids is needed for this task.

The box dimensions create difficulty for carrying the box up the steps of the platform and for tipping. Table 21-2 suggests that the acceptable load taken from Figure 23-2 would probably have to be reduced an additional 15 to 20 percent because the box exceeds each of the recommended dimensions for length, width, and height. Using Figure 23-3 again, one can look at the options for reducing box size, reducing its weight, or decreasing the height of the lift. Each of these approaches would be effective in increasing the number of people who would find the lift acceptable, but all may not be cost effective or feasible.

To accommodate the large majority of the workforce in this task, one would have to reduce the box weight to 5 kg (11 lbm) and bring its dimensions down to the values recommended in Figure 21-6. If it is possible to raise the worker another 30 cm (12 in.) above the floor safely, a total of 100 cm (40 in.), the original weight of 14 kg (31 lbm) would be acceptable, although the box dimensions would still have to be reduced. If box weight is reduced, more lifts will have to be made, although a rate of 12 to 18 lifts per shift is not of concern in terms of workload.

c. Conclusion

Assuming that the worker cannot be safely elevated any higher than the original platform height, one can recommend that the parts be supplied in a smaller box and that the box weight be reduced to 5 kg (11 lbm). This could be accomplished by using plastic liners in the larger box and filling these individually to the 5 kg (11 lbm) limit, or by changing the box and loading it directly. The machine operator would have to load the hopper more often or make two to three trips per load in this new supply configuration, but the opportunity to overexert or to lose control of the load would be considerably reduced.

3. PROBLEM 11—WEIGHT OF A SAMPLE CASE FOR A SALES REPRESENTATIVE

a. Background

A sales representative has a carrying case that measures 50 × 40 × 15 cm (20 × 16 × 6 in.). It has a 3 cm (1.2 in.) cylindrical handle and is balanced when loaded with sales samples. The company employing the sales representative is concerned that the case could be overloaded and has decided that an upper weight limit should be recommended to help prevent sore muscles and joints. Sales representatives in metropolitan areas often have to walk 100 meters (305 feet) between calls. They carry the case in one hand at their side.

You have been asked to determine a maximum weight for the sales sample case.

b. Ergonomic Principles and Information Used to Solve the Problem

Two types of information can be used to determine the maximum weight limit for the sales sample case. Since grip strength limits a person's ability to hold the case, data from Table 26-10 can be used as a starting point. A value of 200 N (45 lbm) would accommodate about 70 percent of the women and most of the men for a short exertion. As a common travel distance is 100 meters which may require one minute of walking and holding, 50 percent of the maximum grip strength can be used, or about 100 N (22.5 lbm) (see Table 11-3). A value of similar magnitude, about 11 kg (24 lbm), can be derived from Figure 24-1.

If this recommended maximum weight is not consistent with the sales representative's need for samples, a second carrying case with similar dimensions and weight could be carried in the other hand, which would balance the load across the body. A second case would be of concern primarily in terms of its potential elevation of blood pressure when holding is sustained for more than three to four minutes. Increasing the size of the carrying case, especially its depth, would drop the maximum weight to approximately half of the value for the more compact load (see Figure 24-1).

c. Conclusion

The recommended upper weight for the sales sample case is 10 to 11 kg (22 to 24 lbm) if it is carried for one minute continuously in one hand. Two cases at this weight are better than one heavier case in terms of reducing the potential for local muscle (arm) fatigue.

E. STRENGTH AND CAPACITY FOR WORK

1. PROBLEM 12—FLOOR SCRUBBER HANDLE DESIGN

a. Background

Cleaning attendants in a building operate a floor scrubber for several hours a shift. The scrubber consists of a large tank with handles on either side located 90 cm (35 in.) above the floor. The handles are 5 cm (2 in.) in diameter and about 50 cm (20 in.) apart. The left handle must be rotated (inward and outward) to control the scrubber's motion. The right handle has a "dead man" switch that is activated at a grip span of 9 cm (3.5 in.). The switch must be closed to use the scrubber, which often requires at least one minute of continuous holding time. The force measured at the point of closing the switch is 135 N (30 lbf). The torque to rotate the left handle is 10 Nm (7 ft lb).

Several cleaners have been taping the "dead man" switch closed, which is against company safety regulations. Their supervisor has asked you to determine if the scrubber is designed well and, if not, to recommend modifications.

b. Ergonomic Principles and Information Used to Solve the Problem

The muscle strength data in Appendix A enable the evaluation of the scrubber handle design. The right handle with the "dead man" switch requires a handspan of 9 cm (3.5 in.), well beyond the optimum span for grip strength (Figure 26-11). One can estimate that grip strength at 9 cm (3.5 in.) is less than 40 percent of that measured at 5 cm (2 in.). Therefore, 445 N (100 lbf) of grip strength at optimum span becomes 180 N (40 lbf) on the "dead man" switch. The required force of 135 N (30 lbf) represents 75 percent of the maximum functional grip strength of the worker, and it will not be possible for him or her to hold the grip for a full minute (see Table 11-3). People with optimum-span grip strength values of 200 N (44 lbf) would be unable to use the "dead man" switch at all. This handle should be modified so that the switch activation point requires a span of 5 to 6 cm (2 to 2.4 in.). This would make the one-minute hold more feasible for people with a maximum grip strength of 270 N (60 lbf) or more, and would thereby accommodate about 50 percent of the female workers and almost all of the male workers.

The left handle is 5 cm (2 in.) in diameter which is greater than the 3 cm (1.2 in.) diameter recommended in Part VI. As the torque required is greater than the average female college student can develop (Table 26-8), it is quite possible that activating this handle switch by pronating and supinating the forearm is difficult for most people. Reduction of handle size alone, even if feasible, would not solve the problem because the torque would still be too large. Reducing the torque to 4 to 5 Nm (3 to 4 ft lb) would make the scrubber easier to operate, but a better method of controlling motion should probably be devised. Rotating a control while using it to direct and stabilize forward motion is awkward at best.

c. Conclusion

The scrubber controls require forces and torques too high for operation, particularly when the control must be held closed for as long as a minute, as is the case for the "dead man" switch. Modifications to reduce the torque of the left handle to 4 to 5 Nm (3 to 4 ft lb) or less or to redesign the control entirely and eliminate rotation, and to reduce the grip span to 5 to 6 cm (2 to 2.4 in.) for activation of the "dead man" switch on the right handle are suggested. Reduction of the required activation force for the right handle should also be considered.

2. PROBLEM 13—WRAPPING AND PACKING JOB SUITABILITY

a. Background

Products are removed from an incoming conveyor, wrapped in brown paper, and placed in a box. The box is sealed and placed in a carton that holds a total of ten boxes. After each carton is full, it is pushed onto a conveyor and is transferred to another part of the building for sealing and palletizing. The wrapping and packing operator completes four cycles per minute, handling each product four times per cycle for a total of 16 transfers in a minute. A pinch grip is used to support the product throughout the cycle. The product is held across its 4 cm (1.6 in.) depth, and it weighs 3 kg (7 lbm) when boxed.

The oxygen consumption of operators doing this job was measured as 10 mL O_2/kg BW • min. The job is done throughout the shift with a 15-minute rest break in the middle of the morning and afternoon periods and a 30-minute lunch break. Some wrappers and packers have requested additional time for recovery from the work, indicating that their arms and shoulders are sore after a full shift on the job.

You have been asked to determine the suitability of the work for men and women in the industrial work force.

b. Ergonomic Principles and Information Used to Solve the Problem

The total workload and local muscle stress of this work must be evaluated to understand why some people find it difficult. The capacity for aerobic work in the wrapping and packing task is limited to the muscles of the upper body, so upper body capacity can be estimated by taking 70 percent of the whole-body aerobic capacity values as measured on the treadmill. Appendix B and Part IV discuss the changes in aerobic work capacity when different amounts of muscle mass are available for work. Since 27 mL O_2/kg BW • min has been used as a design guideline for whole-body maximum aerobic work capacity, 70 percent of this, or 19 mL O_2/kg BW • min, can be used to estimate the suitability of the wrapping and packing workload in this sample.

The measured workload of 10 mL O_2/kg BW • min is 53 percent of the upper body aerobic capacity at the recommended design level. At this workload, the longest continuous duration of work should be about one hour (see Figure 11-1). This explains why some people have requested additional work breaks to recover.

Working backwards from the measured workload, we can determine the percentage of the 50-50 mix population that might have difficulty with the present task. Since the workload (10 mL O_2/kg BW • min) should not be more than 33 percent of the upper body capacity if it is to be sustained for the entire shift, the successful wrapper and packer should have an upper body capacity of 30 mL O_2/kg BW • min. Dividing that by 0.7 to estimate whole-body aerobic capacity, gives 43 mL O_2/kg BW • min. By looking at the cumulative frequency distribution for industrial workers whose maximum aerobic capacities were estimated from submaximal treadmill tests (Figure 26-14), we can see that 80 percent of the population has less aerobic capacity than that needed for the job. Therefore, job modifications would be desirable to reduce the effort level to about 7 mL O_2/kg BW • min. Such modifications could include rotating the wrapping and packing operators every hour with people in less demanding jobs or changing the packaging to require less product handling per cycle.

In addition to the total workload, it is also important to review the local muscle stress of this job. The objects are handled in a pinch grasp that has only 25 percent of the strength of a power grasp. Since the product weighs 3 kg (7 lbm), people with less than 120 N (28 lbf) of grip strength under optimum power grip conditions would have difficulty with this task. As most people have a grip strength greater than that, the interaction of intensity and duration should be examined next (see Table 11-3). The cycle requires 16 grasps per minute, with hand activity in between the grasps as well, so local muscle fatigue of the hand and arm could be important factors in limiting the acceptable work duration. Although the 30 N (7 lbf) grip force is only 15 percent of power grip strength, it is 60 percent of the maximum pinch grip force. At the repetition frequency of 16 per minute, sore arms and hands could be expected in more susceptible workers. Attention to package design, provision of holding fixtures, and other changes in the way the product is handled in wrapping and packing could reduce gripping frequency and reduce local muscle stress levels.

c. Conclusion

The wrapping and packing task was found to be very demanding for 80 percent of the potential work force as measured by upper body aerobic work capacity. It was also difficult to sustain the strength and rate of pinch grasps for long periods of time. Modifications of package design and rotation of workers to less demanding tasks every hour are recommended as ways to reduce job stress.

SELECTED ANNOTATED BIBLIOGRAPHY
RELATING TO ERGONOMICS AND JOB DESIGN

The following are brief descriptions of several books that include subjects discussed in this book. This list is not meant to be all-inclusive, and it is focused especially on books that deal with job design. The interested reader should also review the annotated bibliography in Volume 1 of this series for additional references that deal more specifically with the traditional human factors concerns of workplace, equipment, environmental, and information design.

Some of the older references cited below are now out of print. They remain here because many can be found through libraries and because they are valuable sources of information.

1. General Ergonomics

Edholm, O. G. *The Biology of Work.* New York: McGraw-Hill, 1967, 256 pages.
  An easy-to-read introduction to the basic concepts of the effects of work on the body. Presents a broad but quick introduction to ergonomics and human factors.

Grandjean, E. *Fitting the Task to the Man: An Ergonomic Approach.* 3rd ed. London: Taylor & Francis, 1980, 379 pages.
  Emphasis is placed on the factors that affect people at work. This book provides a summary of some important European ergonomics research that has not been available previously in English. The level of treatment is generally introductory.

Kantowitz, B. H., and R. D. Sorkin. *Human Factors: Understanding People Systems Relationships.* New York: John Wiley and Sons, 1983, 699 pages.
  This is a comprehensive human factors textbook that focuses on the traditional man-machine interface at work. Also included are discussions of environmental factors, such as noise and lighting, and a discussion of computer interactions.

Konz, S. *Work Design.* Columbus, Ohio: Grid Publishing Company, 1979, 592 pages.
  Job and workplace design, from an engineer's perspective. Emphasis on improving work efficiency. The book is organized around design guidelines.

Murrell, K. F. H. *Human Performance in Industry.* New York: Reinhold, 1965, 496 pages.
  Discusses ways in which ergonomics can be used to improve the design of work in industry. The coverage ranges from material of an introductory nature to a few topics that are treated in considerable depth. An excellent introduction to the field.

Osborne, D. J. *Ergonomics at Work.* New York: John Wiley and Sons, 1982, 321 pages.
  A general text on ergonomics with an emphasis on the man-machine interface and workplace and information design. It also includes a discussion on ergonomics and safety and reviews the environmental factors of heat, lighting, noise, and vibration as they affect people at work.

Rohmert, W. and J. Rutenfranz. *Praktische Arbeitsphysiologie.* 3rd. ed. Stuttgart: Georg Thieme Verlag, 1983, 409 pages.

Classical text on industrial ergonomics/work physiology from the Max Planck Gsellschaft in West Germany. Particularly detailed in the measurement of job physical effort demands, the design of work/rest cycles to reduce fatigue, and environmental stressors in the workplace. (In German.)

Shackel, B. *Applied Ergonomics Handbook.* Reprint of *Applied Ergonomics.* Volumes 1 and 2, Nos. 1–3. Surrey, England: IPC Science and Technology Press, Business Press, 1974, 122 pages.

A compilation of articles, appearing in the first two volumes of the journal *Applied Ergonomics,* that are intended to show the state of the art on industrial ergonomics in the early 1970s. Based on earlier booklets produced by the Ministry of Technology on *Ergonomics for Industry.* The articles cover workplace and equipment layout; environmental factors such as noise, light, and thermal conditions; systems design; safety, work organization, and design of work for the disabled.

Singleton, W. T., ed. *The Body at Work. Biological Ergonomics.* Cambridge, England: Cambridge University Press, 1982, 430 pages.

A compilation of information from several specialists in ergonomics that focuses on the physiology of work. The topics covered include metabolism, biomechanics, vibration, climate, vision, and hearing.

2. THE PHYSIOLOGICAL BASIS OF WORK

Astrand, P. O., and K. Rodahl. *Textbook of Work Physiology.* 3rd ed. New York: McGraw-Hill, 1986, 756 pages.

An excellent source of physiological information on people at work. Includes chapters on physical work capacity, physical training, the energy cost of activities, temperature regulation, and factors that affect human performance.

Basmajian, J. V. *Muscles Alive: Their Functions Revealed by Electromyography.* 2nd ed. Baltimore: Williams & Wilkins, 1967, 421 pages.

A source of information on muscle function and electromyography. Suggested for those who need a detailed understanding of the role of individual muscle groups during work. Has applications in the study of light workloads as well as moderate and heavy ones.

Brouha, L. *Physiology in Industry.* 2nd ed. Oxford: Pergamon Press, 1967, 164 pages.

A basic introduction to industrial work physiology. Techniques of measuring the effects of the environment, especially heat, and the physical demands of the job are discussed. There are numerous examples of actual industrial applications.

deVries, H. A. *Physiology of Exercise—For Physical Education and Athletics.* Dubuque, Iowa: William C. Broan Company, 1980, 577 pages.

An introductory text on the physiology of exercise that reviews the circulatory, respiratory, neuromuscular, and metabolic functions of the body. The applications of these data focus on physical fitness and training.

Johnson, W. R., and E. R. Buskirk, eds. *Science and Medicine of Exercise and Sport.* 2nd ed. New York: Harper & Row, 1974, 486 pages.

A contributed volume covering the physiological, structural and mechanical, psychological, and developmental aspects of sports and exercise. Although not on ergonomics, much is directly relevant to the study of work.

Mathews, D. K., and E. L. Fox. *The Physiological Basis of Physical Education and Athletics.* 2nd ed. Philadelphia: W. B. Saunders Company, 1976, 577 pages.
A thorough text that covers the physiology of exercise. It includes basic information about neuromuscular physiology, the cardiopulmonary responses to work, muscle training, nutrition, and responses to environmental heat.

McNaught, A.B., and R. Callander. *Illustrated Physiology.* Baltimore: Williams and Wilkins Company, 1963, 287 pages.
A fully illustrated review of functional physiology. Each plate includes illustrations and text to describe the basic physiological processes of concern. Especially helpful for the student who has no biology background.

Michael, E. D., E. J. Burke, and E. V. Avakian, Jr. *Laboratory Experiences in Exercise Physiology.* Ithaca, N.Y.: Mouvement Publications, 1979, 66 pages.
There are 26 laboratory experiments in this book, most of which focus on exercise physiology. Those that relate heart rate to work, study the strength and endurance relationships of muscles, and show how to use perceived exertion scales are especially useful in supplementing an occupational ergonomics course.

Rasch, P. J. and R. K. Burke. *Kinesiology and Applied Anatomy.* Philadelphia: Lea and Febiger, 1978, 496 pages.
Good basic text on kinesiology and applied anatomy. Reviews muscle, bone, and joint structure and function. A simple presentation of mechanics precedes in-depth discussions of the functions of individual muscle groups. Final chapters are devoted to applications of kinesiology to the study of human performance.

Ricci, B. *Experiments in the Physiology of Human Performance.* Philadelphia: Lea and Febiger, 1970, 208 pages.
The emphasis is on exercise physiology rather than work physiology. A relatively easy-to-understand book that is similar to a laboratory manual and can be used to facilitate self-education. Among the topics included are the cardiac response to exercise, measurement of metabolism, and the analysis of gross muscle function.

Sage, G. H. *Introduction to Motor Behavior: A Neurophysiological Approach.* 2nd ed. Reading, Mass.: Addison-Wesley Publishing Company, 1971, 610 pages.
A source of detailed information on the psychology of exercise or work. It includes basic information on the nervous system and how motor behavior is integrated to produce skilled movements. The interaction of vision with movement and how a motor skill is developed by training are also included.

Stegemann, J. *Exercise Physiology: Physiologic Bases of Work and Sport.* Translated and edited by J. S. Skinner. Chicago: Year Book Medical Publishers, 1981, 345 pages.
A basic text in exercise physiology giving detailed information about muscle, nerve, circulation, respiration, metabolism, and the regulation of bodily functions. From the German school of work and exercise physiology.

3. EVALUATION OF JOB DEMANDS

Frankel, V. H. and M. Nordin. *Basic Biomechanics of the Skeletal System.* Philadelphia: Lea and Febiger, 1980, 297 pages.
Written primarily for medical personnel interested in orthopedic problems. Divided into two sections: The first deals with biomechanics of tissues and structures of the

skeletal system, and the second includes chapters on the biomechanics of specific joints in the body, including the wrist, shoulder, elbow, knee, and back.

Grieve, D. W., D. I. Miller, D. Mitchelson, J. P. Paul, and A. J. Smith. *Techniques for the Analysis of Human Movement*. Princeton, N.J.: Princeton Book Co., 1975, 177 pages. Includes methodologies for photographic and photoelectric measurements of motion, computer simulations, electromyography, and instruments for force measurement.

Miller, D. I., and R. C. Nelson. *Biomechanics of Sport*. Philadelphia: Lea and Febiger, 1973, 248 pages.
Research orientation. Reviews basic equations for static and dynamic biomechanical analyses. Also provides information on instrumentation and on laboratory development.

National Institute for Occupational Safety and Health. *The Industrial Environment—Its Evaluation and Control*. Washington, D.C.: U.S. Department of Health, Education, and Welfare, National Institute for Occupational Safety and Health, 1973, 719 pages. A series of papers by specialists in ergonomics and industrial hygiene relating especially to environmental factors in the workplace and some aspects of job design.

Singleton, W. T., J. G. Fox, and D. Whitefield, eds. *Measurement of Man at Work*. New York: Van Nostrand Reinhold, 1971, 267 pages.
A compilation of 26 symposium papers, most of which describe techniques for obtaining psychological and physiological measurements of man at work. Some of the studies reported were done in an industrial setting. Emphasis is on applications rather than on research.

Sternbach, R. A. *Principles of Psychophysiology*. New York: Academic Press, 1966, 297 pages.
An introductory text that summarizes the variables that can affect physiological measurements of performance. Considers mental tasks and other very light work. Emphasizes the effect of emotion.

Thompson, C. W. *Kranz Manual of Structural Kinesiology*. 8th ed. St. Louis: Mosby, 1977, 159 pages.
An introduction to kinesiology. Clearly describes the anatomy of the muscles and the resulting body movements. Includes numerous illustrations.

Tichauer, E. R. *The Biomechanical Basis of Ergonomics*. New York: Wiley Interscience, 1978, 99 pages.
A summary of the principles of biomechanics as applied to industrial jobs, especially in manual materials handling and workplace and equipment design.

Williams, M., and H. R. Lissner. *Biomechanics of Human Motion*. Philadelphia: W. B. Saunders Company, 1962, 147 pages.
An introductory text to the analysis of human motion using biomechanical principles. Summaries of how to evaluate static and dynamic tasks and how to model the body mechanically are included.

Winter, D. A. *Biomechanics of Human Movement*. New York: John Wiley, 1979, 199 pages. Emphasis is on dynamic biomechanical analysis of human movement. Discusses techniques for measurement and analyses of human movement in more detail than in other texts of biomechanics. Specific topics include kinematics, anthropometry, kinetics, mechanical work and power, muscle mechanics, and electromyography.

4. PATTERNS OF WORK

Belbin, R. M. *Training Methods for Older Workers*. Paris: OECD—McGraw-Hill, 1965, 72 pages.
> Adult learning capacity, the problems of training and retraining, and the selection of training methods for older workers are discussed from a practical viewpoint.

Birren, J. E. *The Psychology of Aging*. Englewood Cliffs, N.J.: Prentice Hall, 1964, 303 pages.
> A general discussion of psychological and some physiological changes with age. Material can be useful in identifying particular needs of older workers in job design.

Fitts, P. M., and M. L. Posner. *Human Performance*. Belmont, Calif.: Wadsworth Publishing Company, 1968, 162 pages.
> Provides a comprehensive basic review of several aspects of human performance. Among these are learning and skilled performance, motivation, sensory capacities and perceptual processing, skills measurement, and perceptual motor and language skills.

Parker, J. F., Jr., and V. R. West, eds. *Bioastronautics Data Book*. 2nd ed. NASA SP-3006. Washington, D.C.: U.S. Government Printing Office, 1973, 930 pages.
> A NASA handbook prepared for the designers of aerospace equipment. The data are presented in a very condensed form. While this book may not be the best choice for a beginner, the large amount of data compiled makes it a worthwhile acquisition for the practicing human factors specialist.

Scherrer, J., H. Monod, A. Wisner, Andlauer, A. Baisset, S. Bouisset, H. Desouile, J. M. Faverage, A. Dubois-Poulson, E. Grandjean, B. Metz, P. Montrasuc, S. Pascoud, M. Pottier, and D. Rohr. *Physiologie du Travail (Ergonomie)—Volume I: Travail Physique Energetique*, 387 pages. *Volume II: Ambiances Physiques, Travail Psycho-Sensoriel*, 362 pages. Paris: Nasson et Cie, 1967.
> A comprehensive review of work physiology with much of the French literature included. Emphasis on the physiological responses to physical effort and environmental stressors, and sensory functions associated with industrial work. (In French.)

Simonson, E., ed. *Physiology of Work Capacity and Fatigue*. Springfield, Ill.: Thomas, 1971, 571 pages.
> Fatigue is the primary subject of this contributed volume. It has a theoretical section on the factors that produce muscle fatigue, a discussion of dynamic and static work and fatigue, and the impact of other factors on the fatigue process.

Welford, A. T. *Aging and Human Skill*. London: Oxford University Press, 1973, 300 pages (reprint of the 1958 edition).
> A summary of the effects of age on performance. Identifies the types of tasks that are most likely to show an age effect.

Welford, A. T. *Fundamentals of Skill*. London: Methuen, 1968, 426 pages.
> A survey of the elements that affect performance on psychomotor tasks. Topics include decision making, short-term memory, stress, and workload.

5. HOURS OF WORK

Colquhoun, W. P., and J. Rutenfranz, eds. *Studies of Shiftwork*. London: Taylor & Francis, 1980, 468 pages.
> A compilation of many of the classic papers on shift work, including sections on:

biological adaptation to shift work; individual and environmental factors in adjusting to shift work; effects on performance efficiency, family life, health, and well-being; the design of shift systems; and the economics of shift work and methods of compensation. An excellent source of primary references about the existing research on human tolerance of, and performance on, shift work.

Johnson, L. C., D. I. Tepas, W. P. Colquhoun, and M. J. Colligan, ed. *Biological Rhythms, Sleep, and Shift Work.* Proceedings of a Conference on Variations in Work-Sleep Schedules: Effects on Health and Performance, September 19–23, 1979, San Diego, Calif. *Advances in Sleep Research,* Volume 7, E. D. Weitzman, series editor. New York: SP Medical and Scientific Books, 1981, 618 pages.
A sequel to the 1975 symposium (see Rentos and Shepard), sponsored by NIOSH and the Department of the Navy. Focuses on work-sleep schedules as they are affected by shift work schedules and as they affect the biological rhythms of the workers. Discusses the health, social, and psychological effects of shift work in light of its increasing use in industry because of around-the-clock service needs, technological developments resulting in continuous processes, and economic factors optimizing equipment use. A thorough, state-of-the-art conference with many of the key researchers in the field contributing to this volume.

Mott, P. E., F. Mann, Q. McLoughlin, and D. Warwick. *Shift Work: Its Social, Psychological, and Physical Consequences.* Ann Arbor: University of Michigan Press, 1965, 351 pages.
A readable summary of the psychosocial and health effects of shift work as they were understood in the middle 1960s. Provides a useful summary of the studies done in the United States in the period between the World Wars. Written more for personnel and supervisory people than for scientists.

Rentos, P. G. and R. D. Shepard, eds. *Shift Work and Health.* HEW Publication Number (NIOSH) 76-203. National Institute for Occupational Safety and Health. Washington, D.C.: U.S. Government Printing Office, 1976, 283 pages.
Proceedings of a 1975 symposium held in Cincinnati, Ohio, to define the state of the art on relationships between shift work and biological rhythms, psychological factors, and health. Reviews results from animal studies of circadian rhythms, laboratory experiments of performance on unusual work/rest schedules, and studies of shift workers. People with different professional backgrounds try to identify what future research is needed to determine if shift work has a negative impact on health. Emphasizes a need for more industrial studies.

Sergean, R. *Managing Shiftwork.* London: Gower Press, Industrial Society, 1971, 242 pages.
A practical discussion of shift work including descriptions of shift systems, techniques for introducing shift work in an industrial setting, and the effects of shift work on the employee. Most of the studies described are from the British literature. The author summarizes most of the information that management needs to make decisions about the appropriateness of different types of shift systems and their potential effects on the worker's performance and satisfaction.

Swensson, A., ed. *On Night and Shift Work.* Stockholm: National Institute of Occupational Health, 1969, 157 pages.
The proceedings of this symposium contains 12 papers. Covers: effects of sleep loss, disruption of diurnal rhythms, attitudes of shift workers, and the relative importance of health problems and social problems in shift work. Several speakers

presented literature surveys of specific aspects of shift work. Provides a comprehensive introduction to the effects of industrial shift work.

Swensson, A., ed. *Proceedings of the Second International Symposium on Night and Shift Work.* Stockholm: National Institute of Occupational Health, 1971, 116 pages.
Contains the proceedings of a symposium held in Bulgaria in 1971. Papers include the effect of shift work on physiological functions, health, accidents, oral temperature, working capacity, and psychological indices.

U.S. Department of Housing and Urban Development. *Work Schedule Design Handbook: Methods for Assigning Employee's Work Shifts and Days Off.* Washington, D.C.: Department of Housing and Urban Development, Capacity Sharing Program, 1978, 365 pages (also available from the Institute for Public Program Analysis, 1328 Baur Boulevard, St. Louis, Missouri 63132).
A handbook designed to help people develop work schedules for operations that require coverage more than eight hours a day, five days a week. It includes: techniques for designing shift schedules given certain requirements for staffing, methods for analyzing existing shift schedules, guidelines for the choice of rotation frequency and lengths of work and rest periods, and practical examples of schedule designs and management practices affecting shift system design.

Webb, W. B., ed. *Biological Rhythms, Sleep, and Performance. Wiley Series on Human Performance,* D. H. Holding, series editor. New York: John Wiley and Sons, 1982, 278 pages.
A series of nine state-of-the-art discussions covering the areas of sleep, biological rhythm, and performance research with the intention of developing their interdependence. Reviews much of the past research in these fields, and makes attempts to reconcile analytical techniques and identify when the data can be extrapolated and when they are specific to the reported study. Does not develop a clear-cut direction, but does show an appreciation for the complexity of biological rhythm, sleep, and performance research.

6. MANUAL MATERIALS HANDLING

Drury, C. G., ed. *Safety in Manual Materials Handling.* Report on international symposium, July 18–20, 1976, Buffalo, N.Y. U.S. Department of Health, Education, and Welfare, National Institute for Occupational Safety and Health (Division of Biomechanical and Behavioral Sciences). Washington, D.C.: U.S. Government Printing Office, 1978, 209.
The proceedings of, and results of discussions at, a symposium on manual handling. Covers the epidemiology, measurement and modeling of human performance in lifting, factors affecting performance, and needs for further research.

National Institute for Occupational Safety and Health. *Work Practices Guide for Manual Lifting.* DHHS (NIOSH) Publication No. 81–122. Department of Health and Human Services, National Institute for Occupational Safety and Health. Washington, D.C.: U.S. Government Printing Office, 1981, 183 pages.
A guide for designing manual lifting tasks so they are within the capacities of more people. It was developed by a special task force and reviews the epidemiological, biomechanical, physiological, and psychophysical evidence for properly designed lifting tasks.

7. HUMAN CAPACITIES

Clarke, H. H. *Muscular Strength and Endurance in Man.* Englewood Cliffs, N.J.: Prentice-Hall, 1966, 211 pages.
A summary of muscle strength testing, methods to measure both dynamic and static muscle work, and information about muscle fatigue and endurance.

Consolazio, F. C., and L. J. Pecora. *Physiological Measurements of Metabolic Functions in Man.* New York: McGraw-Hill, 1963, 505 pages.
The classic discussion of the methodology for measurement of energy expenditure and the factors that affect it. Most useful for those who want to make metabolic measurements of people at work. Not written for the novice.

Damon, A., H. W. Stoudt, and R. A. McFarland. *The Human Body in Equipment Design.* Cambridge: Harvard University Press, 1966, 360 pages.
An explanatory text discussing the proper application of anthropometric and strength data contained in the tables. Additional material covers other factors that affect equipment design.

Dirken, J. M. *Functional Age of Industrial Workers.* Groningen, The Netherlands: Wolters-Noordhoff Printing, 1972, 251 pages.
A study of Dutch industrial workers relating their physiological, sensory, psychological, and social capacities to job demands and performance. The resulting model of "functional" age is a better predictor of performance than is chronological age.

Ellestad, M. H. *Stress Testing. Principles and Practice.* 2nd ed. Philadelphia: F. A. Davis Company, 1980, 424 pages.
An extensive review and discussion of cardiovascular stress testing that includes both protocols and data. It covers stress testing for fitness evaluation and for clinical detection of cardiopulmonary dysfunction.

Jones, N. L., E. J. M. Campbell, R. H. T. Edwards, and D. G. Robertson. *Clinical Exercise Testing.* Philadelphia: W. B. Saunders Company, 1975, 214 pages.
A summary of multiple tests of aerobic and anaerobic work capacities. Includes both the theory and the protocols for such testing as well as some standard data.

Larson, L. A., ed. *Fitness, Health, and Work Capacity. International Standards for Assessment.* Prepared by the International Committee for the Standardization of Physical Fitness Tests. New York: MacMillan, 1974, 593 pages.
A summary of the concensus of an international committee, especially regarding aerobic work capacity and fitness assessment. Includes protocols for testing as well as for measuring physiological changes in metabolism, circulation, and respiration.

NASA. *Anthropometric Source Book—Volume I: Anthropometry for Designers,* 616 pages. *Volume II: A Handbook of Anthropometric Data,* 424 pages. *Volume III: Annotated Bibliography of Anthropometry,* 130 pages. Yellow Springs, Ohio: NASA Scientific and Technical Information Office, 1978.
Compilation of anthropometric data, primarily of Air Force and Army men and women. Most complete and comprehensive source of these data.

Wiener, J. S., and J. A. Lourie, comps. *Human Biology. A Guide to Field Methods.* International Biological Programme Handbook No. 9. Oxford, England: Blackwell Scientific Publications, 1969, 621 pages.
A thorough sourcebook containing details of the testing of many body characteristics, including anthropometry, work capacity, heat tolerance, and muscle strength. Based on information from a large number of consultants.

8. JOURNALS

American Industrial Hygiene Association Journal, published monthly, American Industrial Hygiene Association, 475 Wolf Ledges Parkway, Akron, Ohio 44311.
    The ergonomics group of AIHA frequently publishes work physiology, heat stress, and manual materials handling articles in this journal. The majority of articles relate to chemical and physical workplace exposures, however.

Applied Ergonomics, published quarterly, IPC House, Surrey, England.
    The application of human factors to the solution of problems is emphasized. A very wide range of problem areas is covered. Theoretical research results and lengthy descriptions of experimental technique are avoided. The general style of the articles is such that they can be read easily by persons with little or no background in human factors.

British Journal of Industrial Medicine, published quarterly, British Medical Association House, Tavestock Square, London WCIH 9JR, England, and British Medical Journal, 1172 Commonwealth Avenue, Boston, Massachusetts 02134.
    Articles on environmental stress are frequently included in this journal, especially heat, humidity, and cold stress and noise exposure. Many of the articles deal with chemical and physical environmental exposures.

Ergonomics, published bimonthly, Taylor & Francis, Ltd., 10–14 Macklin Street, London WC2B, 5NF, England.
    The articles published include results of original research, evaluations of experimental techniques and methodologies, and application of human factors to practical problems. The emphasis is on industrial problems, but there are articles related to other areas such as sports, the military, and product design.

Ergonomics Abstracts, published quarterly, Taylor & Francis, Ltd., 10–14 Macklin Street, London WC2B 5NF, England.
    This service regularly scans more than 160 journals for articles related to ergonomics and human factors. Both English and foreign language sources are included. Several thousand abstracts are published and indexed under 40 different topics each year. A comprehensive service.

European Journal of Applied Physiology, published quarterly, Springer-Verlag New York, Inc., 165 Fifth Avenue, New York, N.Y. 10010.
    This journal is a cross between *Ergonomics* and the *Journal of Applied Physiology.* Contains articles on work physiology, most of which are in English, some in French and German. A good source for information on ergonomics work in Germany that is written in English.

Human Factors Journal, published bimonthly, Human Factors Society, Santa Monica, California.
    The types of articles published include results of original research, literature surveys, and applications. The range of topics covered is broad and includes military and aerospace applications, industrial applications, environmental effects, and product and equipment design.

Journal of Applied Physiology, Respiration, Environmental, and Exercise Physiology, published monthly, American Physiological Society, Williams & Wilkins, Baltimore, Maryland.
    Includes papers of physiological applications as well as basic research, with particularly thorough coverage of heat exposure, whole-body physical effort, and local

muscle fatigue. Written more from a research physiology perspective than that of an ergonomist, but much useful information in it.

Journal of Human Ergology, published semiannually, Business Center for Academic Societies Japan, 4–16 Yayoi 2–chome, Bankyo-ku Tokyo 113, Japan.
This is the official journal of the Human Ergology Research Association of Japan. The articles are similar to those in *Ergonomics* and the *European Journal of Applied Physiology*, with a majority coming from Japanese ergonomists.

Journal of Occupational Medicine, published monthly, American Occupational Medical Association, Arlington Heights, Illinois.
Focusses on occupational diseases or injuries and includes many articles on chemical toxicity, airborne particle exposures, heavy physical effort stress, and occupational health programs to detect and reduce occupational illness.

Medicine and Science in Sports and Exercise, published bimonthly, American College of Sports Medicine, P. O. Box 1440, Indianapolis, Indiana 46206.
Covers work physiology and its applications, especially muscle function and capacity. Fitness is a focal point. The information can be used to identify individual capacities in a predominantly young population. Little research on industrial work physiology, but many common interests with ergonomics.

GLOSSARY

The terms listed here are commonly used in industrial human factors work. This glossary is intended primarily to assist the reader in understanding the terms found in this volume. For a more comprehensive listing of ergonomics and human factors terms, see the *Ergonomics Glossary, 1982**.

A Shift—the day shift; usually from 7 a.m. to 3 p.m. or 8 a.m. to 4 p.m. on eight-hour shift schedules.

Abduction—movement of a limb away from the body's midline axis, such as elevating an elbow or raising an arm to the side.

Absenteeism Rate—the number of days lost from work out of the total workdays scheduled, usually measured over a year. It can be calculated for an individual or for a group of workers and is usually expressed as a percentage.

Acceleration—the rate of change of velocity, $A = (V_2 - V_1)/T$.

Accelerometer—an instrument that measures the amount of change in velocity. A triaxial accelerometer identifies velocity changes in three orthogonal planes. Since the velocity of a movement is related to the forces applied in muscular work, accelerometers are used in some situations to assess dynamic lifts.

Acceptable Load—the amount of weight a person chooses to lift in a specific container for a defined time period at a specific location. In psychophysical experiments, the handler determines this amount by adjusting the weight of the container to an amount that appears suitable for the duration of the lifting task.

Accommodation of Workers—the design approach that matches job task demands to the capabilities of the work force so that most people can perform the required work. It can also be the provision of aids, such as platforms or step stools, and tools that permit people to fit the design even if the original design is not suitable for their size or strength.

Accountability—liability for being called on to explain or be responsible for something. A factor that increases the stress on some workers when critical decisions must be made, such as quality inspection decisions prior to the release of a product to a customer.

Acetylcholine—a chemical substance that is the primary transmitter of a nerve fiber's electrical signal to produce muscle contraction. It is released by the nerve's electrical stimulus and changes the characteristics of the muscle protein to form actomyosin and develop tension. Cholinesterase breaks down the acetylcholine to permit recovery to take place.

---

*Anon. 1982. *Ergonomics Glossary: Terms Commonly Used in Ergonomics*. International Publications Service, 114 East 32nd Street, New York, New York 10016.

Acromion—the distal shoulder above the upper arm. It is the bony crest from which anthropometric measurements of the shoulder and arm are often made.

Acrophase—the highest point, or maximum value, of a 24-hour circadian rhythm. It can be used to mark the rhythm's phase or location in time. It will often change its time of occurrence when a person is adjusting to a rotating shift work schedule.

Action Limit (AL)—in the National Institute for Occupational Safety and Health manual lifting guide, the amount of weight that at least 85 percent of the work force can lift with two hands in the sagittal plane at 76 cm (30 in.) above the floor and at different distances in front of the ankles. Weights greater than these values require some selection or training of potential employees or merit consideration of redesign to make them suitable for more people.

Actomyosin—a contractile protein formed from the combination of two muscle proteins, actin and myosin, in the presence of calcium ions and ATP (adenosine triphosphate).

Adaptation—adjustment to conditions in the environment. In biological rhythms, an adaptation is an adjustment in amplitude and phase in response to altered hours of work, specifically night shift work. Adaptation to temperature changes is referred to as acclimatization.

Adduction—movement of a limb toward the midline axis of the body, such as moving an arm across the front of the body.

Aerobic Metabolism—the breakdown of foodstuffs to carbon dioxide and water in the presence of oxygen. Large amounts of ATP (adenosine triphosphate) are produced in this process, which supports muscle activity and body processes such as growth, hormone secretion, and tissue repair.

Alveolar Gas Exchange—the movement of oxygen (O_2) from the small air sacs (alveoli) of the lung into the arterial blood of the capillaries surrounding the alveoli. Also, the movement of carbon dioxide (CO_2) from the venous blood to the alveoli and then to the atmosphere. Alveolar gas exchange in healthy individuals is altered by pulmonary ventilation, blood flow, and the workload.

Amplitude of a Circadian Rhythm—the vertical distance from the lowest to the highest points in the 24-hour periodic wave that indicates changes in the level of a measurable substance or process, such as blood hormones, body temperature, or heart rate. Adaptation to changes in one's hours of sleep, as would be required for people working on C shift, can often be seen as a reduction in amplitude of a circadian rhythm.

Anaerobic Metabolism—the breakdown of starch or sugar molecules to lactic and pyruvic acids in the absence of oxygen. Small amounts of ATP (adenosine triphosphate) are produced in anaerobic metabolism, but the accumulating lactic acid eventually limits useful work by causing severe muscle fatigue.

Anterior Motor Neuron (AMN)—the motor neuron (nerve cell) in the anterior (abdominal side) horn of the spinal cord that is the primary component in a motor unit supplying muscle cells and permitting movement to occur. It is the "final common pathway" for motor control, integrating activity from other parts of the nervous system.

Anthropometry—the study of people in terms of their physical dimensions. It includes the measurement of human body characteristics, such as size, breadth, girth, and distance between anatomical points. It also includes segment masses, the centers of gravity of body segments, and the ranges of joint motion, which are used in biomechanical analyses of work and postures.

Arbitrary Work Breaks—interruptions in work that are not scheduled and do not relate directly to the primary tasks. They are often used to release the worker from prolonged work on a task, especially if that work is externally paced. These breaks are often very short (one to two minutes) and include getting a drink of water, visiting the lavatory, talking to another worker, or taking a short walk in the corridors. If rest breaks are scheduled regularly on a highly repetitive or paced job, arbitrary work breaks are virtually eliminated.

Arm Work—physical effort that takes place in locations that can be reached without bending the knees or the trunk frequently. For example, work on a counter that is at least 90 cm (35 in.) high and does not require reaches more than 38 cm (15 in.) in front of or to the sides of the body will be done primarily by the muscles of the arms, shoulders, and upper trunk. Arm work capacity, or upper body work capacity, is about 70 percent of whole-body aerobic work capacity as measured in a standardized treadmill stress test.

Arthritis—a family of diseases characterized by inflammation of the joints. Thought to have some association with overuse or repetitive trauma to the joints as may occur in some occupational tasks.

Asymmetric Lift—a manual handling task where the hands do not share the load equally, either because the object being lifted is not symmetrical or because the posture assumed does not permit equal use of the hands. Asymmetrical loads usually put more load on the back and upper limb muscles (for the same weight) and can increase the risk for overexertion injuries.

Atlanto-Occipital Joint—the place where the vertebral column joins the cranium.

ATP (Adenosine Triphosphate)—a chemical compound that stores phosphate in high energy bonds and makes it available for chemical reactions in cellular processes such as muscle contraction. When the energy is released, ATP becomes adenosine diphosphate (ADP) or adenosine monophosphate (AMP).

Autocorrelation Analysis—a mathematical approach to evaluating the relationship among several closely related factors. Used to study biological rhythms in order to detect changes that are not explained by chance alone.

Automation—the use of machines or mechanical devices to perform a stereotyped task automatically. In the automation process, the worker becomes machine monitor rather than task performer.

Axis of Rotation—see Fulcrum.

B Shift—the afternoon-evening shift; usually from 3 p.m. to 11 p.m. or 4 p.m. to midnight on eight-hour shift schedules.

Ballistic Lifting—a style of lifting during which large forces are developed early in the load displacement so that the resulting momentum will assist in completion of the lift.

Basal Ganglia—clusters of nerve cells and their projections that organize information from the sensorimotor cortex, the cerebellum, and the senses (such as vision and hearing) and relay it to the extrapyramidal system, which influences the coordination of muscle activity.

Biceps (Biceps Brachii)—a long muscle on the ventral (front) side of the arm that flexes, reducing the elbow angle of the forearm.

Bicycle Ergometer—a stationary bicycle with adjustable and calibrated tension controls. It is used to increase workload incrementally so that an individual's maximum aerobic work capacity can be determined. In most protocols, the pedaling rate is kept constant at 50 cycles per minute and tension on the wheel is increased, requiring more muscle work at each level. This standardized test is used to compare the exercise tolerance and aerobic fitness of an individual to values for the general population.

Biomechanics—the application of mechanical principles, such as levers and forces, to the analysis of body-part structure and movement.

Blood Pressure—the pressure, measured in millimeters of mercury, exerted by the blood on the walls of the peripheral blood vessels. It varies with the muscular efficiency of the heart and the age and health of the subject. Systolic and diastolic blood pressure levels indicate how efficiently the heart is beating and the amount of resistance the vessels offer to flow, respectively. Mean arterial pressure gives an indirect indication of the work of the heart, as does the double product (heart rate times systolic blood pressure). Blood pressure increases significantly with heavy static muscle loading that is sustained for two minutes or more.

Body Segment—a portion of the body that falls between two joints (such as the upper arm, forearm, upper trunk, lower leg) and that can influence the body mechanics in postural or manual handling activities.

Body Weight (BW)—the mass of the body and the force with which it is attracted to the earth by gravity. It is expressed in kilograms in the SI system and pounds in the English system. Oxygen consumption is expressed per kilogram of body weight in order to normalize workload measurements.

Boredom—a state of weariness associated with performance on a tedious or monotonous task. It is an individual response, varying among individuals for work on the same job.

Brain Stem—the lower part of the brain where it joins with the spinal cord. It contains many of the vital centers for the control of breathing, circulation, and other body functions. The reticular activating system (RAS) is found in this area and is important in determining the level of arousal of a person and his or her sensitivity to environmental stimuli.

Bulk Materials—raw materials or products that are not packaged for the trade; includes many construction supplies and large-sized bags or boxes of food or chemicals.

Bursitis—inflammation of a bursa, a sac found near a joint such as the shoulder or knee. The inflammation is attributed in some cases to excessive use of the joint.

C Shift—the late night shift; usually from 11 p.m. to 7 a.m. or midnight to 8 a.m. on eight-hour shift schedules.

Calisthenics—traditionally, a system of bodily exercise to promote strength and gracefulness. In manual handling, it refers to lifting the body each time an object is lifted from a location near the floor. The energy requirements of lifting the body along with the load limit the amount of work a person can do per shift; removing the calisthenics by elevating the load can improve productivity.

Capacity—the maximum ability of a person to perform in a given set of conditions. Aerobic work capacity will vary with the number of muscle groups involved and environmental conditions. A person's strength capacity is known as a maximum voluntary contraction and will change with joint angle and the duration of application, among other factors.

Carbon Dioxide Production—the amount of carbon dioxide (CO_2) produced in the body is determined by the level of aerobic glycolysis. Muscle work receives its energy from the breakdown and oxidation of glycogen, glucose, or fatty acids to CO_2 and water. Carbon dioxide is a waste product and is eliminated through the lungs and kidneys.

Carboy—a large glass or plastic bottle surrounded by a wooden case that protects the bottle during transport. Caustic or corrosive liquids, such as sulfuric acid, are often stored in carboys.

Carpal Tunnel Syndrome—entrapment of the median nerve of the hand and wrist in the passageway (tunnel) through the carpal bones of the wrist. It usually results in numbness in the fingers and pain on gripping and may be accompanied by changes in electromyographic (EMG) patterns and nerve conduction velocities, indicating a pressure block of the nerve.

Caster—a wheel or set of wheels mounted in a swivel frame. They are attached to trucks, furniture, and hand carts to permit easier movement.

Catecholamines—chemical substances in the blood stream that function as hormones and neurotransmitters and influence the activity of the heart and blood vessels and other organs through the autonomic nervous system. They include epinephrine (adrenaline) and norepinephrine (noradrenaline). Their levels in the blood and urine reflect the amount of stress a person experiences.

Center of Gravity—the center of mass of an object that determines its symmetry and ease of handling. The centers of gravity of limb segments are used in biomechanics to determine the torques around joints.

Centimeter (cm)—a metric measure of distance; 1 cm = 1/100 meter, the standard measure of length in the SI system. One inch equals 2.54 cm, or 1 cm equals approximately 0.4 inch.

Central Nervous System—the brain and spinal cord; the integrating and controlling portion of the nervous system.

Cerebellum—a part of the brain that coordinates information from the muscles, joints, and tendons as well as from higher centers in the nervous system so that smooth and efficient motions can be made and body equilibrium can be maintained. It is located just above the brain stem on the dorsal (back) side.

Chiming—turning a cylindrical container, such as a metal drum or a compressed gas tank, on the edge of its bottom rim (chime) to move it between locations. An alternative to using a drum cart.

Chronobiology—the scientific study of biological events and their variations in time. This includes the development of an organism as well as daily (diurnal) and short-cycle (ultradian) rhythms in body temperature, awareness, and other physiological or psychological measures.

Cinematography—in ergonomics, the use of motion picture film to record and measure body motions. This technique is very useful in situations where quantification of joint angles, static muscle loading times, and postures are needed but are difficult to measure directly.

Circadian Rhythm—a physical measurement, such as body temperature, or a chemical response, such as the excretion of catecholamines, that varies periodically over 24 hours.

Combination Shift Schedules—work hours that use more than one type of shift schedule. For example, combining a rapidly rotating shift with weekly rotating shifts or mixing five- and seven-day shift schedules to cover a department that needs more than a five-day shift schedule but less than a full seven-day schedule. Combination shifts are often used instead of fixed overtime schedules.

Comfort Rating—a psychophysical measure of the degree of well-being experienced by a person in a specific set of environmental or task conditions. Used to assess local muscle and joint stress in handling activities or to rate environmental temperature and humidity conditions.

Compact Load—an object that is comfortably handled within 25 cm (10 in.) of the front of the body with the arms spread no more than 50 cm (20 in.) apart. It is usually not more than 30 cm (12 in.) deep.

Compressed Work Weeks—defined as work schedules that are less than five days long, such as four ten-hour days or three 12-hour days.

Compressive Force—a force that is applied perpendicular to a surface; for example, the pressure placed on the intervertebral discs due to forces generated during lifting or maintaining a posture.

Concentric Muscle Contraction—shortening of the muscle as it exerts force against a resistance, as in elbow flexion. See Eccentric Muscle Contraction.

Confounding Factor—a variable that appears at the same time as the effect of one variable on another response is being studied. It makes interpretation of the data collected less easy because the confounding factor may interact with the other variable being studied. Physical work may confound studies of the effect of time of day on body temperature levels, for example. See Hawthorne Effect.

Contingency—a possible but not certain condition or event. Job design must consider contingencies but should not be totally determined by them.

Continuous Work—a workload, such as the exertion of a muscular force, that is sustained and uninterrupted. In dynamic work, it is the sustained pattern of work without rest or light effort breaks. Continuous work, especially when the work is demanding, results in earlier fatigue and less productivity than does intermittent work.

Cosine—a trigonometric relationship for an acute angle in a right triangle; defined as the ratio between the side adjacent to the angle and the hypotenuse.

CP (Creatinine Phosphate)—a storage system for high-energy phosphate bonds that permits very short-duration heavy effort tasks to be done without incurring a large oxygen debt. CP transfers its high-energy phosphate to ADP in order to make more ATP.

Critical Defect—a flaw in a product or subassembly that will make the product function improperly or not at all.

Cross-Sectional Area—the surface exposed when a muscle is cut transversely, perpendicular to its axis. The amount of force a muscle can generate is proportional to its cross-sectional area.

Cumulative Trauma Disorders—see Repetitive-Motion Disorders.

Cycle—a time interval during which a regularly recurring sequence of events is completed. It can be the time to complete a task with many elements or the time to complete a single operation in a repetitive task.

Cylindrical Grip—the contact of the hand with an object where the palm and fingers hold the object securely and the angle of curl of the fingers is similar. The thumb is not essential for a cylindrical grip.

Decrement in Performance—a decrease in human proficiency on a task. It may be associated with fatigue, distraction, or discomfort. It is characterized by increased errors and misjudgments, omission of task elements, and reduced intensity of effort.

Defect—an imperfection, fault, or deficiency in a product or part that will influence the product's performance. There are several levels of defects, from ones that are more cosmetic than functional to ones that are critical and may affect product safety. Inspectors have to learn to recognize defects and to remove that product before it goes on to the next workplace or the customer.

Depolarize—to reduce or reverse the charge, usually on a membrane. Nerve and muscle conduction are accomplished by depolarization of cell membranes produced by the movement of sodium ions into the cells.

Desynchronization of Rhythms—the disruption of a normal rhythmic relationship between a physiological measure and the time of day caused by a changed behavior or activity pattern. For example, a person who rotates weekly among day, afternoon-evening, and night shifts may experience desynchronization of body temperature, heart rate, hormonal, and sleep rhythms for the first part of the week on the night shift.

Diastolic Blood Pressure—the relaxation pressure in the cardiovascular system when the heart is filling with blood, measured at the point when the heart sounds disappear during cuff deflation. The value is also determined by the amount of resistance in the peripheral blood vessels.

Displacement—the difference between the initial position of an object and its position at a later time. In biomechanics, the object may be the body or a particular body segment.

Distal—far from the origin; refers to a point farthest from the body's midline. For example, the distal phalanx of a finger is the finger tip.

Distraction—an event or condition that diverts attention from the primary task. For example, thermal discomfort, either too hot or too cold, may distract a worker from concentrating on the job's activities.

Distribution of Blood Flow—the channeling of blood to different organs or muscles of the body in response to metabolic demands for oxygen or nutrients. It is accomplished by the vasomotor centers of the brain stem that control the tone of the arterioles and by local metabolic effects on the smooth muscle of the small arterial vessels.

Double-Day Shift—a work schedule with weekly alternation between daytime and afternoon-early evening shifts. It is used in five-day work schedules where a late-night shift is not required.

Double Product—an indirect estimate of the work of the heart, found by multiplying the systolic blood pressure, in millimeters of mercury, by the heart rate, in beats per minute.

Drum Cart—a two- or four-wheeled vertical carrier, possessing a hook to grasp the top rim of a metal drum and a narrow platform to slide beneath the drum.

Dynamic Muscle Work—muscle contraction where muscle length changes during activity, resulting in motion around a joint. Most handling and assembly tasks are dynamic. Isotonic contractions can be measured by resisting movement (eccentric) or by pulling against resistance (concentric) and moving an object over distance. Work can be calculated as the force of contraction times the distance moved.

Dynamics—the biomechanical aspects of the human body in motion.

Dynamometer—a device for measuring the force of muscle contraction; for example, a handgrip dynamometer measures power grip strength.

Eccentric Muscle Contraction—muscle force exertion while the muscle is being forcefully lengthened, thereby counteracting its normal shortening. This occurs when a muscle is under an external load that exceeds its ability to exert force, as in trying to lift a too-heavy object.

Efficiency—the effectiveness with which a task or operation is done; usually measured in energy spent, cost, or time required. For muscular work, efficiency is a measure of how much of the energy is translated into useful mechanical work compared to the total amount, mechanical plus that dissipated as heat. Most muscle work is less than 25 percent efficient.

Effort Equivalencies—a way to categorize the physical effort levels of various industrial tasks according to the percentage of maximum work capacity that they usually require. This permits recognition of similarities in effort levels when, because of the amount of muscle mass involved, different amounts of aerobic work capacity are available. For example, the same energy expenditure that would be heavy work in an arm task is only moderate work in a whole-body task because arm work aerobic capacity is only 70 percent of whole-body aerobic capacity.

Electrocardiogram (ECG)—a recorded tracing of the changes in electrical potential that characterize the electrical activity of the heart. There is a small P wave representing depolarization of the atria, a large QRS spike representing depolarization of the ventricles, and a broad T wave representing repolarization of the ventricles. It is used in ergonomics to measure heart rate, through the R-R intervals, and to evaluate job demands.

Electrogoniometer—an instrument used to measure angles. It is placed across a joint, such as the elbow or wrist, with each arm of the instrument aligned over a major bone on either side of the joint. As the joint angle changes, the instrument transforms the change to an electrical signal, which can be recorded throughout the work cycle to measure the biomechanics of the task.

Electromyography—a scientific technique for recording the electrical activity of muscles. In ergonomics, it is used to evaluate the active muscle groups in various job studies, and is especially useful when qualitative, not quantitative, measures are needed.

Endurance—the ability to sustain an activity over time. For example, muscle endurance is described by the length of time a muscle force can be held. Dynamic work endurance is described by the amount of time a given level of aerobic work can be sustained.

Energy Expenditure—the power used during activity or rest. It is usually expressed in watts, in kilocalories per minute or hour, or in milliliters of oxygen per kilogram of body weight per minute (mL O_2/kg BW · min).

Energy Transformation—the transfer of energy from digested foodstuffs to high energy phosphates (ATP, CP) in order to provide energy for cell processes. This is done through anaerobic and aerobic glycolysis for carbohydrates and through similar metabolic pathways for proteins and fats.

Engram—a permanent pathway or trace left on a system as the result of a repeated stimulus or pattern of activity. Motor engrams are established when new skills are learned and practiced, and the coordinated sequences are stored for future use.

Entrainment of Rhythms—the linking of a biological rhythm, such as body temperature, to another rhythm, such as time of day or an activity pattern. Entrainment is best seen in isolation studies where time cues are minimal and the body rhythm of interest can be influenced more directly.

Environment—the circumstances, conditions, and influences that affect the behavior and performance of people in the workplace. Physical factors such as noise, vibration, lighting, temperature, humidity, and air flow are important environmental factors in job design.

Equilibrium—the point at which all forces are balanced and there is no movement. Static equilibrium in the biomechanical analysis of a lifting task is achieved when the torque tending to rotate the body forward is counteracted by a torque in the opposite direction.

Ergonomic Design of Jobs—the use of principles relating peoples' capacities to job demands in the assignment of tasks and levels of effort required in a job. Designing jobs so that a large majority of the potential work force can do them without risk of injury or illness.

Error Rate—the number of mistakes or omissions per unit time or per number of pieces or actions. For example, the number of defects missed per 100 units inspected or the number of improperly recorded readings per eight-hour shift are two types of error rate. Ergonomic job design intends to reduce the opportunities for error, and thus reduce the error rate.

Extended Hours—defined as additional hours worked per shift compared to the standard eight-hour work day. Ten- and 12-hour shifts are extended-hours schedules, even if the total hours per worker per week do not exceed 40. Longer work shifts and overtime have implications for the development of fatigue in physically demanding jobs and for the adjustment of exposure times in environments where chemicals, heat, cold, noise, and other physical factors may be present.

Extension—the straightening of a joint whereby the angle between adjacent bones usually increases. Exceptions are extension of the foot and of the wrist.

Fatigue—the reduction in performance ability caused by a period of excessive activity followed by inadequate recovery time. Muscle fatigue is accompanied by a buildup of lactic acid in the working muscle.

Feedback—the return of the effects of a given action or process to its source. In ergonomics, it refers to information about an activity that is passed back to the operator for evaluation and, if necessary, adjustment of the activity. For example, feedback on product quality can influence the performance of an assembler or inspector.

50–50 Mix—equal numbers of men and women for a hypothetical design population. The capabilities of each separate population are combined statistically to estimate the percentage of the 50–50 mix accommodated by a given design. It is most commonly used in applying anthropometric and strength data to workplace, task, and equipment design.

First-Class Lever—a lever system in which the axis of rotation (fulcrum) is between the force and the resistance. The force arm may be equal to, greater than, or less than the resistance arm.

Flexibility—the ability to adjust to changing conditions. During absences due to illness or vacation, it is easier to cover a job if the population capable of performing it has not been limited too severely by the job design. Management has more options and, therefore, more flexibility if more people have the needed skills or capacities.

Flexion—the bending of a joint whereby the angle between adjacent bones usually decreases.

Flextime—a work schedule, usually on the day shift, that requires employees to be at work for the core hours between 10 a.m. and 3 p.m., with the remaining work hours being chosen according to individual needs or work preferences. Each person is expected to work 40 hours per week. The schedule has been successfully applied in white collar jobs in Europe and the United States.

Foot (ft)—a measure of distance in the English system of measurement; one foot has 12 inches and is equivalent to approximately 30 cm.

Force—a push or pull, defined as mass times acceleration, that an object exerts on another object. It is measured in newtons (N) or pounds of force (lbf). See Torque.

Force Arm—in a lever system, that part of the lever between the axis of rotation (fulcrum) and the point of application of muscle force.

Force Transducer—an instrument that, most commonly, converts the deformation of a piece of metal into an electrical signal or a needle movement, thereby quantifying the deforming force. Strain gauges and push-pull gauges are force transducers.

Force-Velocity Relationship for Muscle—the characteristic of muscles that describes their reduced ability to generate tension or force at higher contraction velocities. Because force varies with the velocity of movement, isokinetic strength-testing devices have been developed to measure dynamic forces.

Forklift Truck—a powered, open vehicle having two horizontal prongs, often about 90 cm (36 in.) long, which are adjustable vertically and, sometimes, horizontally. The forklift truck is used to handle pallets of materials or products and can be adapted to handle metal drums, large containers, or other materials.

Fourier Analysis—a mathematical technique that uses wave mechanics to determine the spectrum of energies associated with a complex signal. For example, a surface electromyogram can be subjected to Fourier analysis in order to determine whether muscle fibers are fatiguing with sustained effort.

Free-Running Rhythm—the inherent biological rhythm set free from external entrainers; usually accomplished only in isolation experiments. In nonexperimental conditions, rhythms are entrained by psychosocial, environmental, and work cues.

Frequency of Lifting—the number of lifts made per minute or other short time period.

It should be related to the distribution of rest or recovery periods in order to determine the intensity of the workload.

Fulcrum—the axis of rotation for a lever. In the human body, the joints serve as fulcrums for the body's lever system.

Gastrocnemius—the largest and most external muscle of the lower leg's calf.

Glycolysis—the breakdown of carbohydrates in the presence (aerobic) or absence (anaerobic) of oxygen. The anaerobic process results in an accumulation of lactic acid, while the aerobic process, or Kreb's cycle, results in total degradation to carbon dioxide (CO_2) and water (H_2O). ATP is formed in each process, but aerobic glycolysis is about 18 times more effective.

Gripping Block—a 5 to 8 cm (2 to 3 in.) wide modification to the sides of a tray that permits a person to get a stable grip for handling. It is used in designing plastic trays because molding requirements make it difficult to use other handle styles.

Gussets—triangular sections in the corners of a bag that give it strength and provide a place for a hook grasp during bag handling.

Handhold—the part of an object that enables it to be lifted or handled manually, It should be designed to provide adequate clearance for the hand and should have rounded edges that do not concentrate the object's weight on a small part of the fingers.

Handling—lifting, lowering, conveying, pushing, pulling, or sliding an object in order to move it from one place to another. If the motion is powered by a person's muscles, it is termed manual handling.

Handling Aids—devices that aid a person in moving an object from one location to another. Examples are hoists, scissors tables, conveyors, hooks and clamps, and hand carts or trucks.

Hawthorne Effect—refers to a study at the Hawthorne plant of Western Electric Company where a study of lighting levels and productivity illustrated a confounding factor in work studies. The attention paid to the workers was considered to have a profound effect on the success of the workplace intervention.

Heart Rate—the frequency with which the heart contracts, expressed in beats per minute. It is obtained from electrocardiogram traces or telemetry. It is used to estimate workload and other job stresses.

Heavy Effort—physical work that can be sustained for only one hour or less; also the handling of objects weighing more than 18 kg (40 lbm) and the application of forces greater than 250 newtons (56 lbf).

Hemoglobin—the iron-containing blood protein that transports oxygen molecules to the body tissues via the circulatory system. In the high oxygen partial pressure of the lung capillaries, the hemoglobin molecule becomes saturated with the gas; it releases the oxygen to working muscles where its partial pressure is low.

High-Energy Phosphate Compounds—chemical substances that store energy in the body in phosphate bonds and release that energy, as needed, to fuel chemical reactions such as muscular contraction.

Hormone—a chemical substance that is formed by specific cells and that exerts a specific effect on a part or parts of the body remote from the place of origin. For example, hormones from the adrenal glands are released in response to stress and affect blood circulation, muscles, and metabolism.

Humidity—the atmospheric water vapor pressure. It can be measured with a natural wet bulb thermometer, a sling or electric psychrometer with a wetted thermometer covering, or with a relative humidity meter.

Hyperextension of the spine—extension of the trunk beyond the upright position, forming a more extreme backward arch and changing the distribution of pressure on the spinal discs. Seen for instance, in work done above shoulder height. May aggravate back pain symptoms in susceptible people.

Hypotenuse—the side of a right triangle opposite the right angle.

Incentive—in work situations, it usually refers to a pay plan whereby performance above the standard level for a job is financially rewarded up to some fixed level, often 115 percent. It can be an individual incentive, as on some piecework tasks where performance is measured by the number of good units produced per shift by an individual worker, or it can be a group incentive, where the performance of a production team or a department is rewarded.

Inch (in.)—a measure of distance in the English system of measurement; equal to 2.54 centimeters.

Incident—an event, action, or situation whose occurrence or near-occurrence is noted. In safety research, it may be an accident, a near-accident, or an illness.

Innervate—to supply nerves to a part of the body.

In-Process Inventory—the buildup of products or parts on or just off a production line, permitting individual workers some control of their work pace. For example, two to three minutes of in-process inventory space at the supply and take-off ends of a production machine can give the operator more flexibility in loading and unloading the machine.

Intensity-Duration Relationship—the observed pattern that the longer a physical workload is sustained, the less percentage of maximum work capacity or strength is available for use. This is important both for static work where the available percentage of maximum voluntary strength is reduced as holding time increases and for dynamic work where the available aerobic work capacity is affected.

Intermittent Work—physical effort (usually moderately to highly demanding) that is interrupted regularly by short rest or light work periods lasting a few seconds to a few minutes. These rest periods permit the muscles to replenish their oxygen and energy stores and to reduce their accumulation of lactic acid compared to that measured with continuous work.

Isokinetic—exerting muscle force at a constant velocity in a dynamic task. Isokinetic strength-testing devices measure a person's capability for dynamic lifting.

Isometric Muscle Work—force that is developed without significantly changing the length of muscle fibers. Because there is no motion associated with an isometric contraction, no external work is done. Static muscle work, such as maintaining a posture or holding onto an object, are examples of isometric muscular contraction. The workload can be estimated by measuring the force required, relating it to the maximum force available, and determining how long it must be sustained.

Isotonic Muscle Work—see Dynamic Muscle Work.

Job Demands—the physiological, psychological, and perceptual requirements of a job that determine the suitability of a given workload for the potential work force.

Job Design—the arrangement of tasks over a work shift. Good job design reduces the opportunities for fatigue and human error.

Job Restriction—a medical response to helping an injured or chronically ill person return to work. Specific tasks or jobs that would tend to aggravate the illness or injury are designated as not suitable for that person. The restriction exists as long as the condition remains.

Job Satisfaction—a multidimensional psychophysical measure that compares and rates a person's opinions about job requirements to individual goals for meaningful work.

Job Sharing—an agreement between the employer and two employees that a job can be accomplished by having two part-time workers instead of one full-time one. This is often done to accommodate people with young children or who are attending classes.

Joule—a unit of work or energy in the SI system; equal to approximately 0.25 small calories, 10^7 ergs, or 0.7376 foot-pounds.

Kilocalorie—the amount of heat required to raise one kilogram of water one degree Celsius between 14 and 15°C; 1,000 small calories. It is used to express workload and the energy value of food when it is oxidized in the body. One kilocalorie per minute equals approximately 70 watts or 0.2 liters of oxygen per minute.

Kilogram (kg)—the unit of mass in the SI system of measurement; equal to approximately 2.2 pounds. Although the newton is the appropriate measure of force in the SI system, some measurement equipment quantifies force in kilograms (kgf); one kgf equals approximately 9.8 newtons.

Kilojoule (kJ)—the measure of work or energy in the SI system; 1,000 joules or approximately 0.2 kilocalories.

Kilopascal (kPa)—the measure of pressure in the SI system of measurement. It is equal to 1,000 newton-meters or 0.15 pounds per square inch.

Kinematics—a dynamic biomechanical analysis dealing with the descriptive aspects of movement without considerations of mass or force.

Kinetics—in biomechanics, the study of the forces that influence movement of the human body.

Lactic Acid—a three-carbon acid ($C_3H_6O_3$) formed in the breakdown of a glucose molecule if sufficient oxygen is not available to complete the degradation to carbon dioxide and water. In the blood, lactic acid takes the form of sodium lactate. High muscle and blood levels of lactate accompany exhaustive exercise.

Lateral—toward the side of the body away from the midline.

Learning—any change in behavior or performance that occurs as a result of teaching, practice, or experience.

Length-Tension Relationship for Muscle—the characteristic of muscle fibers that describes their ability to generate maximum tension at 1.2 times their resting lengths. Tension drops off at both longer and shorter lengths. This inverted-U-shaped curve is thought to be accounted for by the formation of cross-bridges between the actin and myosin proteins in the muscle fibers.

Light Assembly Tasks—work with low energy expenditure demands that is often performed in a seated position. Muscles of the hands, arms, and shoulders are usually most actively involved in these tasks, and the repetitiveness of the work can be high.

Light Effort—physical work that can be easily sustained for at least eight hours a day; also the handling of objects weighing less than 5 kg (11 lbm) and the application of forces less than 100 newtons (22.5 lbf).

Liter—the unit of volume in the SI system of measurement; equals approximately 0.26 gallons in the English system.

Long-Term—occurring over or lasting a relatively long period of time. It can be a change that takes years to become apparent, such as an illness associated with psychological stress.

Lost Time Illness or Injury—time lost from work that is associated with an illness or injury sustained at work. Lost time usually indicates a more serious injury or illness, but it may also reflect a lack of alternative work. Low-back pain and wrist soreness are among the leading occupational contributors to lost time in industry.

Lumbar Disc—the intervertebral discs between the lumbar vertebrae in the back. They are usually under the greatest stress when a person is lifting, bending forward, or sitting in a slumped posture without back support.

Maneuvering a Hand Cart or Truck—moving a vehicle in a limited space and turning it, using changes in direction rather than pushing or pulling in a straight line. This method of maneuvering usually limits the acceptable forces for moving a hand cart or truck.

Manual Dexterity—the ability to manipulate objects with the hands, ranging from the very fine finger dexterity required in traditional watchmaking tasks to the gross hand dexterity of packing operations.

Materials Handling—the movement of parts, raw supplies, chemicals, subassemblies, finished products, or other materials between sections of a manufacturing system or through distribution systems to the customer or client. The movement may be done by hand, as in lifting cases and pushing hand trucks and carts, or with automated equipment or aids, as in using forklift trucks, storage and retrieval systems, or conveyors.

Maximum Aerobic Work Capacity—the highest oxygen consumption rate that can be achieved in a given work situation or on a standardized test. For most healthy people, it is determined by their cardiovascular fitness level.

Maximum Grip Span—the largest distance between the thumb and fingers that permits the exertion of force on an object in a power grip. Grip strength is reduced to less than 50 percent of the strength at an optimal grip span of about 5 cm (2 in.).

Maximum Heart Rate Range— the difference between the predicted maximum heart rate (220 − age) and the resting level as measured under controlled conditions. The range represents the reserve capacity for elevating blood flow and thereby oxygen delivery to the muscles during work. Elevations in heart rate above resting values can be expressed as a percent of the heart rate range in order to estimate the percent of aerobic work capacity being used on the job.

Maximum Permissable Limit (MPL)—in the NIOSH manual lifting guide, the recommended upper limit of weight to be handled with two hands in the sagittal plane at 76 cm (30 in.) above the floor and at different locations in front of the ankles. Less than 15 percent of the work force has the capacity to do this type of lifting without increased risk of musculoskeletal injury.

Maximum Voluntary Contraction—the largest force that can be developed by a muscle or muscle group under a given set of conditions. Joint angle, available muscles, degree of worker motivation, and duration of holding all determine the maximum voluntary contraction strength.

Mean Arterial Pressure (MAP)—the root mean square of the difference between the systolic and the diastolic blood pressures plus the diastolic reading. It is about 67 percent of the distance between the systolic and diastolic pressures, expressed in millimeters of mercury.

MET—a conventional way of describing workloads in relation to the resting metabolic rate; used especially in cardiac rehabilitation. One MET is resting metabolism, usually about 3.5 mL O_2/kg BW • min. Aerobic work capacities of 10 and 15 METs are average values for industrial women and men; an average workload of 4 to 5 METs is a moderately heavy to heavy job demand for an eight-hour shift.

Metabolism—the sum of the physical and chemical processes by which a living organism is produced and maintained. In muscular work, metabolism is the chemical process through which foodstuffs are broken down to form carbon dioxide, water, and high-energy phosphates (ATP) for muscle contraction.

Meter—the standard unit of length or distance in the SI system of measurement; equal to approximately 39.37 inches.

Micropauses—a very short (0.5 to 3 seconds) recovery period during a physically demanding task that allows some regeneration of the energy supply for the working muscles. Heavy work can be sustained for longer periods if it is interrupted by micropauses than if it is sustained continuously until completion. See Intermittent Work.

Milliliter (mL)—0.001 liter; used to quantify the oxygen used, or the carbon dioxide produced, per minute during rest or work.

Moderate Effort—physical work that can be sustained for about two hours without a major work break; also the handling of objects weighing up to 18 kg (40 lbm) and the application of forces up to 250 newtons (56 lbf) for short periods.

Moment—see Torque.

Momentum—the product of the mass of an object and its velocity.

Monitoring—observing, keeping tack of, listening to, and exercising surveillance over a process or activity. Monitoring of the quality of the product being manufactured, the amount of inventory, the progress of a chemical reaction, or the performance of an individual on a new job are examples in industry.

Monotony—lack of variety, sameness; sometimes applied to highly repetitive tasks that require little decision making and that might be done better by machines. Individuals can overcome inherent job monotony with creative "games playing" provided that the games do not detract from the primary task.

Motor Unit—the functional unit of neuromuscular control of movement. It consists of an anterior motor neuron in the spinal cord, the nerve fibers that innervate muscle cells, and the junctions between the nerve and muscle cells.

MTM (Methods Time Measurement)—a technique for evaluating individual efficiency of motion, especially on highly repetitive jobs; work is broken down into elemental tasks such as "move, grasp, turn." It is used to set production standards in some plants and to identify where new methods can increase productivity. MTM is based on the work of F. Gilbreth in the early 1920s.

Muscle Spindles—proprioceptors that lie parallel to the muscle fibers and respond to changes in tension in the muscle. They form a feedback system that is very important in coordinating and sensing the force required to perform a muscle movement.

Musculoskeletal—pertaining to the muscles, bones, and joints.

Myoglobin—a muscle protein similar to hemoglobin that stores and releases oxygen molecules as needed to keep the energy supply at appropriate levels. It is particularly important in the initiation of muscle work when the cardiovascular system has not yet delivered enough oxygen to supply the needs of the working muscles.

Myoneural Junction—the specialized area on a muscle fiber where a nerve fiber innervates it. The junction has small packets or vesicles of transmitter substance (usually acetylcholine) that transform the electrical stimulus of the nerve fiber into a chemical activation of the muscle fiber, resulting in contraction.

Myositis—muscle inflammation often associated with heavy exertion or repeated use of a muscle group with inadequate recovery time. It is sensed as a "sore" muscle.

"Natural" Selection on Heavy Jobs—the process whereby people of lower capacity, whether strength, endurance, or visual ability, move out of the more demanding jobs because they cannot sustain them for the required durations. The difficulty of the job will determine what percent of the people starting it will eventually be able to do it on a full-time basis. In most instances, natural selection is an expensive alternative to ergonomic job design.

Nerve Entrapment Syndromes—any of a group of neuromuscular problems where a nerve is partially blocked as it passes through or across a bony structure. For example, carpal tunnel syndrome is an entrapment of a median nerve in the wrist bones.

Neuromuscular—pertaining to the muscles and the motor side of the nervous system.

Newton—the unit of force in the SI system; equal to approximately 0.22 lbf or 0.10 kgf.

Newton-meter (Nm)—a measure of torque or rotational force in the SI system. It is used in biomechanical analyses of movements around a joint.

Newton's Second Law—the law of motion and energy that describes the relationship between force (F), mass (m), and acceleration (a), F = ma.

NIOSH—the National Institute for Occupational Safety and Health, a research institute of the Department of Health and Human Services. Provides information to the Occupational Safety and Health Administration (OSHA).

Noninvasive Measurement Techniques—methods for measuring the effects of work that do not require penetrating the skin or causing significant discomfort. Examples are the electrocardiogram (ECG), motion detection techniques, and measurements of oxygen consumption.

Nutrition—the processes by which a person takes in food, breaks it down, and utilizes it for body activities.

Oblique Grip—a variant of the cylindrical grasp in which the object is held in the palm along the base of the thumb and the fingers take different degrees of flexion.

Overtime—defined as time worked that exceeds the standard hours required of the employee in a week. If more than 40 hours are worked in a seven-day period, the additional hours are overtime. These hours are usually paid at 1.5 to 3 times standard pay, depending on when they occur; weekend and holiday hours are paid at a higher rate than additional hours per weekday.

"Overuse" Syndromes—see Repetitive-Motion Disorders.

Oxygen Consumption—the rate at which the body's tissues and cells use oxygen; measured as the amount of oxygen entering the body, limb, or organ, minus the amount leaving. In assessing physical workload, researchers measure oxygen consumption as the respiratory exchange of oxygen across the body. This quantifies the active muscle mass involved in the task and, if corrected for body weight, is quite consistent from person to person for a given task.

Oxygen Debt—the excess oxygen used (above resting values) that is measured upon completion of physical work. It represents the amount of anaerobic work done at the initiation of a work task or during a sustained work period. It must be "repaid" in the recovery period before resting values are reestablished. Oxygen debts are higher at high workloads with rapid onsets and accumulate more during prolonged work of a person is working at more than 50 percent of aerobic capacity for the task.

Pacing—controlling a worker's rate of movement through external means, such as a continuous conveyor moving at a fixed speed, production pressures, peer pressure, or pay incentives. Too-rigid pacing can have a negative effect on individual productivity.

Pallet—a wooden or plastic double-sided platform, often 14 cm (5.5 in.) high and 102 × 122 cm (40 × 48 in.) in width and length (US), on which materials are stored and transported.

Part-Time Employment—working less than 35 hours a week (or the contracted minimum hours) and thereby often being ineligible for certain employee benefits such as full medical coverage or a pension plan. There is usually partial coverage for vacation time.

Pay Differential for Shift Work—the difference in hourly pay between day, afternoon-evening, and late-night shifts. The late-night shift generally has the highest pay rate.

Peak Load—the heaviest work done or weight lifted during the shift. See Short Duration Heavy Effort.

Perceived Exertion—a psychophysical measure of the amount of effort required for a given action or task. Measured on a three-, seven-, or 15-point scale, using word descriptions such as "heavy, very heavy, light" or "uncomfortable, slightly uncomfortable, and so on." The work of G. Borg on whole-body and local muscle perceived exertion is the most extensive.

Perceptual Work—tasks that are done using the senses to gather information and to determine what action should be taken. For example, using auditory or visual information to identify a product or part that is defective, not running to specification, or tending to move out of the acceptable quality range.

Performance Measurement—measuring the effectiveness of an individual on a task or job. It can be done objectively using special tests, error or productivity measurements, or evaluation checklists, and subjectively through peer, self-, and supervisory ratings.

Performance Curve—a measure of the accomplishment of a given task against a variable such as time or the number of trials. The task can be measured in a number of ways: the number of acceptable units produced, the time to complete one cycle or to make one unit, the number of errors made, and others. It could also be a measure of the physiological response to a given workload over several weeks of physical training.

Period—the reciprocal of frequency (f), or $1/f$, measured in units of time per event. For biological rhythms, the period is the time it takes to complete one cycle, usually 24 hours.

Peripheral Nervous System—the nerves, receptors, and neuromuscular connections outside the brain and spinal cord that sense the environment and effect motor activity.

Permanent Night Shift—a work schedule in which people are assigned to the late-night/early morning shift and do not rotate to daytime duty.

Phase Shift—movement of a circadian rhythm's acrophase (maximum value) in time indicating that physiological adaptations are occurring. The amount of change is measured as a phase angle, and this value is used to measure the person's adaptation to external factors, such as shift work.

Physical Effort—the use of muscles to accomplish a task. The amount of effort depends on the number of muscles involved, the intensity of their activity, and the duration of the task. It is measured by the amount of oxygen used (above resting levels) or by the elevation of the heart rate above resting values if there are no major environmental or emotional stressors present.

Pinch—applying pressure between the thumb and the ends of the fingers or the side of the hand (lateral pinch). The strength is about 25 percent that of a power grip. This grip does not involve the palm of the hand.

Postprandial—after a meal; relates to the digestion of food and the accompanying physiological changes associated with digestive processes. For example, the heart rate is elevated five to ten beats per minute above the resting heart rate within one hour after a substantial meal.

Posture—the relative arrangement of body parts, specifically the orientation of the limbs, trunk, and head during a work task. Posture can influence productivity since static muscle loading reduces the amount of continuous work a person can do.

Potential Work Force—the distribution of male and female, younger and older workers that could appear at the company's employment office on any given day to apply for work. For convenience, it is assumed that there is an equal probability of men and women (a 50-50 mix) applying, with a wide age distribution. Job design should attempt to accomodate a large percentage of this potential work force.

Pound (lbm, lbf)—a measure of mass (lbm) or force (lbf) in the English system of measurement; equal to approximately 0.45 kg or 4.45 newtons, respectively.

Predicted Maximum Heart Rate—the highest rate a person's heart can be expected to attain during maximum physical whole-body work levels; it is estimated by subtracting a person's age from 220 beats per minute. The accuracy of this prediction is about plus or minus 10 percent. Submaximal aerobic capacity testing is used to estimate the work capacities of people who are also studied as they do their job.

Production Machine—a piece of equipment or system of interlocking machines that performs a specific function, such as manufacturing or packaging a product. There are often several work stations on the machine where supplies are loaded, inspection done, or jams cleared to keep it running smoothly.

Productivity—the amount of good product completed during a shift in relation to the number of people and amount of money needed to produce it. In manufacturing operations, it is often expressed as the number of good parts produced per operator or the time required per piece assembled. Productivity is affected by worker characteristics and motivation, workplace and job design, supervisory style, and environmental factors.

Pronation—rotation of a joint forward and towards the midline of the body; for the hand and arm, palm down and thumb next to the body.

Proprioceptors—receptors located in the muscles, joints, tendons, and ear that sense the position of the limbs, head, and trunk. They convey that information to the cerebellum and brain cortex, allowing coordinated and smooth movements to be made. Proprioceptors include the semicircular canals of the inner ear, Golgi tendon organs, and muscle spindles.

Proximal—nearest the origin; refers to a point nearest the midline of the body. The proximal phalanx of a finger is the joint nearest the palm.

Psychology—the study of the human mind and behavior.

Psychomotor Task—a muscle activity requiring skill and coordination, and often spatial perception as well, such as hand-eye coordination. Most light assembly tasks require psychomotor skills.

Psychophysical Measures—data collection instruments that permit a person to evaluate the heaviness of an object or its importance by trying it. Its weight is adjusted upward and downward until it meets some set of criteria, such as the ability to lift it four times per minute throughout the shift. The selection of an acceptable load depends on the integration of many factors.

Psychosocial—referring to factors that produce both psychological and social effects. For example, prolonged time on the afternoon-evening shift can isolate a worker from his or her children because they are often either in school or asleep when the shift worker is home; this produces social isolation and is often accompanied by psychological responses of concern or guilt. Psychosocial factors are the primary determinants of the acceptability of various shift work schedules.

Radial Deviation—movement of the hand (with the palm outstretched) toward the thumb side. See Ulnar Deviation.

Radius of a Handle—the amount of bend in a handle that determines the surface area in contact with the hands during lifting or holding. The smaller the radius, the more uncomfortable the object is to support or hold as its weight is increased.

Rapidly Rotating Shift Work Schedules—hours of work wherein the shift workers work all three shifts within the course of a week. The usual rotation schedule is 3-2-2 or 2-2-3 on day, evening, and late-night shifts or late-night, evening, and day shifts, respectively, for a seven-day schedule. These schedules are advantageous in that they increase social interactions among the shift workers and their families and friends during the work week as compared to full weeks on the afternoon-evening or late-night shifts in weekly rotating schedules. These schedules are used more frequently in Europe than in the United States.

Receptor—the sensory ending of a nerve cell that responds to physical or chemical stimuli from the environment.

Recovery Time—work periods when task demands are light or when rest breaks are scheduled, permitting a person to recover from heavy effort work or exposure to an environmental extreme, such as high temperatures.

Redesign—recommended changes to an existing workplace or to production equipment to make it suitable for more workers. Also, the reexamination of job requirements and their pattern of occurrence. Redesign is a more expensive alternative to incorporation of ergonomic principles in the initial design.

REM (Rapid Eye Movement) Sleep—the stage of sleep where rapid movements of the eyes are noted and appear as intermittent electrical activity in the electroencephalogram (EEG). People whose sleep is interrupted, and who do not get enough REM sleep, tend to feel much less rested than those who have normal amounts of REM sleep. Sleep disturbances for workers on the early morning shift (C) are thought to have a negative effect on REM activity.

Repetitive-Motion Disorders—a family of musculoskeletal or neurological illnesses or symptoms that appear to be associated with repetitive tasks in which forceful exertions of the fingers, or deviations or rotations of the hand, wrist, elbow, or shoulder are required. Also called cumulative trauma disorders (CTD). Examples are tendonitis, tenosynovitis, carpal tunnel syndrome, epicondylitis (elbow), and bursitis.

Residual Time—the amount of time spent doing activities that are not primary physical effort. These activities are recognized through supplementary effort requirements analysis.

Resistance—a counteractive force, such as the force a muscle develops to counteract the weight of an object being held or lifted.

Resistance Arm—in a lever system, that part of the lever between the axis of rotation (fulcrum) and the point of application of resistive forces.

Rest Allowances—recovery time in addition to regularly scheduled work breaks. They are usually provided in jobs where heavy physical work or exposure to environmental extremes occurs. Rest allowances are built into the job standard so that productivity ratings recognize the need for additional recovery time in these jobs.

Robotics—the use of computer-programmed machines to simulate humans in work tasks that are highly repetitive or must be done in hostile environments.

Sagittal—any plane parallel to the midsagittal plane dividing the body into right and left halves; used in describing symmetrical lifting tasks using two hands.

Sarcomere—the functional unit of a muscle cell. A sarcomere contains actin and myosin proteins that combine to form actomyosin, and it is characterized by bands formed by the orientation of the muscle proteins and their interactions.

Second-Class Lever—a lever system in which the resistance lies between the axis of rotation (fulcrum) and the force. The force arm is always longer than the resistance arm; consequently, a mechanical advantage always exists.

Secondary Work—activities that are related to the main job activity but not directly linked to productivity on the job. They may include the procurement of supplies, discussions with supervision or colleagues about product quality, or adjustments to the equipment.

Selection Testing—the use of performance or capacity measurements to determine the suitability of a worker for a job requiring above-average capacity. If it is used to select people for initial employment or progression, a selection test must be validated to the job demands.

Sensorimotor Integration—a process by which information from both the peripheral and central nervous systems is channeled to the cerebral cortex, resulting in coordinated, appropriate motor action.

Shear Force—a force that is applied tangentially to a surface. See Compressive Force.

Sheet Materials—products that are very wide and long with very little depth, such as plywood, glass plates, paper sheets, and cardboard. Their dimensions make it necessary to handle them with a pinch grip and often require the arms to be spread widely apart.

Shift—a period of work, most often eight hours long, through which workers may rotate to provide 16- or 24-hour coverage of a production process or service.

Shift Work—the rotation of a worker's hours over a specified time period, often one week on each eight-hour shift. Shift work is done in order to provide 24-hour service to cover operations that cannot be easily shut down, or to make use of manufacturing equipment to the fullest extent so that the unit cost of the product can be reduced. Many shift work schedules have been developed to try to match the manufacturing and service needs to human behavior and needs.

Shipping Case—a cardboard container containing single or multiple units of product for distribution to a customer or to a regional warehouse and distribution center.

Short-Duration Heavy Effort—a period less than 20 minutes in length when very demanding physical effort work is done. This effort level often requires more than 70 percent of a person's maximum aerobic capacity for that type of work.

Short-Term—lasting or requiring a relatively short period of time.

SI Units—abbreviation for Le Système Internationale d'Unités (International System of Units), based on the metric system. It is a coherent measurement system based on units of 10, and has been adopted for worldwide scientific and commercial measurements.

Sine—a trigonometric relationship. For an acute angle within a right triangle, it is defined as the ratio between the side opposite the angle and the hypotenuse.

Siphon Pump—a method of transferring liquid from one container to another using gravity and air pressure.

Situational Factors in Job Analysis—job characteristics that are not inherent in the jobs themselves but are associated with external factors; for example, supervisory style, management policies, and production deadlines.

Skewness—the characteristic of a frequency distribution that is associated with the assymetric clustering of cases at one extreme, or tail, of the distribution.

Skid—a metal platform, similar in dimensions to a U.S. wooden pallet, that is most often handled by a forklift truck. It can hold more weight than a pallet and is generally used in storage and transport situations where a wooden pallet is not sufficiently rugged.

Skill Acquisition—the development of improved performance on a psychomotor task that is associated with practice, experience, or other learning. A motor pattern (engram) is produced in the central nervous system that permits the skill to be accessed as needed.

Sleep Satisfaction—a psychophysical rating of how rested a person feels after a period of rest. Satisfaction ratings are generally lower for people who work the late-night shift and have to sleep during the day than for people on the other two shifts who can sleep during the night.

Slip Sheet—a plastic or cardboard sheet, approximately the size of a pallet, that can be used in some applications to transfer materials into a truck without using wooden pallets.

Social Isolation—a possible result of shift work schedules that require long periods of work on the afternoon-evening and night shifts. The worker is not free during times when many people and his or her children are usually free, so interindividual contacts are reduced.

Stability—the property of a system to return to equilibrium after it has been disturbed. In biomechanics, stability is the return of the torques around the joints to dynamic equilibrium after a change in posture.

Stacker-Retriever System—an automated warehousing system that permits a worker to store or take out a truck or pallet that holds a particular product using a computer-controlled handling system. It reduces the amount of manual handling and is especially useful for the storage of large supply rolls or the intermediate stages of a manufacturing process that requires several steps.

Standard Deviation—a measure of the variability of values around the mean, or average, value; expressed as plus and minus ($\pm$) values.

Static Muscle Work—see Isometric Muscle Work.

Statics--the biomechanical aspects of bodies at rest or forces in equilibrium.

Steady State—a state or condition that does not change over time; equilibrium. In work physiology, it means that the muscles' demands for oxygen are met by the appropriate adjustments in respiration, heart rate, and blood flow. A steady physiological state is characterized by a steady heart rate over time unless the workload is very heavy.

Strain—indices of stress, such as heart rate and oxygen consumption. Also, the deformation of part of the body, such as a finger, in response to increased force per unit area.

Strain Gauge—see Force Transducer.

Stress—physiological, psychological, environmental, or mental effects that may produce fatigue or degrade a person's performance.

Submaximal Aerobic Capacity Testing—the prediction of a person's maximal aerobic capacity from a multi-staged test that does not include a maximum workload.

Susceptibility—the tendency of a person to respond to a stress with a specific type of illness; a weakness. For example, some people are more susceptible to low back pain or repetitive-motion disorders than others. The reasons for greater susceptibility are not clear but many have genetic, nutritional, or development bases.

Supination—rotation of a joint backward and away from the midline of the body; for the hand and arm, palm up and thumb away from the body.

Systems Approach—the design of manufacturing, materials handling, or other systems through consideration of the whole process and not just its separate parts; for example, designing materials handling systems to eliminate or reduce multiple rehandling of items, using conveyor systems to move products between work stations, and consistent record keeping to permit tracking the product with minimal paper work.

Systolic Blood Pressure—the pressure in the cardiovascular system when the heart contracts to send blood to the rest of the body, measured at the point when the heart sounds are first heard during cuff deflation.

Tangent—a trigonometric relationship. For an acute angle in a right triangle, it is defined as the ratio between the side opposite the acute angle and the side adjacent to the same angle.

Task Analysis—an analytical process that measures behavior on a job against time to determine the physiological and psychological demands of the job on the workers.

Telemetry—the measurement of a response, such as heart rate, by means of an electrical device that transmits the measurement by radiowaves (or a wire) to a recorder or other receiver, permitting the information to be collected at a remote site. This reduces the amount of interference with the person being studied.

Tendonitis—inflammation of a tendon; usually associated with repetitive, forceful exertions, often involving rotation around a joint, such as the wrist or elbow.

Tenosynovitis—inflammation of a tendon sheath; similar to tendonitis.

Third-Class Lever—a lever system in which the applied force lies between the axis of rotation (fulcrum) and the resistance. The resistance arm is always longer than the force arm so a mechanical disadvantage always exists.

Three-Shift Discontinuous Schedules—three eight-hour shifts on a five-day schedule with weekly rotation between the shifts. The discontinuity is produced by the intervening weekend, which most people live on a daytime schedule, regardless of the shift worked.

Tier—a row, layer, or level.

Timed Activity Analysis—a technique for identifying the patterns of work on a job, especially in analyzing the distribution of physically demanding activities in relation to recovery periods. It is used to identify potentially fatiguing job tasks and the impacts of external pacing on the worker.

Torque—a force that produces or tends to produce rotation; quantified as the product of the perpendicular force times its distance from the axis of rotation.

Tote Tray—a container, usually more than 25 cm (10 in.) deep, that is used to transport or store large items or to hold waste materials within a production area.

Tray—a rectangular container, usually less than 15 cm (6 in.) deep, commonly used in industry to transport multiple units of products or parts. See Tote Tray.

Treadmill—a moving belt that can be varied in speed and slope to produce a graded exercise test for determining whole-body maximum aerobic work capacities; also used to diagnose abnormalities in the electrocardiogram (ECG) during physical work.

Trigonometry—the study of the properties of triangles. Trigonometric relationships are used in biomechanical analyses.

True Grip Span—the relationship of the thumb and fingers in a power grip where near maximum force can be exerted, usually at about 5 to 6 cm (2.5 in.) of separation.

Ulnar Deviation—movement of the hand (with the palm outstretched) towards the little finger side. See Radial Deviation.

Ultradian Rhythm—a pattern of human response or behavior that usually has a periodicity of less than two hours. Often around 90 minutes.

Unit Cost—the amount of money needed to produce each product that leaves a manufacturing department or division. This includes the cost of raw materials, labor, overhead, waste, and associated distribution and storage in most cases. Some costs are relatively fixed and others are subject to modification through better use of the worker's time and skills, better equipment, earlier detection of defective materials or parts, and more effective organizational controls.

Variability—having the characteristics of changeability; having different values under different conditions.

Vasomotor Control of Blood Flow—nervous regulation of the smooth muscles in small arterial blood vessels (arterioles) to increase (vasodilatation) or decrease (vasoconstriction) the amount of blood flowing through them. The centers for this control system are in the brain stem and their influence is exerted through the autonomic nervous system.

Velocity (V)—the rate of change of displacement (D); $V = \Delta D/\Delta T$, where T is time.

Ventilation—the amount of air exchanged between the environment and the alveoli (air sacs) of the lungs in a given time period, usually one minute; expressed as liters of gas per minute (L/min) and corrected for body temperature and water vapor saturation.

Watt—a unit of power in the SI system; equals one joule per second, approximately 0.014 kilocalories per minute, or 0.0013 horsepower.

Weekly Rotation—changing the shift worked every seven days.

Weight—a measure of an object's mass, given in kilograms (kg) or pounds (lbm). If the biomechanics of the lift are known, the weight can determine the force, in newtons (N) or pounds (lbf), needed to counteract it.

Whole-Body Work—using most of the body's muscles to accomplish a task. The large muscles of the legs and buttocks as well as the muscles of the trunk, arms, and shoulders are involved. Work that is located lower than 75 cm (30 in.) above the floor requires whole-body effort.

Work/Rest Cycles—the job pattern that defines how more demanding work is organized with respect to lighter tasks or rest. High work/rest ratios, measured as continuous time on each type of activity, have higher potential for fatigue.

Workplace—the physical area in which a person performs job activities; includes tables or counters, chairs, any controls and displays necessary, the lighting, and other environmental controls.

Work Station—a workplace that is included in a production system or on a piece of manufacturing equipment and at which the operator may spend only a portion of the working shift. One operator may work at several work stations, but may have only one workplace for other types of work. Both work station and workplace should be designed according to ergonomic principles, but workplace design is more critical because of the amount of time spent there per shift.

Index